Plant Biomechanics

Anja Geitmann · Joseph Gril
Editors

Plant Biomechanics

From Structure to Function at Multiple Scales

Springer

Editors
Anja Geitmann
Faculty of Agricultural and Environmental Sciences
McGill University, Macdonald Campus
Sainte-Anne-de-Bellevue, QC
Canada

Joseph Gril
CNRS, Université Clermont Auvergne
SIGMA Clermont, Institut Pascal
Clermont-Ferrand
France

and

Université Clermont Auvergne, INRA
UMR PIAF
Clermont-Ferrand
France

ISBN 978-3-030-07727-3 ISBN 978-3-319-79099-2 (eBook)
https://doi.org/10.1007/978-3-319-79099-2

Printed on acid-free paper

This Springer imprint is published by the registered company Springer International Publishing AG part of Springer Nature
The registered company address is: Gewerbestrasse 11, 6330 Cham, Switzerland

Contents

Part I Organ and Tissue Mechanics

Wood Cell Wall Structure and Organisation in Relation to Mechanics 3
Lennart Salmén

Modelling, Evaluation and Biomechanical Consequences of Growth Stress Profiles Inside Tree Stems 21
Tancrède Alméras, Delphine Jullien and Joseph Gril

Bending Stress in Plant Stems: Models and Assumptions 49
Christopher J. Stubbs, Navajit S. Baban, Daniel J. Robertson, Loay Alzube and Douglas D. Cook

Tree Mechanics and Wind Loading 79
John Moore, Barry Gardiner and Damien Sellier

Part II Growth, Morphogenesis and Motion

The Mechanics of Leaf Growth on Large Scales 109
Eran Sharon and Michal Sahaf

Twisting Growth in Plant Roots 127
Hirofumi Wada and Daichi Matsumoto

Plants at Bodybuilding: Development of Plant "Muscles" 141
Tatyana Gorshkova, Polina Mikshina, Anna Petrova, Tatyana Chernova, Natalia Mokshina and Oleg Gorshkov

Modeling Plant Morphogenesis: An Introduction 165
Anne-Lise Routier-Kierzkowska and Adam Runions

Mechanical Conflicts in Growth Heterogeneity 193
Nathan Hervieux and Olivier Hamant

Folding, Wrinkling, and Buckling in Plant Cell Walls 209
Dorota Borowska-Wykręt and Dorota Kwiatkowska

Structural Principles in the Design of Hygroscopically Moving Plant Cells 235
Rivka Elbaum

Using Modeling to Understand the Hygromechanical and Hysteretic Behavior of the S2 Cell Wall Layer of Wood 247
Dominique Derome, Karol Kulasinski, Chi Zhang, Mingyang Chen and Jan Carmeliet

Part III Molecular Underpinnings of Cell Wall Mechanics

Calcium–Pectin Chemistry and Biomechanics: Biological Background and Mathematical Modelling 273
Mariya Ptashnyk and Henry R. Allen

Cell Wall Expansion as Viewed by the Creep Method 305
Dmitry Suslov and Kris Vissenberg

Tensile Testing of Primary Plant Cells and Tissues 321
Amir J. Bidhendi and Anja Geitmann

Part IV Water Transport, Mechanosensing and Biomimetics

Water Motion and Sugar Translocation in Leaves 351
Tomas Bohr, Hanna Rademaker and Alexander Schulz

Molecular Mechanisms of Mechanosensing and Mechanotransduction 375
Masatsugu Toyota, Takuya Furuichi and Hidetoshi Iida

Biomechanics and Functional Morphology of Plants—Inspiration for Biomimetic Materials and Structures 399
Thomas Speck, Georg Bold, Tom Masselter, Simon Poppinga, Stefanie Schmier, Marc Thielen and Olga Speck

Index 435

About the Editors

Dr. Anja Geitmann is the Dean of the Faculty of Agricultural and Environmental Sciences and Associate Vice-Principal (Macdonald Campus) of McGill University in Montreal, Canada. She holds the Canada Research Chair in Biomechanics of Plant Development and leads an interdisciplinary team of biologists and engineers. Her research group is affiliated with the Department of Plant Science located at the Macdonald Campus. Dr. Geitmann is actively involved with several learned societies. In 2013–2015, she served as President of the Microscopical Society of Canada and she was the President of the Canadian Society of Plant Biologists in 2015–2017 following mandates as the Vice-President and Education Director. She is the Vice-President of the International Association of Plant Reproduction Research and she serves on the editorial boards of multiple scientific journals including *Plant Physiology* and *The Cell Surface*.

Dr. Geitmann obtained her Ph.D. in Environmental Biology in 1997 from the University of Siena (Italy), following undergraduate and graduate studies in biology at the University of Constance (Germany), Oregon State University (USA), and Stockholm University (Sweden). From 1997 to 2001, she performed postdoctoral research at the *Université Laval*, Québec, and at the University of Wageningen, the Netherlands. In 2001, she was recruited as a faculty member by the Department of Biological Sciences of the *Université de Montréal* and the *Institut de recherche en biologie végétale*. She taught Plant Anatomy, Microscopy, and Scientific Communication at undergraduate and graduate levels, respectively, before moving to McGill University in 2015. Dr. Geitmann has organized numerous international conferences and workshops including a NATO Workshop on Tip Growth, joint meetings between the American and Canadian Societies of Plant Biology, and an international Workshop on Cell Wall Mechanics at the Banff International Research Station. In 2018, she organized the 9th International Plant Biomechanics Conference in Montreal.

Dr. Geitmann investigates the mechanical aspects of processes involved in plant growth and reproduction. Her research combines cell biology with engineering methodology and focuses on the intracellular processes that lead to the formation of differentiated cell shapes, plant organs, and functional tissues. These studies aim at understanding the interrelation between the primary cell wall and turgor, as well as the intracellular transport processes regulating cell wall assembly and mechanical behavior. One of the model cell types investigated is the pollen tube, a very rapidly growing plant cell that has the capacity to penetrate tissues, direct its growth following guidance cues, and to deliver the sperm cells during plant fertilization. Dr. Geitmann's research group has developed sophisticated experimental and computational tools enabling researchers to answer questions such as how the pollen tube turns and how it exerts invasive forces. In other research projects, her team investigates how plant cells undergo morphogenesis to generate intricate geometrical shapes and mechanical tissues with specific material properties.

Dr. Joseph Gril holds a CNRS senior scientist position in mechanical engineering science and currently leads a French consortium of institutions and researchers involved in wood science. Dr. Gril graduated from *Ecole Polytechnique* in 1981 and from *Ecole du Génie rural des Eaux et des Forêts* in 1983. In 1988, he obtained a Ph.D. in mechanical engineering from Paris VI University. He was hired as a CNRS researcher in 1989 and worked in the Laboratory of Mechanics and Civil Engineering (LMGC), where in 2004 he succeeded Bernard Thibaut as leader of a research group devoted to basic and applied knowledge on wood, tree biomechanics, support of research in developing countries, introduction of wood culture in university curricula. In 2008 and 2009, he worked 5 months in the Research Institute of Sustainable Humanosphere as Invited Professor of Kyoto University. In 2017, he moved to Clermont-Ferrand where he joined *Institut Pascal* with a part-time position in the group "*Physique et Physiologie Intégrative de l'Arbre en Environnement Fluctuant (PIAF)*".

Joseph Gril established collaborations with wood scientists in Europe, Japan, China, Morocco, Iran, etc., where he contributed through data analysis and modeling. He contributed significantly to the progress of wood mechanics in Europe through active networking activity through the COST system, for instance, as Vice-Chair of Actions E50 "Cell-wall macromolecules and reaction wood" (2005–2009) and IE0601 "Wood science for conservation of cultural heritage" (2007–2011). He won the CNRS silver medal in 2003 and since 2004 leads the mechanical study of the wooden support of Leonardo da Vinci's famous painting Mona Lisa. He has organized several international conferences mostly in connection with COST Actions. He co-organized the 8th International Plant Biomechanics Conference in Nagoya and served on the International Advisory Committee for the 9th Conference in Montreal.

Dr. Gril studies how wood performs its mechanical functions both in man-made objects and in living trees. He models the time-dependent deformation of wood subjected to changing climate and the relationships between structure and properties. He studies how the process of wood formation contributes to the material variability by generating mechanical interactions between tissues and collaborates with restorers to improve the conservation of ancient wooden objects. In 2012, he obtained the support of several CNRS Institutes to launch GDR3544, a multidisciplinary network aiming at the promotion of wood research and education through the organization of annual meetings, workshops, training schools, and individual scientific missions. Renewed once until 2020, this network currently groups about 500 staff and 150 students from 100 French laboratories, as well as 100 scientists from 25 countries.

Part I
Organ and Tissue Mechanics

Wood Cell Wall Structure and Organisation in Relation to Mechanics

Lennart Salmén

Abstract The wood cell wall, as well as the entire wood structure, is a highly intermixed assembly of biopolymers building up various structural elements. The understanding of the organisation of these wood polymers and their interaction is a key to be able to better utilise wood materials. The complexity of the wood cell wall is here discussed regarding the cellulose fibrillar network, the cellulose aggregate structure and the arrangement of the matrix polymers of hemicelluloses and lignin. The ability to model the wood cell wall properties, based on the structural organisation within different cell wall structures, and the difficulties in relating predictions to actual measurements of cell wall properties are described. The deficiencies regarding our structural knowledge in relation to mechanical properties are also being defined.

Keywords Cell wall · Cellulose · Hemicellulose · Lignin · Microfibrils
Microfibril angle · Micromechanics · Humidity · Temperature
Secondary cell wall

Introduction

Wood is a material that has been utilised by mankind since the beginning of civilization, not only as an energy resource but also as a building material and as a source for creating various utilities. The advantages of this material come to a high degree from its availability and excellent mechanical performance in relation to its density, which relates to the structural organisation of the material. Plants have a remarkable organisation with a highly hierarchical structure from the level of cellulose microfibrils to the cell wall organisation and the tissue level (Fratzl 2003; Gibson 2012).

Wood is the secondary xylem of trees, i.e. dicotyledons, and provides both strength and water transport for the plant. In gymnosperms, softwood, tracheids are the dominating cell type providing both for transport and strength. Parenchyma cells in radial

L. Salmén (✉)
RISE Bioeconomy, Stockholm, Sweden
e-mail: lennart.salmen@ri.se

A. Geitmann and J. Gril (eds.), *Plant Biomechanics*,
https://doi.org/10.1007/978-3-319-79099-2_1

rays functioning for storage for gums, starch, resin, tannins, latex and other materials. In angiosperms, hardwood, fibres provide the mechanical rigidity while larger thin-walled vessel elements, joined end-to-end function as water-transporting channels also here with parenchyma cells in rays for storage. The mechanical features of wood fibres or tracheids are dominated by the structure of the secondary cell wall. Understanding the cell wall structure and its properties are the key not only to properties of wood or plant materials themselves, but also to the way these materials may be utilised in modern material applications and how to dismantle them in order to utilise individual polymeric components. In addressing questions of the relationships between structure and mechanics of a biological material like wood, one should consider the hierarchical levels involved. In many cases where an engineering property of the material is requested, a homogenisation of properties at a higher hierarchical level may be adequate (Hofstetter and Gamstedt 2009; de Borst and Bader 2014). When it comes to questions regarding the roles of different components of the cell wall and the underlying phenomena governing structure/property relations, knowledge of the structural organisation on the molecular level may be necessary. This issue is here addressed primarily with regard to the structure of the wood cell wall, as well as the properties of the constituent biopolymers building up this structure.

The Fibre Cell Wall

At the wood cell wall level, the structural organisation is a composite involving primarily cellulose microfibrils, amorphous hemicelluloses and lignin. This is illustrated in Fig. 1, which shows the build-up of the main secondary wall, the S_2 wall, with the cellulose aggregates undulating in the length direction of the fibre. The variability is still large on this nanoscale, and makes it difficult to get a deeper insight into the structural arrangement, as the variability is at the limit of resolution of imaging techniques presently available. Thus, understanding of the arrangement of the wood-polymer components of the cell wall still relies on indirect observations of their response to specific techniques or methods. Here, modelling plays an important role as a way of testing different theories against each other to assess the importance of different factors. At the present time, too many of the variables are nevertheless not sufficiently well determined to permit a complete understanding of the detailed structure of the cell wall.

It is well known that cellulose dominates many of the properties of the plant cell wall, especially its strength properties. For example, the microfibril angle (MFA), relative to the long axis of the secondary S_2 wall (which constitutes 80% of the fibre volume) to a high extent determines the properties in the longitudinal fibre direction such as stiffness, strength and toughness. The fact that the MFA is usually less than 30° and is often in the range of 5–10° with fibrils having a right-handed chirality, makes the fibre highly anisotropic. Numerous modelling approaches have been carried out based on estimated properties of the wood components and have shown a sufficiently good correlation with the measured longitudinal fibre strength

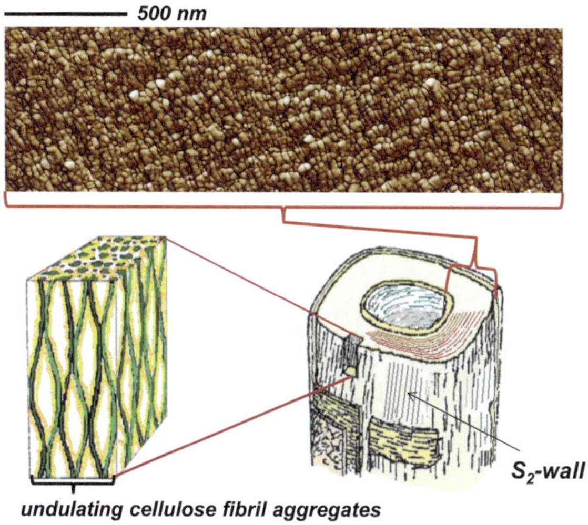

Fig. 1 Schematic representation of the complex organisation of the tracheid cell wall polymers with cellulose microfibrils arranged in undulating aggregates (after Boyd 1982) with hemicelluloses and lignin interspaced. The structure across the secondary S_2 wall (as viewed by Atomic Force Microscopy, AFM Fahlén and Salmén 2005) shows a clear variability of aggregate sizes equally dispersed over the entire cross section. An indication of a lamellar structure may be seen

properties. However, modelling transverse fibre properties is still far from satisfactory (Bergander and Salmén 2002; Wang et al. 2014a, b). The problem relates both to the fact that measurement of transverse fibre properties in itself is extremely difficult and to the fact that the properties of the constituent polymers in the transverse direction are much more uncertain. In early models of the cell wall, only the main layers, the primary wall, P, and the three secondary walls, S_1, S_2 and S_3 (with different microfibril angles), were considered (Cave 1968; Mark 1972; Salmén and de Ruvo 1985; Bergander and Salmén 2002). Later models increased the complexity by adding transition layers between S_1 and S_2 and between S_2 and S_3 (Wang et al. 2013, 2014a, b). The existence of such layers has been confirmed in a number of studies, although the variation of the MFA and extent of each layer has not been fully quantified; there may even be large variations from fibre to fibre. However, it seems clear that there is no cross-laminar structure of the S_1 wall (Brändström et al. 2003; Reza et al. 2014) but rather that it forms a gradual transition towards the S_2 (Reza et al. 2014; Reza 2016) as indicated in Fig. 2.

Incorporating such a structure into the modelling of the cell wall properties has to some extent improved the modelling accuracy. However, despite this progress, there are still rather substantial uncertainties with respect to an overestimated transverse rigidity (Wang et al. 2013). The difficulties in measurements of transverse fibre properties (Bergander and Salmén 2000a, b), may have led to an underestimation of these properties. The experimentally determined stiffness values are lower than modelled values (Wang et al. 2013, 2014a, b) indicating that both structural information and a

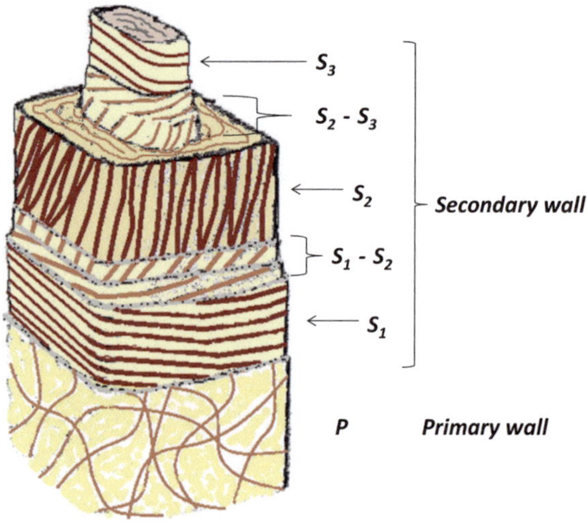

Fig. 2 Schematic representation of the structure of a wood fibre indicating the orientation of the cellulose microfibril aggregates in the different layers. The existence of transition layers between the S_1 and S_2 layers as well as between the S_2 and S_3 is indicated (Wang et al. 2013; Reza et al. 2014)

precise knowledge of the properties of the constituents and their interaction are still lacking.

The following information is most urgently required for a better understanding of the secondary wall of wood cells:

- The three-dimensional structure of the cellulose fibrillar structure
- The elastic, viscoelastic and expansion properties of the cellulose fibril aggregate (considering the structural arrangement of non-crystalline areas)
- The degree of interaction between the wood components.

The Cellulose Fibrillary Network

The cellulose microfibrils of the secondary wall are generally regarded as consisting of concentric lamellae (Kerr and Goring 1975) as indicated in Figs. 1 and 2. However, such a strict organisation has been questioned, using modelling and Transmission Electron Microscopy, TEM, studies of the cell wall. Instead, it has been proposed that microfibrils are arranged more randomly in clusters, with varying degrees of tangential or radial organisation (Donalson 2001). Such a more radial arrangement of microfibril aggregates was also suggested by Sell and Zimmermann (1993) based on fracture analysis. This fractured structure was later disputed as being only a reflection of the process of dissipation of fracture energy (Fahlén and Salmén 2002). All of these studies have been based on transverse sections of the fibre wall whereas in reality a three-dimensional imaging is necessary to be able to resolve the structure. An attempt to perform such measurements was recently made by using TEM images taken at angles to the cell wall (Reza et al. 2014) and using image processing to construct tomograms (Reza 2016). The data suggest that microfibril aggregates run transverse

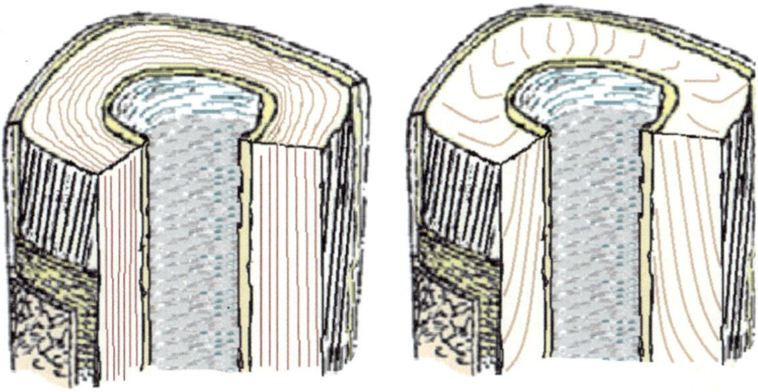

Fig. 3 Schematic illustration of the organisation of the lamellar structure of the secondary S_2 cell wall. Left, the conventional concentric lamellar structure (Kerr and Goring 1975). Right, a radial lamellar structure proposed based on TEM tomographs (Reza et al. 2014; Reza 2016). The latter structure is difficult to reconcile with the mechanism of the cell wall assembly process

to the fibre axis in the S_2-wall layer going from the outer S_1-wall continuously towards the inner S_3-wall as illustrated in Fig. 3. This configuration is very different from the conventional concept. At present times, it is difficult to understand how such a fibrillar structure could be developed during cell growth, considering that the microfibrils are continuously produced from the plasma membrane in a layer-by-layer deposition during the formation of the cell wall. The lamellar structure of the cell wall is also corroborated by studies of the shrinkage of the cell wall when lignin is removed (Stone et al. 1971). Lignin removal results in a distinct shrinkage of the thickness of the cell wall suggesting that the lignin is organised mostly as tangential lamellae within the S_2-wall layer.

In the longitudinal direction, images clearly show that the cellulose microfibrils are not straight but are undulating at regular intervals. They apparently connect to each other forming a network with differently sized aggregates (Bardage et al. 2004), as schematically indicated in Fig. 1. The size distribution is large with a mean aggregate size of around 16 nm (Bardage et al. 2004; Fahlén and Salmén 2005). Lenticular spaces, with a larger longitudinal axis where matrix material of hemicellulose and lignin are deposited, are clearly visible in Scanning Electron Microscopy, SEM, images (Bardage et al. 2004). This structure resembles that of the models of an undulating cellulose network originally suggested by Boyd (1982). This undulating structure may be the reason for the puzzling orientation interpreted from tomography by Reza et al (2014). Indeed, in order to fully answer the question of the three-dimensional organisation of the cellulose microfibrils and aggregates, more accurate imaging of the cell wall is needed, probably requiring synchrotron facilities to provide the necessary resolution.

Cellulose Aggregate Properties

The cellulose microfibrils are clearly the cell wall units that have the largest impact on mechanical properties of the cell wall, but the exact properties of these nanostructured filaments are still the subject of discussion. Most modelling and experimental data converge towards a longitudinal elasticity of crystalline cellulose of 140–150 GPa (Sakurada et al. 1962; Kroon-Batenburg et al. 1986; Iwamoto et al. 2009). However, there is a large range in moduli determined from molecular modelling and experiments with values as low as 120 and as high as 170 GPa (Matsuo et al. 1990; Tashiro and Kobayashi 1991; Nishino et al. 1995; Tanaka and Iwata 2006; Bergenstråhle et al. 2007; Iwamoto et al. 2009). Considering that the cellulose microfibril is not composed of 100% crystalline cellulose, one must take into account the organisation of disturbed, non-crystalline portions of the cellulose. A large part of these non-crystalline areas have been assigned to the surfaces of the microfibrils (Wickholm et al. 1998) but the presence of non-crystalline parts along the length of the microfibril in the form of periodical defects have been noticed using various methods (Nishiyama 2009). The existence of non-crystalline portions combined with the observation of undulations (Fig. 1) explains why cellulose aggregates have a lower effective elasticity than an entirely crystalline structure. The possible twisting of the microfibrils (Fernandes et al. 2011) could also contribute to non-regular structures in the length direction. The influence on the longitudinal cellulose stiffness may though be rather small (Fernandes et al. 2011). Most of the non-crystalline regions may be associated with the surface areas of the microfibrils rather than interrupting the crystalline regions along the length of the microfibril (Salmén and Bergström 2009). The effects of moisture on the cellulose stiffness indicates, together with the fact that moisture affects only non-crystalline structures of the cellulose (Nishiyama et al. 2002), that the deformation follows a relation of parallel amorphous and crystalline structures (Salmén and Bergström 2009). This is corroborated by the fact that for low MFA (<10°) the strain of small wood elements exposed to a tensile stress is to more than 95% given by the strain of the crystalline cellulose (Nakai et al. 2006). From this, it can be concluded that non-crystalline structures arranged in series with crystalline sections must have a rather limited extension, possibly because of adjacent crystalline microfibrils which prevent them from being highly strained (Fig. 4). Due to the uncertainty of the estimated values for the cellulose aggregate stiffness as well as of the experimental cell wall data it is presently not possible to back calculate the true value of the cellulose microfibril stiffness. However, given that the transverse properties seem to be highly overestimated, it seems to be unlikely that the tensile elasticity of the cellulose aggregates exceeds 140–150 GPa.

It is even more difficult to assess the cellulose properties in the transverse direction of the cellulose aggregates. Two factors may be considered; the effect of the undulations and the connections between microfibrils in aggregates. The cellulose crystals are inaccessible to moisture but it is now clear that the surfaces, between the fibrils in the aggregates, do adsorb moisture (Lindh and Salmén 2017;

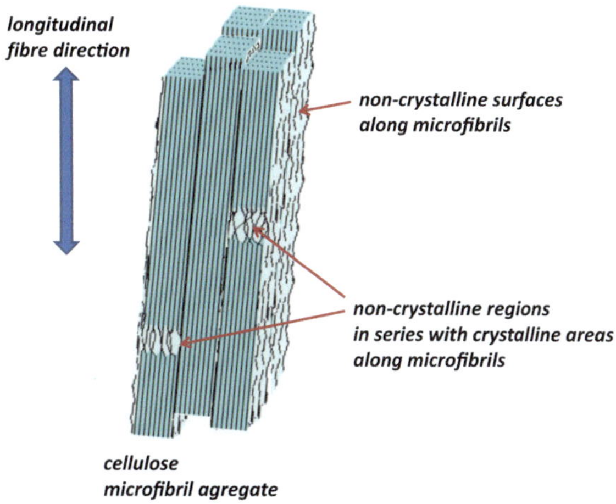

Fig. 4 Schematic representation of the organisation of non-crystalline cellulose areas in a cellulose microfibril aggregate. The largest part of the non-crystalline areas is associated with the surface of the microfibrils while a smaller portion is indicated as disturbances along the microfibrils

Lindh et al. 2017), and cellulose microfibrils have both hydrophilic and hydrophobic surfaces (Matthews et al. 2006; Fernandes et al. 2011).

Due to the asymmetry of the building block of cellulose, the glucose unit, the cellulose molecule has two chain ends of different characters, where one end may form an aldehyde function, the reducing end. During its synthesis at the plasma membrane, by the rosette terminal complex, several cellulose chains are extruded at the same time thus having similar direction. It has also been shown that the formed microfibril is composed of molecules pointing in the same direction, i.e. that the reducing end of the cellulose is pointing only upwards or downwards in the same microfibril (Hieta et al. 1984; Chanzy and Henrissat 1985). The fact that the rosettes during cellulose synthesis are travelling both up- and downwards in relation to the fibre direction implies that cellulose microfibrils may be pointing in different directions with the reducing end either up or down in relation to the microfibril axis, a fact verified for the cell wall of Valonia (Revol and Goring 1983). Conversion of cellulose, in the cell wall, to cellulose II involves a blending of microfibrils with opposite orientations of the chains forming a regular crystal structure with alternating chains with up- and downward reducing end groups. In order that this may be possible microfibrils close to each other must expose opposite molecular directions. However, if these opposite microfibrils exist in the same aggregate or in different aggregates close to each other, see Fig. 5, is an unsolved question. How the connections within the aggregate should be mediated by the presence of hemicellulose molecules, glucomannan in the case of softwood (Tokoh et al. 1998), is also unclear. The implications with regard to the forces holding together the fibrils in aggregates are thus highly uncertain.

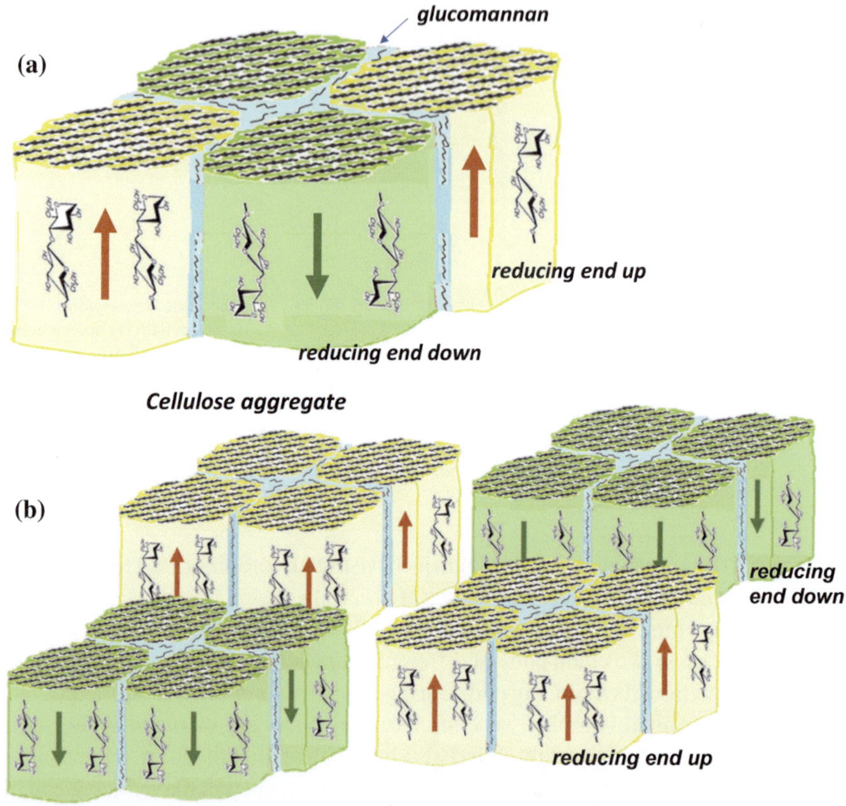

Fig. 5 Schematic representation of arrangement of cellulose microfibrils in **a** aggregate composed of microfibrils with different cellulose chain directions, the reducing end pointing upwards (marked yellow) or downwards (marked green) **b** uniform aggregates with different cellulose chain direction in different aggregates. Hemicellulose, glucomannan in softwood, separates the individual microfibril surfaces within the aggregate

The Matrix Polymers

The matrix polymers consist of hemicelluloses (xylan and glucomannan) and lignin. The properties of these polymers are difficult to determine precisely since they are intimately mixed within the structure of the cell wall. Furthermore, extraction of any of the components probably alters the properties of the component of interest or at least alters its environment. Since all of these matrix polymers are amorphous, the approximate range of properties can be estimated based on their glass transition temperature (Salmén 1982; Salmén et al. 1985). Many questions remain though to be answered concerning interrelation between lignin and hemicelluloses and their organisation within the cell wall. It is unclear, for example, whether or not lignin and xylan are intermixed so that they act as a single component. Measurements based on lignin softening tend to indicate that at least some part of the lignin acts as a separate component (Olsson and Salmén 1997).

Recent data indicate that all of the matrix polymers are to some extent aligned in relation to the fibre axis. In softwood, the glucomannan is closely associated with the cellulose microfibrils (Åkerholm and Salmén 2001) and also has a preferred orientation, albeit less prominently than the microfibrils. It may be that the presence of glucomannan controls the organisation of the microfibrils and prevents the formation of larger ribbon structures (Tokoh et al. 1998) Similarly, xylans in both softwood and hardwood exhibit a preferred orientation in the direction parallel to the cellulose microfibrils (Simonovic et al. 2011).

In the case of lignin, it is difficult to talk about a real orientation. In the middle lamella, the lignin is clearly arranged in an isotropic manner (Salmén et al. 2012). In the secondary wall of both hardwoods and softwoods, a weakly detectable preferred direction parallel to the cellulose seems to exist (Simonovic et al. 2011; Salmén et al. 2012). It has also been shown that the phenylpropane units of the lignin molecule have a tangential orientation i.e. following the tangential lamellae of the S_2 wall (Atalla and Agarwal 1985).

The anisotropy of the amorphous biopolymers (hemicelluloses and lignin) means that the transverse elasticity will be somewhat lower than the elasticity in the polymer chain direction. Since they have a preferred orientation along the microfibrils, the matrix will also display a slightly reduced stiffness in the transverse fibre direction. The tight interaction between the wood components is also reflected by the fact that during moisture absorption the cellulose crystalline structure of the microfibrils has been observed to be compressed, presumably as a result of the swelling of the matrix substances (Abe and Yamamoto 2005).

The impact of lignin on the stiffness of the wood fibre structure has also been discussed at the molecular deformation level, since no molecular deformation of lignin has been detected using spectroscopic techniques (Gierlinger et al. 2006; Salmén and Bergström 2009). Recent studies indicate, however, that such deformations of lignin may be recorded but only when wood fibres are highly deformed (Salmén et al. 2016). The conclusion drawn by the authors is that in native wood lignin contributes only marginally to the tensile strength properties in line with earlier suggestions based on Raman-spectroscopy studies on single wood fibres (Burgert 2006; Gierlinger et al. 2006).

The amorphous wood polymers are highly hygroscopic, and even native lignin absorbs sufficient moisture to affect its properties although not at ambient temperatures. Consequently, the properties of the cell wall depend on the environment and it is essential to control testing conditions. Transient changes in the testing environment such as variations in humidity or temperature, therefore, compromise measurement of fibre properties.

At higher degrees of deformation, especially for fibres with a larger MFA, sliding of the matrix material along microfibrils probably contributes to plastic flow (Spatz et al. 1999; Keckes et al. 2003). The involvement of hemicelluloses in the stick-slip mechanism of the deformation is highly plausible (Altaner and Jarvis 2008). However, the role of hemicellulose in the sliding mechanisms may not be essential, as it has been demonstrated that during mechano-sorptive creep the contents of neither lignin nor of the two types of hemicelluloses, glucomannan and xylan in softwood, play a role (Olsson and Salmén 2014). Indeed, the mechano-sorptive creep of fibres seems to be governed exclusively by the MFA of the cellulose microfibrils (Dong et al. 2010). Given that moisture may access the interfaces between microfibrils (Lindh and Salmén 2017; Lindh et al. 2017) it is reasonable to hypothesize that water molecules may act as a lubricant promoting sliding between microfibrils.

Modelling Cell Wall and Fibre Properties

Modelling structure–property relations may serve several purposes from that of predicting the service properties of a construction down to understanding molecular phenomena of the material. In the concept here addressed, modelling serves the purpose of increasing our understanding of the role of the different wood-polymer constituents and their arrangement for the mechanical properties of the cell wall. These models relay on knowledge of thermomechanical properties of the wood polymers and the structural arrangement of them in different cell wall elements. The fact that most data, as well as the more theoretically simpler relations, refer to elastic properties has led to that most models are focused on various aspects of cell wall elasticity. Thus, the elastic properties under both dry and moist conditions, the hygroexpansive properties and the relation to the MFA have been modelled. Both angle-ply models and concentric cylinder models of the cell wall have been used (Hofstetter and Gamstedt 2009), where both types of models give good agreement with regard to the sensitivity of the longitudinal stiffness variation to the MFA and moisture (Yamamoto and Koijima 2002; Joffre et al. 2014; Wang et al. 2014a, b). Regarding hygroexpansivity, the models estimate values in the proper range of data (Joffre et al. 2014; Wang et al. 2014a, b) but there are not enough measurements under critical conditions to validate the predictions. Data for the transverse elasticity, are also scarce, but so far it seems that all models overestimate the stiffness (Bergander and Salmén 2002; Wang et al. 2014), see Fig. 6. However, it is well demonstrated that the extent of softening is well captured by the assumed softening of the hemicelluloses, indicating the essential role of hemicellulose properties for the transverse cell wall properties (Salmén 2004).

It could be noted that modelling, outgoing from wood polymer properties, utilising homogenisation techniques in several steps may provide reasonable stiffness data on the tissue level for all directions (de Borst and Bader 2014). However, in this case the influence of the structural organisation at the level of the cell wall layer may not be easily assessed. On the other hand, it is shown that also by utilising the measured transversal fibre stiffness (Bergander and Salmén 2000a, b) incorporated into a tissue structure containing ray cells transverse stiffness values of the wood tissue may be reasonably estimated (Salmén 2007).

With the development of molecular dynamic (MD) simulations, the possibility of predicting cell wall properties based on atomistic properties of the constituent

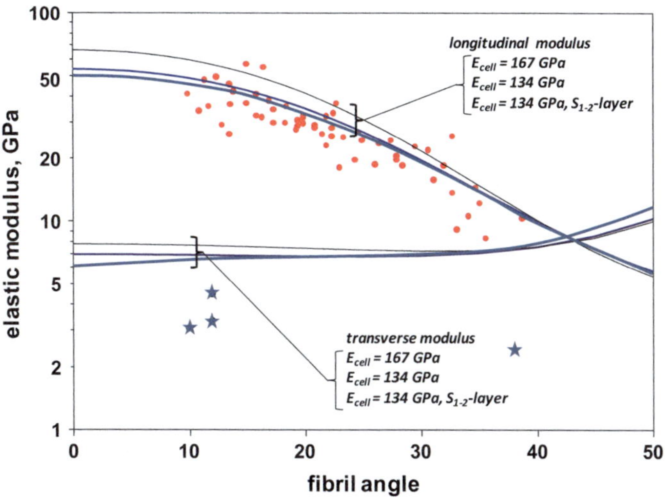

Fig. 6 The elastic modulus in the longitudinal and transverse direction of a softwood fibre cell wall. Calculations are based on micromechanics and laminate theory considering oriented hemicelluloses and lignin in the S_2 wall (Bergander and Salmén 2002), and in one case the incorporation of a transition layer (Wang et al. 2014a, b). Curves are displayed from the top and down with cellulose modulus = 167, GPa; = 134 GPa; = 134 GPa + $S_{1\text{-}2}$ layer. Measured data in the longitudinal direction (dots) are those of Cave (1969) and in the transverse direction (stars) those of Bergander and Salmén (2000a, b)

polymers increases. Such MD simulations of the action of moisture on cellulose clearly demonstrate that it is possible to model the loss of stiffness and hygroexpansive properties of amorphous cellulose (Keckes et al. 2003). Consequently, hemicelluloses may not be the exclusive agent responsible for softening of the wood fibre. Knowledge of the distribution and organisation of the non-crystalline cellulose portion is, therefore, essential for understanding how the properties are affected by moisture changes in the cell wall. Because of the dominant role of the cellulose, particular interest should be dedicated to the effect of moisture on microfibrils.

Cell Wall Property Measurements

Measurements of the mechanical properties at the fibre or cell wall level represent a challenge, not only because of the small dimensions but also because of biological variation. Any measured property has to be related to the appropriate hierarchy level. In general, due to the dominance in the volume of the fibre in relation to the connecting middle lamella, there is a good agreement between the longitudinal elastic modulus at the fibre level and that at the tissue level (Wang et al. 2014a, b). However, when it comes to failure properties, the ultimate strain and breaking stress are lower in the tissue compared to the fibre level because fibres are debonded at high stresses. These levels are probably also highly dependent on the environmental conditions, as these conditions influence the properties of the middle lamella.

One of the much-used experimental strategies to assess and understand cell wall properties is based on a nanoindentation technique. For large MFA, the elastic modulus values are close to those measured in tensile fibre testing, but for low MFA, the modulus measured through indentation often greatly underestimates the properties (Gindl et al. 2004; Eder et al. 2013).

The contribution of the different wood polymers to the properties of the cell wall has often been measured by selective removal of part of the component. Nanoindentation techniques typically yield lower moduli and lower hardness values than the native cell wall when lignin and/or hemicelluloses are removed (Wang et al. 2016; Zhang et al. 2016). However, this result may to some extent reflect the drop in material density rather than truly lower material properties of the remaining polymers.

To assess the ability of different parameters to predict structure–property relations of the wood fibre, it is not enough to be able to estimate a single value by some specific measurement. The model and structure may only be considered reliable if it is possible to predict some variation in important cell wall parameters. The most important parameters include MFA, wood-polymer composition and the effect of moisture. Differences in the effect of moisture on the stiffness of the S_2 wall of different wood samples were noted by nanoindentation measurements (Wagner et al. 2015), which could perhaps be related to variations in chemical composition and MFA differences. However, as indicated long ago, the relative degree of moisture softening is independent of the amount of lignin and hemicellulose (Salmén et al. 1985; Kolseth and Ehrnrooth 1986; Zhang et al. 2016). Nanoindentation has shown that the effect of moisture on the hardness of the cell wall is greater than the effect on the elastic modulus (Yu et al. 2011). To some extent, this could be related to the complex deformation of the nanoindenter. However, if one considers the wood fibre to be a more ductile material, at least at the higher moisture contents, the result is highly plausible. The influence of the MFA was insignificant for hardness (Li et al. 2014).

Another interesting approach for mechanical testing is micropillar compression. In this technique micrometre-sized pillars are eroded from the cell wall using a focused ion-beam. The pillar is then uniaxially compressed with a flat indenter with a diameter of a few micrometres (Adusumalli et al. 2010; Raghavan et al. 2012) thus deforming a somewhat larger area as compared to the nanoindentation technique. This gives a more uniaxial deformation of the material and less controversial data may be obtained. It has been possible to determine the yield point in compression as a function of MFA, and it appears to be in accordance with theoretical predictions (Schwiedrzik et al. 2016). At higher strains, buckling affects the results but this may, on the other hand, give information regarding the integrity of the cell wall.

In order to obtain stiffness values on an even higher level of resolution, i.e. for the in-situ wood polymers, a variety of AFM techniques may be promising. These include contact resonance atomic force (CR-AFM) (Arnould and Arinero 2015). However, the measured properties are generally lower than those obtained from nanoindentation, and they are not yet fully correlated to the structure and polymers of the cell wall.

X-ray tomography tools offer increasingly greater resolution, allowing structures and properties to be monitored. Current technology allows resolution at the fibre level, but hopefully in the future subcellular levels can be reached. Synchrotron X-ray tomography has been used to measure the hygroexpansion coefficients of a single wood fibre in different directions (Joffre et al. 2016). As an example, the values of a longitudinal hygroexpansivity of 0.014/RH and a transverse hygroexpansivity of 0.17/RH can be regarded as material properties, but they are in the lower range of previous estimates which emphasises the difficulties in obtaining reliable experimental data. So far this is a rather exclusive technique, but future studies are certain to benefit from technological progress in this field.

Future Perspective

From the picture of cell wall properties presented here, it is clear that the principal gap in knowledge is related to the difficulty in making precise measurements of properties at the cell wall level. In order better to evaluate the importance of various parameters for modelling, critical measurements particularly with regard to transverse fibre properties must be made. To enhance our understanding of the cell wall the following structural information is necessary;

- The three-dimensional structure of the cellulose aggregates within the different cell wall layers. This would presumably require synchrotron X-ray tomography investigations in order to reach the proper level of resolution.
- Elastic properties of the cellulose microfibril or aggregate structure including the non-crystalline regions. Due to the dominance of the cellulose, an unambiguous data set for cellulose would provide a solid base for all calculations.
- Transverse cell wall properties, elastic and hygroexpansional as a function of MFA. With data as a function of MFA, it would be possible better to match different modelling approaches to assess the influence of different variables.
- A better knowledge of the interactions of the cell wall components in the transverse direction. This may be a way in order to clarify the reasons for the discrepancy between modelled and measured transverse fibre properties.

References

Abe K, Yamamoto H (2005) Mechanical interaction between cellulose microfibril and matrix substance in wood cell wall determined by X-ray diffraction. J Wood Sci 51:334–338

Adusumalli R-B, Raghavan R, Ghislni R, Zimmermann T, Michler J (2010) Deformation and failure mechanism of secondary cell wall in Spruce late wood. Appl Phys A 100:447–452

Åkerholm M, Salmén L (2001) Interactions between wood polymers studied by dynamic FT-IR spectroscopy. Polymer 42(3):963–969

Altaner CM, Jarvis MC (2008) Modelling polymer interactions of the 'molecular Velcro' type in wood under mechanical stress. J Theor Biol 253:434–445

Arnould O, Arinero R (2015) Towards a better understanding of wood cell wall characterisation with contact resonance atomic force microscopy. Compos A 74(69–76)

Atalla RH, Agarwal UP (1985) Raman microprobe evidence for lignin orientation in the cell walls of native woody tissue. Sience 227:636–639

Bardage S, Donalson L, Tokoh C, Daniel G (2004) Ultrastructure of the cell wall of unbeaten Norway spruce pulp fibre surfaces. NPPJ 19(4):448–452

Bergander A, Salmén L (2000a) The transverse elastic modulus of the native wood fibre wall. J Pulp Pap Sci 26(6):234–238

Bergander A, Salmén L (2000b) Variations in transverse fibre wall properties: relations between elastic properties and structure. Holzforschung 54(6):655–661

Bergander A, Salmén L (2002) Cell wall properties and their effects on mechanical properties of fibers. J Mater Sci 37(1):151–156

Bergenstråhle M, Berglund lA, Mazeau K (2007) Thermal response in crystalline Iâ cellulose: a molecular dynamics study. J Phys Chem B 111:9138–9145

Boyd JD (1982) An anatomical explanation for visco-elastic and mechanosorptive creep in wood, and effects of loading rate on strength. New perspective in wood anatomy. P

Brändström J, Bardage SL, Daniel G, Thomas N (2003) The structural organisation of the S1 cell wall layer of Norway spruce tracheids. IAWA J 24(1):27–40

Burgert I (2006) Exploring the micromechanical design of plant cell walls. Am J Bot 93(10):1391–1406

Cave ID (1968) The anisotropic elasticity of the plant cell wall. Wood Sci Technol 2:268–278

Cave ID (1969) The longitudinal Young's modulus of Pinus radiata. Wood Sci Technol 3:40–48

Chanzy H, Henrissat B (1985) Unidirectional degradation of Valonia cellulose micrystals subjected to cellulase action. FEBS Lett 184:285–288

de Borst K, Bader TK (2014) Structure-function relationships in hardwood—insight from micromechanical modellig. J Theor Biol 345:78–91

Donalson LA (2001) A three-dimensional computer model of the tracheid cell wall as a tool for interpretations of wood cell wall ultrastructure. IAWA J 29(4):345–386

Dong F, Olsson A-M, Salmén L (2010) Fibre morphological effects on mechano-sorptive creep. Wood Sci Technol 44(3):475–483

Eder M, Arnould O, Dunlop JWC, Hornatowska J, Salmen L (2013) Experimental micromechanical characterisation of wood cell walls. Wood Sci Technol 47:163–182

Fahlén J, Salmén L (2002) On the lamellar structure of the tracheid cell wall. Plant Biol 4:339–345

Fahlén J, Salmén L (2005) Pore and matrix distribution in the fibre wall revealed by atomic force microscopy and image analysis. Biomacromol 6(1):433–438

Fernandes AN, Thomas LH, Altaner CM, Callow P, Forsyth VT, Apperley DC, Kennedy CJ, Jarvis MC (2011) Nanostructure of cellulose microfibrils in spruce wood. PNAS 108(47):1195–1203

Fratzl P (2003) Cellulose and collagen: from fibres to tissues. Currrent opin. colloid interface sci. 8:32–39

Gibson LJ (2012) The hierarchical structure and mechanics of plant materials. J R Soc Interface 9:2749–2786

Gierlinger N, Schwanninger M, Reinecke A, Burgert I (2006) Molecular changes during tensile deformation of single wood fibres followed by Raman microscopy. Biomacromol 7:2077–2081

Gindl W, Gupta HS, Schöberl T, Lichtenegger HC, Fratzl P (2004) Mechanical properties of spruce wood cell walls by nanoindentation. Appl Phys A 79:2069–2073

Hieta K, Kuga S, Usuda M (1984) Electron staining of reducing ends evidences a parallel-chain structure in Valonia cellulose. Biopolymers 23:1807–1810

Hofstetter K, Gamstedt EK (2009) Hierarchical modelling of microstructural effects on mechanical properties of wood. A review. Holzforschung 63:130–138

Iwamoto S, Kai W, Isogai A, Iwata T (2009) Elastic modulus of single cellulose microfibrils from tunicate measured by atomic force microscopy. Biomacromol 10(9):2571–2576

Joffre T, Isaksson P, Dumont PJJ, Roscoat SR, Sticko S, Orgéas L, Gamstedt EK (2016) A method to measure moisture induced swelling properties of a single wood cell. Experimental Mechanics On-line

Joffre T, Neagu RC, Bardage SL, Gamstedt EK (2014) Modelling of the hygroelastic behaviour of normal and compression wood tracheids. J Struct Biol 185:89–98

Keckes J, Burgert I, Frûhmann K, Mûller M, Kölln K, Hamilton M, Burghammer M, Roth SV, Stanzl-Tschegg SE, Fratzl P (2003) Cell-wall recovery after irreversible deformation of wood. Nat Mater 2:810–814

Kerr AJ, Goring DAI (1975) The ultrastructural arrangement of the wood cell wall. Cell Chem Technol 9(6):563–573

Kolseth P, Ehrnrooth EML (1986) Mechanical softening of single wood pulp fibers. In: Bristow JA, Kolseth P (eds) Paper structure and properties. Marcel Dekker Inc, New York, pp 27–50

Kroon-Batenburg LM, Kroon J, Norholt MG (1986) Chain modulus and intramolecular hydrogen bonding in native and regenerated cellulose fibers. Polym Commun 27:290–292

Li W, Wang H, Wang H, Yu Y (2014) Moisture dependence of indentation deformation and mechanical properties of mason pine (Pinus Massoniana Lamb) cell walls as related to microfibrilar angle. Wood Fiber Sci 46(2):228–236

Lindh EL, Salmén L (2017) Surface accessibility of cellulose fibrils studied by hydrogen-deuterium exchange with water. Cellulose 24:21–33

Lindh EL, Terenzia C, Salmén L, Furó I (2017) Water in cellulose: evidence and identification of immobile and mobile adsorbed phases by 2H MAS NMR. PCCP

Mark RE (1972) Mechanical behaviour of the molecular components of fibers. In: Jayne BA (ed) Theory and design of wood and fiber composite materials. Syracuse University Press, Syracuse, pp 49–82

Matsuo M, Sawatari C, Iwai Y, Ozaki F (1990) Effect of orientation distribution and crystallinity on the measurement by X-ray diffraction of the crystal lattice moduli of cellulose I and II. Macromolecules 23(13):3266–3275

Matthews JF, Skopec CE, Mason PE, Zuccato P, Torget RW, Sugiyama J, Himmel ME, Bradya JW (2006) Computer simulation studies of microcrystalline cellulose Ib. Carbohydr Res 341:138–152

Nakai T, Yamamoto H, Nakano T, Hamatake M (2006) Mechanical behavior of the crystal lattice of natural cellulose in wood under repeated uniaxial tensile stress in the fiber direction. Wood Sci Technol 40:683–695

Nishino T, Takano K, Nakamae K (1995) Elastic modulus of the crystalline regions of cellulose polymorphs. J Polym Sci, Part B: Polym Phys 33(11):1647–1651

Nishiyama Y (2009) Structure and properties of the cellulose microfibril. J Wood Sci 55:241–249

Nishiyama Y, Langan P, Chanzy H (2002) Crystal structure and hydrogen-bonding system in cellulose Ib from synchrotron X-ray and neutron fiber diffraction. J Am Chem Soc 124:9074–9082

Olsson A-M, Salmén L (1997) The effect of lignin structure on the viscoelastic properties of wood. Nordic Pulp Pap Res J 12(3):140–144

Olsson A-M, Salmén L (2014) Mechano-sorptive creep in pulp fibres and paper. Wood Sci Technol 48(3):569–580

Raghavan R, Adusumalli R-B, Buerki G, Hansen S, Zimmermann T, Michler J (2012) Deformation of the compound middle lamella in spruce latewood by micro-pillar compression of double cell walls. J Mater Sci 47:6125–6130

Revol J-F, Goring DAI (1983) Directionality of the fibre c-axis of cellulose crystallites in microfibrils of Valonia ventricosa. Polymer 24:1547–1550

Reza M (2016) Study of Norway spruce wall structure with microscopy tools. Applied physics. Helsinki, Aalto University. PhD

Reza M, Ruokolainen J, Vourinen T (2014) Out-of-plane orientation of cellulose elementary fibrils on spruce tracheid wall based on imaging with high-resolution transmission electroc microscopy. Planta 240:565–573

Sakurada I, Nukushima Y, Ito T (1962) Experimental determination of the elastic modulus of crystalline regions in oriented polymers. J Polym Sci 57:651–660

Salmén L (1982) Temperature and water induced softening behaviour of wood fibre based materials. PhD thesis, KTH, Stockholm

Salmén L (2004) Micromechanical understanding of the cell-wall structure. CR Biologies 337:873–880

Salmén L (2007) The mechanical deformation of wood—relation to ultrastructure. In: Entwistle KM, Walker CF (eds) The comprised wood workshop 2007. University of Canterbury, Christchurch, New Zeeland, pp 143–157

Salmén L, Bergström E (2009) Cellulose structural arrangement in relation to spectral changes in tensile loading FTIR. Cellulose 16(6):975–982

Salmén L, de Ruvo A (1985) A model for the prediction of fiber elasticity. Wood Fiber Sci 17(3):336–350

Salmén L, Kolseth P, de Ruvo A (1985) Modeling the softening behavior of wood fibers. J Pulp Pap Sci 11(4):J102–J107

Salmén L, Olsson A-M, Stevanic JS, Simonovic J, Radotic K (2012) Structural organisation of the wood polymers in the wood fibre structure. Bioresources 7(1):521–532

Salmén L, Stevanic JS, Olsson A-M (2016) Contribution of lignin to the strength properties in wood fibres studied by dynamic FTIR spectroscopy and dynamic mechanical analysis (DMA). Holzforschung 70(12):1155–1163

Schwiedrzik J, Raghavan R, Rüggeberg M, Hansen S, Wehrs J, Adusumalli RB, Zimmermann T, Michler J (2016) Identification of polymer matrix yield stress in the wood cell wall based on micropillar compression and micromechanical modelling. Philos Mag 96(32–34):3461–3478

Sell J, Zimmermann T (1993) Radial fibril agglomerations of the S2 on transverse-fracture surfaces of tracheids of tension-loaded spruce and white fir. Holz Roh Werkstoff 51:384

Simonovic J, Stevanic J, Djikanovic D, Salmen L, Radotic K (2011) Anisotropy of cell wall polymers in branches of hardwood and softwood: a polarized FTIR study. Cellulose 18(6):1433–1440

Spatz H-C, Köhler L, Niklas KJ (1999) Mechanical behaviour of plant tissues: composite materials or structures? J Exp Biol 202:3269–3272

Stone J, Scallan AM, Ahlgren PAV (1971) The ultrastructural distribution of lignin in tracheid cell walls. Tappi 54:1527–1530

Tanaka F, Iwata T (2006) Estimation of the elastic modulus of cellulose crystal by molecular mechanics simulation. Cellulose 13:509–517

Tashiro K, Kobayashi M (1991) Theoretical evaluation of three-dimensional elastic constants of native and regenerated celluloses: role of hydrogen bonds. Polymer 32(8):1516–1526

Tokoh C, Takabe K, Fujita M, Saiki H (1998) Cellulose synthesized by acetobacter xylinum in the presence of acetyl glucomannan. Cellulose 5:249–261

Wagner L, Bos C, Bader TK, de Borst K (2015) Effect of water on the mechanical properties of wood cell walls—results of a nanoindentation study. Bioresources 10(3):4011–4025

Wang N, Liu W, Peng Y (2013) Gradual transition zone between cell wall layers and its influence on wood elastic modulus. J Mater Sci 48(14):5071–5084

Wang N, Wangyn L, Lai J (2014a) An attempt in model the influence of gradual transition between cell wall layers on cell wall hygroelastic properties. J Mater Sci 49:1984–1993

Wang X, Keplinger T, Gierlinger N, Burgert I (2014b) Plant material features responsible for bamboo's excellent mechanical performance: a comparison of tensile properties of bamboo and spruce at the tissue, fibre and cell wall levels. Ann Bot 8:1627–1635

Wang X, Li Y, Deng Y, Yu W, Xie X, Wang S (2016) Contribution of basic chemical components to the mechanical behavior of wood fiber cell walls as evaluated by nanoindentation. Bioresources 11(3):6026–6039

Wickholm K, Larsson PT, Iversen T (1998) Assignment of non-crystalline forms in cellulose I by CP/MAS carbon 13 NMR spectroscopy. Carbohydr Res 312(3):123–129

Yamamoto H, Koijima Y (2002) Properties of cell wall constituents in relation to longitudinal elasticity of wood. Wood Sci Technol 36:55–74

Yu Y, Fei B, Wang H, Tian G (2011) Longitudinal mechanical properties of cell wall of Masson pine (Pinus massoniana Lamb) as related to moisture content: A nanoindentation study. Holzforschung 65:121–126

Zhang S-Y, Fei B-H, Wang C-G (2016) Effects of chemical extraction treatment on nano-scale mechanical properties of the wood cell wall. Bioresources 11(3):7365–7376

Modelling, Evaluation and Biomechanical Consequences of Growth Stress Profiles Inside Tree Stems

Tancrède Alméras, Delphine Jullien and Joseph Gril

Abstract The diameter growth of trees occurs by the progressive deposition of new wood layers at the stem periphery. These wood layers are submitted to at least two kinds of mechanical loads: maturation stress induced in wood during its formation, and the effect of the increasing self-weight. Interaction between growth and these loads causes mechanical stress with a particular distribution within the stem, called growth stresses. Growth stresses have technical consequences, such as cracks and deformations of lumber occurring during sawing, and biological consequences through their effect on stem strength. The first model for computing the field of stress inside a growing stem was set long ago by Kübler. Here, we extend these analytical formulations to cases with heterogeneous wood properties, eccentricity and bending stresses. Simulated profiles show reasonable agreement with measured profiles of released strains in logs. The particular shape of these profiles has consequences on stem bending strength. During bending in response to transient loads such as wind, most of the load is supported by outer parts of a stem cross section. The tensile maturation stress at this level increases the bending strength of the stem by delaying compression failure. Compressive stress in reaction to this tension does not reduce the bending strength because it is located near the centre of the stem and thus not loaded during bending, except if growth is strongly eccentric. Permanent bending stresses are concentrated at the mid-radius of the section, so that they do not cumulate with above-mentioned sources of stress. This smart distribution of stresses makes it possible that the stem is stronger than the wood it is made of, and that a growing stem can bend considerably more than its non-growing beam equivalent without breaking.

T. Alméras (✉) · D. Jullien
LMGC, CNRS, Université of Montpellier, Montpellier, France
e-mail: tancrede.almeras@umontpellier.fr

J. Gril
CNRS, Université Clermont Auvergne, SIGMA Clermont, Institut Pascal, Clermont-Ferrand, France

J. Gril
Université Clermont Auvergne, INRA, UMR PIAF, Clermont-Ferrand, France

A. Geitmann and J. Gril (eds.), *Plant Biomechanics*,
https://doi.org/10.1007/978-3-319-79099-2_2

Introduction

Like most plants, tree growth occurs by the progressive accumulation of new elements, either the formation of new stem and root portions (primary growth) or their thickening by deposition of new layers of wood or bark (secondary growth). Tree growth stress refers to the distribution of mechanical stress present in the wood of tree stems and branches, as a result of their growth (hereafter we will always refer to stems, but all considerations exposed here apply to both stems and branches). It is caused by the combined effect of permanent weight accumulation and dimensional changes induced by the maturation process during wood formation in the cambial zone (Archer 1986). The mechanisms involved during the maturation process, at the cellular and molecular levels, will be ignored in this chapter; a discussion on that subject is available, for instance in Alméras and Clair (2016). The viewpoint will be essentially that of stem portions, considered as slender structures subjected to normal or bending loads at all stages of development. At the macroscopic level that will be considered here, wood can be considered as a continuous solid medium, characterized by a high degree of anisotropy with an extreme dominance of the fibre direction in terms of rigidity and strength (the fibre directions will be assumed aligned with the stem axis). As a result, the stress distribution within a cross section of the stem can be analysed in this direction only.

The existence of tree growth stress has been initially recognized through its technological consequences, such as checks at log ends after crosscutting, or lumber distortion after sawing (Jacobs 1945). The strain release resulting from any cutting operation, combined with information on wood constitutive equations, is also the usual way to estimate the growth stress. Later biomechanical functions of maturation stress have been identified, and can be classified as 'skeletal' and 'motor' by analogy with animals (Moulia et al. 2006). The motor function refers to the control of tree posture, (i.e. the shape and orientation of its axes), enabled by an asymmetric distribution of maturation stress around the stem circumference. The skeletal function refers to its effect on stem strength (Bonser and Ennos 1998), as will be illustrated in the present paper.

Kübler (1959a, b) was the first to propose a mechanical model of growth stress distribution, which was very useful to analyse either technological or biomechanical aspects. Although the general mechanical framework of stress generation in a growing stem has been well/extensively described in later works (Archer 1986; Fournier et al. 1991a, b), including general numerical formulations (Fourcaud and Lac 2003; Ormarsson et al. 2010). There is a need to return to analytical formulations extending from the initial one by Kübler as they allow straightforward links with parameters describing the growth conditions. The aim of this work is to summarize a number of available models corresponding to typical growth histories, such as radial variations of maturation stress or elastic modulus, eccentric growth with tangential variations of maturation stress (as due to reaction wood formation), bending stresses induced in an inclined stem by passive bending under the self-weight or active uprighting.

Stress profiles computed for these different situations will be shown and compared to measured profiles, and their implication on stem strength will be discussed.

Model: Computing the Stress Profile Inside Growing Stems

In this section, we state hypotheses on which the calculations are based, derive related equations, and show case studies of application to different situations. The first section ("General Principles of the Mechanics of Growing Stems") is dedicated to general assumptions underlying the mechanics of a growing structure. Hypotheses formulated in this section will be kept in all case studies. Next sections present case studies of a vertical axisymmetric stem loaded either by its self-weight (section "Case of a Straight and Vertical Axisymmetric Stem Loaded by Its Self-weight") or by maturation (section "Case of a Straight and Vertical Axisymmetric Stem Loaded by Wood Maturation"), with different patterns of radial distribution of wood properties. Next, the condition of axisymmetry will be released by considering the effect of axial maturation load for an eccentrically growing stem with circumferential variations in wood properties (section "Case of an Actively Reacting Stem with Stationary Orientation"). Finally, the case of a stem bending under the self-weight will be considered (section "Case of a Stem Passively Bending Under Its Self-weight"), as well as the case where the stem both bends and reacts (section "Case of a Stem Bending While Reacting").

General Principles of the Mechanics of Growing Stems

The radial growth of a stem is achieved by the cambium. Cambial cells divide and enlarge, increasing the volume of the stem. The cambium is located at the outermost surface of the wood and its thickness is generally very small compared to the section's diameter. From a macroscopic perspective, the growth of the stem can be viewed as the deposition of new wood layers on the existing wood. The increase in stem volume is thus not described as occurring through deformations of the existing material, but through the accretion of new material at its surface. As a consequence, large strains of existing material are not necessary to achieve large volume changes. The problem will be set in the framework of small strains and linear elasticity.

Modelling Additive Growth with Pre-stresses

***(H1)** The material is loaded only since it was created.*

The mechanical stress σ at position x and time t is equal to the sum of an internal stress developing in response to the formation process (the maturation stress σ_0) and external stress increments that have occurred since its formation

$$\sigma(x,t) = \sigma_0(x,t) + \int_{t_x}^{t} \frac{\partial\sigma(x,\tau)}{\partial\tau} d\tau \tag{1}$$

where t_x is the time when the material point located at x was created.

***(H2)** The material has a linear elastic behaviour.*

In an elastic context, this pre-stress can be accounted for by using Hooke's law with prescribed strain

$$\sigma = E(\varepsilon - \alpha_0) = E\varepsilon + \sigma_0 \tag{2}$$

where σ is the elastic stress, α_0 is the prescribed strain, ε is the deformation relative to the configuration of the material at its creation, i.e. before maturation and subsequent loading occur, $(\varepsilon - \alpha_0)$ is the elastic strain, $\sigma_0 = -E\alpha_0$ is the initial stress and E is the material stiffness (or elastic modulus for 1D formulations).

After a material point is formed (i.e. a new elementary volume is added at the stem surface), it undergoes a phase of maturation during which its density, stiffness and state of stress change and reach a final value. This phase is relatively fast (between a few of weeks and a couple of month) compared with the characteristic time of growth, implying that the ring of immature wood at the stem surface has a negligible thickness compared to the diameter of the stem and its increase during growth. As a consequence, we assume that

***(H3)** The induction of pre-stress is instantaneous.*

$$\sigma_0(x,t) = \sigma_0(x,t_x) = \sigma_0(x) \tag{3}$$

Equation (1) becomes

$$\sigma(x,t) = \sigma_0(x) + \int_{t_x}^{t} \frac{\partial\sigma(x,\tau)}{\partial\tau} d\tau \tag{4}$$

The differential form of Eq. (2) is

$$\frac{\partial\sigma(x,t)}{\partial t} = E(x)\frac{\partial\varepsilon(x,t)}{\partial t} + \frac{\partial\sigma_0(x,t)}{\partial t} = E(x)\frac{\partial\varepsilon(x,t)}{\partial t} \tag{5}$$

(H4) *Size can replace time to describe growth in a monotonously growing structure.*

Because the volume always increases during growth, any volume of the structure is associated to a unique time (when the structure reached this volume). As time by itself is not involved in the constitutive Eq. (2), the volume can be used to monitor the succession of events, i.e. the progressive changes in stress

$$\int_{t_x}^{t} \frac{\partial\sigma(x,\tau)}{\partial\tau} d\tau = \int_{V_x}^{V} \frac{\partial\sigma(x,u)}{\partial u} du \tag{6}$$

Equation (4) then becomes

$$\sigma(x,V) = \sigma_0(x) + \int_{V_x}^{V} \frac{\partial\sigma(x,u)}{\partial u} du \tag{7}$$

where V_x is the volume of structure when the material point located at x was created.

Assumptions for a Growing Stem: Beam Theory

(H5) *As the stem is a slender structure, beam theory will be used to derive the model.*

We are here concerned only by the distribution of longitudinal stress, so that a 1D formulation will be used, where mechanical parameters mentioned above ($\varepsilon, \sigma, \alpha_0, \sigma_0, E$) are scalar. In the context of beam theory, the field of stress is computed for a cross section far enough from the stem ends. The location of a point within the section is (x, y). The increment in strain field within the section is plane. Here, we assume that the change in curvature involves a rotation around the Y axis. The strain increment $\delta\varepsilon$ in response to the addition of a new layer of material between t and $t + \delta t$ is

$$\delta\varepsilon(x,y,t) = \delta\varepsilon_O(t) - (x - x_O)\delta C(t) \tag{8}$$

where $\delta\varepsilon_O$ is the strain at a reference point of the section (typically the pith or the geometrical centre) located at abscissa x_O and δC is the change in curvature.

Case of a Straight and Vertical Axisymmetric Stem Loaded by Its Self-weight

Here we assume that the tree is submitted only to the effect of self-weight, neglecting maturation stress. This situation is virtual since wood maturation always occurs

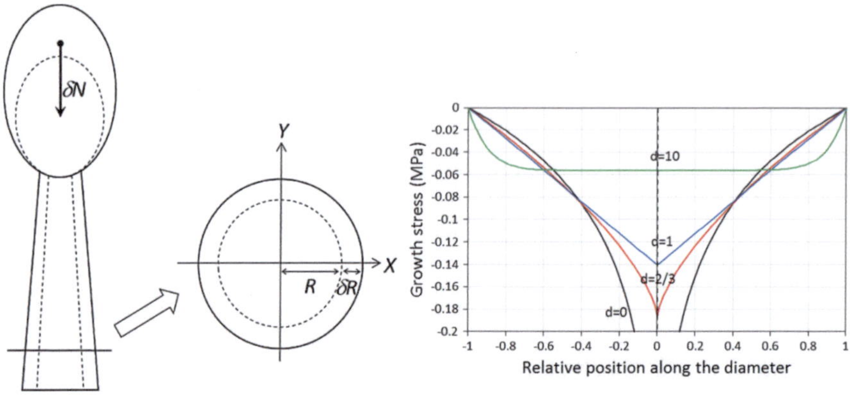

Fig. 1 Case of a vertical and straight stem with circular and homogeneous cross section, loaded by its own weight. In this case, wood maturation stress is neglected. Calculations are made for a stem growing allometrically, exponents d representing the relation between stem radius and length (Eq. 18) related to self-weight (Eq. 19). The case $d = 0$ represents a stem growing only in diameter, yielding a log profile with asymptotic behaviour at the centre. The infinite value at this level is an artefact of the model, as commented in section "Case of a Straight and Vertical Axisymmetric Stem Loaded by Wood Maturation". The case $d = 2/3$ represents the allometry for elastic stability (Greenhill 1881), and the case $d = 1$ isometric growth. In these two cases, the profile has a V shape. Large values of the exponent (here $d = 10$) represent the case where extension growth is much faster than radial growth. In this case, the profile is uniform in most of the section, tending to the uniform profile expected for a beam without radial growth. The computation has been made for a given stem dimension and weight, but the magnitude of the profile (not its shape) is actually depending on stem size. The mean stress in the section depends on its height (Eq. 25). This mean stress has very low magnitude (for reference, the strength of wood is several tens of MPa)

during tree growth, but it will enable the comparison between these two sources of stress. The tree is assumed straight and vertical and therefore submitted to purely axial loads, with circular and homogeneous cross section (Fig. 1). The first two subsections state hypotheses suitable for this case, but also for some cases presented in sections "Case of a Straight and Vertical Axisymmetric Stem Loaded by Wood Maturation" and "Case of an Actively Reacting Stem with Stationary Orientation".

Assumptions for a Stationary Stem: Pure Axial Loading

We define as "stationary" a stem whose shape (curvature) does not change during growth. This is the case, for example of a vertical stem that remains vertical. It is also the case of a leaning stem for which lean and shape remain constant, because the effect of weight increment is compensated by the active reaction (asymmetry of maturation stress within the section), or, for example because bending is prevented by a stalk.

Let us assume that the stem is submitted to a normal force increment $\delta N(t)$, as due, for example to weight or maturation load. As we assume that only axial contraction occurs ($\delta C = 0$), the strain increment is uniform over the section

$$\delta\varepsilon(x, y, t) = \delta\varepsilon_O(t) \tag{9}$$

The strain increment due to the normal force increment is

$$\delta\varepsilon_O(t) = \delta N(t) / \big(\bar{E}(t) A(t)\big) \tag{10}$$

where $A(t)$ is the section area and $\bar{E}(t)$ is the homogenized modulus of elasticity given by

$$\bar{E}(t) = \iint_{\delta S} E(x, y) dx dy \big/ A(t) \tag{11}$$

Considering the incremental form of the constitutive Eq. (2), $\delta\sigma = E\delta\varepsilon$, together with Eq. (10), the stress increment in response to the load increment is

$$\delta\sigma(x, y, t) = \frac{\delta N(t)}{A(t)} \frac{E(t)}{\bar{E}(t)} \tag{12}$$

Formulation for a Circular Homogeneous Cross Section

For a circular cross section, the section geometry can be characterized by its outer radius R and this variable can be used to monitor time. Because the loading is purely axial, the problem is axisymmetric, and the position of a point is characterized by its radial position r. Therefore, coordinates (x, y, t) become (r, R), where R is the radius of the stem at time t. Neglecting maturation stress, Eq. (1) becomes

$$\sigma(r, R) = \int_r^R \frac{\partial\sigma(r, u)}{\partial u} du \tag{13}$$

As the elastic modulus is homogeneous, we have $\bar{E} = E$, so that Eq. (12) becomes

$$\delta\sigma(r, R) = \frac{\delta N(R)}{A(R)} \tag{14}$$

For a circular cross section, we have $A(R) = \pi R^2$, so that Eq. (14) becomes

$$\delta\sigma(r, R) = \frac{\delta N(R)}{\pi R^2} \tag{15}$$

Dividing by δR and setting $\delta R \to 0$, we obtain the following differential equation

$$\frac{\partial \sigma(r,R)}{\partial R} = \frac{1}{\pi R^2}\frac{dN(R)}{dR} \tag{16}$$

Finally, the relation between growth stresses and the evolution of axial loading is

$$\sigma(r,R) = \frac{1}{\pi}\int_r^R \frac{dN(u)}{du}u^{-2}du \tag{17}$$

Application to Stem Allometric Growth

Let us assume that the tree grows in radius R and length L following an allometric law given by

$$L = cR^d \tag{18}$$

The normal force induced by self-weight N is assumed proportional to the volume of the stem and expressed as

$$N = -k\pi R^2 L = -k\pi c R^{d+2} \tag{19}$$

where $k = N/(\pi R^2 L)$ is a constant linking stem weight to stem dimensions, accounting for the effect of taper, branch distribution and density.

Deriving with respect to R, we obtain

$$\frac{dN(R)}{dR} = -k\pi c(d+2)R^{d+1} \tag{20}$$

Substituting into (17), we obtain the field of stress within the section

$$\sigma(r,R) = -kc(d+2)\int_r^R u^{d-1}du \tag{21}$$

Which integrates as

$$\sigma(r,R) = -\frac{kc(d+2)}{d}\left(R^d - r^d\right) \quad \text{for } d \neq 0 \tag{22}$$

$$\sigma(r,R) = 2kc\ln(r/R) \quad \text{for } d = 0 \tag{23}$$

We define the relative radius ρ

$$\rho = r/R \tag{24}$$

and the mean support stress over the section $\bar{\sigma}$

$$\bar{\sigma} = \frac{N(R)}{A(R)} = \frac{-kc\pi R^{d+2}}{\pi R^2} = -kcR^d = -kL \tag{25}$$

Equations (22) and (23) can be rearranged as

$$\sigma(r, R) = \bar{\sigma}\frac{d+2}{d}\left(1 - \rho^d\right) \quad \text{for}\, d \neq 0 \tag{26}$$

$$\sigma(r, R) = -2\bar{\sigma}ln\rho \quad \text{for}\, d = 0 \tag{27}$$

Results are shown and commented on Fig. 1.

Case of a Straight and Vertical Axisymmetric Stem Loaded by Wood Maturation

Here we compute the field of growth stress of a straight and vertical tree with circular cross section submitted to maturation stress, neglecting the self-load (Fig. 2). The problem is axisymmetric and equations derived in section "Assumptions for a Stationary Stem: Pure Axial Loading" still apply. First (Section "Kübler's Model") we will derive Kübler's model assuming uniform properties (maturation stress and elastic modulus) within the section. Next, the assumption of uniformity will be released, by considering radial variations in either maturation stress (Section "Case of Polynomial Radial Variations of Maturation Stress") or elastic modulus (Section "Case of Linear Radial Variations of Elastic Modulus"). As only radial variations are considered, the problem is axisymmetric and coordinates (r, R) will be used to refer to radial position r when the stem has radius R.

Kübler's Model

Because the problem is axisymmetric and maturation stress is uniform Eq. (4) becomes

$$\sigma(r, R) = \sigma_0 + \int_r^R \frac{\partial\sigma(r, u)}{\partial u} du \tag{28}$$

The normal force resulting from maturation stress in the section increment between R and $R + \delta R$ is

$$\delta N(R) = -\iint_{\delta S} \sigma_0 du = -\sigma_0 \iint_{\delta S} du = -2\pi R\delta R\sigma_0 \tag{29}$$

Dividing by δR and setting $\delta R \to 0$, we obtain

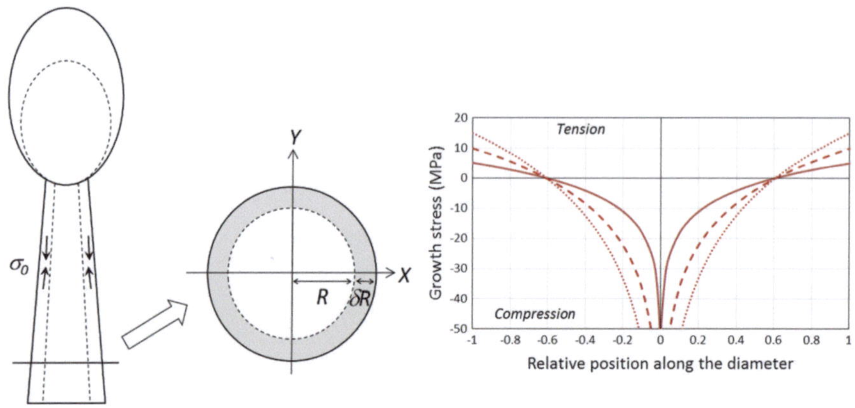

Fig. 2 Case of a straight and vertical stem with circular and homogeneous cross section, loaded by wood maturation (Kübler' model). Here, the stem self-weight is neglected. The figure shows the growth stress profile, for different values of maturation stress: 5 MPa (continuous line), 10 MPa (dashed line) and 15 MPa (dotted line). Profiles have a typical log shape, with the value of maturation stress in the periphery (tension), a null stress located at 0.6 times the radius, and strong compression in the core. Profiles are self-similar (i.e. the pattern does not depend on stem size), and balanced (i.e. the integral over the whole section is null). The magnitude of stress is several orders of magnitude larger than that of self-weight (Fig. 1), providing an a posteriori verification of the assumption that self-weight has a negligible effect. The infinite value at the centre of the section is an artefact due to the mathematical formulation of the model, because the profile is integrated since $R = 0$, whereas the stem has finite size at the very beginning of radial growth, with tissues other than wood having a dominant contribution to stem stiffness

$$\frac{dN(R)}{dR} = -2\pi R\sigma_0 \tag{30}$$

Using Eq. (16), we obtain

$$\frac{\partial\sigma(r,R)}{\partial R} = \frac{1}{\pi R^2}\frac{dN(R)}{dR} = -2\frac{\sigma_0}{R} \tag{31}$$

Substituting in Eq. (28) and defining $\rho = r/R$, we obtain

$$\sigma(r,R) = \sigma_0 - \int_r^R 2\frac{\sigma_0}{u}du = \sigma_0(1 + 2\ln(\rho)) \tag{32}$$

Results are shown and commented on Fig. 2.

Case of Polynomial Radial Variations of Maturation Stress

Here, we still consider the axisymmetric problem of a vertical and straight tree with circular cross section (Fig. 2), but account for possible radial variations in maturation stress. This problem may represent the case of a tree with ontogenic or environmental changes inducing variable level of maturation stress during growth.

We assume that the radial change in maturation stress during growth has polynomial form

$$\sigma_0(r) = \sum_{0}^{N} a_n r^n \tag{33}$$

This can describe any pattern of variation.

For a non-uniform pattern of maturation stress in axisymmetric case, Eq. (4) becomes

$$\sigma(r, R) = \sigma_0(r) + \int_{r}^{R} \frac{\partial\sigma(r, u)}{\partial u} du \tag{34}$$

Substituting Eqs. (33), (31) becomes

$$\frac{\partial\sigma(r, R)}{\partial R} = -2\sum_{0}^{N} a_n R^{n-1} \tag{35}$$

The growth stress profile is obtained by combining Eqs. (33), (34) and (35)

$$\sigma(r, R) = \sum_{0}^{N} a_n r^n - 2\sum_{0}^{N} a_n \int_{r}^{R} u^{n-1} du \tag{36}$$

Which integrates as

$$\sigma(r, R) = \sum_{1}^{N} \left(1 + \frac{2}{n}\right) a_n r^n - 2\sum_{1}^{N} \frac{a_n}{n} R^n + a_0\left(1 + 2\ln\left(\frac{r}{R}\right)\right) \tag{37}$$

Defining the relative radius $\rho = r/R$, this can be rearranged as

$$\sigma(r, R) = a_0(1 + 2\ln\rho) + \sum_{1}^{N} a_n R^n \left(\left(1 + \frac{2}{n}\right)\rho^n - \frac{2}{n}\right) \tag{38}$$

Results are shown and commented on Fig. 3.

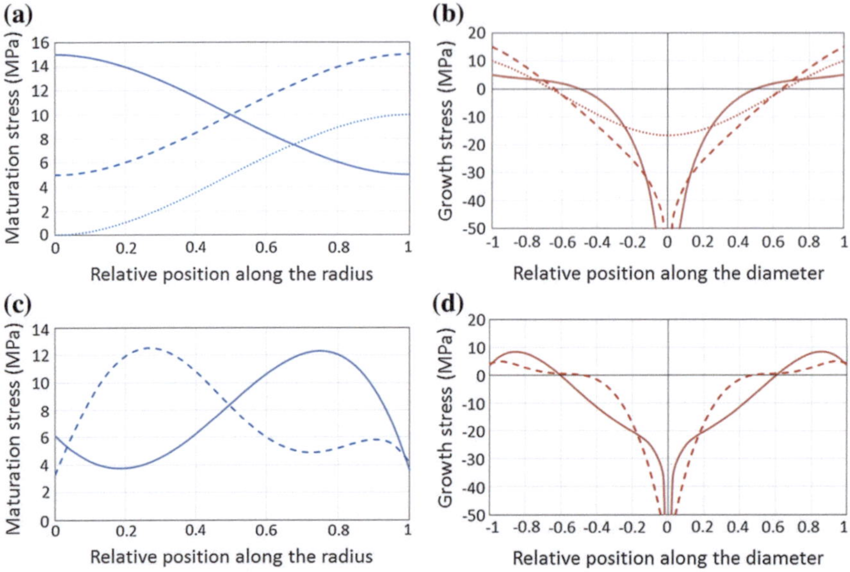

Fig. 3 Results for the same situation as Fig. 2, but with non-homogeneous patterns of radial variations in maturation stress (**a** monotonous variations; **c** non-monotonous variations). The maturation stress patterns represent possible scenario of response to ontogenic or environmental changes (Jullien et al. 2013; Dassot et al. 2015). Simulations (**b**, **d**) show that the shape of growth stress profiles can differ substantially from Kübler's model. The shape generally presents asymptotic infinite negative value on the middle of the section, which is a mathematical artefact of the model. The case where maturation stress is small in early stages (**a**, dotted lines) don't show this asymptotic behaviour and is convex with a U shape. For non-monotonous variations in maturation stress growth stress profiles show a variety of shape, either convexo-concave or even non-monotonous along a radius

Case of Linear Radial Variations of Elastic Modulus

Here, we still consider the axisymmetric problem of a vertical and straight tree with circular cross section (Fig. 2), but account for possible radial variations in elastic modulus. This problem may represent the case of a vertical and straight tree with ontogenic variations of the modulus of elasticity, as typically happens for conifers in juvenile stages.

We assume that the change in elastic modulus during growth has linear form

$$E(r) = ar + b \tag{39}$$

The maturation strain is assumed uniform. The maturation stress is then non-uniform and is

$$\sigma_0(r) = -\alpha_0 E(r) = -\alpha_0 (ar + b) \tag{40}$$

For nonuniform elastic modulus, Eq. (11) becomes

$$\bar{E}(R) = \int_0^{2\pi}\int_0^{R} uE(u)dud\theta \Big/ \int_0^{2\pi}\int_0^{R} udud\theta = \frac{2}{3}aR + b \tag{41}$$

Applying the same principle as (H1) to the growth strain profile, and reminding that ε is the strain since initial configuration (i.e. the initial strain is 0 by definition of ε) Eq. (1) is replaced by

$$\varepsilon(r, R) = \int_r^R \frac{\partial\varepsilon(r, u)}{\partial u} du \tag{42}$$

Equation (10) becomes

$$\delta\varepsilon(r, R) = \frac{\delta N(R)}{\bar{E}(R)A(R)} = \frac{2\pi R\alpha_0 E(R)\delta R}{\pi R^2 \bar{E}(R)} = 2\alpha_0 \frac{aR + b}{\frac{2}{3}aR^2 + bR}\delta R \tag{43}$$

Therefore,

$$\frac{\partial\varepsilon(r, R)}{\partial R} = 2\alpha_0 \frac{aR + b}{\frac{2}{3}aR^2 + bR} \tag{44}$$

The strain profile is obtained by combining Eqs. (42) and (44)

$$\varepsilon(r, R) = 2\alpha_0 \int_r^R \frac{au + b}{(2/3)au^2 + bu} du \tag{45}$$

Which integrates as

$$\varepsilon(r, R) = -\alpha_0 \left(2\ln(r/R) + \ln\left(\frac{(2/3)ar + b}{(2/3)aR + b} \right) \right) \tag{46}$$

Combined with Eqs. (2) and (39), the stress profile can be deduced

$$\sigma(r, R) = -\alpha_0(ar + b)\left(1 + 2\ln(r/R) + \ln\left(\frac{(2/3)ar + b}{(2/3)aR + b} \right) \right) \tag{47}$$

Results are shown and commented on Fig. 4.

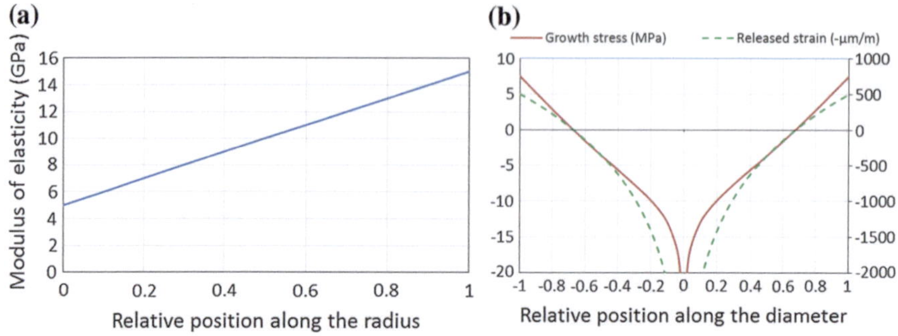

Fig. 4 Results for the same situation as Fig. 2, but with non-homogeneous patterns of radial variations in modulus of elasticity (**a**) representing, for example the case of a conifer in juvenile stage. The stress profile (**b**, solid red line) is close to linear along a radius. However, because the MOE is heterogeneous, the shape of the released strain profile (**b**, green dashed line) is different and closer to Kübler's model

Case of an Actively Reacting Stem with Stationary Orientation

Here, we consider a non-axisymmetric problem where tangential variations of maturation stress and growth increments occur (Fig. 5). Particular cases (only tangential variations in maturation stress or only eccentric growth) are encompassed by this formulation. This problem is no more axisymmetric. The stem is assumed stationary, i.e. not submitted to changes in curvature during growth, and thus loaded only axially. This corresponds to a tilted stem that remains tilted at constant angle, and, therefore, undergoes no bending. This situation typically happens for a staked stem or for branches where the reaction exactly compensates for the effect of increasing weight.

Let us consider a circular cross section growing by addition of eccentric layers, but remaining circular during growth. We also assume homogeneous elastic properties within the cross section. Here the position of the pith is not at the centre of the section. Let x_O be position of the centre of the section relative to pith at a given time. During growth, this position moves relatively to the pith. The pith is taken as a reference because the position of a material point is fixed in this reference. Let δx_O be the displacement of the centre during the addition of a growth ring increasing the section's diameter by δD. Let δR be the half of diameter increment. The eccentric growth is characterized by parameter k_O such that

$$k_O = \delta x_O / \delta R \tag{48}$$

Assuming that the section grows with constant eccentricity since the beginning of growth, the position of the geometric centre can be obtained by integration

$$X_O = k_O R \tag{49}$$

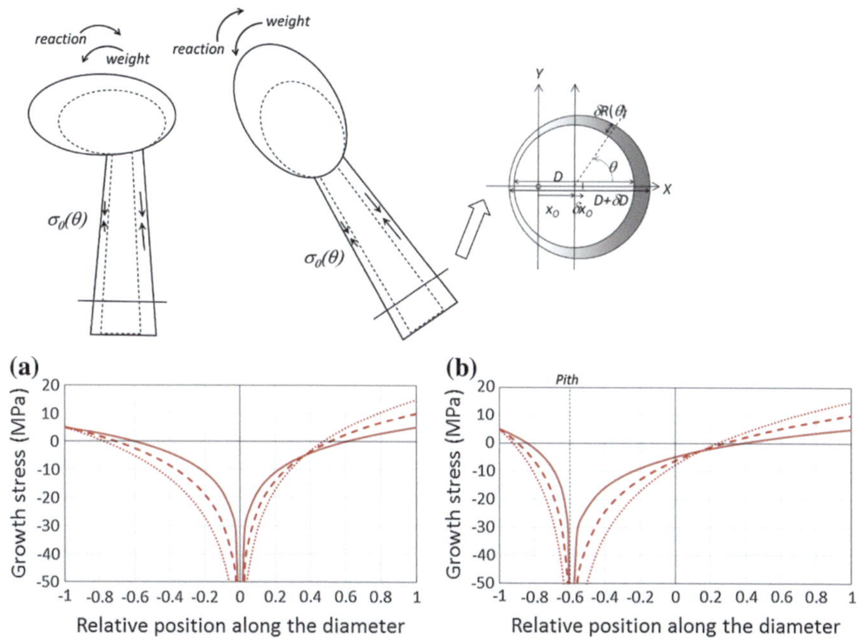

Fig. 5 Case of a "stationary" stem (i.e. a stem keeping constant shape) reacting to eccentric growth and the production of tangential variations in maturation stress. The stem can be vertical or tilted, provided the load caused by the self-weight is exactly compensated by the load caused by the reaction. Cases with no eccentricity and various levels of maturation stress on the tension wood side are shown on chart (**a**). Profiles have a log shape, with possible shift between the two sides, due to the asymmetry in maturation stress. Cases with eccentricity are shown on chart (**b**). Profiles still have a log shape, but with marked asymmetry between the two sides. Note that the maximal compressive stress is located at the pith, not at the geometric centre of the section

The tangential distribution of the radius increment is approximated as

$$\delta R(\theta) = \delta R(1 + k_O \cos\theta) \tag{50}$$

The tangential distribution of the maturation strain and stress is given as

$$\alpha_O(\theta) = \bar{\alpha} + \frac{1}{2}\Delta\alpha\cos\theta \tag{51}$$

$$\sigma_O(\theta) = \bar{\sigma} + \frac{1}{2}\Delta\sigma\cos\theta \tag{52}$$

With $\bar{\sigma} = -E\bar{\alpha}$ and $\Delta\sigma = -E\Delta\alpha$.

Here, the modulus E is assumed uniform.

Let us consider a point of the section located in (x, y) in a reference attached to the pith. Equation (7) becomes

$$\sigma(x, y, R) = \sigma_0(x, y) + \int_{r_{xy}}^{R} \frac{\partial \sigma}{\partial u} du \tag{53}$$

where r_{xy} is the mean radius of the section when the point located in (x, y) was created.

Equation (14) becomes

$$\delta\sigma(x, y, R) = -\frac{\delta N(R)}{A(R)} \tag{54}$$

Equation (29) becomes

$$\delta N(R) = \delta R \int_{0}^{2\pi} \sigma_0(\theta)(1 + k_O \cos\theta) R d\theta \tag{55}$$

Substituting Eq. (52), we obtain

$$\delta N(R) = R\delta R \int_{0}^{2\pi} \left(\bar{\sigma} + \frac{1}{2}\Delta\sigma \cos\theta\right)(1 + k_O \cos\theta) d\theta \tag{56}$$

Which integrates as

$$\delta N(R) = R\delta R\left(2\pi\bar{\sigma} + \frac{\pi}{2}k_O \Delta\sigma\right) \tag{57}$$

Reminding that $A(R) = \pi R^2$ and combining with (54), Eq. (31) becomes

$$\frac{\partial\sigma(x, y, R)}{\partial R} = \frac{2\bar{\sigma} + \frac{1}{2}k_O \Delta\sigma}{R} \tag{58}$$

Let us compute the stress profile at the level of the axis of symmetry ($y = 0$). For $x > 0$, $y = 0$, the maturation stress is

$$\sigma_0(\theta) = \bar{\sigma} + \frac{1}{2}\Delta\sigma \tag{59}$$

From Eq. (48), it can be deduced that

$$r_{xy} = x/(k_O + 1) \tag{60}$$

Substituting Eqs. (58), (59) and (60) into (53)

$$\sigma(x, 0, R) = \bar{\sigma} + \frac{1}{2}\Delta\sigma + \int\limits_{x/(k_O+1)}^{R} \frac{2\bar{\sigma} + \frac{1}{2}k_O\Delta\sigma}{u} du \tag{61}$$

Which integrates as

$$\sigma(x, 0, R) = \bar{\sigma} + \frac{1}{2}\Delta\sigma + \left(2\bar{\sigma} + \frac{1}{2}k_O\Delta\sigma\right)\ln\left(\frac{\frac{x}{R}}{k_O + 1}\right) \tag{62}$$

For $x < 0$, $y = 0$, the maturation stress is

$$\sigma_0(\theta) = \bar{\sigma} - \frac{1}{2}\Delta\sigma \tag{63}$$

From Eq. (48), it can be deduced that

$$r_{xy} = x/(k_O - 1) \tag{64}$$

Substituting Eqs. (58), (63) and (64) into (53)

$$\sigma(x, 0, R) = \bar{\sigma} - \frac{1}{2}\Delta\sigma + \int\limits_{x/(k_O-1)}^{R} \frac{2\bar{\sigma} + \frac{1}{2}k_O\Delta\sigma}{u} du \tag{65}$$

Which integrates as

$$\sigma(x, 0, R) = \bar{\sigma} - \frac{1}{2}\Delta\sigma + \left(2\bar{\sigma} + \frac{1}{2}k_O\Delta\sigma\right)\ln\left(\frac{\frac{x}{R}}{k_O - 1}\right) \tag{66}$$

Results are shown and commented on Fig. 5.

Case of a Stem Passively Bending Under Its Self-weight

Here, we consider a stem that grows at the same time it bends. In this section, maturation stress is neglected on purpose (Fig. 6). The general problem of a stem submitted to both bending and maturation will be studied in next section. As will be shown, the stress profile only depends on the changes in curvature, not on their cause. The model is therefore valid whether this change in curvature occurs in response to sagging under the self-weight, to uprighting due to maturation stress asymmetry or to any external forcing. The first section is dedicated to general assumptions for a stem changing in curvature. Then, two cases will be studied. First, an axisymmetric section will be considered. This corresponds to the situation where a stem does not react during growth, so that it bends under its self-weight. Note that even if the

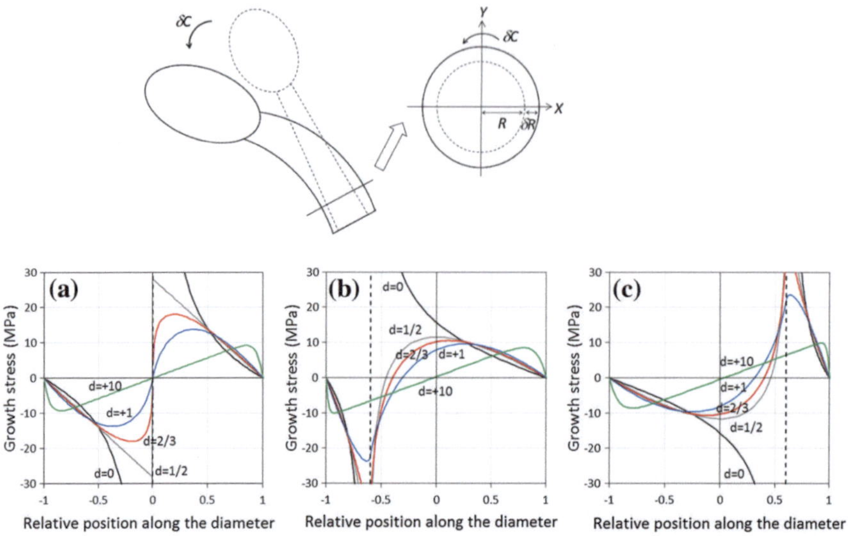

Fig. 6 Case of a tilted stem bending under its own weight. The section is circular with homogeneous modulus of elasticity, and possibly eccentric growth. Here only bending stresses are computed, and maturation stresses are neglected on purpose. The calculation was made for different allometric laws of the stem, defined by an allometric exponent between radius and length (Eq. 18). Results are shown for different patterns of eccentricity and different allometric exponents d. **a** Case with no eccentricity. **b** Case with eccentricity on the upper side of the tilted stem. **c** Case with eccentricity on the lower side of the tilted stem. Values of d represent different remarkable cases of allometry: only radial growth ($d = 0$); allometry associated to constant stress (McMahon 1976) or biomechanical stability (Alméras and Fournier 2009) ($d = 1/2$); allometry associated to elastic stability (Greenhill 1881) (d = 2/3); isometric growth ($d = 1$); situation where radial growth is much slower than weight increase ($d = 10$). Case $d = 0$ shows asymptotic behaviour near the pith, which is an artefact of the model. Maximal bending stress is located near the pith, tensile on the upper side, and compressive on the lower side. This location of maxima is also observed for $d = 1/2$, where linear variations occur from the periphery to the pith in the concentric case. For $d = 2/3$ and $d = 1$, the profile is sigmoid, with maximal magnitude located in the middle part of the radius. The case $d = 10$ shows a linear pattern in most of the section, tending to the usual bending stress profile expected for a non-growing beam. The case where eccentricity occurs on the upper part of the section shows similar but distorted results. For the case $d = 10$, the profile is almost exactly the same, and the stress is null at the geometric centre. For other cases, compression is concentrated on the lower side, and its maximal magnitude is larger than in the concentric case. Tension occurs in the part of the section located above the pith, and it maximal value is lower than in the concentric case. When eccentric growth occurs on the lower side, the pattern is inverse: tension is concentrated above the pith and its maximal value is larger than in the concentric case, and compression is concentrated under the pith and its maximal value is lower than in the concentric case

section is axisymmetric, the problem is not axisymmetric because bending occurs. Second, we will consider a non-axisymmetric section with eccentric growth. This corresponds to the situation of a stem in active reaction and uprighting movement (although only the effect of changes in curvature will be accounted for, not the stress directly due to maturation).

Assumption for a Stem Submitted to Changes in Curvature

In both situations of passive bending and active uprighting, the change in curvature associated to an increment in section radius is depending on section's radius. For the active uprighting, it has been shown (Fournier et al. 2006; Alméras and Fournier 2009) that the change in curvature scale at power −2 with the radius. For the passive bending case, we will show that changes in curvature and radius are also related by a power law when assuming a power law for allometric growth (Section "Case of an Eccentric Section"). We assume that the stem undergoes a bending movement during growth, the change in curvature δC that occurs during growth from R to $R + \delta R$ following a power law

$$\frac{\delta C}{\delta R} = C'(R) = aR^b \quad \text{with}\, b \neq -1 \tag{67}$$

The problem is no more axisymmetric because of the distribution of bending stresses. Let (x, y) be the position of a point relative to the centre of the section. As the stem is assumed submitted to pure bending around its neutral line, Eq. (8) becomes

$$\delta\varepsilon(x, y, R) = (x - x_O)\delta C(R) \tag{68}$$

where x_O is the position of the neutral line of the section.

Case of an Axisymmetric Section

Here we assume that the section is circular and homogeneous, and compute the stress profile along the diameter parallel to X. From Eqs. (66) and (67), and reminding that the geometric centre and the position of neutral line are equal for an homogeneous section we deduce that

$$\frac{\partial\varepsilon(x, y, R)}{\partial R} = xC'(R) \tag{69}$$

Combining with Eq. (5) we have

$$\frac{\partial\sigma(x, y, R)}{\partial R} = ExC'(R) \tag{70}$$

Neglecting initial stress in the section, Eq. (7) becomes

$$\sigma(x, y, R) = \int_{r_{xy}}^{R} \frac{\partial\sigma(x, y, u)}{\partial u} du \tag{71}$$

where r_{xy} is the radius of the section when point located at (x, y) was created.

Combining Eqs. (66), (69) and (70), we obtain

$$\sigma(x, y, R) = xE \int_{r_{xy}}^{R} C'(u)du = xE \int_{r_{xy}}^{R} au^b du = \frac{axE}{b+1}\left(R^{b+1} - r_{xy}^{b+1}\right) \tag{72}$$

In particular, for the diameter parallel to X $(y = 0)$ we have $r_{xy} = x$ if $x > 0$ and $r_{xy} = -x$ if $x < 0$.

Defining $\rho = x/R$, this can be rearranged as

$$\sigma(\rho, R) = \frac{aER^{b+2}}{b+1}\rho\left(1 - |\rho|^{b+1}\right) \tag{73}$$

Case of an Eccentric Section

Let us consider an eccentric section characterized by parameter k_O (Fig. 5). Combining Eqs. (67), (68) and (49), Eq. (54) becomes

$$\frac{\partial \varepsilon(x, y, R)}{\partial R} = (x - k_O R)C'(R) \tag{74}$$

Equation (54) then becomes

$$\frac{\partial \sigma(x, y, R)}{\partial R} = E(x - k_O R)C'(R) \tag{75}$$

Combining Eqs. (71) and (75), the final stress is obtained by integration

$$\sigma(x, y, R) = E \int_{r}^{R} C'(u)du - Ek_O \int_{r}^{R} RC'(u)du \tag{76}$$

Injecting Eq. (67)

$$\sigma(x, y, R) = aE\left(x \int_{r}^{R} u^b du - k_O \int_{r}^{R} u^{b+1} du\right) \tag{77}$$

Which integrates as

$$\sigma(x, y, R) = aE\left(\frac{x\left(R^{b+1} - r^{b+1}\right)}{b+1} - \frac{k_O\left(R^{b+2} - r^{b+2}\right)}{b+2}\right) \tag{78}$$

In particular, for the diameter parallel to X ($y = 0$) we have $r = x/(1 + k_O)$ for $x > 0$ and $r = x/(1 - k_O)$ for $x < 0$. Let ρ be the reduced position defined as

$$\begin{aligned} \rho &= (x/R)/(1 + k_O) \quad \text{if}\, x > 0 \\ \rho &= (x/R)/(1 - k_O) \quad \text{if}\, x < 0 \end{aligned} \tag{79}$$

Equation (78) can be rearranged as

$$\begin{aligned} \sigma(\rho, R) &= aER^{b+2}\left(\frac{\rho(1 + k_O)\left(1 - \rho^{b+1}\right)}{b+1} - \frac{k_O\left(1 - \rho^{b+2}\right)}{b+2}\right) \quad \text{if } \rho > 0 \\ \sigma(\rho, R) &= aER^{b+2}\left(\frac{\rho(1 - k_O)\left(1 - \rho^{b+1}\right)}{b+1} - \frac{k_O\left(1 - \rho^{b+2}\right)}{b+2}\right) \quad \text{if } \rho < 0 \end{aligned} \tag{80}$$

Application of the Bending Model to Allometric Growth

Let us assume that the tree grows following the same allometric law ($L = cR^d$) as in section "Application to Stem Allometric Growth" and compute the change in curvature due to the increase in self-weight. The normal force induced by self-weight N is given by Eq. (19).

Assuming that the centre of mass is located at mid-length of the stem load, the bending moment applied at the base of the stem can be computed

$$M = PL/2 = kc^2 R^{2d+2}/2 \tag{81}$$

The bending moment increment in response to a growth increment dR is

$$\frac{dM}{dR} = kc^2(2d + 2)R^{2d+1}/2 \tag{82}$$

The change in curvature is then

$$\frac{dC}{dR} = \frac{dM}{EIdR} = aR^b \tag{83}$$

Reminding that for a circular cross section $I = \pi R^4/4$, we have

$$b = 2d - 3 \tag{84}$$

$$a = \frac{2kc^2(2d + 2)}{\pi E} \tag{85}$$

Results are shown and commented on Fig. 6.

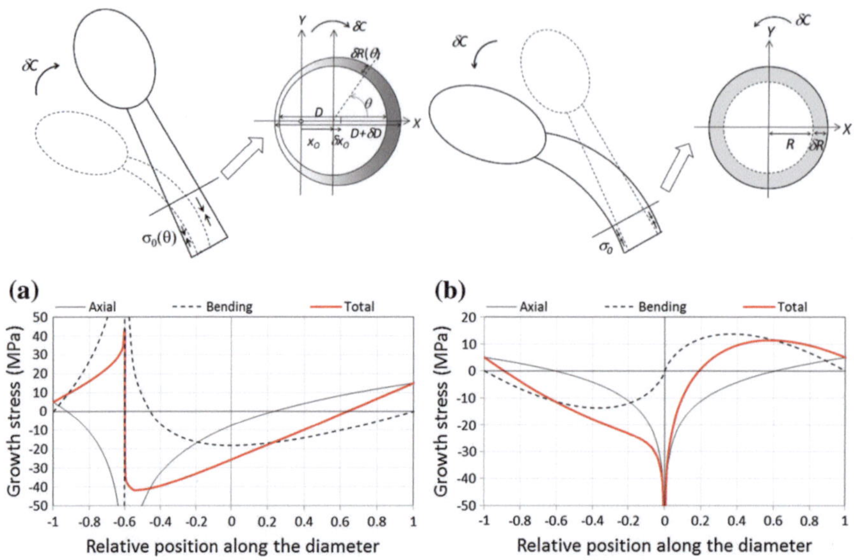

Fig. 7 General case of a stem bending while producing maturation stress. Two cases are illustrated: case of active up-righting due to asymmetric maturation (**a**), and case of passive bending under the self-weight due to isometric growth with production of axisymmetric maturation stress (**b**). Figures show separately stresses due to axial load resulting from maturation (continuous thin black line), stresses due to bending (dotted line) and total stress (thick red line). Stress profiles are asymmetric, with a "ν"-shape (Greek letter nu). In the case of active up-righting, the maximal magnitude of total stress is reduced compared to its individual components (axial and bending stresses)

Case of a Stem Bending While Reacting

In the general case, a stem always produces maturation stress and is sometimes also submitted to bending stresses. Whether the bending stresses are due to self-weight, active reaction or any combination of these the causes does not change the stress profile, only the variation of curvature during the course of growth matters. For the given section's parameters, the growth stress profile is simply obtained by adding axial and bending stress profiles. Results are shown and commented in Fig. 7.

Experiments: Growth Stress Measurements and Model Validation

Experiments

The study was conducted on four tilted poplar trees located near Montpellier (France). Tree diameter at breast height ranged between 6 and 14 cm. Trees were felled and cut

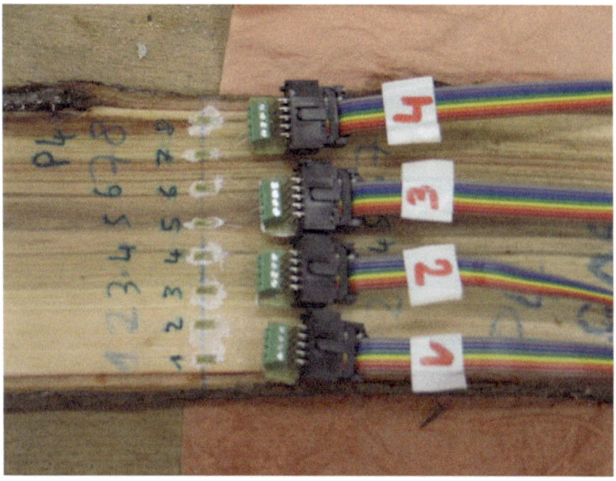

Fig. 8 Measurements of residual strains of a freshly sawn poplar board. 8 strain gauges are pasted along the grain across the diameter. Growth stresses are later released by operating transverse cuts below and above the gauges, and recording released strains

into logs 10–20 longer than the diameter, before being transported to the laboratory. Logs were stored in a plastic sheet in a refrigerated room, and a diametral board was sawn within a few weeks after felling to avoid drying. The board was oriented along the axis of symmetry of the tilted tree. Across the diameter located at mid-length of the board, 7–11 5 mm-long strain gauges were pasted in the grain direction and connected to a data logger (Fig. 8). Residual strains of growth stresses were released and measured by operating transversal cuts close to each side of the gauges.

Results and Assessment of the Model

Measured released strain profiles are shown in Fig. 9. The opposite of the strain is plotted, so that a contraction (negative strain corresponding to tensile growth stress) is represented by positive values. Stems had variable degrees of eccentricity (the Y axis corresponds to the pith). In all cases, tension is observed at the periphery and compression near the core, consistently with simulation. For two specimens (P1 and P4), a marked asymmetry of maturation strain (peripheral values of strain) is observed. In contrast with simulated growth stress profiles, we observe finite negative values of strains near the core, less negative and sharp than expected from simulations. This may be due to some nonlinear behaviour near the core, where large compression may be released by viscoelastic and/or plastic behaviour. Consistently with simulations, the maximal compression is mostly located near the pith (except for P4) rather than near the geometric centre of the section.

The variety of profiles observed are likely due to variable patterns of heterogeneity (of maturation stress or elastic modulus) and growth history, as attested by the variety of profiles obtained from simulations. Local increase in stress away from the pith as observed in P1 and P5 are likely due to past temporary reaction or bending movement of the stem.

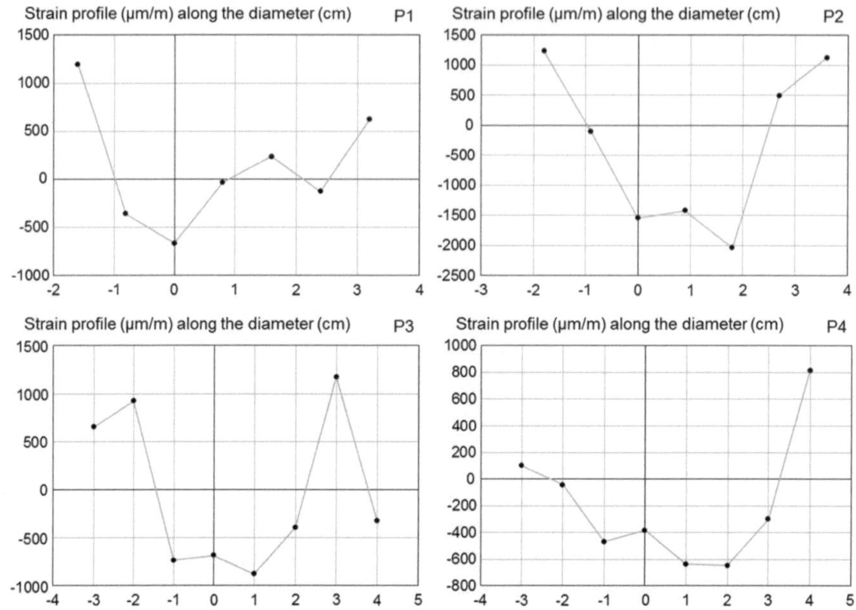

Fig. 9 Profiles of residual strains (10^{-6} m/m) measured across the diameter (cm) of four poplar trees. Different shapes are observed, more or less in accordance with simulated profiles, with tensile stress at the periphery and compressive stress near the core. The absence of strongly negative value near the core is likely due to plastic deformations occurring in this area, not taken into account by the models

Discussion

Limits of the Model and Possible Extensions

In this section, we will review possible extensions of the preceding formulations. In section "Extension of the Formulations", the case of different geometry or loading history, or the refinement of the mechanical analysis will be considered while keeping the general principles stated in part 2.1. In section "Reconsidering the Basic Equations", their partial release will be considered.

Extension of the Formulations

Using a numerical formulation, the model can be easily extended to any section shape or distribution of mechanical properties or maturation stress within the section. It is, however, still possible to obtain analytical expressions in some cases that were not detailed here

- The constant allometry assumed in section "Application to Stem Allometric Growth" or "Application of the Bending Model to Allometric Growth" can be generalized to more complex situations. Based on the incremental formulation giving the stress increment $\delta\sigma$ at a given position in the cross section for an increase of stem size δR, the partial derivative $\partial\sigma/\partial R$ can be obtained and the integration of Eq. (1) can be made. Possible extensions that could also lead to analytical equations, include a change—either progressive or sudden—of the allometry exponent, corresponding to a change of growth condition or of the limiting factor causing the allometry.
- The equations obtained for a circular cross section can be extended to the ellipsoid case by an appropriate change of variable in the computation of surface integrals. Here also a change of the growth conditions, resulting in a progressive or sudden change of circularity, can be introduced without much difficulty.
- The variables appearing in section "Modelling Additive Growth with Pre-stresses" and Eqs. (1–7)—stress σ, strain ε, stiffness E …—have been so far considered as scalars for a 1D formulation. The equations apply equally to 3D formulations, with all the variables becoming tensors. The beam theory for a slender structure, stated in section "Assumptions for a Growing Stem: Beam Theory" and Eq. (8), remains applicable using the formulation of generalized plane deformation as shown for example by Archer (1986) or Fournier et al. (1991a, b) for the case of a circular cross section with concentric growth rings (no eccentricity). These authors assumed orthotropic elasticity and also included in their formulation the torsion resulting from inclined grain. A model of transverse isotropy that neglects the differences between the transverse directions can be also used to simplify the formulations.

Reconsidering the Basic Equations

Except for (H1) (material loaded since created), all basic assumptions made in part 2.1 can be questioned. Linear elasticity (H2) obviously needs to be questioned considering the excessive stress levels predicted by the models near the biological centre of the stem. Elastoplastic models reducing the stress increase above a certain limit (the yield stress) can be introduced (Archer 1986), as well as the apparition of damage characterized by a decrease of stiffness. The generation to hollow stems can be addressed through such an approach, by considering that the core of the stem is totally damaged. Such time-independent formulations remain compatible with (H4).

When the viscoelastic behaviour is introduced, (H4) is not valid anymore: the time cannot be replaced by stem size, as there is an explicit dependence on time in the constitutive equations of the material. (H3) no longer applies when the effect of progressive maturation is studied as in Coutand et al. (2007). Viscoelasticity can be used to account for the progressive stiffening of the material during the process of maturation (Gril and Fournier 1993).

Beam theory (H5) can be used to model the movement of the whole stem under the action of internal or external forces. In that case, slow variations of properties along the stem axis, such as conicity, can be taken into account using the current description applicable to each stem portion. In case slenderness (length to diameter ratio) is not large enough, the effect of shear must be taken into account using the theory developed by Timoshenko (1940). It results in an additional contribution to the stem movement, in proportion to the wood anisotropy characterized by the ratio of longitudinal Young's modulus (E_L) to longitudinal shear modulus (G_{LT} or G_{LR}). In the long term, due to the higher relaxation in shear compared to that along the fibre, the effective level of anisotropy is likely to increase and the contribution of shear likely to become significant.

Consequences of Growth Stress Patterns for Stem Strength

Patterns of growth stress always exhibit very nonlinear variations across the diameter. This pattern is in sharp contrast with the linear profile expected for a beam submitted to axial and bending loads. These non-classical patterns are a direct consequence of the interaction between growth and loading: although the stress increment associated to a radius increment has linear variations within the section, the integration of this stress overgrowth is nonlinear. The profile is nonlinear because new wood layers are progressively added to structure, so that each layer has its own loading history: old layers have undergone more load increments than younger peripheral layers.

The state of pre-stressing has direct consequences on stem strength. If a transient bending load is applied to the stem, a linear stress profile is added to the growth stress profile. This additional stress field is maximal at the periphery, compressive on one side and tensile on the opposite side. Peripheral pre-tension reduces compression on the compressive side of the stem, and increases tension on the tensile side. As fresh wood is weaker in compression than in tension, the pre-stress is beneficial for the stem strength. The stem as a structure has larger strength than the wood it is made of. Since peripheral tension is typically 5–10 MPa (compared to a compressive strength typically ranging between 20 and 50 MPa), this strength increase amounts several tens of percent.

The peripheral tensile pre-stress is balanced by a strong compression near the core of the stem. Usually, this does not significantly reduce stem strength, since central parts of the stem bear only negligible bending loads. However, in the case where eccentricity is strong, large compression near the pith is located away from the neutral line of the stem. The pith, pre-stressed in compression, may then be additionally loaded in compression during transient bending. A non-classical behaviour may happen in this case, where the mechanical failure starts inside the stem (at the level of the pith) rather than at its periphery. Because stress profiles are often asymmetric, another non-classical behaviour may happen: the strength of the beam is not symmetric with respect to the bending direction. Figure 7 illustrates a case where strength is lower for downward movement (where compression away from the pith may add with transient compression) than for an upward movement. Permanent bending stress profiles (Fig. 6) have a particular sigmoid shape. An important point is that although it has undergone large changes in curvature, the stress is null at the periphery. As a consequence, a growing stem can withstand considerably larger changes in curvature without breaking than its non-growing equivalent. To illustrate this let us compare the case of a stem of diameter R, submitted to a curvature increment C either during the course of growth or at final state. If the change occurs at final state, then the maximal stress in the section is located at the periphery and can be computed using usual formulae, namely $\sigma_{max} = ERC$, where E is the elastic modulus. For the growing stem, we assume that the rate of change in curvature growth is equal to the rate of change of diameter ($b = 0$ in Eq. 67). Then, the location of the maximal stress can be obtained by deriving Eq. 73, and its magnitude can be expressed as $\sigma_{max} = \frac{1}{4}ERC$. Therefore, the maximal curvature that can be achieved by the growing stem is fourfold larger than the non-growing stem.

The state of stress of a stem can be viewed as the sum of three kinds of stress: pre-stress due to maturation at the periphery and associated growth stress in the core; stress due to permanent bending in response to self-weight and/or asymmetric maturation; stress due to transient bending when the stem is submitted to instantaneous external action. It is noteworthy that these stresses are smartly distributed in the section: maturation stresses are tensile at the periphery, growth stresses are concentrated near the core of the stem, bending stresses are concentrated in the mid-parts of the radius, and transient stresses are located in the peripheral part. Thanks to this smart distribution, a stem that has already undergone large permanent bending movements can also withstand transient bending without loss of strength.

References

Alméras T, Clair B (2016) Critical review on the mechanisms of maturation stress generation in trees. J R Soc Interface 13:20160550

Alméras T, Fournier M (2009) Biomechanical design and long-term stability of trees: morphological and wood traits involved in the balance between weight increase and the gravitropic reaction. J Theor Biol 256(3):370–381

Archer RR (1986) Growth stresses and strains in trees. Springer, New York

Bonser RHC, Ennos AR (1998) Measurement of prestrain in trees: implications for the determination of safety factors. Funct Ecol 12(6):971–974

Coutand C, Fournier M, Moulia B (2007) The gravitropic response of poplar trunks: key roles of prestressed wood regulation and the relative kinetics of cambial growth versus wood maturation. Plant Physiol 144(2):1166–1180

Dassot M, Constant T, Ningre F, Fournier M (2015) Impact of stand density on tree morphology and growth stresses in young beech (*Fagus sylvatica* L.) stands. Trees 29(2):583–591

Fourcaud T, Lac P (2003) Numerical modelling of shape regulation and growth stresses in trees: I. An incremental static finite element formulation. Trees 17(1):23–30

Fourcaud T, Blaise F, Lac P, Castéra P, De Reffye P (2003) Numerical modelling of shape regulation and growth stresses in trees. Trees 17(1):31–39

Fournier M, Stokes A, Coutand C, Fourcaud T, Moulia B (2006). Tree biomechanics and growth strategies in the context of forest functional ecology. In: Ecology and biomechanics—a mechanical approach to the ecology of animals and plants, pp 1–33

Fournier M, Chanson B, Thibaut B, Guitard D (1991a) Mécanique de l'arbre sur pied: modélisation d'une structure en croissance soumise à des chargements permanents et évolutifs. 2. Analyse tridimensionnelle des contraintes de maturation, cas du feuillu standard. Annales des Sciences forestières 48(5):527–546

Fournier M, Chanson B, Guitard D, Thibaut B (1991b) Mécanique de l'arbre sur pied: modélisation d'une structure en croissance soumise à des chargements permanents et évolutifs. 1. Analyse des contraintes de support. Annales des sciences forestières 48(5):513–525

Gril J, Fournier M (1993) Contraintes d'élaboration du bois dans l'arbre: un modèle multicouche viscoélastique, 11ème Congrès Français de Mécanique, vol 4. Association Universitaire de Mécanique, Lille-Villeneuve d'Ascq, pp 165–168

Greenhill A-G (1881) Determination of the greatest height consistent with stability that a vertical pole or mast can be made, and of the greatest height to which a tree of given proportions can grow. Proc Camb Philos Soc IV(Part II)

Jacobs RR (1945) The growth stresses of woody stem. Commonw For Bur Aust Bull 28:1–64

Jullien D, Widmann R, Loup C, Thibaut B (2013) Relationship between tree morphology and growth stress in mature European beech stands. Ann For Sci 70(2):133–142

Kübler H (1959a) Studien über Wachstumsspannungen des Holzes I. Die Ursache der Wachstumsspannungen und die Spannungen quer zur Faserrichtung. (Studies of growth stresses in trees Part 1. The origin of growth stresses and the stresses in the transverse direction). Holz Roh-Werk 17:1–9

Kübler H (1959b) Studien über Wachstumsspannungen des Holzes II. Die Spannungen in Faserrichtung. (Studies of growth stresses in trees Part 2. Longitudinal stresses). Holz Roh-Werk 17:44–54

McMahon TA, Kronauer RE (1976) Tree structures: deducing the principle of mechanical design. J Theor Biol 59(2):443–466

Moulia B, Coutand B, Lenne C (2006) Posture control and skeletal mechanical acclimation in terrestrial plants: implications for mechanical modeling of plant architecture. Am J Bot 93:1477–1489

Ormarsson S, Dahlblom O, Johansson M (2010) Numerical study of how creep and progressive stiffening affect the growth stress formation in trees. Trees 24(1):105–115

Timoshenko S (1940) Strength of materials. D. Van Nostrand Company, Inc

Bending Stress in Plant Stems: Models and Assumptions

Christopher J. Stubbs, Navajit S. Baban, Daniel J. Robertson, Loay Alzube and Douglas D. Cook

Abstract Analytic expressions for bending stress can be used to predict mechanical stresses and failure of plant stems. However, the nonuniform shape and anisotropic material properties of plant stems contradict several assumptions that are typically used in the derivation of bending stress equations. The purpose of this chapter is to analyze each of these assumptions to determine the accuracy with which beam theory can predict stresses in plant stems. Finite element models of plant stems were used to investigate and quantify the effect of each assumption. Finally, experimental case-study data was used to illustrate the applications of these equations. The goal of this work is to enable researchers to make informed decisions regarding mechanical models of plant stems used to predict of measure mechanical behavior of plants and plant tissues.

Keywords Plant stems · Stress · Modeling · Assumptions · Bending · Beams Maize

Introduction

Plant stems typically fail due to bending loads. When a failure occurs, it typically manifests in one of the four modes: buckling or kinking, splitting along longitudinal fibers, compressive tissue collapse, or tensile fracture of the stem tissue (Wegst and Ashby 2007). This chapter focuses on the techniques for modeling tensile, com-

N. S. Baban · L. Alzube · D. D. Cook (✉)
New York University, Abu Dhabi Campus, PO BOX 129188, Abu Dhabi, United Arab Emirates
e-mail: douglascook@nyu.edu

D. D. Cook
Brigham Young University, Provo, UT 84602, USA

D. J. Robertson
University of Idaho, 875 Perimeter Drive, Gauss-Johnson 234 G, Idaho, MS 0902, USA

C. J. Stubbs
New York University, 6 MetroTech Center, Brooklyn, NY 11201, USA

A. Geitmann and J. Gril (eds.), *Plant Biomechanics*,
https://doi.org/10.1007/978-3-319-79099-2_3

pressive, and shear stresses within the stem, which are responsible for the latter two failure modes.

Equations from engineering beam theory are frequently used to predict mechanical stresses that occur in plant stems due to external loading. Beam theory has been applied to studies on the thigmomorphogenesis of tomato stems (Coutand et al. 2000), the mechanics of tree trunks, (Dean and Long 1986; Speck et al. 1990), flexural strength of crops (Kokubo et al. 1989), and plant growth mechanisms such as gravitropism (Zandomeni and Schopfer 1994). More detailed information about bending stress equations are available in books such as *Plant Biomechanics* (Niklas 1992), *Plant Physics* (Niklas and Spatz 2012), and in the engineering literature (Beer et al. 2012; Boresi and Schmidt 2003; Crandall 2012; Timoshenko 1940).

In each of these books, certain assumptions are required to arrive at a set of equations describing structural stresses under bending. These assumptions are typically made as a means to an end, with the focus being on the final equations. But Karl Niklas alluded to a deeper understanding in the preface to *Plant Biomechanics* (Niklas 1992). He stated that the purpose of that book was "to illustrate representative principles rather than to delve deeply into them." The purpose of this chapter is to undertake the deeper work referred to by Niklas by dissecting and evaluating the derivation process in order to provide the reader with a clearer understanding of the accuracy and limitations of bending stress equations as applied to plant stems.

The chapter consists of several sections. Bending stress equations are derived in section "Derivation of Bending Stress Equations" using a relatively small set of assumptions. This section gives an explicit mathematical definition of each assumption used in the derivation process. Alternative formulations are also provided where traditional assumptions are not appropriate for the modeling of plant stems. In section "Evaluating Model Assumptions", the validity of each assumption made during the derivation is investigated by comparing analytic results with those obtained from a finite element model. Section "Structural Characteristics of Plant Stems" addresses structural aspects of the plant stems that are not explicitly addressed within the modeling assumptions. Section "Case Studies" provides several experimental case studies in which bending equations were applied to plant stems. Finally, the chapter concludes with section "Summary and Future Work": a brief summary of the chapter and identification of areas in which further work is needed.

Background Information

To aid the reader, several essential terms and concepts will be briefly defined and described before proceeding with the derivation of bending equations. Table 1 provides a list of terms, symbols, and nomenclature.

When an external force is applied to a plant stem, internal stresses develop within the stem to resist the externally applied load. These stresses act in addition to internal stresses that may be present in the absence of external loading (i.e., residual internal stresses Archer 1987; Vandiver and Goriely 2009). Mechanical stresses can be

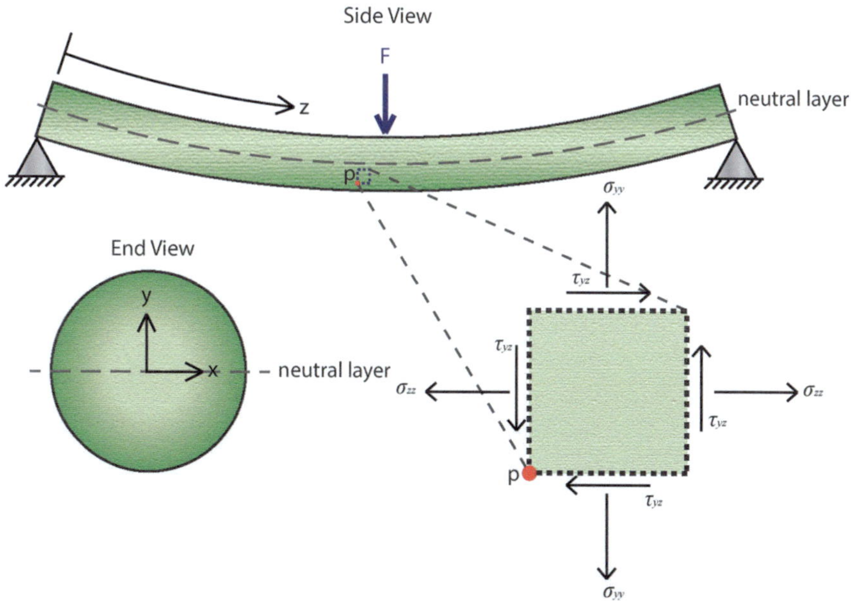

Fig. 1 An infinitesimally small cube with unit thickness taken within the plant stem under three-point bending used to quantify the stresses at a point P

characterized into two general categories: normal stresses and shear stresses. Normal stresses act perpendicular to the surface of interest and cause tension or compression. Shear stresses act parallel to the surface of interest. They do not induce tension or compression but cause shear deformation or torsion.

When an external load causes a plant stem to bend, internal normal and shear stresses are developed in the plant tissue. In this chapter, bending loads will primarily be illustrated using the common three-point bending load configuration depicted in Fig. 1. As shown in this figure, the top of the beam will be compressed and bottom of the beam will be stretched. Thus, the top of the beam will experience normal compressive stresses, while the bottom of the beam experiences normal tensile stresses. The normal stress changes linearly from maximum compressive stress at the top of the beam, to maximum tensile stress at the bottom of the beam (Byars et al. 1983). The location inside the beam at which the normal stress changes from compressive to tensile is termed as the neutral layer. The neutral layer is neither stretched nor compressed, thus the normal stress at this layer is equal to zero. The neutral layer is typically, though not always located at or near the geometric centroid of the stem (Boresi and Schmidt 2003; Niklas 1992). Methods for determining the precise location of the neutral layer are beyond the scope of this study, but are available in the literature (Boresi and Schmidt 2003).

Table 1 Terminology and symbols used within this chapter

Symbol(s) or Terms	Description
x, y, z	Cartesian coordinates. The z axis is aligned with the apical–basal direction of the stem. The x and y coordinates originate from the geometric cross section of the plant stem (see Fig. 1)
L	Length
D	Diameter of the plant stem
σ_{ii}	Normal stresses in the i direction
ε_i	Normal strain in the i direction
τ_{ij}	Shear stress on the i face, in the j direction
ρ	Radius of the curvature
E_i	Stiffness (Young's Modulus) in the i direction
G_{ij}	Shear modulus between the i direction and j direction
V	Internal shear load
M	Internal bending moment
I	Area moment of inertia
w(y)	Width of the cross section as a function of y coordinates
I_E	Material stiffness weighted second moment of area
σ_1, σ_2	Principal normal stresses
σ_v	von Mises stress
[ε]	Strain matrix
[K]	Stiffness matrix
[σ]	Stress matrix
S_{rind}	Section modulus of the rind
ν_{ij}	Poisson's ratio between the i and j directions
Beam	A slender structure that is primarily subjected to transverse loading (Riley and Zachary 1989)
Neutral layer	The layer in a beam where the axial stress is zero. This layer corresponds with $y=0$
Sample slenderness (L/D)	Ratio of beam's length to its diameter (or height)
Diameter distance	The nondimensional distance from a support or load to a point of measurement, measured in terms of the number of stem diameters
Isotropy	A material which has the same properties in every direction (Callister and Rethwisch 2012)
Anisotropy	Material which has different material properties in different directions (Callister and Rethwisch 2012)
Orthotropy	Material that possesses material symmetry along orthogonal axes (e.g., wood) (Shukla et al. 1989)

Table 2 List of assumptions used to derive bending stress equations

Assumptions	
A	Orthotropic linear elasticity (Eq. 1)
B	Normal stress σ_{zz} is much larger than σ_{xx} & σ_{yy} (Eq. 2)
C	Bending is locally approximated by an arc (Eq. 3)
D	Homogeneous cross-sectional stiffness (no equation)
E	Shear stress τ_{yz} is much larger than shear stresses τ_{xz} & τ_{xy} (Eq. 11)
F	Shear stress is a function of y only (Eq. 11)

Derivation of Bending Stress Equations

Many sources that describe the derivation of bending stress equations introduce more assumptions than are actually necessary. The derivation in this chapter utilizes just 6 assumptions, which are briefly described in Table 2. As mentioned previously, the validity of each assumption will be evaluated in section "Evaluating Model Assumptions" along with the effect each assumption has on the accuracy of the bending equations.

Normal Stresses

We begin by assuming plant tissue can be modeled using a three-dimensional, orthotropic, linear elastic material (Assumption A). This type of material model is often used to describe the properties of wood and other fibrous composites and is, therefore, a good approximation of plant tissue (Agarwal et al. 2006; Niklas 1992; Shukla et al. 1989). Orthotropic, linear elastic material theory states that material strains are the product of a compliance matrix and a stress vector as shown below (refer to Table 1 for the interpretation of subscripts).

$$\begin{bmatrix} \varepsilon_{xx} \\ \varepsilon_{yy} \\ \varepsilon_{zz} \\ \varepsilon_{yz} \\ \varepsilon_{zx} \\ \varepsilon_{xy} \end{bmatrix} = \begin{bmatrix} \frac{1}{E_x} & -\frac{\upsilon_{yx}}{E_y} & -\frac{\upsilon_{zx}}{E_z} & 0 & 0 & 0 \\ -\frac{\upsilon_{xy}}{E_x} & \frac{1}{E_y} & -\frac{\upsilon_{zy}}{E_z} & 0 & 0 & 0 \\ -\frac{\upsilon_{xz}}{E_x} & -\frac{\upsilon_{yz}}{E_y} & \frac{1}{E_z} & 0 & 0 & 0 \\ 0 & 0 & 0 & \frac{1}{2G_{yz}} & 0 & 0 \\ 0 & 0 & 0 & 0 & \frac{1}{2G_{zx}} & 0 \\ 0 & 0 & 0 & 0 & 0 & \frac{1}{2G_{xy}} \end{bmatrix} \begin{bmatrix} \sigma_{xx} \\ \sigma_{yy} \\ \sigma_{zz} \\ \sigma_{yz} \\ \sigma_{zx} \\ \sigma_{xy} \end{bmatrix} \tag{1}$$

Next, we assume normal stresses in the z direction are much larger than the normal stresses in the x and y directions (Assumption B). The accuracy of this assumption primarily depends upon local slenderness (see section "Evaluating Model Assumptions"). If σ_{xx} and σ_{yy}, are neglected, and, therefore, assumed to be zero, Eq. 1 reduces to a much simpler form:

$$\sigma_{zz} = E_z \varepsilon_{zz}, \tag{2}$$

Next, we assume that local deformation of a plant stem in bending can be approximated by an arc (Assumption C). Under this assumption, the strain is inversely proportional to local arc radius (ρ) and is linearly related to distance from the neutral layer (y). In other words, tensile and compressive normal forces occur on opposing sides of the neutral layer and smaller arc radii induce larger strains than bigger arc radii. The assumption is stated mathematically by the following equation:

$$\varepsilon_{zz} = \frac{y}{\rho} \tag{3}$$

Combining Eq. 2 and Assumption C, we have the following:

$$\sigma_{zz} = E_{zz} \frac{y}{\rho} \tag{4}$$

By Newton's first law, the total internal moment, *M,* of each cross section must be equal to the bending moment created by internal normal forces. Thus, at each point in the stem cross-section, the internal moment with respect to the neutral layer is equal to the sum of all internal normal forces multiplied by their distance from the neutral layer. Mathematically,

$$M = \int_A y\, \sigma_{zz}\, dA \tag{5}$$

Substituting Eq. 4 into Eq. 5 we have the following:

$$M = \int y \left(E_z \frac{y}{\rho} \right) dA \tag{6}$$

At this point, E_z is assumed to be constant across the cross section (Assumption D) and can, therefore, be placed outside of the integral. This is the first assumption which appears to be inappropriate for plant stems. Many plant stems, particularly grasses, utilize stiffness gradients to optimize structural integrity (Rüggeberg et al. 2008; Speck and Burgert 2011). Assumption D is used at this point to illustrate the traditional formulation and is later eliminated. Carrying on with an assumption of constant stiffness we obtain:

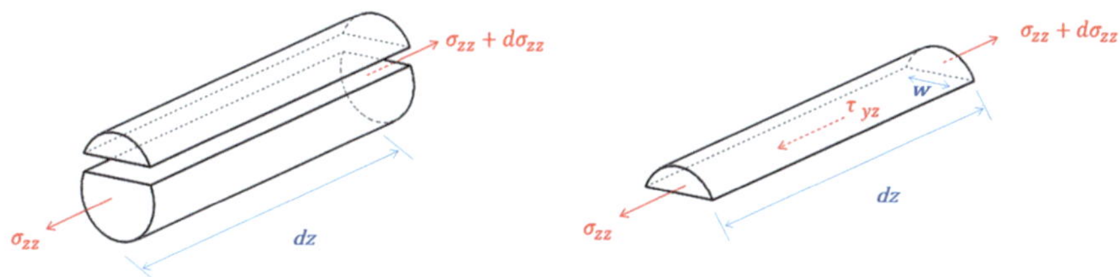

Fig. 2 Free-body diagrams of the differential segment of the plant stem showing the generation of shear stress due to changing normal stress along the longitudinal direction. The differential section sliced along a horizontal plane (left); the upper section in isolation (right)

$$M = \frac{E_z}{\rho} \int_A y^2 dA. \tag{7}$$

The integral term in Eq. 9 is recognized as the area moment of inertia (I):

$$I = \int_A y^2 dA \tag{8}$$

Thus, the internal bending moment depends upon material properties, area moment of inertia, and radius of curvature:

$$M = \frac{E_z I}{\rho} \tag{9}$$

Solving for the radius of curvature, ρ from Eq. 4 and substituting into Eq. 9 we find that the material effect is canceled yielding the classic equation for stress as a function of the internal bending moment, distance from the neutral layer, and area moment of inertia:

$$\sigma_{zz} = \frac{M\,y}{I} \tag{10}$$

Shear Stresses

None of the assumptions made thus far prohibit variation in the internal bending moment and normal stresses, σ_{zz}, along the longitudinal axis (z direction) of the stem. When internal normal stresses do vary along the length of the stem additional stress components are required to achieve internal static equilibrium. The magnitude of these stresses can be determined by creating a free body diagram of a small region of excised tissue (Fig. 2).

Next, we assume that shear stress τ_{yz} is much larger than shear stresses τ_{xz} and τ_{xy} (Assumption E). Furthermore, we assume τ_{yz} does not vary in the x direction (Assumption F). Thus, τ_{yz} becomes a function only of y. When the τ_{yz} shear stress varies only in the y direction, the free body diagram of Fig. 2 can be used to determine the value of shear stresses by summing forces in the z direction.

The net force in the z direction due to normal stress, σ_{zz}, is the differential increase in σ_{zz}, integrated across xy cross section. This is the term on the right-hand side of Eq. 11. The net normal force is balanced by the net shear force, which is the integral of τ_{yz} across the horizontal plane. Since the shear stress is assumed to only vary in y, the integral reduces to the shear stress times the width as a function of, $w(y)$, which is the left-hand side of Eq. 11. The limits of the integration y_i and y_f denote y direction distances from the neutral layer to the considered and the outermost layer respectively.

$$\tau_{yz} w(y) dz = \int_{y_i}^{y_f} d\sigma_{zz} \cdot dA \tag{11}$$

Solving for the shear stress we have the following:

$$\tau_{yz} = \frac{1}{w(y)} \int_{y_i}^{y_f} \frac{d\sigma_{zz}}{dz} \cdot dA \tag{12}$$

Equation 10 can be substituted into Eq. 12 to obtain following equation:

$$\tau_{yz} = \frac{1}{w(y)} \int_{y_i}^{y_f} \frac{d}{dz} \left(\frac{M(z)y}{I} \right) \cdot dA \tag{13}$$

Because the derivative of the internal bending moment is equal to the internal shear, *v(z)* (Beer et al. 2012), a simplified form is obtained:

$$\tau_{yz} = \frac{V(z)}{I\, w(y)} \int_{y_i}^{y_f} y\, dA \tag{14}$$

For illustrative simplicity, if we assume a *rectangular* cross section, the width as a function of y becomes a simple constant and the shear stress is found as a function of y for a given value of z:

$$\tau_{yz} = \frac{V(z)}{I} \left(\frac{h^2}{8} - \frac{y^2}{2} \right) \tag{15}$$

This parabolic distribution of shear stress is the classic solution presented in most engineering textbooks, but because it required an assumption of the rectangular cross section, it is not appropriate for plant stems.

Adjusted Bending and Shear Stress Equations

Bending Stresses

Not all assumptions used in the previous derivation are valid for plant stems. The most erroneous assumption in Table 1 is constant cross-sectional stiffness (Assumption D). While some woody plants may be approximated as having a constant cross-sectional stiffness, this is not generally true (Niklas 1992). Instead of assuming that E_z is constant, we more accurately model a plant stem by assuming that stiffness varies as a function of y and x. Equation 7 is then rewritten as:

$$M = \int_A y \left(E_z(x, y) \frac{y}{\rho} \right) dA \tag{16}$$

We can remove ρ from the integral because it is constant in the plane of integration. After doing so, we define a new quantity: the material-weighted area moment of inertia. It is analogous to the traditional area moment of inertia, but each differential element *dA* is weighted by the material stiffness, $E_z\ (x, y)$:

$$I_E = \int_A E_z(x, y)\, y^2\, dA \tag{17}$$

We now progress as before, with

$$M = \frac{I_E}{\rho} = \frac{I_E}{E_z(x, y)\frac{y}{\sigma}} \tag{18}$$

and finally,

$$\sigma = \frac{M\, E_z(x, y) y}{I_E} \tag{19}$$

When variable stiffness is considered, we see that the material stiffness remains in the equation for bending stress, thus leading to a slightly modified form of the traditional equation. As in the traditional form, I_E, and M act as a scaling constant in the determination of stress. Both of these terms affect the stress magnitude, but do not affect the spatial distribution of stress. The spatial stress distribution is found to be the first moment material stiffness about the stem's neutral layer: $E_z\ (x, y)y$.

Shear Stresses

A similar process can be followed to obtain shear stress for variable cross-sectional stiffness. By assuming E_z to be a function of x and y:

$$\tau_{yz} = \frac{1}{w(y)} \int_{y_i}^{y_f} \frac{d}{dz} \left(\frac{M(z)\, E(x, y)\, y}{I_E} \right) \cdot dA \tag{20}$$

By using Eq. 2, the above equation simplifies to

$$\tau_{yz} = \frac{V(z)}{I_E\, w(y)} \int_{y_i}^{y_f} E(x, y)\; y\; dA \tag{21}$$

Combined Stress States

Cartesian stresses provide mathematical convenience, but have certain limitations. For structures involving multiple stress components, the stresses responsible for failure are rarely aligned with the Cartesian axes. Three common methods for predicting failure stresses are principal stresses, maximum shear stress, and von Mises stresses. Readers that are not familiar with these terms may refer to *Advanced Mechanics of Materials* (Boresi and Schmidt 2003). A brief explanation of principal stresses and von Mises stresses are given below.

For two-dimensional stress states, any combination of σ_{zz} and τ_{yz} are equivalent to two principal normal stresses, σ_1 and σ_2 which are (a) invariant (i.e., independent of Cartesian coordinate orientation) and (b) orthogonal to each other (Beer et al. 2012). A state of pure stress, τ_{max} also exists for every stress state. For the two-dimensional bending stress states described above, the two principal normal stress (σ_1 and σ_2) can be obtained by the following relation (Beer et al. 2012):

$$\sigma_1\,,\, \sigma_2 = \frac{\sigma_{zz}}{2} \pm \sqrt{\frac{\sigma_{zz}^2}{4} + \tau_{yz}^2} \tag{22}$$

Maximum shear stress can be obtained by the following relation (Beer et al. 2012):

$$\tau \text{max} = \sqrt{\frac{\sigma_{zz}^2}{4} + \tau_{yz}^2} \tag{23}$$

Angle between σ_{zz} and σ_1 can be obtained by the following relation (Beer et al. 2012):

$$\tan(2\theta) = \frac{2\tau_{yz}}{\sigma_{xx}} \tag{24}$$

The maximum shear stress is always oriented at a 45° angle from the principal stresses.

Another common method for combining stresses is the von Mises failure criterion. For the two-dimensional bending stresses derived above, von Mises stress is given as (Beer et al. 2012):

$$\sigma_v = \sqrt{\sigma_{zz}^2 + 3\,\tau_{yz}^2} \tag{25}$$

These stresses provide a means of assessing total stress that is independent of the choice of coordinate system.

Evaluating Model Assumptions

We now proceed to analyze each assumption used in section "Derivation of Bending Stress Equations". In particular, the stresses predicted by the equations developed in section "Derivation of Bending Stress Equations" will be compared to stresses derived from a computational finite element model of a plant stem. A finite element model is used by the authors because finite element methods allow for a generalized strain and stress solution based on internal potential energy minimization, and therefore do not inherently rely on any of the assumptions made in section "Derivation of Bending Stress Equations" (Kim and Sankar 2009). Finite element modeling also enables parameterization of individual assumptions, loading conditions, and boundary conditions.

In the current work, all finite models of plant stems possessed circular cross sections. Both solid and hollow cross sections were analyzed. Hollow models did not include nodes. In general, discrepancies between the analytic equations of section "Derivation of Bending Stress Equations" and the solid cross-section finite element model were smaller than discrepancies between the analytic equations and the hollow cross-section finite element model. To provide conservative assessments of assumption validity, many results below only show comparisons with the hollow cross-section finite element model. A complete description of each finite element model can be found at the end of this chapter.

Assumption A: Material Model

Both the analytic equations of section "Derivation of Bending Stress Equations" and the finite element models presented in this section assume linearly elastic, transversely isotropic material properties. Although alternatives to this approach are possible, such formulations are beyond the scope of this chapter. In addition, all finite element models developed as part of this work possessed material stiffness in the *z* direction was 20 times higher than the material stiffness in the *x* and *y* directions (Gibson 2012).

Assumption (B): Negligible Normal Stresses in the x *and* y *Directions*

Assumption (B) dictates that the stress state of the stem is dominated by normal stress in the *z* direction (σ_{zz}). A comparison of the stresses predicted by the analytic bending equations and the finite element model is provided in Fig. 3. Because normal stress (σ_{zz}) increases linearly along the stem, each stress component was normalized by σ_{zz}. As shown in Fig. 3, the analytic model predicts two stress components, while the finite element model predicts all six stress components. As seen in the right-hand side of the figure, stress components σ_{xx}, σ_{yy}, and τ_{xy} reduce to less than 0.1% of the bending stress within two diameters of the support ($z/D < 2$), and within one diameter ($z/D < 1$) of the point of loading. In the span where σ_x, σ_y, and τ_{xy} are not depicted along the bottom edge of the chart ($2 < z/D < 24$), their values are much less than 0.1% of the bending stress. This measure is referred to as *diameter distance*, defined as the distance from a support or load point divided by the stem diameter, and will be discussed further in the following sections. Shear stresses τ_{yz} and τ_{xz} remain significant throughout, but reduce to below 10% of the bending stress at five diameters from the support. The shear components will be discussed in more detail later in this section.

Assumptions (C): Normal Strains Approximated by an Arc

The analytical stress equations assume that normal strains are linearly distributed about the neutral layer, and inversely proportional to the radius of curvature (Assumption C). The validity of this assumption was checked by examining the stress distributions in the *xy*-plane of both the hollow and the solid finite element model. Figure 4 shows the distribution of σ_{zz} predicted by the bending equations as well as the σ_{zz} distribution for the finite element model.

Figure 4 provides a comparison of normal stresses for a single cross section. A broader assessment of Assumption C is possible by examining the validity of this assumption as a function of diameter distance. If Assumption C is valid, each planar cross section of the stem will remain planar as the stem bends. Many textbooks use the phrase "plane sections remain plane" to describe this phenomenon (Bauchau and Craig 2009).

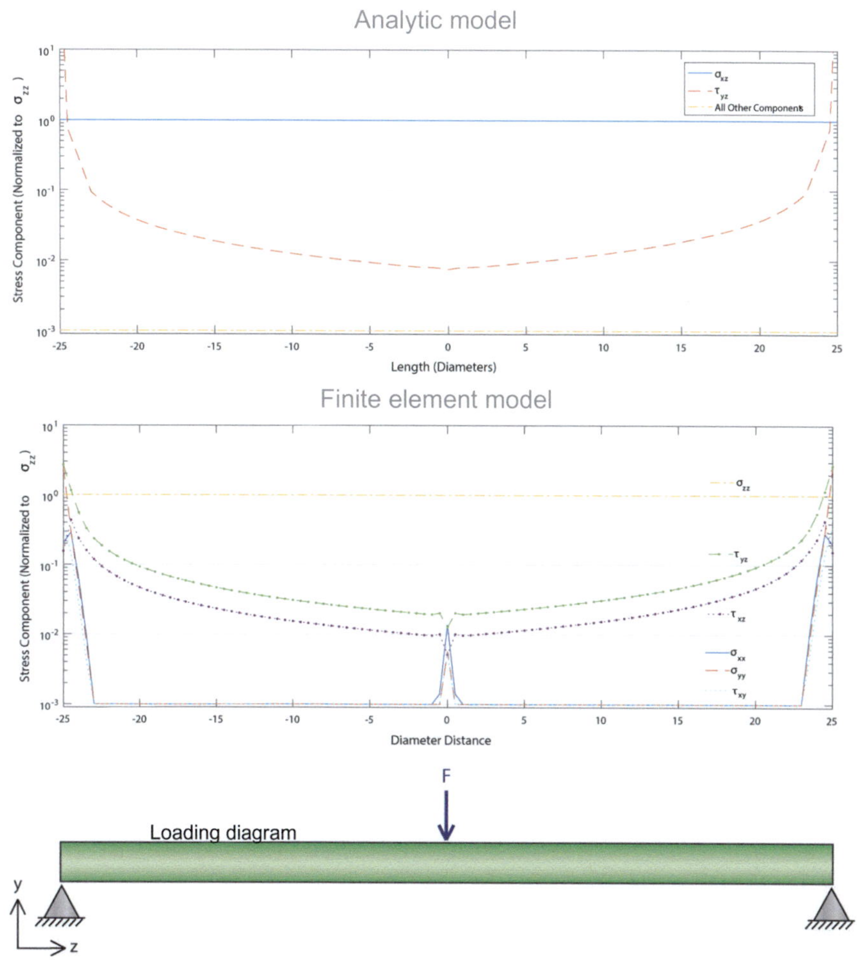

Fig. 3 Comparison of the stresses predicted by analytic bending equations (top) and the finite element model (middle). Both models are based on modeling a hollow circular cross section. Stresses have been normalized to σ_{zz}. All distances are measured in terms of diameters from the point of loading. For convenience, a loading diagram is provided at the bottom of the figure

Figure 5 illustrates the "nonplanarity" as a function of diameter distance based on data from the finite element model. Nonplanarity is defined as the maximum deviation from planar deformation, divided by the maximum deformation within the same plane. A nonplanarity value of zero thus implies perfectly planar deformation, while a nonplanarity value of 0.1 would imply that the maximum out-of-plane deformation is one-tenth of the maximum deformation of the same plane.

Out-of-plane deformation is directly related to the ratio between shear stresses and σ_{zz}. When σ_{zz} dominates, plane sections remain plane. This assumption is also

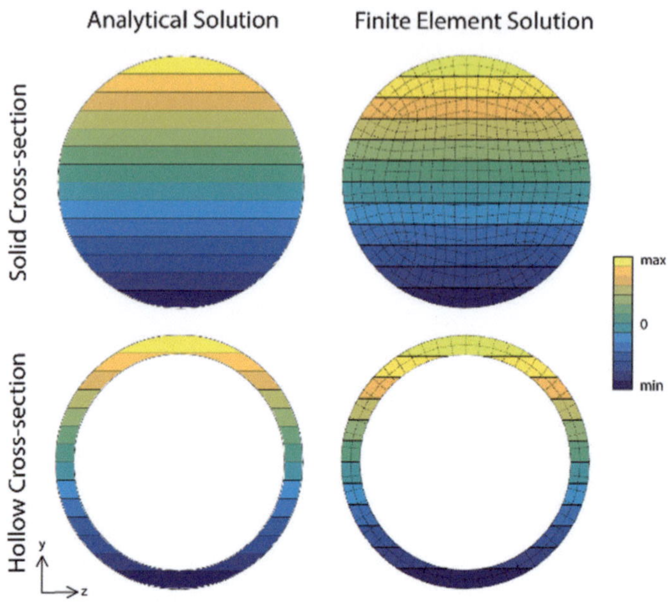

Fig. 4 Validation of assumption C. Bending stress distributions for a solid and hollow cross section, for the analytical solution (left) and finite element model (right)

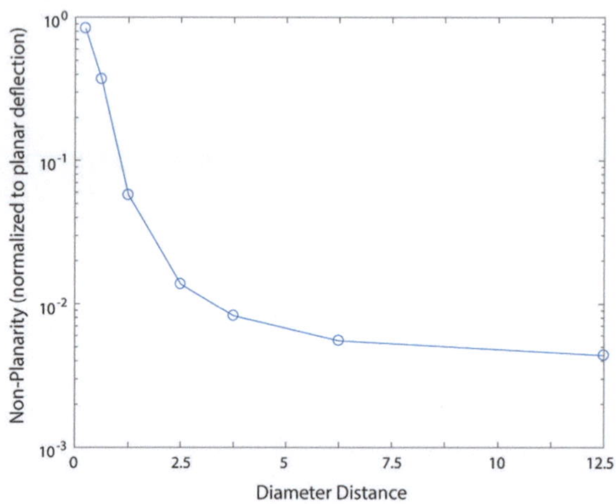

Fig. 5 Nonplanarity as a function of diameter distance from the point of loading. Non-planarity is defined as the maximum deviation from planarity divided by the maximum planar deformation of the cross-section

related to the diameter distance of the plant stem. For positions above 5 diameters from the left support, non-planarity is significantly less than 1%.

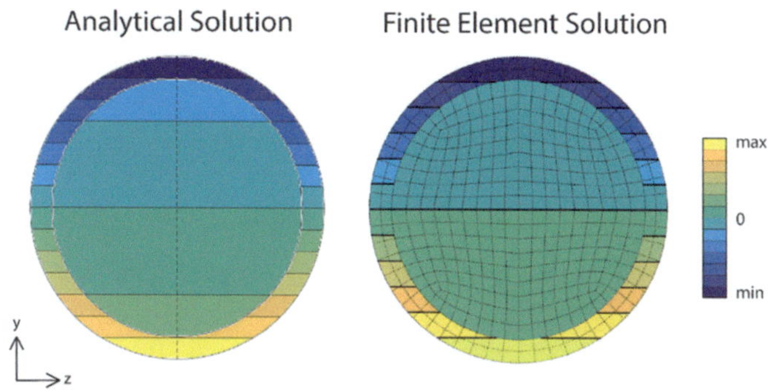

Fig. 6 Bending stress in a two-material cross section, analytical model (left) using Eq. 19 and the finite element model (right)

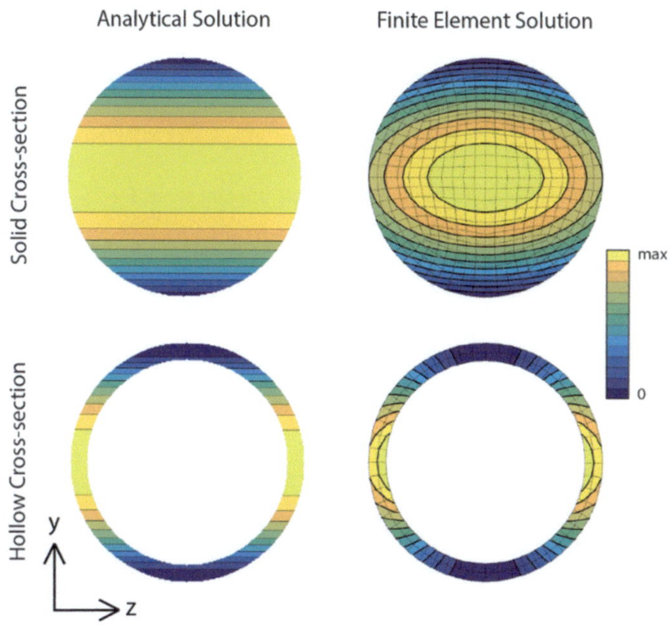

Fig. 7 Shear stress distributions for solid and hollow models of plant stems for τ_{yz}. Analytical model (left) using uniform cross-sectional stiffness and the finite element model (right)

Assumption (D): Constant Material Properties

Assumption (D) is that the stem is made of a single homogenous material. This assumption is not valid for most plant stems. Equation 19, utilizes the material-weighted area moment of inertia, which incorporates varying material properties for the calculation of bending stress. This generalized formula can be accurately used

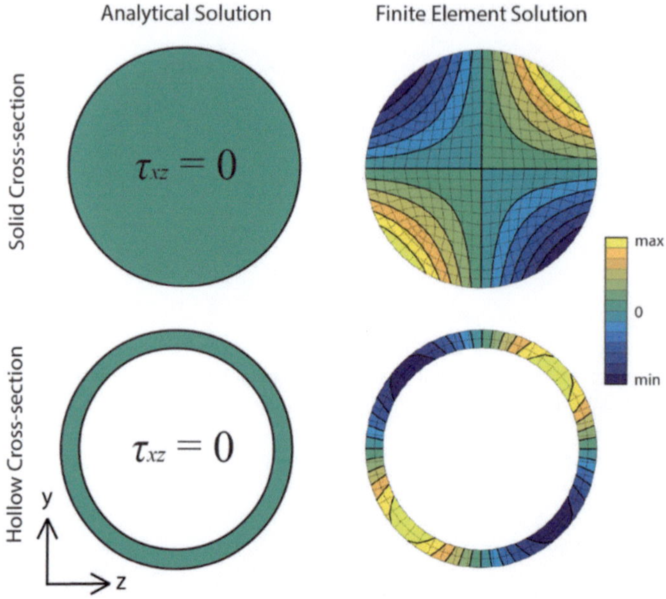

Fig. 8 Stress distributions for solid and hollow models of plant stems for τ_{xz}. Analytical model (left) and the finite element model (right)

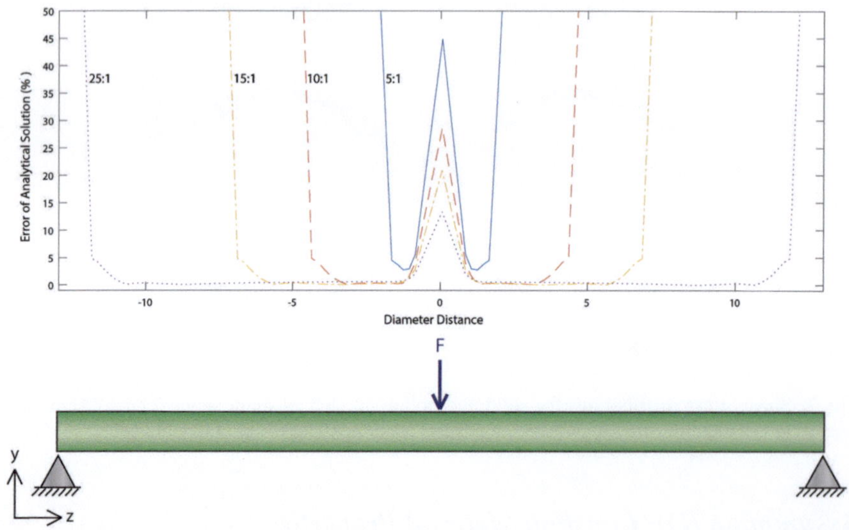

Fig. 9 Relative error of the analytic solution for σ_{zz} in hollow stem models having sample slenderness values of 25:1, 15:1, 10:1 and 5:1. All lengths are measured in number of diameters from the point of loading

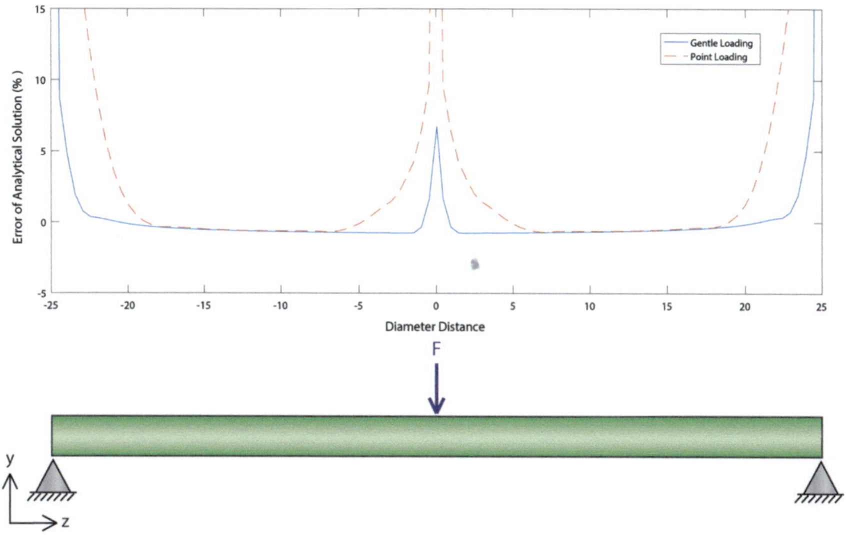

Fig. 10 Relative error of the analytic solution for σ_{zz} in a hollow stem model for gentle loading and point loading. All lengths are measured in number of diameters from the point of loading

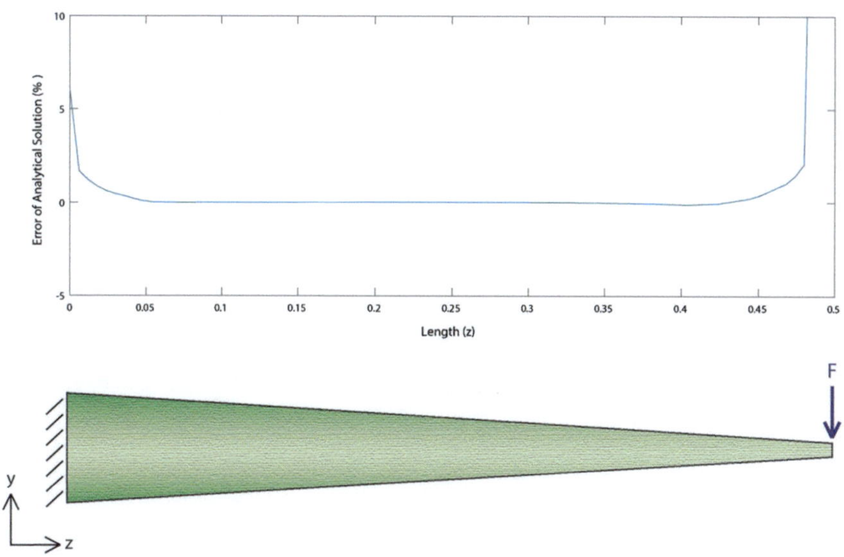

Fig. 11 Relative error of the analytic solution for σ_{zz} in a 10:1 tapered hollow stem model for cantilever point loading

for cross sections with nonconstant material stiffness or for two-material models (constant material properties within the two domains like rind and pith). The following figure shows stress distributions for rind–pith plant stems in which the rind

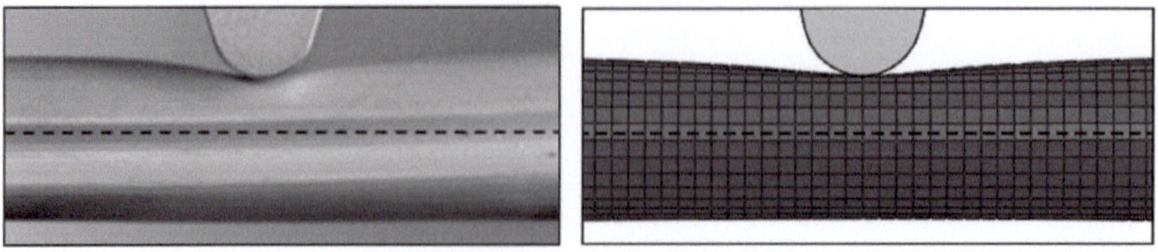

Fig. 12 Cross-sectional deformation caused by intermodal loading for (Robertson et al. 2015b) laboratory testing (left) and finite element solution (right)

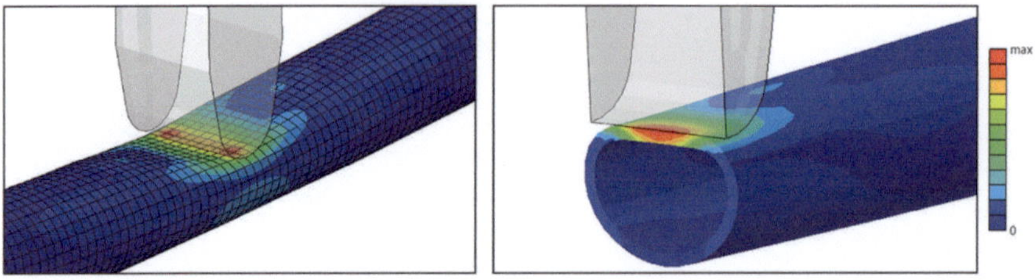

Fig. 13 Finite element model, depicting von Mises stresses for an isometric view (left) and a view cut through the center of the loading (right)

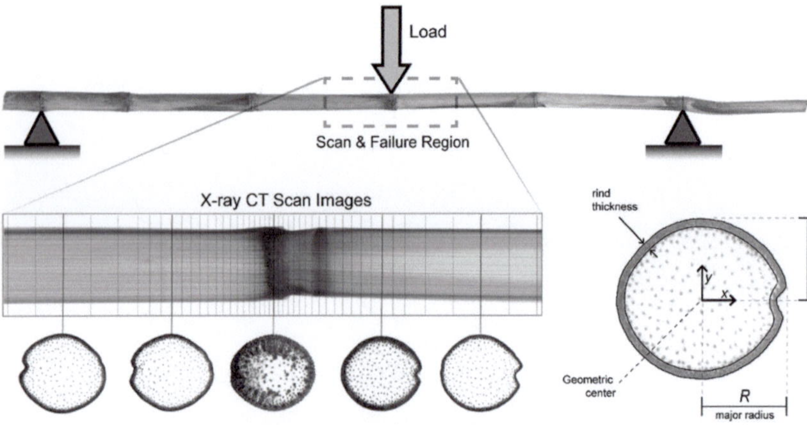

Fig. 14 Diagram depicting three-point bending test arrangement and region of CT scan (top), axial and transverse CT scan images (lower left), and an annotated cross section indicating some features identified through image processing (lower right)

is 20 times stiffer than the pith material. For such situations, Eq. 19 can be used to show that the ratio of stresses on either side of the pith/rind interface is exactly equal to the stiffness ratio across the interface. For the case shown below, σ_{zz} near the interface in the pith is 20x lower than the stresses in the rind. Equation 19 can also be used to assess stresses for continuously varying material stiffness, as shown later (see Fig. 16).

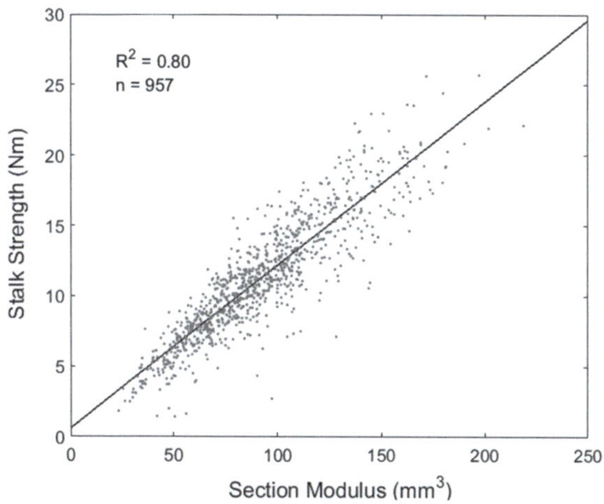

Fig. 15 Scatter plot and regression line between the stalk strength (maximum internal bending moment) and the section modulus of the rind

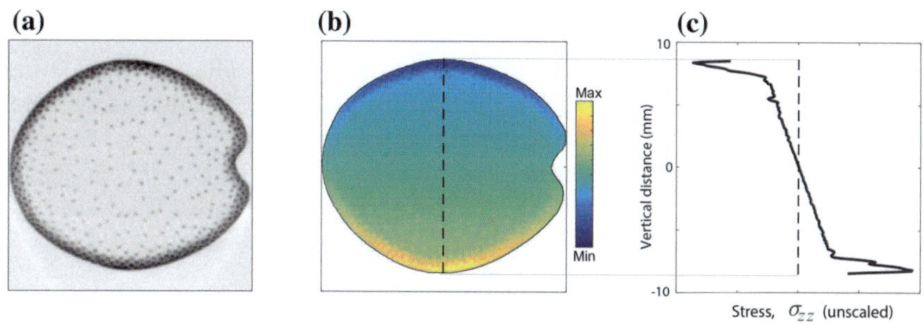

Fig. 16 Left: CT cross-section image; Center: σ_{zz} as determined by Eq. 19; Right: A line plot of σ_{zz} along the dashed line in the center image

Assumptions (E) and (F): Constant Shear in the X Direction

Assumption (F) states that τ_{yz} is constant in the x direction. This assumption is valid for rectangular cross sections in which there is no geometric or material variation along x direction, but it is not necessarily valid for cross sections that do not meet these criteria.

Figure 7 shows the distribution of τ_{yz} for hollow and solid circular cross sections. While the analytic bending equations generate horizontal bands of shear stress, the finite element model does not. This is because the circular geometry results in distributions of τ_{yz} that differ from the Assumption F. It should be noted that this model uses uniform cross-sectional stiffness. Stiffness variation in the cross section, which is common in plant stems would likely cause even further deviation from this assumption.

Assumption E is that two shear components can be neglected ($\tau_{xy} = \tau_{xz} = 0$). As shown in Fig. 3, although τ_{xy} is negligible, τ_{xz} is comparable in magnitude to τ_{yz}. Figure 8 shows that this shear component is zero in the analytic model, but not in the finite element model.

As seen in Figs. 7 and 8, the analytic bending equations do not accurately predict shear stresses. However, these inaccuracies may not be severe limitations for the prediction of failure. Recall that σ_{zz} is generally much larger than either stress component, and that the discrepancy between normal and shear stresses decreases with increasing distance from an applied load or a support. This is because stress formulations such as principal stress, maximum shear stress, and von Mises stress tend to amplify the influence of σ_{zz} (see Eqs. 22, 23, and 25). In addition, maximum σ_{zz} occurs at the top center and bottom center of the cross section, but shear stress components are very small at these points. As a result, analytic bending equations tend to be quite accurate, as will be discussed in the following section.

Structural Characteristics of Plant Stems

This section addresses aspects of plant stems that are not typically addressed in traditional mechanics textbooks, such as orthotropy, error as a function of diameter distance, sample slenderness, inhomogeneity, taper, and loading effects. The influence of each of these factors on the accuracy of bending stress equations is discussed below.

Orthotropy

Orthotropic materials are those in which material properties are orthogonal to each other (Nagarajan and Zak 1985). Wood is a common example of an orthotropic material. Wood typically has one stiffness in the longitudinal direction, another stiffness in the radial direction, and yet another stiffness in the circumferential direction.

Derivations of bending stress equations in textbooks vary in their treatment of orthotropy. Introductory level textbooks generally do not address this issue. As shown above, equations for bending stress (Eqs. 10 and 19) are valid for orthotropic materials (Hashin 1968; Silverman 1964). Thus, the primary material property for modeling bending stresses within plant stems is the longitudinal elastic stiffness or longitudinal Young's modulus, E_z.

Assumptions About the Stem Cross Section

The derivation of bending stress equations was made without restrictions on cross-sectional geometry. Finite element results provided above show that the analytic equations are reliable for both hollow and solid cross sections. However, loads that do not pass through the shear center of a beam induce torsional stresses in addition to bending stress (Boresi and Schmidt 2003). Thus, the equations above should only be applied when loading is aligned with the shear center of a stem. The determination of the shear center is beyond the scope of this chapter, but for circular and elliptical stems with symmetrical or axisymmetrical material properties, the shear center is collocated with the geometric center of the stem.

Most textbook derivations of bending stress equations assume a homogeneous cross-section. While this may be a reasonable approximation for woody stems, most plant stems exhibit some spatial variation in cross-sectional stiffness (Niklas 1992). Equations 16–21 show how the traditional equations can be modified to account for variation in cross-sectional stiffness. Figures 4, 6, 7 and 8 show distributions for the analytical and finite element models for various cross sections relevant to plant stems. In most cases, the bending equations provide relatively good estimates of stress.

Slenderness

Many sources indicate the accuracy of the bending stress equations is related to sample slenderness, defined as the ratio of length to diameter (*L/D*) of the sample (Boresi and Schmidt 2003; Niklas and Spatz 2012). This is not completely correct. The accuracy of the analytic equations is primarily dependent upon *diameter distance*. This is illustrated in Fig. 9, which contains data for several hollow stem models, each having a different sample slenderness. Observe that the accuracy of the analytic model at a diameter distance of 3 is identical for models of varying sample slenderness. Model accuracy is, therefore, highly dependent upon diameter distance, but only weakly dependent on sample slenderness.

Errors in Fig. 9 above 1–2% are due to the complex stress states that exist near loading points. These complex stresses cannot be predicted by the analytic model because several stress components have been neglected (Assumptions B and E). However, the finite element can predict stress and deformation near a loading point. Similar studies can be found in the engineering field that demonstrates the limitations of the elementary analytical equations to predict stresses near the loading points and boundary conditions (Sandorff 1980; Sullivan and Van Oene 1986; Williams 1989; Yu 1973).

Loading Effects

The analytic bending equations do not capture the complex stress states that occur near applied loads. Within the finite element method, similar loading conditions can be specified in a number of ways. Figure 9 was generated with distributed loading configurations that most closely mimic the analytical model. However, these boundary conditions underestimate loading effects. An alternative involves applying loads and supports at individual points within the finite element model, which tends to overestimate loading effects. Figure 10 shows curves for both gentle and point loading cases, illustrating the range of possible loading effects. The finite element model used to generate this figure was a hollow model with a rind thickness that was 15% of the radius. Because hollow stems are most susceptible to loading effects, the "Point Loading" curve in Fig. 10 is a reasonable "worst-case scenario" for bending in plant stems. For distances of 5 diameters from any loading point, the analytical model has an error of less 1%. Similar studies were conducted to assess the loading effects on the orthotropic beams under concentrated loads pointing toward considerable departure from classical beam theory (Sandorff 1980; Sullivan and van Oene 1986; Whitney 1985).

Taper

Plant stems are often tapered, but traditional derivations of bending equations often state an assumption of constant cross-sectional geometry (Beer et al. 2012; Crandall 2012; Niklas 1992; Timoshenko 1940). The derivation presented in this chapter does not make use of any assumptions about cross-sectional variation. However, the derivation does assume that the stress state is such that σ_{zz} is much larger than any other components (Assumption B). While abrupt variations in cross-sectional area will activate σ_{xx} and σ_{yy} stress components, gradual taper generally will not violate this assumption.

Figure 11 depicts the σ_{zz} predictions for a tapered (10:1), hollow plant stem. In this figure, the diameter decreases by one unit for every 10 units of length along the stem, as shown in the diagram below the chart. A relatively moderate taper actually provides a fairly accurate prediction of internal stresses. Excluding end effects, the finite element model is within 1% of the analytical solution. The effect of taper is addressed further in the following case study.

In general, gradual variation in stem properties is accurately predicted by the bending stress equations (Maki and Kuenzi 1965; Shahba et al. 2013). This is true for gradual changes in shape as well as material properties. However, rapid changes in shape and/or stiffness may not be predicted accurately.

Case Studies

This section provides an experimental perspective on the equations and results presented above. Laboratory testing of maize stems in three-point bending are described and the bending stress equations are used to aid in interpreting the experimental data.

A number of studies have investigated the strength of maize stem using three-point bending studies (Davis and Crane 1976; Robertson et al. 2014, 2015b; Tongdi et al. 2011). However, in a number of these studies, the load was applied at the internode. It was recently demonstrated that transverse loads, when applied at internode locations, cause deformation of the cross section, leading to premature failure (Robertson et al. 2015b, 2016; Tongdi et al. 2011). As mentioned above, bending equations cannot be trusted to produce accurate predictions of internal stresses at loading points.

When transverse loads are applied to a non-solid (i.e., hollow or pith-filled) plant stem, significant transverse deformation of the stem cross section can occur. This is depicted both experimentally and computationally in Fig. 12. A more detailed depiction of the deformation and stresses that result from this loading configuration are shown in Fig. 13. Robertson et al. recommended that loads and support points should be located at the nodes, and that the overall stem length should be maximized (Robertson et al. 2014, 2015b). Node-loading reduces transverse deformation, while the longer stem length reduces the transverse load required to achieve a given internal bending moment.

The node-loading approach was applied to a set of 1000 maize stems to determine the connection between stem morphology and strength (Robertson et al. 2015a, 2016). X-ray computed tomography was used to obtain three-dimensional images of maize stems. After imaging, the stems were tested in three-point bending. Displacement was increased and the applied load was measured until failure of the stem occurred.

Image processing techniques were used to extract various features of the stem morphology from X-ray CT data. This information was used to calculate the area moment of inertia of the entire stem cross section, as well as the area moment of inertia of the pith region and the rind region. Each region was assumed to possess homogeneous material properties.

Because the rind tissue is the primary load-bearing component of the stem, the pith tissue was neglected. The rind tissue was assumed to be homogeneous (Assumption D). Equation 10 was rearranged as follows to model these stems:

$$M\text{max} = \sigma\text{max}\,\frac{I_{rind}}{y} = \sigma\text{max}\,S_{rind} \tag{26}$$

In the above equation, S is referred to as the section modulus. The term σ_{max} refers to the maximum tissue strength. Figure 15 shows the relationship between the section modulus of the rind (S_{rind}) and stem strength (M_{max}).

The slope of the curve fit line in Fig. 15 can be interpreted as representing the maximum tissue strength (Eq. 26). The slope of the line in Fig. 15 is 120 MPa, which represents the average value of maximum tissue strength for this sample. Individual values of σ_{max} ranged from 28 to 210 MPa, which is within the range spanned by many species of wood (Gibson et al. 2010; Forest Products Society 2011). It should be noted that the use of Eq. 26 neglects the role of the pith. The tissue strength values should, therefore, be interpreted cautiously. Equation 19 would be a more appropriate choice, given the architecture of the maize stalk shown in Fig. 14. But that approach would require assessment of both rind and pith stiffnesses, which is relatively difficult.

Although rind and pith stiffnesses were not measured, an exploration of the stress distribution within the stem is still possible. In human biomechanics, the intensity of CT scans has been shown to be correlated to material stiffness (Zioupos and Currey 1998). In addition, X-ray absorption is closely related to density, and for plant tissues, density is often correlated with tissue stiffness (Gibson and Ashby 1999; Nakagawa et al. 1986). It is, therefore, reasonable to assume that pixel intensity of maize stems is correlated with tissue stiffness. With this assumption in place, Eq. 19 was used to obtain an estimate of the distribution of bending stresses within the maize cross section. This was accomplished by assuming (1) a linear relationship between CT intensity and tissue stiffness; (2) equal tissue stiffness in tension and compression; and (3) the neutral layer coincident with the geometric center of the cross section. Figure 16 shows one CT cross section and an estimate of the σ_{zz} distribution within the cross section.

It should be noted that the stresses in the above figure are qualitative rather than quantitative. Because the mapping between CT intensity and material stiffness is not well defined, this approach provides the shape of stress distributions rather than stress magnitudes.

Nevertheless, this analysis can provide some interesting insights. First, although the pith is composed of parenchyma and vascular bundles, it generally seems to act as a nearly homogeneous material. Second, an assessment of the percentage of total internal bending moment supported by the pith versus the rind can be performed. For this cross section, the rind was measured using an image processing algorithm (Robertson et al. 2016) and the stresses within the pith and rind were assessed separately. The average stiffness of the rind was assumed to be 20 times higher than the average stiffness of the pith. Under this assumption, the rind provided 87% of the total internal bending moment supported by the cross section, with the pith providing the remaining 13% of support. Third, although stress is related to stiffness, strength and stiffness are different mechanical properties. Tissue failure will, therefore, occur whenever/wherever the local tissue stress exceeds the local tissue strength, which may not be in regions of highest stress. Finally, examination of model results often raises new questions. For example, the line graph above indicates that the highest stresses are developed just inside the outer surface of the stem. But for models with a homogeneous rind, the highest stresses are developed at the outer surface. There may be structural advantages when highest stresses are interior rather than at the surface. On the other hand, this result may be due to limitations in CT technology. Boundaries are never exact in CT images but are often blurred. This blurring effect may artificially decrease CT pixel intensity and therefore decrease the calculated material stiffness near the edge of the stem thus lowering stresses. In either case, the use of such models can generate research questions that may not have been asked without insights from model results.

Summary and Future Work

This chapter has provided a derivation of bending stress equations that take into consideration many issues that are relevant to modeling bending stresses in plant stems. It should be noted that bending stresses developed in response to external loads are in addition to internal stresses that may exist within the plant stem prior to the application of bending loads. The equations presented in this chapter, therefore, address stresses due to external loads only.

Although plant stems are complex structures, relatively few assumptions are required to obtain bending stress equations that are applicable to plant stems. Equations for stems with variation in cross-sectional stiffness were also developed. The assumptions used to develop these equations were examined and the results of these investigations should allow researchers to make well-informed decisions when modeling plant stems in this manner. The examination of additional issues such as diameter distance, sample slenderness, taper, and loading effects provided additional insights into issues that are of interest when modeling plant stems.

For easy reference, the most commonly used bending stress equations are presented in Table 3 along with their accompanying assumptions and notes regarding applications of the respective equations.

Future Work

Nodes in plant stems serve as mechanical bulkheads and, therefore, act to minimize cross-sectional ovalization which in turn increases bending stiffness (Robertson et al. 2015a; Schulgasser and Witztum 1992). Due to the complexities involved in creating realistic finite element models of stems with nodes, this issue was not investigated in this study. Based on the results shown in Fig. 15 and other studies, bending stress equations may be suitable in the presence of nodes. A more detailed analysis will be required to adequately address this issue.

The primary limitation of the bending equations appears to be the difficulty of assessing spatial variations in material properties. Without information about the distribution and value of material properties, only the traditional form of the bending stress equation can be used. This equation assumes a single material stiffness, which may be a rather crude assumption for many plant stems. Techniques for assessing spatial distribution of material properties would enable a much more detailed analysis of stress distributions within plant stems.

At the same time, it is not yet clear how the material properties of the pith and rind vary with moisture content. In a wet plant stem, water may actually play a

Table 3 Normal and shear stress equations with their accompanying assumptions and notes regarding application of these equations

	Assumptions required	Equation(s)	Notes
Normal stresses	A. Stress state is dominated by the normal stress σ_{zz} B. The material is linearly elastic C. Bending is locally approximated by an arc D. Material tissue is homogeneous in the cross section	$\sigma = \frac{My}{I}$	•Diameter distance>5:1 typically ensures that assumptions A & C are valid •Material may be orthotropic •Not valid near loading points or in regions where the cross-sectional shape changes rapidly
	All of the above except D	$\sigma = \frac{M\,E(x, y)\,y}{I_E}$ $I_E = \int_A E(x, y) y^2 dA$	•Same notes as above •Captures stress variation due to cross-sectional variation in material stiffness
Shear stresses	Assumptions A—D and: A. τ_{yz} is constant in the x direction B. τ_{xy} and τ_{xz} are negligible	$\tau_{yz} = \frac{V(z)}{I\,w(y)} \int_{y_i}^{y_f} y\,dA$	•This assumption introduces errors when applied to non-rectangular cross sections
	Assumptions A—C and: E. τ_{yz} is constant in the x direction F. τ_{xy} and τ_{xz} are negligible	$\tau_{yz} = \frac{V(z)}{I_E\,w(y)} \int_{y_i}^{y_f} E(x, y)\; y\; dA$	•This assumption introduces errors when applied to non-rectangular cross sections
Combined stresses	The only significant stresses are σ_{zz} and τ_{yz}	$\sigma_1, \sigma_2 = \frac{\sigma_{zz}}{2} \pm \sqrt{\frac{\sigma_{zz}^2}{4} + \tau_{yz}^2}$ $\tau_{\max} = \sqrt{\frac{\sigma_{zz}^2}{4} + \tau_{yz}^2}$ $\sigma_v = \sqrt{\sigma_{zz}^2 + \frac{1}{3}\tau_{yz}^2}$	•For diameter distance values above 5, σ_{xx}, σ_{yy} τ_{xy}, and τ_{xz} are negligible

role in mechanical support (Niklas 1992).This is because the water is essentially incompressible—in a confined space it could potentially resist compression as well as tension (Crum 1979). More work is needed in order to determine the effect of water on bending strength of stems.

The lack of quantitative material properties is also a limitation for predicting other modes of failure. Equations for plant stem buckling have been provided by Spatz et al. (1990) and Schulgasser and Witztum (1992). Whereas bending stress depends upon just one material stiffness, buckling equations typically include several additional material properties, including transverse stiffness. Connecting these equations with experiments is very difficult due to limitations in assessing materials properties. Methods for assessing multiple material properties would enable a great deal of additional work on the mechanisms of stem failure.

In general, further progress in the prediction of plant stem failure is limited by a lack of methods for quickly assessing mechanical properties of plant samples. More research is needed in this area to support advances in plant biomechanics.

Acknowledgments This work was supported by funding from the National Science Foundation (Award #1400973), the United States Department of Agriculture (Award #2016-67012-24685) and the New York University Abu Dhabi Global Ph.D. Fellowship.

Finite Element Model Description

The finite element model used throughout this chapter was developed in Abaqus/CAE 16.4-1. The analysis consisted of a single part, meshed with three-dimensional continuum 20-noded full integration hexahedral elements, with the mesh densities as shown in the figures throughout the chapter. The material represents an average maize stem as described in (Von Forell et al. 2015). For the three-point bending analysis, half the length of the stem was modeled, with symmetry boundary conditions applied at the point of loading. For the gentle loading analyses, the load was applied through the entire cross section as a distributed traction pressure, and the simple support was modeled as constrained vertical deflection at the neutral layer nodes. For the point loading analysis, the load was applied at the top node as a point load, and the simple support was modeled as constrained vertical deflection at the bottom node. The model was run as a linear general static analysis, using a full Newton direct solver in Abaqus/Standard 6.14–1.

References

Agarwal BD, Broutman LJ, Chandrashekhara K (2006) Analysis and performance of fiber composites. Wiley, New York

Archer RR (1987) Internal residual stress patterns in tree stems. In: Growth stresses and strains in trees. Springer, pp 66–113

Bauchau O, Craig J (2009) Euler-bernoulli beam theory. In: Structural analysis. Springer, pp 173–221

Beer F Jr, Russell Johnston E, DeWolf J, Mazurek D (2012) Mechanics of materials. McGraw-Hill, New York

Boresi A, Schmidt R (2003) Advanced mechanics of materials. Wiley, New York

Byars EF, Snyder RD, Plants HL (1983) Engineering mechanics of deformable bodies. Harpercollins College Div

Callister WD, Rethwisch DG (2012) Fundamentals of materials science and engineering: an integrated approach. Wiley, New York

Coutand C, Julien J, Moulia B, Mauget J, Guitard D (2000) Biomechanical study of the effect of a controlled bending on tomato stem elongation: global mechanical analysis. J Exp Bot 51(352):1813–1824

Crandall SH (2012) An introduction to mechanics of solids. Tata McGraw-Hill Education, New York

Crum LA (1979) Tensile strength of water. Nature 278(5700):148–149

Davis SM, Crane PL (1976) Recurrent selection for rind thickness in maize and its relationship with yield, lodging, and other plant characteristics. Crop Sci 16(1):53–55

Dean T, Long JN (1986) Validity of constant-stress and elastic-instability principles of stem formation in pinus contorta and trifolium pratense. Ann Bot 58(6):833–840

Forest Products Society (2011) Wood Handbook: Wood as an engineering material
Gibson Lorna J et al. (2010) Cellular materials in nature and medicine. Cambridge University Press
Gibson LJ (2012) The hierarchical structure and mechanics of plant materials. J R Soc Interface:rsif20120341
Gibson LJ, Ashby MF (1999) Cellular solids: structure and properties. Cambridge University Press, New York
Hashin Z (1968) Erratum: "plane anisotropic beams" (journal of applied mechanics, 1967, 34, pp 257–262). J Appl Mech 35 (3):623–623
Kim N-H, Sankar BV (2009) Introduction to finite element analysis and design. Wiley, New York
Kokubo A, Kuraishi S, Sakurai N (1989) Culm strength of barley correlation among maximum bending stress, cell wall dimensions, and cellulose content. Plant Physiol 91(3):876–882
Maki A, Kuenzi EW (1965) Deflection and stresses of tapered wood beams
Nagarajan S, Zak AR (1985) Finite element model for orthotropic beams. Comput Struct 20(1–3):443–449
Nakagawa F, Kobayashi S, Takemae T, Sugita K (1986) Aneurysms protruding from the dorsal wall of the internal carotid artery. J Neurosurg 65(3):303–308
Niklas KJ (1992) Plant biomechanics: an engineering approach to plant form and function. University of Chicago press, Chicago
Niklas KJ, Spatz H-C (2012) Plant physics. University of Chicago Press, Chicago
Riley WF, Zachary LW (1989) Introduction to mechanics of materials. Wiley, New York
Robertson D, Smith S, Gardunia B, Cook D (2014) An improved method for accurate phenotyping of corn stalk strength. Crop Sci 54(5):2038–2044
Robertson DJ, Julias M, Gardunia BW, Barten T, Cook DD (2015a) Corn stalk lodging: a forensic engineering approach provides insights into failure patterns and mechanisms. Crop Sci 55(6):2833–2841
Robertson DJ, Lee SY, Julias M, Cook DD (2016) Maize stalk lodging: flexural stiffness predicts strength. Crop Sci 56(4):1711–1718
Robertson DJ, Smith SL, Cook DD (2015b) On measuring the bending strength of septate grass stems. Am J Bot 102(1):5–11
Rüggeberg M, Speck T, Paris O, Lapierre C, Pollet B, Koch G, Burgert I (2008) Stiffness gradients in vascular bundles of the palm washingtonia robusta. Proc R Soc Lond B Biol Sci 275(1648):2221–2229
Sandorff PE (1980) Saint-venant effects in an orthotropic beam. J Compos Mater 14(3):199–212
Schulgasser K, Witztum A (1992) On the strength, stiffness and stability of tubular plant stems and leaves. J Theor Biol 155(4):497–515
Shahba A, Attarnejad R, Hajilar S (2013) A mechanical-based solution for axially functionally graded tapered euler-bernoulli beams. Mech Adv Mater Struct 20(8):696–707
Shukla A, Agarwal B, Bhushan B (1989) Determination of stress intensity factor in orthotropic composite materials using strain gages. Eng Fract Mech 32(3):469–477
Silverman IK (1964) Orthotropic beams under polynomial loads. J Eng Mech Div 90(5):293–320
Spatz H, Speck T, Vogellehner D (1990) Contributions to the biomechanics of plants. ii. Stability against local buckling in hollow plant stems. Plant Biol 103(1):123–130
Speck T, Burgert I (2011) Plant stems: Functional design and mechanics. Annu Rev Mater Res 41:169–193
Speck T, Spatz H, Vogellehner D (1990) Contributions to the biomechanics of plants. I. Stabilities of plant stems with strengthening elements of different cross-sections against weight and wind forces. Botanica Acta 103 (1):111–122
Sullivan J, Van Oene H (1986) An elasticity analysis for the generally and specially orthotropic beams subjected to concentrated loads. Compos Sci Technol 27(2):133–155
Timoshenko S (1940) Strength of materials. D. Van Nostrand Company Incorporated, New York
Tongdi Q, Yaoming L, Jin C (2011) Experimental study on flexural mechanical properties of corn stalks. In: 2011 international conference on new technology of agricultural engineering (ICAE). IEEE, pp 130–134

Vandiver R, Goriely A (2009) Differential growth and residual stress in cylindrical elastic structures. Philos Trans R Soc Lond A Math Phys Eng Sci 367(1902):3607–3630

Von Forell G, Robertson D, Lee SY, Cook DD (2015) Preventing lodging in bioenergy crops: a biomechanical analysis of maize stalks suggests a new approach. J Exp Bot 4367–4371

Wegst UG, Ashby MF (2007) The structural efficiency of orthotropic stalks, stems and tubes. J Mater Sci 42(21):9005–9014

Whitney JM (1985) Elasticity analysis of orthotropic beams under concentrated loads. Compos Sci Technol 22(3):167–184

Williams J (1989) End corrections for orthotropic dcb specimens. Compos Sci Technol 35(4):367–376

Wood Handbook: Wood as an Engineering Material. Forest Products Society, (2011)

Yu JC (1973) Local effects of a concentrated load applied to orthotropic beams. J Franklin Inst 296(3):191–206

Zandomeni K, Schopfer P (1994) Mechanosensory microtubule reorientation in the epidermis of maize coleoptiles subjected to bending stress. Protoplasma 182(3–4):96–101

Zioupos P, Currey J (1998) Changes in the stiffness, strength, and toughness of human cortical bone with age. Bone 22(1):57–66

Tree Mechanics and Wind Loading

John Moore, Barry Gardiner and Damien Sellier

Abstract The response of trees to applied wind loads ranges from minor movement of leaves, branches and stems through to catastrophic failure in the form of stem breakage and uprooting. Catastrophic wind damage is a major source of economic loss in managed forests but is also an important ecological process in natural forests. Exposure to chronic wind stress results in a number of thigmomorphogenic responses including changes in tree shape and internal wood properties. In order to better understand the impacts of wind on trees, knowledge is required on wind loading of trees and their response to these loads. In this chapter, we provide an overview of the mechanics of wind loading of trees, starting with the drag force acting on trees for a given wind speed and how this wind speed varies within forest canopies. We then discuss how this load is resisted by the stem and root system, including tree dynamic response to fluctuating wind loads. Throughout the chapter, we focus on advances in instrumentation and modelling techniques that have aided our understanding of this complex phenomenon. We also highlight some of the key gaps in our knowledge and suggest future directions where research advances could be made. An improved knowledge of the mechanics of wind loading on trees aids the better management of the risk of damage to forests and a better understanding of the thigmomorphogenic responses of trees to wind stress and the biomechanical benefits these confer. It can also aid our understanding of the effects of wind exposure on wood properties and the potential consequences for the wood products sector.

J. Moore (✉) · D. Sellier
Scion, Rotorua, New Zealand
e-mail: john.Moore@scionresearch.com

B. Gardiner
UMR 1391 ISPA, INRA, Bordeaux Sciences Agro, 33140 Villenave D'Ornon, France

Present Address
B. Gardiner
EFI Atlantic, 69 Route de Arcachon, F-33612 Cestas cedex, France

A. Geitmann and J. Gril (eds.), *Plant Biomechanics*,
https://doi.org/10.1007/978-3-319-79099-2_4

Introduction

Wind is a persistent force in the natural environment which is primarily generated by differential heating of the Earth's surface by the sun (Stull 1988). Regional differences in wind direction and the distribution of wind speeds are primarily due to latitude and differences in meteorological conditions, but are also the result of differences in topography and surface roughness. Wind imposes a stress on trees to which their response can range from relatively minor leaf flutter within canopies to mild wind sway of limbs and trunks well within the elastic limits of the wood material to physical damage manifesting as micro-failures internally within the trunk branches and roots, and in the most extreme cases to catastrophic failure of branches, stems and roots (Fig. 1) (Grace 1977; de Langre 2008; Gardiner et al. 2016).

Catastrophic wind damage in the form of uprooting and breakage of individual trees through to whole stands is a major disturbance in both managed and natural forests in many regions of the world (Foster and Boose 1995; Everham and Brokaw 1996; Martin and Ogden 2006; Gardiner et al. 2011; Mitchell 2012). In forests managed for timber production, wind damage can result in significant financial losses due to reduced revenues and the increased cost of salvaging trees (Manley and Wakelin 1989). It can also result in substantial disruption to regional wood supply, particularly when the amount of wind damaged timber is a substantial proportion of the annual harvest levels (Hanewinkel and Peyron 2013). While the effects of wind are generally considered to be negative for managed forests, wind is an important disturbance agent in natural forests that is responsible for ecological succession and soil mixing (Schaetzl et al. 1989; Ennos 1997; Ulanova 2000). A better understanding of the catastrophic effects of wind on trees is needed as forests represent the major terrestrial sink for carbon and are a key component in the global carbon cycle (Grace et al. 2013) and an increased frequency of disturbances (including wind damage) under projected future climate change may result in reduced carbon storage by forests (Aber et al. 2001; Bonan 2008). Catastrophic wind damage can also have a large impact on trees in the urban environment (Duryea and Kampf 2007; Lopes et al. 2009). Broken and uprooted trees can pose a hazard to infrastructure and the human population, with partial or complete tree failures being one of the main causes of injury and death during strong wind events.

Exposure to chronic wind loading can result in a number of morphological changes in trees—a phenomenon known as thigmomorphogenesis—as well as affect their physiology (Telewski and Jaffe 1986; Telewski 1995, 2012; Bonnesoeur et al. 2016; Gardiner et al. 2016). In extreme cases, where trees are exposed to persistent strong winds they exhibit highly deformed growth which has terms such as 'krummholz' or 'elfin' associated with it (Lawton 1982; Grace 1989). In natural forests, these responses are important for understanding forest structure and species distribution, while in managed forests they can have important, and often adverse, impacts on log quality which in turn has implications for the forest products sector (Telewski and Moore 2016).

Fig. 1 Examples of wind effects on trees in the form of catastrophic damage (top) and flag form growth (bottom). Photographs from John Moore and Scion photo collection

Because of the economic and ecological importance associated with wind impacts on trees, there have been a large number of studies that have addressed this issue. These studies have often sought to understand and quantify the response of trees to different levels of wind loading and the contribution that different factors (i.e. tree size and shape, stand structure, soil type, etc.) have on this response. This is particularly true for catastrophic wind damage where researchers have attempted to better understand the underlying risk factors. One approach has been to use data collected from catastrophic events to identify empirical relationships between the probability of wind damage and various tree, stand and site factors (Quine 1995; Mitchell et al. 2001; Schmidt et al. 2010; Albrecht et al. 2012). However, two of the challenges with this empirical approach are that the scope of inference is limited to the same underlying population that was used to develop the models, and they do not necessarily identify underlying causal mechanisms (Quine 1995; Lanquaye-Opoku and Mitchell 2005). As a result, there has been an increasing focus on developing hybrid models (i.e. containing both mechanistic and empirical elements) to predict the probability of wind damage to individual trees and stands (see Gardiner et al. 2008 for a summary of models). These models combine plant biomechanics and wind engineering to calculate the critical wind speed at which trees will either uproot or break. However, a biomechanical approach has also enabled a better understanding of the non-catastrophic effects of wind on trees, including the impacts of tree thigmomorphogenic responses to chronic wind stress that improve the resistance of trees to applied loading (Nicoll et al. 2008; Fournier et al. 2013; Gardiner et al. 2016).

In this chapter, we review recent progress in understanding and quantifying wind–tree interactions. We particularly focus on the development of new measurement and modelling approaches that have facilitated a greater understanding of these interactions, highlight where gaps exist in our current knowledge and identify future directions for research. The chapter starts with an overview of tree responses to wind, including a summary of thigmomorphogenic responses and the biomechanical benefits these confer. We then describe the interaction between wind and trees and how advances in instrumentation and modelling techniques have enabled data to be more readily collected in the field and more realistic simulations of the response of trees to fluctuating wind loads. Finally, we discuss some of the gaps in our understanding of wind–tree interactions, which are often manifest as simplifications in simulation models, and provide some perspectives on future directions for research.

Tree Responses to Applied Wind Loading

Wind induces a variety of responses in trees which depend on the magnitude, duration and frequency of loading. The mechanical stability of trees subjected to strong winds depends on the magnitude of the applied force from the wind acting on the crown and the relative capacity of the stem and root–soil system to resist this applied loading (Fig. 2) (Wood 1995; Peltola 2006; Gardiner et al. 2008; Peltola et al. 2013). The threshold (or critical) wind speed required to cause uprooting or stem failure can then be defined as the wind speed required to produce an applied bending moment equal to the maximum resistive moment that can be provided by either the stem or the root–soil system. This is the fundamental basis for mechanistic models, such as GALES and HWIND that have been developed to predict the risk of wind damage to forests (Gardiner et al. 2008). The magnitude of both the applied loads and resistive forces depends on a number of site, tree and stand factors, including (but not limited to) wind climate, tree species, tree height and diameter, wood stiffness and strength, stand density (inter-tree spacing), soil type and rooting depth. Many of these factors are able to be manipulated by forest managers through the selection of sites for establishing forests, choice of species and stand structure (Ruel 1995), and mechanistic models provide a useful basis for evaluating the impacts of different factors on the risk of wind damage (Gardiner and Quine 2000).

There are a variety of growth responses of trees to chronic wind exposure and these are described in more detail in recent reviews (de Langre 2008; Badel et al. 2015; Telewski 2016; Gardiner et al. 2016). It is hypothesised that these thigmomorphogenic responses increase the ability of a tree to withstand wind stress. The earliest observations of these responses were made around 300 B.C. (Telewski 2012) and the first experiments examining the effects of restricting trees from swaying were conducted over 200 years ago (Knight 1803). Within the linear elastic range of material behaviour, thigmomorphogenic responses are characterised by an increase in stem taper, mostly resulting from a reduction in the rate of height growth (Telewski 1995), increase in branch diameter and a reduction in the rate of branch extension (Watt et al. 2005), the formation of flexure wood with a lower modulus of elasticity but higher strength (Telewski 2016), increased radial growth of roots (Urban et al. 1994) and changes in the cross-sectional shape of structural roots (Nicoll and Ray 1996). Wind stress has also been proposed as a significant factor influencing the shape of tree stems (Metzger 1893; Larson 1963, 1965). Under this theory, the cambium is hypothesised to produce new wood in order to equalise the distribution of stresses along the outer surface of the stem (Morgan and Cannell 1994). This theory is known as the mechanical theory of uniform stress and has been the subject of considerable debate in the scientific literature (Mattheck 2000; Niklas and Spatz 2000a, b).

Exposure to higher magnitude chronic wind stress can result in plastic deformation of the stem, branches and wood cells (Telewski 1995, 2012, 2016). Examples of woody tissues responses to plastic deformation include the formation of 'Wulstholz' in response to fibre buckling near the cambium (Koch et al. 2000), resin pockets and shake (Moore et al. 2014). The most obvious and striking whole-tree response to strong prevailing winds is the windswept growth form in which branches and in some cases stems curve away from the wind (Fig. 1) (Telewski 2012). Where wind stress causes the stem to be displaced for a period of time much longer than the presentation time for gravitropism, reaction wood will form in order to help the tree recover an upright position (Fournier et al. 2014).

The thigmomorphogenic responses of trees to chronic wind stress increase the resistance of trees to wind stress by reducing the wind forces acting on the tree for a given wind speed and increasing the mechanical strength of the stem and root–soil system. The degree to which this acclimation to chronic wind stress can affect the

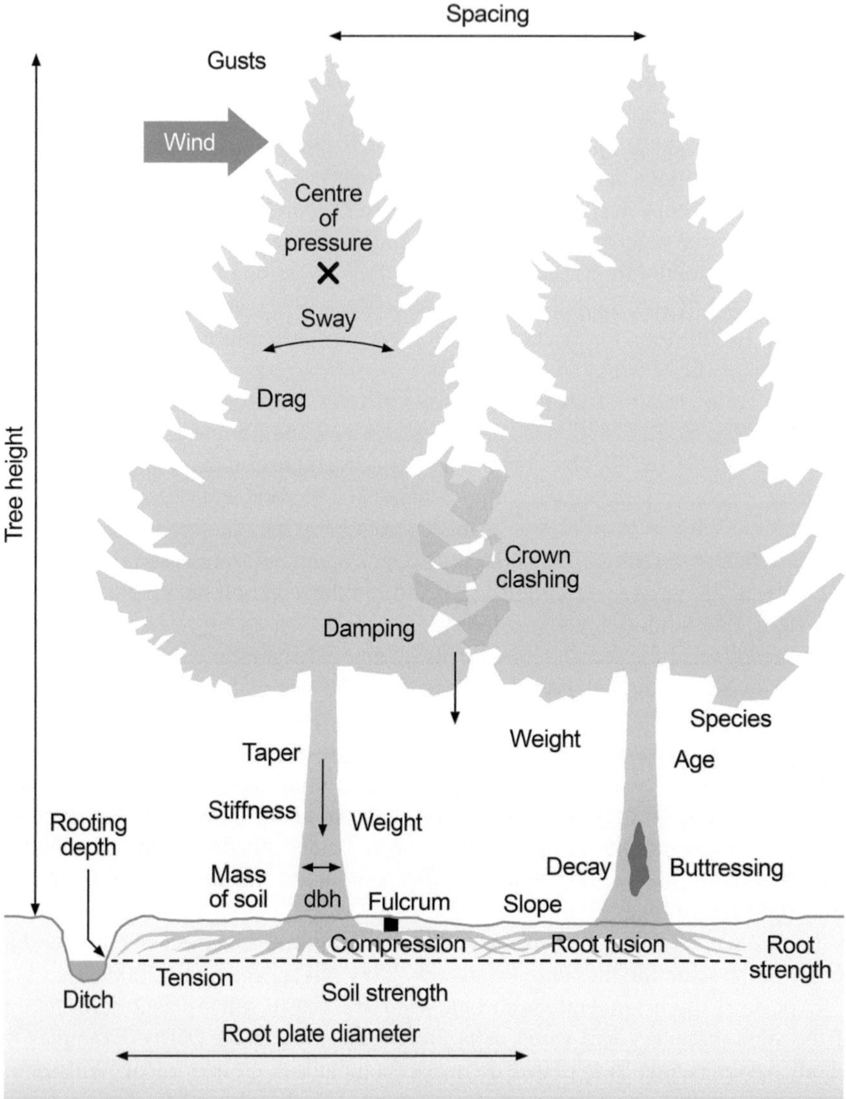

Fig. 2 Factors affecting the mechanical stability of a single tree subjected to wind loading (after Quine et al. (1995) © Crown Copyright Forestry Commission. Used with permission

mechanical stability of trees under strong wind conditions can be determined through analysis of the mechanics of wind loading on trees. This type of analysis is also a key component of understanding how the tree acclimation responses to wind stress are actually produced (Telewski 2006; Coutand 2010; Moulia et al. 2015) and can potentially be used by tree breeders to select for traits that confer greater wind firmness (Telewski and Moore 2016).

The Mechanics of Wind Loading on Trees

Wind Forces Acting on the Tree

The wind force acting on a tree is due to a combination of viscous drag and pressure drag (Patton and Finnigan 2012). Viscous drag (or skin friction) is the friction of the air flowing over the surfaces of the stem, branches and needles. At the interface between these surfaces and the air, the velocity of the air is zero and thus the viscosity of the air imparts a drag force on the tree. The second component of drag is pressure drag (or form drag) which is due to the pressure difference that occurs between the front and back of the elements (i.e. stem, branches and foliage) of the tree as the flow moves around these elements. As the wind speed increases, the flow becomes more turbulent and the pressure drag dominates. The wind force (F) acting on a tree for a given wind speed (u) can be determined through empirical approaches in which wind speed data and the tree response data are collected simultaneously (e.g. Hale et al. 2012; Butler et al. 2012) or by using the drag equation:

$$F = \frac{1}{2} C_D \rho A u^2 \tag{1}$$

where C_D is a non-dimensional drag coefficient, A is the projected crown frontal area and ρ is the mass density of air (1.226 kg m^{-3} at standard atmospheric pressure and temperature). This equation requires information on the drag coefficient of tree crowns, which can be obtained from wind tunnel tests. However, one of the challenges is that the working section of most wind tunnels only allows small sections of tree crowns to be tested (Ennos 1999). Wind tunnel tests on larger trees (diameter at breast height = 10–15 cm, height = 6–8 m) were undertaken by the British Forestry Commission in the 1960 and 1970s to determine the drag force acting on different species as a function of wind speed and to calculate the drag coefficients for the species tested (Raymer 1962; Mayhead 1973). Reanalysis of data from these wind tunnel tests showed that wind drag was not proportional to the square of wind speed, but to the power of between 1.24 and 1.41 (Fig. 3). As a result, the drag coefficients calculated from these tests decreased with increasing wind speed as the crown frontal area measured under still air conditions was applied in Eq. (1). Because tree crowns are flexible structures and as the wind speed acting on them increases as the foliage and branches move into alignment with the direction of the applied wind loading (Fig. 4). This process of streamlining acts to reduce the crown frontal area and more recent wind tunnel experiments found that the crown frontal area of trees decreased by between 20 and 54% at a wind speed of 20 ms^{-1} depending on the species (Rudnicki et al. 2004; Vollsinger et al. 2005).

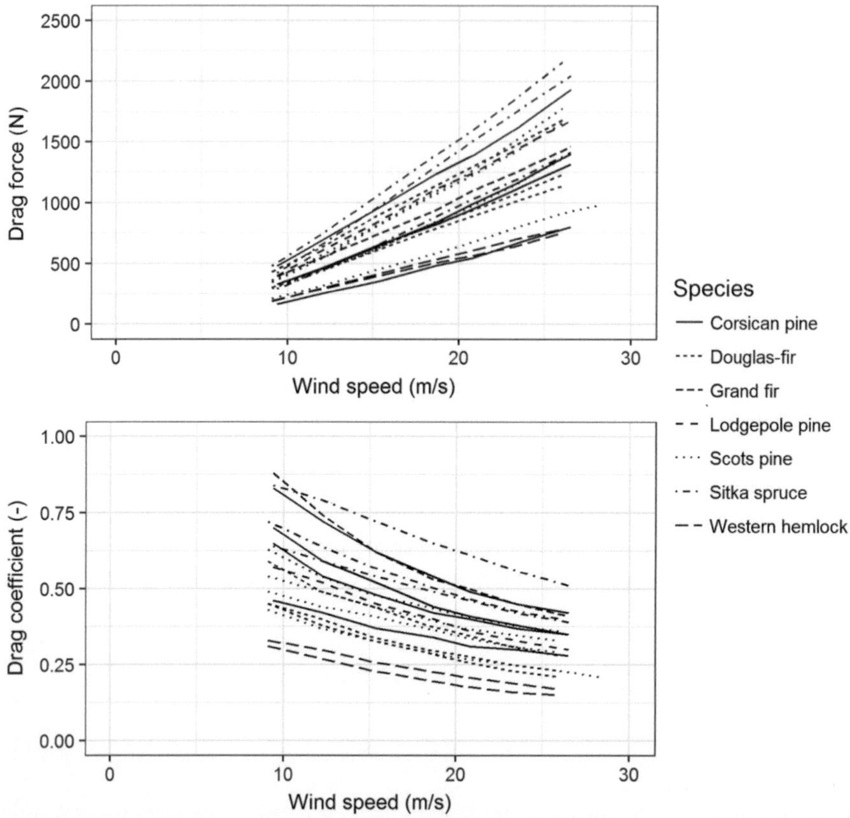

Fig. 3 Relationship between drag force and wind speed based on wind tunnel testing of conifer trees. The bottom panel shows the dependence of drag coefficient on wind speed. Relationships are based on data presented in Raymer (1962) and Mayhead (1973)

The degree of reconfiguration of the crown in high winds can be quantified through the value of E, which is calculated through the following equation (Vogel 1994):

$$\frac{F}{u^2} \propto u^E \tag{2}$$

Based on the data from wind tunnel studies undertaken by the British Forestry Commission, the average value of E across all trees tested was −0.65.

The total mean wind-induced force, F, acting on the tree is the sum of the wind forces acting at each point on the stem and crown, i.e. $F = \sum_{z=0}^{h} F(z)$. To calculate the wind load acting at height z on a tree, the drag equation can be used in combination with an equation to represent the vertical profile of mean horizontal wind speed, i.e.

Fig. 4 Reconfiguration of the crown of a Sitka spruce tree in the Royal Aircraft Establishment wind tunnel at a constant wind speed of 24.9 m/s. Photograph from Graham Mayhead © Crown Copyright Forestry Commission. Used with permission. For details of tests see Mayhead (1973)

$$F(z) = \frac{1}{2} C_D \rho A(z) u(z)^2 \tag{3}$$

Over open ground containing a few isolated trees, the vertical profile of mean horizontal wind speed can be represented by a logarithmic function of the following form (Stull 1988):

$$u(z) = \frac{u_*}{k} \ln\left(\frac{z}{z_0}\right) \tag{4}$$

where u_* is the friction velocity (ms^{-1}), k is von Karman's constant (0.4), z is the height above ground (m) and z_0 is the roughness length (m). As the density of trees increases, trees are sheltered by other trees and the wind speed u at height z is no longer linearly related to ln(z) but instead is linearly related to $\ln(z - d)$ where d is the zero-plane displacement (m). The wind speed profile **above** the forest canopy is described by the following equation:

$$u(z) = \frac{u_*}{k} \ln\left(\frac{z - d}{z_0}\right) \tag{5}$$

The zero-plane displacement is the height at which zero wind speed occurs due to the presence of a tree canopy and is also considered to be the mean level of momentum absorption by the tree canopy (Thom 1971). Theoretical equations for predicting z_0 and d based on tree dimensions and inter-tree spacing have been developed by Raupach (1992, 1994), although approximate values are given by

$$d = 0.78\ \mathrm{h} \tag{6a}$$

$$z_0 = 0.075\ \mathrm{h} \tag{6b}$$

where h is the height to the top of the canopy (Jones 1992). Within a forest canopy, the wind speed profile can be approximated by an exponential function of the following form:

$$\frac{u(z)}{u_h} = e^{-\nu_e(1-z/h)} \tag{7}$$

where ν_e is an extinction coefficient that is dependent on leaf area density. The canopy-top wind speed u_h can be obtained from Eq. 5 with $z = h$.

The wind force acting on trees within a forest canopy can also be calculated by partitioning the shear stress acting on the canopy among the trees within the forest. The shear stress on the surface of the canopy is given by

$$\tau = \rho u_*^2 \tag{8}$$

The shear stress is the drag per unit area imposed on the canopy surface by the wind. If D represents the average spacing between trees within the forest then τD^2 is the average drag force acting on each tree. Both the 'profile' method (Eqs. 3, 5 and 7) and the 'roughness' method (Eq. 8) are used within the GALES model to calculate wind loading on trees (Gardiner et al. 2000, 2008). Using the roughness method, the mean bending moment BM_{mean} acting at any height z on the stem below the zero-plane displacement height is given by $(d - z)\,\tau D^2$ or as a function of the canopy-top wind speed u_h as

$$BM_{mean}(z) = (d - z)\,\rho \left(\frac{D u_h k}{\ln\left(\frac{h-d}{z_0}\right)} \right)^2 \quad (9)$$

There is also an additional moment exerted on the tree due to the offset mass of stem and crown from the wind-induced stem bending and this is discussed in a subsequent section.

Effects of Canopy Structure and Forest Layout on Wind Loads Acting on Trees

Wind flow in and above forests is highly turbulent and hence the wind loading on trees is not simply a function of the mean wind speed. A detailed description of canopy turbulence is beyond the scope of this chapter (for reviews see Finnigan 2000; Patton and Finnigan 2012) and the following is a brief and simplified overview of what is a complex process. There is strong wind shear (change in wind speed with height) at the canopy top with an associated inflection point in the vertical profile of mean horizontal wind speed. This inflection point is an instability that leads to the formation of Kelvin–Helmholtz waves that quickly become transformed into across-wind vortices. These vortices are unstable and begin to distort to produce coherent gusts at the top of the canopy. Each gust consists of a rapid increase in horizontal wind speed together with downward movement of air into the canopy (Finnigan 2000). These sweep events are responsible for much of the momentum transfer into the canopy and the strongest gusts can exert forces on trees that are up to ten times larger than those due to the mean wind speed (Gardiner et al. 1997). There are relatively few experimental studies of the influence of canopy morphology and forest structure on turbulent statistical fields and coherent structures because of the difficulties in implementing such experiments, although new techniques such as LiDAR anemometry offer new possibilities to obtain measurements of the three-dimensional wind field at multiple heights above the ground without the alignment errors that can affect sonic anemometers (Dellwik et al. 2010; Boudreault et al. 2016). However, considerable insight into turbulence in and above forest canopies has been gained through both wind tunnel studies (e.g. Stacey et al. 1994; Gardiner et al. 1997;

Harman et al. 2016; Poëtte et al. 2017) and large eddy simulation (LES) techniques (Yang et al. 2006a, b; Dupont and Brunet 2008, 2009; Dupont et al. 2008). One of the challenges in these studies is realistically representing the forest structure, although LiDAR provides a means for rapidly capturing information on forest structure that can be used in numerical simulations (Boudreault et al. 2015).

In forests, particularly those managed for timber production, there may be a mosaic of forest edges created by differences in age (and thus height) between adjacent stands and also by harvesting activities. Wind tunnel studies on model trees have been used to determine the relative wind load acting on a tree as a function of the distance downwind from the stand edge (*X*) and the relative spacing of trees (*D/h*) in the stand (Gardiner et al. 1997, 2005). These experiments were also used to determine the ratio of maximum loading to mean loading as a function of these two variables (Fig. 5). These studies have shown that while the mean wind load acting on a tree is much higher at the stand edge than inside the forest, the ratio of maximum load to mean load increases with distance back from the edge. These two factors can be added to Eq. 9 in order to estimate the maximum bending moment acting on a tree at a known distance downwind from the stand edge, i.e.

$$BM_{max}(z) = BM_{mean}(z)\, G f_{edge} \tag{10}$$

where G is the ratio of the maximum bending moment to the mean bending moment as a function of relative spacing and distance downwind from the stand edge and f_{edge} is the ratio of the mean bending at a known distance downwind from the forest edge relative to the mean bending moment well inside the stand (normally taken as a distance of 9 h downwind from the edge). The use of this empirical gust factor is a major simplifying assumption in mechanistic wind damage models and understanding the relationship between peak loading and mean loading under actual forest conditions is a key area for future research (see section on Gaps in knowledge and suggestions for future research).

Bending of the Stem Caused by Applied Wind Loads

The wind force acting on the stem creates an applied bending moment which causes the stem to bend. Euler–Bernoulli beam theory can be used to determine the deflection of the stem under applied loading as well as the stress distribution in the outer fibres in the stem. For an initially straight beam or beam section, Euler–Bernoulli beam theory states that

$$\frac{1}{R} = \frac{M}{EI} \tag{11}$$

where *R* is the radius of curvature of the beam, *M* is the bending moment acting on the beam, *E* is modulus of elasticity and *I* is the second area moment of inertia.

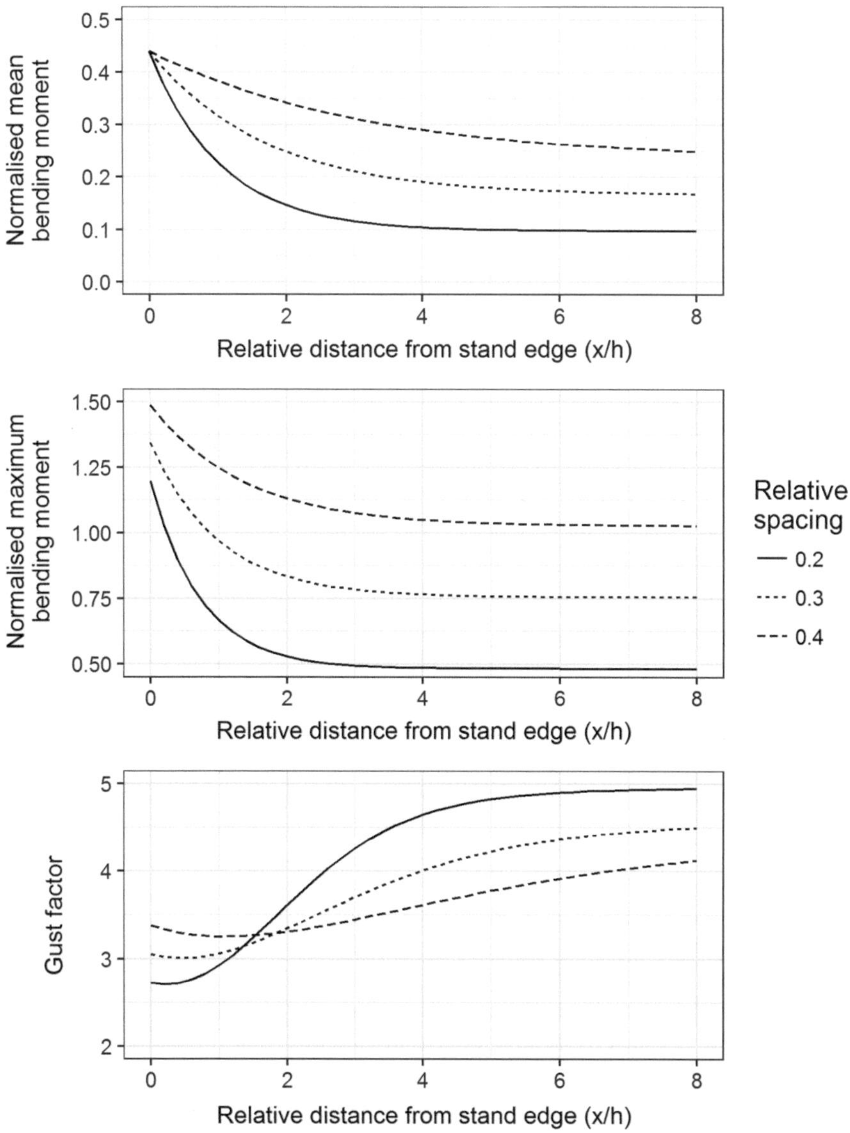

Fig. 5 Change in mean and maximum bending moment and gust factor with relative distance (normalised by tree height) from the windward stand edge. Based on the equations presented in Gardiner et al. (2000)

If a tree is assumed to have a circular cross section then $I = \frac{\pi d^4}{64}$ where d is the diameter of the stem, taken as constant over length. Various analytical approaches have been developed to calculate stem deflection that account for the taper of the stem (Petty and Worrell 1981; Petty and Swain 1985; Blackburn et al. 1988; Gardiner

1992), variation in E along the stem (Wood 1995), the relative flexibility of the stem and root system (Neild and Wood 1999), and bending through large deflections (Morgan and Cannell 1987). Predictions from these equations have generally been in good agreement with deflection measurements made on trees in the field that have been bent through the application of artificial loads (Milne and Blackburn 1989; Neild and Wood 1999). If the bending stress in the tree is in the linear elastic region, then the tensile or compressive stress σ in a section of the tree as a result of a bending moment M will vary with distance y from the neutral axis, i.e.

$$\sigma = \pm \frac{My}{I} \tag{12}$$

An additional compressive stress will occur because of the self-weight of the tree, but this stress is small compared with the bending stress in a deflected tree. The maximum tensile and compressive stresses occur in the fibres at the surface of the stem (where $y = d/2$). Substitution of $I = \frac{\pi d^4}{64}$ into Eq. 12 gives

$$\sigma = \pm \frac{32M}{\pi d^3} \tag{13}$$

Failure of the stem is assumed to occur when the stress in the outer fibres exceeds the strength of the wood. Therefore, by calculating the vertical profile of maximum outer fibre stress due to wind-induced bending and comparing this with the modulus of rupture of green wood, the wind speed required to break the stem can be calculated (Gardiner et al. 2000). Values for the strength of green wood often come from tests made on small defect-free samples (Lavers 1983; USDA 1999) and need to be adjusted to account for the presence of strength reducing defects such as knots and decay. Studies that have measured the modulus of rupture on whole sections taken from living trees have found that values are between 80 and 100% of values obtained from tests on small defect-free specimens (Ruel et al. 2010).

In addition to determining the wind speed required to cause the stem to break, quantifying the distribution of the stresses and strains along the stem is important for understanding the thigmomorphogenic responses of trees to wind loading. Models based on the Finite Element Method (FEM) (Zienkiewicz et al. 2014) have more realistic descriptions of stem geometry and internal wood properties and have been used to simulate the static behaviour under wind and snow loads (Gaffrey and Kniemeyer 2002) and growth mechanics (Fourcaud et al. 2003). These models have particular utility for quantifying the wind-induced displacement and stresses in trees whose shape is not readily approximated by a single tapered cantilever beam.

Mechanics of Root Anchorage

Root anchorage strength is dependent on root system topology as well as the structure and characteristics of the soil that the tree is growing in. Tree winching studies in which a simulated wind load is applied to the stem have been used to assess the root anchorage strength of a wide range of tree species growing on different soil types (Fraser 1962; Fraser and Gardiner 1967; Fredericksen et al. 1993; Moore 2000; Peltola et al. 2000; Cucchi et al. 2004; Achim et al. 2005; Nicoll et al. 2006; Byrne and Mitchell 2007). In these studies, root anchorage strength was found to positively scale with some measure of tree size, e.g. tree height, diameter at breast height, stem volume or stem mass (see summary table in Peltola 2006). These empirical relationships between root anchorage strength and tree size are used in mechanistic wind damage models to calculate the critical wind speed required to uproot a tree (Gardiner et al. 2008).

While these relationships are useful for predicting the overall root anchorage strength, they do not explain how root system and soil characteristics affect anchorage. Several winching studies have been undertaken to identify the contribution of different root types to the overall anchorage strength (Coutts 1983; Crook and Ennos 1996; Stokes 1999; Stokes et al. 2005). In shallowly rooted Sitka spruce (*Picea sitchensis* (Bong.) Carr) trees without a tap root, Coutts (1983) found that the windward lateral roots held in tension were the most important component of root anchorage. However, on free draining soils that afforded deeper rooting, Crook and Ennos (1996) found that approximately 75% of the root anchorage strength was provided by the taproot and the windward sinker roots and Stokes et al. (2005) found that the most resistant species to overturning were those possessing a 'heart' root system, i.e. with many branched, oblique, horizontal and lateral roots.

One of the challenges of studying the factors affecting root anchorage is describing root system topology. Root systems are complex, they are located underground and are difficult to extract without damaging them. Three-dimensional digitisation techniques have been used to study excavated root systems and to identify some of the key features that contribute to anchorage (Danjon et al. 2005). Other non-invasive techniques such as ground penetrating radar may offer the possibility of capturing information on the root system topology without the need to excavate trees. However, further development of these systems is needed in order to be able to provide accurate information in a wide range of conditions (Barton and Montagu 2004). The complex topology of root systems also presents challenges for modelling their behaviour under applied load. Since the first mechanistic anchorage model developed by Blackwell et al. (1990), most models of root system behaviour have been based on the FEM (Dupuy et al. 2005, 2007; Fourcaud et al. 2008; Yang et al. 2014). The development of these models has involved increasingly refined descriptions of the root system topology, soil mechanics and the inclusion of progressive root breakage under increasing loads and they are able to produce results that are in close agreement with those obtained from tree-pulling tests.

Dynamic Response of Trees to Fluctuating Loads

By fluctuating over time the wind exerts a dynamic force on trees and sets them in motion. Under such conditions, the magnitude of a tree's response depends on both the magnitude of the applied wind load and also the frequency of loading, with the greatest response occurring when the frequency of applied loading coincides with the natural frequency of the tree (Holbo et al. 1980; Mayer 1987; Gardiner 1992). Structural dynamics approaches have been used to study the behaviour of trees subject to dynamic loads by considering them as a damped harmonic oscillator. The response of a damped harmonic oscillator depends on the natural frequency and damping ratio and a number of field studies have measured these properties on a range of species (for a summary see Moore and Maguire 2004; James et al. 2006). Such studies have often involved pull-and-release tests in which the tree returns to its rest position under damped free vibration. A number of different sensors including stopwatches, displacement transducers, accelerometers, tilt sensors, strain gauges and lasers have been used to collect data on the frequency and magnitude of tree oscillations in these tests. For trees with simple architecture (i.e. conifer species with a single stem), the natural frequency is strongly and linearly related to the ratio of diameter at breast height (DBH—typically measured at 1.3–1.4 m above ground) to total tree height (H) squared (i.e. DBH/H^2) in accordance with theory for the oscillation of a beam with circular cross section. Damping of tree oscillations is due to a number of factors including: aerodynamic drag on foliage; collisions between neighbouring trees; internal friction; and structural damping (Milne 1991; Spatz et al. 2007; Spatz and Theckes 2013). Published values of damping ratios from free vibration tests on conifer trees range from 0.05 up to 0.23 (Moore and Maguire 2004). Few studies have attempted to model damping due to a lack of knowledge about the energy loss mechanisms involved although recent field experiments and numerical simulations have provided considerable additional insight into these mechanisms, particularly structural damping due to the multiple oscillation frequencies of different parts of a tree (Theckes et al. 2011; Spatz and Theckes 2013; Theckes et al. 2015) and the effects of crown clashing between adjacent trees (Schindler et al. 2011; Webb et al. 2013).

Several theoretical models of tree dynamics have been developed based on the equations of motion for a harmonic oscillator. In these models, many representations of a tree as a dynamical system are possible. It can be represented by a simple oscillator or inverted pendulum (Flesch and Wilson 1999), a system of masses coupled by spring-dashpot connectors (James 2003), a rigid body system (Kerzenmacher and Gardiner 1998; Murphy and Rudnicki 2012), and a vertical cantilever beam (Papesch 1974; Mayer 1987; Saunderson et al. 1999; England et al. 2000). Independently of the representation, a set of equations must be written to describe mechanical equilibrium between inertia, elastic and applied forces. Tree motion is predicted by solving those equations which, for systems with distributed properties, can be achieved using numerical methods (Saunderson et al. 1999).

In the most detailed representations of tree structure, the system is modelled by coupled cantilevered beams oriented in three-dimensional space. Typically, the mechanical behaviour of the system is predicted using the FEM (Zienkiewicz et al. 2014). The FEM relies on a division of the geometric domain into elements with finite properties and a weak formulation of mechanical equilibrium equations for which the displacement field (i.e. the solution) can be numerically approximated. It has been designed to simulate large and complex problems.

The first FEM model of tree dynamic behaviour was developed to investigate the vibration modes in two tree species, *Pinus pinaster* and *Hevea braziliensis*, with distinct architectural forms (Fournier et al. 1993). The model was instrumental in documenting branch deformation and mode shapes in the crown for higher frequency modes. FEM models were later developed to simulate the free sways of real trees (Sellier et al. 2006) as well as wind-induced ones (Moore and Maguire 2008; Sellier et al. 2008). Including branch motion improved the accuracy of predicted displacements if compared to models with the branches described as lumped masses (Kerzenmacher and Gardiner 1998). In angiosperm trees, especially open-grown individuals with a decurrent form, the architecture plays an even more prominent role in tree sways. A FEM-based approach is valuable to understand the complex interplay of branch dynamics for these trees (Ciftci et al. 2013). Building detailed geometrical models of real trees is an expensive task. However, access to remote sensing technologies such as terrestrial LiDAR makes it now possible to digitise tree geometry (Dassot et al. 2012) and to predict the response to wind using FEM with a semi-automated approach (Jackson et al. 2015). FEM modelling is not limited to characterising real structures. It has also become a de facto investigation tool to test hypotheses and isolate mechanisms relative to the branched system and its theoretical organisation. For example, it has been used to research evidence for branching scaling laws and mode compartmentation (Rodriguez et al. 2008), irreversible crown reconfiguration in high winds (Lopez et al. 2011), or branch function in dissipating kinetic energy (Theckes et al. 2011, 2015).

Most FEM models predict tree vibrations by modal analysis using linearised equations of motion. This approach is valid while structural displacements remain small, which is reasonable under chronic wind loads but less so under acute ones. Nonlinear analyses, suitable for large displacements, can be performed using a direct integration technique such as a modified Newton–Raphson algorithm (Sellier and Fourcaud 2009). In that case, the drag forces are calculated for the deformed crown shape, and by doing so account for the reduction in crown area during loading, i.e. streamlining. If a stress–strain analysis is not required, rigid body dynamics provides an alternative to FEM as a less computationally intensive numerical method (Murphy and Rudnicki 2012).

Gaps in Knowledge and Suggestions for Future Work

There have been considerable advances in our understanding of the interactions between wind and trees, particularly the mechanical stability of trees subjected to strong winds, over the past 20 years. These advances have come about through a combination of factors, including: improvements in instrumentation that have enabled better data on the mechanical behaviour of real trees to be collected; improved models of the turbulence processes within and above forest canopies; improved models of the wind-induced responses of trees to dynamic loading that include more realistic representations of above- and below-ground tree architecture; and the tremendous increase in computing power that has enabled complex simulations of different components of wind and tree interactions to be performed.

Understanding and modelling turbulent wind flow over complex forested terrain remains one of the key challenges for better understanding the damaging effects of strong winds. In many empirical studies examining patterns of wind damage across the landscape, information on the spatial variation in wind speed across the landscape is often missing. In the absence of such information, surrogate parameters such as topographic exposure are often used in an attempt to explain this variation (Quine 2000; Quine and White 1994). Recently, however, there has been a marked improvement in our ability to model airflow across forested landscapes (Blennow and Sallnäs 2004; Lopes da Costa et al. 2006; Belcher et al. 2012) but continued improvement in predicting airflow at the landscape level is still an important area for further research.

Despite improvements in instrumentation, we are not aware of any situations in which an instrumented tree has failed under extreme wind loading. All our empirical knowledge about the maximum resistance of trees to applied loads comes from simulated loads (generally point loads applied by a winch and cable system). The manner in which these simulated loads are applied differs from that of actual wind loads as the simulated loads are concentrated while the actual ones are non-uniformly distributed over the length of the crown. This may result in different stress profiles along the stem (Wood 1995). Having instrumented trees loaded to failure would also provide important insight into their dynamic behaviour under extreme wind loading. Anecdotal evidence from video footage of trees that have failed suggests that that failure occurs due to quasi-static loading rather than resonant oscillations (i.e. incremental failure due to a series of strong gusts rather than due to an increasing magnitude of oscillation) but empirical evidence is required to confirm this.

The FEM has resulted in models of the mechanical behaviour of trees in response to wind loading that better represent the architecture of trees and have offered new insights into structural effects such as the coupling of branch and stem oscillations. These models rely on the beam theory in which the wood properties are elastic, isotropic and homogeneous. A better representation of internal stresses and strains is likely to be obtained by using more elaborate material models (Schwager et al. 2013). In materials, failure is often evaluated on the basis of criteria combining all components of the stress tensor. While orthotropic elastic constants are still relatively scarce for green wood (Davies et al. 2016), including orthotropic material behaviour in models would represent a significant improvement in the characterisation of the process of stem failure, in particular to anticipate non-catastrophic damage such as wind-induced compression failures (Hocking 1949; Arnold and Steiger 2006). Wood is also a prestressed material; internal growth stresses are progressively generated during the tree growth. Growth stresses can be modelled using an incremental formulation of mechanical equilibrium (Fourcaud and Lac 2003; Ormarsson et al. 2010). As growth stresses can be quite large (c. 5 MPa), accounting for them may be a significant advance for evaluating the point at which stem breakage occurs and also to better comprehend the role played by physical forces in the developmental processes of wood formation (mechanobiology). Accounting for the radial and longitudinal variations of wood properties that exists within tree stems is another aspect that mechanical models of trees could benefit from. Better material models are an overarching priority to better understand the mechanical design of trees (Fournier et al. 2013). In particular, these models would enable us to better understand the various thigmomorphogenic responses to wind as they would show where regions of high stress and strain occur under wind loading (Moulia et al. 2015) and can also help us to understand the benefits from acclimation. Using highly simplified assumptions about stem shape and wood properties in mechanistic models of tree behaviour can result in their inability to discern key differences between species and individuals within a species (Virot et al. 2016; Albrecht et al. 2016).

The FEM has also facilitated a greater understanding of the mechanics of root anchorage. However, there are still a number of gaps in our knowledge, particularly around the effects of soil moisture on root anchorage, the adhesion between roots and soil (i.e. root resistance to sliding through the soil) and the response of root systems to repeated cyclic loading. These gaps could be addressed through field measurements, e.g. tree-pulling experiments on wet and dry soils, and monitoring of root system behaviour under cyclic loading (O'Sullivan and Ritchie 1993; Kamimura et al. 2011).

Mechanistic wind damage models such as GALES have provided useful insights into the effect of different factors on the risk of wind damage. One of the key gaps in these models is the acclimation of trees to wind, particularly following abrupt changes in conditions such as the creation of a new edge by harvesting an adjacent stand or by thinning. It is assumed that trees acclimate to these new conditions and that stand stability recovers over time (Mitchell 2000). However, the rate to which trees acclimate to these new conditions is not known, nor is the impact of this acclimation on tree stability. Mechanistic models such as GALES were originally developed to predict the critical wind speed required to damage the mean tree within a stand and, therefore, a better understanding of the effects of variability in tree dimensions and relative wind loading on the critical wind speed for damage is needed. The work of Hale et al. (2012) is an important step in this direction but more research is needed across a wider variety of species and stand conditions (e.g. Kamimura et al. 2016, 2017). This improved understanding of the variation in critical wind speeds within a stand will also lead to a better understanding of the relationship between wind speed and the proportion of a stand that is damaged. Dupont et al. (2015) have linked a LES model and a simple mechanistic model to explore damage propagation in uniform stands and Byrne and Mitchell (2012) provide an example of how mechanistic models can be applied to heterogeneous stands in order to model the progression of wind damage. Such models allow us to estimate the effect of forest structure, wind speed intensity and storm duration on the overall level of damage.

One of the key factors affecting the predictions of critical wind speed from the GALES model is the empirical gust factor, which relates the maximum loading to the mean loading. This is an attempt to represent the dynamic response to trees to fluctuating wind loads. Ultimately, a coupled wind flow and tree response model would alleviate the need for this gust factor. Progress towards this goal has been made with the development of a tree sway model that is suitable for use in large eddy simulations (Pivato et al. 2014) and more recently Dupont (2016) has used large eddy simulations to develop a stochastic model to predict extreme wind speeds and wind loading in forest canopies.

References

Aber J, Neilson RP, McNulty S, Lenihan JM, Bachelet D, Drapek RJ (2001) Forest processes and global environmental change: predicting the effects of individual and multiple stressors. Bioscience 51(9):735. https://doi.org/10.1641/0006-3568(2001)051(0735:fpagec)2.0.co;2

Achim A, Ruel J-C, Gardiner BA, Laflamme G, Meunier S (2005) Modelling the vulnerability of balsam fir forests to wind damage. Forest Ecol Manag 204:35–50

Albrecht A, Badel E, Bonnesoeur V, Brunet Y, Constant T, Défossez P, de Langre E, Dupont S, Fournier M, Gardiner B, Mitchell SJ, Moore JR, Moulia B, Nicoll BC, Niklas KJ, Schelhaas M-J, Spatz H-C, Telewski FW (2016) Comment on "Critical wind speed at which trees break". Phys Rev E 94(6). https://doi.org/10.1103/physreve.94.067001

Albrecht A, Hanewinkel M, Bauhus J, Kohnle U (2012) How does silviculture affect storm damage in forests of south-western Germany? Results from empirical modeling based on long-term observations. Eur J Forest Res 131:229–247

Arnold M, Steiger R (2006) The influence of wind-induced compression failures on the mechanical properties of spruce structural timber. Mater Struct 40:57–68

Badel E, Ewers FW, Cochard H, Telewski FW (2015) Acclimation of mechanical and hydraulic functions in trees: impact of the thigmomorphogenetic process. Front Plant Sci 6:266. https://doi.org/10.3389/fpls.2015.00266

Barton CVM, Montagu KD (2004) Detection of tree roots and determination of root diameters by ground penetrating radar under optimal conditions. Tree Physiol 24(12):1323–1331. https://doi.org/10.1093/treephys/24.12.1323

Belcher SE, Harman IN, Finnigan JJ (2012) The wind in the willows: flows in forest canopies in complex terrain. Ann Rev Fluid Mech 44(1):479–504. https://doi.org/10.1146/annurev-fluid-120710-101036

Blackburn P, Miller KF, Petty JA (1988) An assessment of the static and dynamic factors involved in windthrow. Forestry 61:29–43

Blackwell PG, Rennolls K, Coutts MP (1990) A root anchorage model for shallowly rooted Sitka spruce. Forestry 63(1):73–91

Blennow K, Sallnäs O (2004) WINDA—a system of models for assessing the probability of wind damage to forest stands within a landscape. Ecol Modell 175(1):87–99. https://doi.org/10.1016/j.ecolmodel.2003.10.009

Bonan GB (2008) Forests and climate change: forcings, feedbacks, and the climate benefits of forests. Science 320(5882):1444–1449. https://doi.org/10.1126/science.1155121

Bonnesoeur V, Constant T, Moulia B, Fournier M (2016) Forest trees filter chronic wind-signals to acclimate to high winds. New Phytol 210(3):850–860. https://doi.org/10.1111/nph.13836

Boudreault L-É, Bechmann A, Tarvainen L, Klemedtsson L, Shendryk I, Dellwik E (2015) A LiDAR method of canopy structure retrieval for wind modeling of heterogeneous forests. Agric Forest Meteorol 201:86–97. https://doi.org/10.1016/j.agrformet.2014.10.014

Boudreault L-É, Dupont S, Bechmann A, Dellwik E (2016) How forest inhomogeneities affect the edge flow. Bound-Layer Meteorol 162(3):375–400. https://doi.org/10.1007/s10546-016-0202-5

Butler DW, Gleason SM, Davidson I, Onoda Y, Westoby M (2012) Safety and streamlining of woody shoots in wind: an empirical study across 39 species in tropical Australia. New Phytol 193(1):137–149. https://doi.org/10.1111/j.1469-8137.2011.03887.x

Byrne KE, Mitchell SJ (2007) Overturning resistance of western redcedar and western hemlock in mixed-species stands in coastal British Columbia. Can J Forest Res 37(5):931–939. https://doi.org/10.1139/x06-291

Byrne KE, Mitchell SJ (2012) Testing of WindFIRM/ForestGALES_BC: a hybrid-mechanistic model for predicting windthrow in partially harvested stands. Forestry 86(2):185–199. https://doi.org/10.1093/forestry/cps077

Ciftci C, Brena SF, Kane B, Arwade SR (2013) The effect of crown architecture on dynamic amplification factor of an open-grown sugar maple (*Acer saccharum* L.). Trees 27(4):1175–1189. https://doi.org/10.1007/s00468-013-0867-z

Coutand C (2010) Mechanosensing and thigmomorphogenesis, a physiological and biomechanical point of view. Plant Sci 179(3):168–182. https://doi.org/10.1016/j.plantsci.2010.05.001

Coutts MP (1983) Root architecture and tree stability. Plant Soil 71(1–3):171–188. https://doi.org/10.1007/bf02182653

Crook MJ, Ennos AR (1996) The anchorage mechanics of deep rooted larch, *Larix europea* × *L. japonica*. J Exp Bot 47(10):1509–1517. https://doi.org/10.1093/jxb/47.10.1509

Cucchi V, Meredieu C, Stokes A, Berthier S, Bert D, Najar M, Denis A, Lastennet R (2004) Root anchorage of inner and edge trees in stands of Maritime pine (*Pinus pinaster* Ait.) growing in difference podzolic soil conditions. Trees 18:460–466

Danjon F, Fourcaud T, Bert D (2005) Root architecture and wind-firmness of mature *Pinus pinaster*. New Phytol 168(2):387–400. https://doi.org/10.1111/j.1469-8137.2005.01497.x

Dassot M, Colin A, Santenoise P, Fournier M, Constant T (2012) Terrestrial laser scanning for measuring the solid wood volume, including branches, of adult standing trees in the forest environment. Comput Electron Agric 89:86–93. https://doi.org/10.1016/j.compag.2012.08.005

Davies NT, Altaner CM, Apiolaza LA (2016) Elastic constants of green *Pinus radiata* wood. N Z J For Sci 46(1). https://doi.org/10.1186/s40490-016-0075-x

de Langre E (2008) Effects of wind on plants. Ann Rev Fluid Mech 40(1):141–168. https://doi.org/10.1146/annurev.fluid.40.111406.102135

Dellwik E, Mann J, Bingöl F (2010) Flow tilt angles near forest edges—part 2: Lidar anemometry. Biogeosciences 7(5):1759–1768. https://doi.org/10.5194/bg-7-1759-2010

Dupont S (2016) A simple wind–tree interaction model predicting the probability of wind damage at stand level. Agric Forest Meteorol 224:49–63. https://doi.org/10.1016/j.agrformet.2016.04.014

Dupont S, Brunet Y (2008) Influence of foliar density profile on canopy flow: a large-eddy simulation study. Agric Forest Meteorol 148(6–7):976–990. https://doi.org/10.1016/j.agrformet.2008.01.014

Dupont S, Brunet Y (2009) Coherent structures in canopy edge flow: a large-eddy simulation study. J Fluid Mech 630:93. https://doi.org/10.1017/s0022112009006739

Dupont S, Brunet Y, Finnigan JJ (2008) Large-eddy simulation of turbulent flow over a forested hill: validation and coherent structure identification. Q J R Meteorol Soc 134(636):1911–1929. https://doi.org/10.1002/qj.328

Dupont S, Pivato D, Brunet Y (2015) Wind damage propagation in forests. Agric For Meteorol 214–215:243–251. https://doi.org/10.1016/j.agrformet.2015.07.010

Dupuy L, Fourcaud T, Stokes A (2005) A numerical investigation into the influence of soil type and root architecture on tree anchorage. Plant Soil 278(1–2):119–134. https://doi.org/10.1007/s11104-005-7577-2

Dupuy LX, Fourcaud T, Lac P, Stokes A (2007) A generic 3D finite element model of tree anchorage integrating soil mechanics and real root system architecture. Am J Bot 94(9):1506–1514. https://doi.org/10.3732/ajb.94.9.1506

Duryea M, Kampf E (2007) Wind and trees: lessons learned from hurricanes. University of Florida, Gainsville, FL

England AH, Baker CJ, Saunderson SET (2000) A dynamic analysis of windthrow of trees. Forestry 73(3):225–237

Ennos A (1997) Wind as an ecological factor. Trends Ecol Evol 12(3):108–111. https://doi.org/10.1016/s0169-5347(96)10066-5

Ennos AR (1999) The aerodynamics and hydrodynamics of plants. J Exp Biol 202:3281–3284

Everham EM, Brokaw NVL (1996) Forest damage and recovery from catastrophic wind. Bot Rev 62(2):113–185. https://doi.org/10.1007/bf02857920

Finnigan JJ (2000) Turbulence in plant canopies. Ann Rev Fluid Mech 32:519–571. https://doi.org/10.1146/annurev.fluid.32.1.519

Flesch TK, Wilson JD (1999) Wind and remnant tree sway in forest cutblocks. II. Relating measured tree sway to wind statistics. Agric Forest Meteorol 93:243–258

Foster DR, Boose ER (1995) Hurricane disturbance regimes in temperate and tropical forest ecosystems. In: Coutts MP, Grace J (eds) Wind and trees. Cambridge University Press, pp 305–339

Fourcaud T, Blaise F, Lac P, Castéra P, de Reffye P (2003) Numerical modelling of shape regulation and growth stresses in trees. II. Implementation in the AMAPpara software and simulation of tree growth. Trees 17(1):31–39. https://doi.org/10.1007/s00468-002-0203-5

Fourcaud T, Ji JN, Zhang ZQ, Stokes A (2008) Understanding the impact of root morphology on overturning mechanisms: a modelling approach. Ann Bot 101(8):1267–1280. https://doi.org/10.1093/aob/mcm245

Fourcaud T, Lac P (2003) Numerical modelling of shape regulation and growth stresses in trees: I. An incremental static finite element formulation. Trees 17(1):23–30

Fournier M, Almeras T, Clair B, Gril J (2014) Biomechanical actions and biological function. In: Gardiner B, Barnett J, Saranpaa P, Gril J (eds) Biology of reaction wood. Springer-Verlag, Berlin, Germany, pp 139–169

Fournier M, Dlouha J, Jaouen G, Almeras T (2013) Integrative biomechanics for tree ecology: beyond wood density and strength. J Exp Bot 64(15):4793–4815. https://doi.org/10.1093/jxb/ert279

Fournier M, Rogier P, Costes E, Jaeger M (1993) Modélisation mécanique des vibrations propres d'un arbre soumis aux vents, en fonction de sa morphologie. Ann Des Sci For 50(4):401–412. https://doi.org/10.1051/forest:19930407

Fraser AI (1962) The soil and roots as factors in tree stability. Forestry 35(2):117–127. https://doi.org/10.1093/forestry/35.2.117

Fraser AI, Gardiner JBH (1967) Rooting and stability in Sitka spruce, vol 40. HMSO, London

Fredericksen TS, Hedden RL, Williams SA (1993) Testing loblolly pine wind firmness with simulated wind stress. Can J Forest Res 23(9):1760–1765. https://doi.org/10.1139/x93-222

Gaffrey D, Kniemeyer O (2002) The elasto-mechanical behaviour of Douglas fir, its sensitivity to tree-specific properties, wind and snow loads, and implications for stability—a simulation study. J Forest Sci 48(2):49–69

Gardiner B, Berry P, Moulia B (2016) Review: wind impacts on plant growth, mechanics and damage. Plant Sci 245:94–118. https://doi.org/10.1016/j.plantsci.2016.01.006

Gardiner B, Blennow K, Carnus J-M, Fleischer P, Ingemarson F, Landmann G, Linder M, Marzano M, Nicoll B, Orazio C, Petron J-L, Reviron M-P, Schelhaas M-J, Schuck A, Spielmann M, Usbeck T (2011) Destructive storms in European forests: past and forthcoming impacts. European Forest Institute, Atlantic European Regional Office

Gardiner B, Marshall B, Achim A, Belcher R, Wood C (2005) The stability of different silvicultural systems: a wind-tunnel investigation. Forestry 78(5):471–484. https://doi.org/10.1093/forestry/cpi053

Gardiner BA (1992) Mathematical modelling of the static and dynamic characteristics of plantation trees. In: Franke J, Roeder A (eds) Mathematical modelling of forest ecosystems. Sauerlander's Verlag, Frankfurt am Main, pp 40–61

Gardiner BA, Byrne K, Hale SE, Kamimura K, Mitchell SJ, Peltola H, Ruel J-C (2008) A review of mechanistic modelling of wind damage risk to forests. Forestry 81(3):447–463. https://doi.org/10.1093/forestry/cpn022

Gardiner BA, Peltola H, Kellomaki S (2000) Comparison of two models for predicting the critical wind speeds required to damage coniferous trees. Ecol Modell 129:1–23

Gardiner BA, Quine CP (2000) Management of forests to reduce the risk of abiotic damage—a review with particular reference to the effects of strong winds. Forest Ecol Manag 135:261–277

Gardiner BA, Stacey GR, Belcher RE, Wood CJ (1997) Field and wind tunnel assessments and the implications of respacing and thinning for tree stability. Forestry 70(3):233–252

Grace J (1977) Plant responses to wind. Academic Press, London

Grace J (1989) Tree lines. Philos Trans R Soc Lon Ser B 324:233–245

Grace J, Morison JIL, Perks MP (2013) Forests, forestry and climate change. In: Fenning TM (ed) Challenges and opportunities for the world's forests in the 21st century. Forestry sciences, vol 81. Springer, Dordrecht, pp 241–266. https://doi.org/10.1007/978-94-007-7076-8_11

Hale SE, Gardiner BA, Wellpott A, Nicoll BC, Achim A (2012) Wind loading of trees: influence of tree size and competition. Eur J Forest Res 131(1):203–217. https://doi.org/10.1007/s10342-010-0448-2

Hanewinkel M, Peyron J-L (2013) The economic impact of storms. In: Gardiner B, Schuck A, Schelhaas M-J, Orazio C, Blennow K, Nicoll B (eds) Living with storm damage to forests: what science can tell us. European Forest Institute, Joensuu, pp 55–63

Harman IN, Böhm M, Finnigan JJ, Hughes D (2016) Spatial variability of the flow and turbulence within a model canopy. Bound-Layer Meteorol 160(3):375–396. https://doi.org/10.1007/s10546-016-0150-0

Hocking GH (1949) Compression failure in *Pinus radiata* stems exposed to strong wind. N Z J For 9(1):65–66

Holbo HR, Corbett TC, Horton PJ (1980) Aeromechanical behaviour of selected Douglas-fir. Agric Meteorol 21:81–91

Jackson T, Raumonen P, Shenkin A, Malhi Y (2015) Modelling trees response to wind forcing using terrestrial LiDAR data. In: 8th Plant biomechanics international conference, Nagoya, Japan, 30 Nov–4 Dec 2015, p 253

James K (2003) Dynamic loading of trees. J Arboric 29(3):165–171

James KR, Haritos N, Ades PK (2006) Mechanical stability of trees under dyanmic loads. Am J Bot 93(10):1522–1530. https://doi.org/10.3732/ajb.93.10.1522

Jones HG (1992) Plants and microclimate. A quantitative approach to environmental plant physiology, 2nd edn. Cambridge University Press, Cambridge

Kamimura K, Gardiner B, Dupont S, Guyon D, Meredieu C (2016) Mechanistic and statistical approaches to predicting wind damage to individual maritime pine (*Pinus pinaster*) trees in forests. Can J Forest Res 46(1):88–100. https://doi.org/10.1139/cjfr-2015-0237

Kamimura K, Gardiner B, Koga S (2017) Observations and predictions of wind damage to *Larix kaempferi* trees following thinning at an early growth stage. Forestry. https://doi.org/10.1093/forestry/cpx006

Kamimura K, Kitagawa K, Saito S, Mizunaga H (2011) Root anchorage of hinoki (*Chamaecyparis obtuse* (Sieb. Et Zucc.) Endl.) under the combined loading of wind and rapidly supplied water on soil: analyses based on tree-pulling experiments. Eur J Forest Res 131(1):219–227. https://doi.org/10.1007/s10342-011-0508-2

Kerzenmacher T, Gardiner BA (1998) A mathematical model to describe the dynamic response of a spruce tree to the wind. Trees 12:385–394. https://doi.org/10.1007/s004680050165

Knight TA (1803) Account of some experiments on the descent of the sap in trees. Philos Trans R Soc Lond 93:277–289

Koch G, Bauch J, Puls J, Schwab E (2000) Biological, chemical and mechanical characteristics of "Wulstholz" as a response to mechanical stress in living trees of Picea abies (L.) Karst. Holzforschung 54(2):137–143

Lanquaye-Opoku N, Mitchell SJ (2005) Portability of stand-level empirical windthrow risk models. Forest Ecol Manag 216(1–3):134–148. https://doi.org/10.1016/j.foreco.2005.05.032

Larson PR (1963) Stem form development of forest trees. Forest Sci Monogr 5:1–42

Larson PR (1965) Stem form of young Larix as influenced by wind and pruning. Forest Sci 11:412–421

Lavers GM (1983) The strength properties of timber, 3rd edn. Building Research Establishment, London

Lawton RO (1982) Wind stress and elfin stature in a montane rainforest tree: an adaptive explanation. Am J Bot 69(8):1224–1230

Lopes A, Oliveira S, Fragoso M, Andrade JA, Pedro P (2009) Wind risk assessment in urban environments: the case of falling trees during windstorm events in Lisbon. In: Střelcová K, Mátyás C, Kleidon A et al. (eds) Bioclimatology and natural hazards. Springer, Dordrecht, pp 55–74. https://doi.org/10.1007/978-1-4020-8876-6_5

Lopes da Costa JC, Castro FA, Palma JMLM, Stuart P (2006) Computer simulation of atmospheric flows over real forests for wind energy resource evaluation. J Wind Eng Ind Aerodyn 94(8):603–620. https://doi.org/10.1016/j.jweia.2006.02.002

Lopez D, Michelin S, de Langre E (2011) Flow-induced pruning of branched systems and brittle reconfiguration. J Theor Biol 284(1):117–124. https://doi.org/10.1016/j.jtbi.2011.06.027

Manley B, Wakelin S (1989) Modelling the effect of windthrow at the estate level. In: Somerville A, Wakelin S, Whitehouse L (eds) Workshop on wind damage in New Zealand exotic forests. FRI Bulletin 146. Ministry of Forestry, Forest Research Institute, Rotorua, pp 66–72

Martin TJ, Ogden J (2006) Wind damage and response in New Zealand forests: a review. N Z J Ecol 30:295–310

Mattheck C (2000) Comments on "Wind-induced stresses in cherry trees: evidence against the hypothesis of constant stress levels" by KJ Niklas, H-C Spatz. Trees (2000) 14:230–237. Trees 15:63

Mayer H (1987) Wind-induced tree sways. Trees 1:195–206

Mayhead GJ (1973) Some drag coefficients for British forest trees derived from wind tunnel studies. Agric Meteorol 12:123–130

Metzger C (1893) Der wind als massgebender Faktor fur das Wachstum der Baume. Mundener forstl Hefte 3:35–86

Milne R (1991) Dynamics of swaying *Picea sitchensis*. Tree Physiol 9:383–399

Milne R, Blackburn P (1989) The elasticity and vertical distribution of stress within stem of *Picea sitchensis*. Tree Physiol 5:195–205

Mitchell SJ (2000) Stem growth responses in Douglas-fir and Sitka spruce following thinning: implications for assessing wind-firmness. Forest Ecol Manag 135(1–3):105–114. https://doi.org/10.1016/s0378-1127(00)00302-9

Mitchell SJ (2012) Wind as a natural disturbance agent in forests: a synthesis. Forestry 86(2):147–157. https://doi.org/10.1093/forestry/cps058

Mitchell SJ, Hailemariam T, Kulis Y (2001) Empirical modeling of cutblock edge windthrow risk on Vancouver Island, Canada, using stand level information. Forest Ecol Manag 154(1–2):117–130. https://doi.org/10.1016/s0378-1127(00)00620-4

Moore JR (2000) Differences in maximum resistive bending moments of Pinus radiata trees grown on a range of soil types. Forest Ecol Manag 135:63–71

Moore JR, Cown DJ, Lee JR, McKinley RB, Brownlie RK, Jones TG, Downes GM (2014) The influence of stem guying on radial growth, stem form and internal resin features in radiata pine. Trees 28(4):1197–1207. https://doi.org/10.1007/s00468-014-1030-1

Moore JR, Maguire DA (2004) Natural sway frequencies and damping ratios of trees: concepts, review and synthesis of previous studies. Trees—Struct Funct 18(2):195–203. https://doi.org/10.1007/s00468-003-0295-6

Moore JR, Maguire DA (2008) Simulating the dynamic behavior of Douglas-fir trees under applied loads by the finite element method. Tree Physiol 28(1):75–83

Morgan J, Cannell MGR (1987) Structural analysis of tree trunks and branches: tapered cantilever beams subject to large deflections under complex loading. Tree Physiol 3:365–374

Morgan J, Cannell MGR (1994) Shape of tree stems—a re-examination of the uniform stress hypothesis. Tree Physiol 14:49–62

Moulia B, Coutand C, Julien JL (2015) Mechanosensitive control of plant growth: bearing the load, sensing, transducing, and responding. Front Plant Sci 6:52. https://doi.org/10.3389/fpls.2015.00052

Murphy KD, Rudnicki M (2012) A physics-based link model for tree vibrations. Am J Bot 99(12):1918–1929. https://doi.org/10.3732/ajb.1200141

Neild SA, Wood CJ (1999) Estimating stem and root-anchorage flexibility in trees. Tree Physiol 19:141–151

Nicoll BC, Gardiner BA, Peace AJ (2008) Improvements in anchorage provided by the acclimation of forest trees to wind stress. Forestry 81(3):389–398. https://doi.org/10.1093/forestry/cpn021

Nicoll BC, Gardiner BA, Rayner B, Peace AJ (2006) Anchorage of coniferous trees in relation to species, soil type, and rooting depth. Can J Forest Res 36(7):1871–1883. https://doi.org/10.1139/x06-072

Nicoll BC, Ray D (1996) Adapative growth of tree root systems in response to wind action and site conditions. Tree Physiol 16:891–898

Niklas KJ, Spatz H-C (2000a) Response to Klaus Mattheck's letter. Trees 15:64–65

Niklas KJ, Spatz H-C (2000b) Wind-induced stresses in cherry trees: evidence against the hypothesis of constant stress levels. Trees 14:230–237

O'Sullivan MF, Ritchie RM (1993) Tree stability in relation to cyclic loading. Forestry 66(1):69–82

Ormarsson S, Dahlblom O, Johansson M (2010) Numerical study of how creep and progressive stiffening affect the growth stress formation in trees. Trees 24(1):105–115. https://doi.org/10.1007/s00468-009-0383-3

Papesch AJG (1974) A simplified theoretical analysis of the factors that influence windthrow of trees. In: 5th Australasian conference on hydraulics and fluid dynamics, Christchurch, New Zealand. University of Canterbury

Patton EG, Finnigan JJ (2012) Canopy turbulence. In: Fernando HJS (ed) Handbook of environmental fluid dynamics, volume one. CRC Press, pp 311–328. https://doi.org/10.1201/b14241-28

Peltola H, Gardiner B, Nicoll B (2013) Mechanics of wind damage. In: Gardiner B, Schuck A, Schelhaas M-J, Orazio C, Blennow K, Nicoll B (eds) Living with storm damage to forests: what science can tell us. Eur Forest Inst, Joensuu, pp 31–38

Peltola H, Kellomäki S, Hassinen A, Granander M (2000) Mechanical stability of Scots pine, Norway spruce and birch: an analysis of tree-pulling experiments in Finland. Forest Ecol Manag 135(1–3):143–153. https://doi.org/10.1016/s0378-1127(00)00306-6

Peltola HM (2006) Mechanical stability of trees under static loads. Am J Bot 93(10):1501–1511. https://doi.org/10.3732/ajb.93.10.1501

Petty JA, Swain C (1985) Factors influencing stem breakage of conifers in high winds. Forestry 58(1):75–84. https://doi.org/10.1093/forestry/58.1.75

Petty JA, Worrell R (1981) Stability of coniferous tree stems in relation to damage by snow. Forestry 54(2):115–128

Pivato D, Dupont S, Brunet Y (2014) A simple tree swaying model for forest motion in windstorm conditions. Trees 28(1):281–293. https://doi.org/10.1007/s00468-013-0948-z

Poëtte C, Gardiner B, Dupont S, Harman I, Böhm M, Finnigan J, Hughes D, Brunet Y (2017) The impact of landscape fragmentation on atmospheric flow: a wind-tunnel study. Bound-Layer Meteorol. https://doi.org/10.1007/s10546-017-0238-1

Quine CP (1995) Assessing the risk of wind damage to forests: practices and pitfalls. In: Coutts MP, Grace J (eds) Wind and trees. Cambridge University Press, pp 379–403
Quine CP (2000) Esimation of mean wind climate and probability of strong winds for wind risk assessment. Forestry 73(3):247–258
Quine CP, Gardiner BA, Coutts MP, Pyatt DG (1995) Forests and wind: management to minimise damage. For Comm Bull 114. HMSO, London
Quine CP, White IMS (1994) Using the relationship between rate of tatter and topographic variables to predict site windiness in upland Britain. Forestry 67(3):245–256
Raupach MR (1992) Drag and drag partition on rough surfaces. Bound-Layer Meteorol 60:375–395
Raupach MR (1994) Simplified expressions for vegetation roughness length and zero-plane displacement as functions of canopy height and area index. Bound-Layer Meterol 71:211–216
Raymer WG (1962) Wind resistance of conifers. National Physical Laboratory, Aerodynamics Division, Middlesex, U.K
Rodriguez M, Langre E, Moulia B (2008) A scaling law for the effects of architecture and allometry on tree vibration modes suggests a biological tuning to modal compartmentalization. Am J Bot 95(12):1523–1537. https://doi.org/10.3732/ajb.0800161
Rudnicki M, Mitchell SJ, Novak MD (2004) Wind tunnel measurements of crown streamlining and drag relationships for three conifer species. Can J Forest Res 34(3):666–676. https://doi.org/10.1139/x03-233
Ruel J-C (1995) Understanding windthrow: silvicultural implications. For Chron 71(4):434–445
Ruel J-C, Achim A, Herrera RE, Cloutier A (2010) Relating mechanical strength at the stem level to values obtained from defect-free wood samples. Trees 24(6):1127–1135. https://doi.org/10.1007/s00468-010-0485-y
Saunderson SET, England AH, Baker CJ (1999) A dynamic model of the behaviour of Sitka spruce in high winds. J Theor Biol 200(3):249–259
Schaetzl RJ, Johnson DL, Burns SF, Small TW (1989) Tree uprooting: review of terminology, process, and environmental implications. Can J Forest Res 19(1):1–11. https://doi.org/10.1139/x89-001
Schindler D, Fugmann H, Schönborn J, Mayer H (2011) Coherent response of a group of plantation-grown Scots pine trees to wind loading. Eur J Forest Res 131(1):191–202. https://doi.org/10.1007/s10342-010-0474-0
Schmidt M, Hanewinkel M, Kändler G, Kublin E, Kohnle U (2010) An inventory-based approach for modeling single-tree storm damage—experiences with the winter storm of 1999 in southwestern Germany. Can J Forest Res 40(8):1636–1652. https://doi.org/10.1139/x10-099
Schwager H, Masselter T, Speck T, Neinhuis C (2013) Functional morphology and biomechanics of branch-stem junctions in columnar cacti. Proc Biol Sci 280(1772):20132244. https://doi.org/10.1098/rspb.2013.2244
Sellier D, Brunet Y, Fourcaud T (2008) A numerical model of tree aerodynamic response to a turbulent airflow. Forestry 81(3):279–297. https://doi.org/10.1093/forestry/cpn024
Sellier D, Fourcaud T (2009) Crown structure and wood properties: influence on tree sway and response to high winds. Am J Bot 96(5):885–896. https://doi.org/10.3732/ajb.0800226
Sellier D, Fourcaud T, Lac P (2006) A finite element model for investigating effects of aerial architecture on tree oscillations. Tree Physiol 26(6):799–806. https://doi.org/10.1093/treephys/26.6.799
Spatz HC, Bruchert F, Pfisterer J (2007) Multiple resonance damping or how do trees escape dangerously large oscillations? Am J Bot 94(10):1603–1611. https://doi.org/10.3732/ajb.94.10.1603
Spatz HC, Theckes B (2013) Oscillation damping in trees. Plant Sci 207:66–71. https://doi.org/10.1016/j.plantsci.2013.02.015
Stacey GR, Belcher RE, Wood CJ, Gardiner BA (1994) Wind flows and forces in a model spruce forest. Bound-Layer Meteorol 69:311–334

Stokes A (1999) Strain distribution during anchorage failure of *Pinus pinaster* Ait. at different ages and tree growth response to wind-induced root movement. Plant Soil 217(1/2):17–27. https://doi.org/10.1023/a:1004613126353

Stokes A, Salin F, Kokutse AD, Berthier S, Jeannin H, Mochan S, Dorren L, Kokutse N, Ghani MA, Fourcaud T (2005) Mechanical resistance of different tree species to rockfall in the French Alps. Plant Soil 278(1–2):107–117. https://doi.org/10.1007/s11104-005-3899-3

Stull RB (1988) An introduction to boundary layer meteorology. Kluwer Academic Publishers, Dordrecht, The Netherlands. https://doi.org/10.1007/978-94-009-3027-8

Telewski FW (1995) Wind-induced physiological and developmental responses in trees. In: Coutts MP, Grace J (eds) Wind and trees. Cambridge University Press, pp 237–263

Telewski FW (2006) A unified hypothesis of mechanoperception in plants. Am J Bot 93(10):1466–1476. https://doi.org/10.3732/ajb.93.10.1466

Telewski FW (2012) Is windswept tree growth negative thigmotropism? Plant Sci 184:20–28

Telewski FW (2016) Flexure wood: mechanical stress induced secondary xylem formation. In: Kim YS, Funada R, Singh AP (eds) Secondary xylem biology: origins, functions and applications. Elsevier, Oxford, U.K

Telewski FW, Jaffe MJ (1986) Thigmomorphogenesis: field and laboratory studies of Abies fraseri in response to wind or mechanical pertubation. Physiol Plant 66:211–218

Telewski FW, Moore JR (2016) Trait selection to improve windfirmness in trees. CAB Rev 11(50). https://doi.org/10.1079/pavsnnr201611050

Theckes B, Boutillon X, de Langre E (2015) On the efficiency and robustness of damping by branching. J Sound Vib 357:35–50. https://doi.org/10.1016/j.jsv.2015.07.018

Theckes B, Langre E, Boutillon X (2011) Damping by branching: a bioinspiration from trees. Bioinspir Biomim 6(4):046010. https://doi.org/10.1088/1748-3182/6/4/046010

Thom AS (1971) Momentum absorption by vegetation. Q J R Meteorol Soc 97:414–428

Ulanova NG (2000) The effects of windthrow on forests at different spatial scales: a review. Forest Ecol Manag 135(1–3):155–167. https://doi.org/10.1016/s0378-1127(00)00307-8

Urban ST, Lieffers VJ, MacDonald SE (1994) Release in radial growth in the trunk and structural roots of white spruce as measured by dendrochronology. Can J Forest Res 24:1550–1556

USDA (1999) Wood handbook: wood as an engineering material. Forest Products Laboratory General Technical Report FPL-GTR-113, 486pp

Virot E, Ponomarenko A, Dehandschoewercker E, Quere D, Clanet C (2016) Critical wind speed at which trees break. Phys Rev E 93(2):023001. https://doi.org/10.1103/PhysRevE.93.023001

Vogel S (1994) Life in moving fluids—the physical biology of flow. Princeton University Press

Vollsinger S, Mitchell SJ, Byrne KE, Novak MD, Rudnicki M (2005) Wind tunnel measurements of crown streamlining and drag relationships for several hardwood species. Can J Forest Res 35(5):1238–1249. https://doi.org/10.1139/x05-051

Watt MS, Moore JR, McKinlay B (2005) The influence of wind on branch characteristics of *Pinus radiata*. Trees 19:58–65

Webb VA, Rudnicki M, Muppa SK (2013) Analysis of tree sway and crown collisions for managed *Pinus resinosa* in southern Maine. Forest Ecol Manag 302:193–199. https://doi.org/10.1016/j.foreco.2013.02.033

Wood CJ (1995) Understanding wind forces on trees. In: Coutts MP, Grace J (eds) Wind and trees. Cambridge University Press, pp 133–164

Yang B, Morse AP, Shaw RH, Paw UKT (2006a) Large-eddy simulation of turbulent flow across a forest edge. Part II: momentum and turbulent kinetic energy budgets. Bound-Layer Meteorol 121(3):433–457. https://doi.org/10.1007/s10546-006-9083-3

Yang B, Raupach MR, Shaw RH, Paw UKT, Morse AP (2006b) Large-eddy simulation of turbulent flow across a forest edge. Part I: flow statistics. Bound-Layer Meteorol 120(3):377–412. https://doi.org/10.1007/s10546-006-9057-5

Yang M, Defossez P, Danjon F, Fourcaud T (2014) Tree stability under wind: simulating uprooting with root breakage using a finite element method. Ann Bot 114(4):695–709. https://doi.org/10.1093/aob/mcu122

Zienkiewicz OC, Taylor RL, Fox DD (2014) The finite element method for solid and structural mechanics, 7th edn. Butterworth-Heinemann, Oxford, U.K. https://doi.org/10.1016/b978-1-85617-634-7.00018-1

Part II
Growth, Morphogenesis and Motion

The Mechanics of Leaf Growth on Large Scales

Eran Sharon and Michal Sahaf

Abstract Leaf growth is a process during which the leaf expands by many orders of magnitude, while maintaining a proper shape. Complex mechanisms which are not fully understood regulate the local, individual expansion of multiple cells into an organ with a well-defined shape. Various factors of mechanical and biochemical nature play a role in the growth process. The mechanical factors include the cell wall properties, osmotic pressure, and interactions between neighboring cells. This happens under the strict restrictions stemming from geometric considerations, which are dictated by the three-dimensional shape of the leaf. In this chapter, we review this topic focusing on its geometric and mechanical aspects, such as stress–strain response, viscous and elastic moduli, and typical statistical properties of the growth field. We view the leaf as a sheet of active matter, capable of adjusting and responding to signals originating both from its natural internal growth processes and from its surrounding. We aim at a better understanding of the growth patterns and mechanical properties of the tissue as a whole.

Keywords Leaf growth · Remodeling · Mechanical stress · Morphogenesis

Large-Scale Behavior

When treating the tissue as a whole, tools which are often used to study materials come in handy. The whole tissue can be described as an active material, whose properties may depend on a large number of factors. These factors can be related to the plant itself (such as age or genetics) or its environment (like water, light, or temperature). Treating biological entities as active materials is becoming more common (Prost et al. 2015), and this approach allows for a simplified yet accurate phenomenological description of the tissue. Global biological entities which are described as active materials range from biofilaments and molecular motors, through

---

E. Sharon (✉) · M. Sahaf
The Racah Institute of Physics, The Hebrew University of Jerusalem, Jerusalem, Israel
e-mail: erans@vms.huji.ac.il

A. Geitmann and J. Gril (eds.), *Plant Biomechanics*,
https://doi.org/10.1007/978-3-319-79099-2_5

collections of motile microorganisms, to animal flocks (Marchetti et al. 2013). In this approach, typical behaviors and their spatiotemporal scales are identified and formulated mathematically. In this chapter, we will lay down a basis for describing the process of leaf growth within this conceptual framework, treating the tissue as a thin sheet of active material.

Growth and Shape in Differential Geometry

We are all aware of the existence of a connection between growth and shape. A rod that grows uniformly (evenly in all parts) will become longer, a flat circular disc, which grows uniformly and isotropically (evenly in all directions) will form a bigger circle, while if it grows anisotropically (grows differently in different directions), it may become oval, as seen in Fig. 1a. However, when we consider nonuniform growth, the connection between growth and shape becomes highly nontrivial, leading to shapes which are not necessarily flat, as seen in Fig. 1b (Thompson 1942). In some cases, our well-developed mechanical intuition provides good guesses. For example, if the upper layer of a strip expands more than its lower layer (as in the case of a bimetallic strip used for measuring temperature), we can guess that the strip will bend downwards. However, even for smooth and symmetric growth distributions, there are cases which are highly contra-intuitive, and when the growth is highly nonuniform it is practically impossible to guess the emergent shape.

In recent years, there has been a significant progress in the study of shaping via growth of soft solid thin sheets. Differential geometry (O'Neill 2001) was found to be an adequate tool for formulating the connection between growth distribution and large-scale deformation. In differential geometry, the *shape* of a surface in space is described by the *curvature* field; the in-plane distances between points on the surface, which can be tracked to provide a measurement of its growth, are described by the *metric* field. Using these exact mathematical entities, the theory of non-Euclidean plates can describe two separate entities: the active growth is described using the *target* metric and curvature, which are the metric and curvature dictated to the sheet by the growth pattern; while the shape the sheet takes in practice is described using the *actual* metric and curvature (Efrati et al. 2009; Sharon and Efrati 2010). A more elaborate explanation on the difference and the connection between the two will be given in the following section. This mathematical decoupling between the target geometry and the actual one opened the way for extended theoretical and experimental study, in which different "growth fields" were generated and the emergent 3D surface shapes were measured or calculated. It was shown that simple growth distributions can lead to fractal shapes (Audoly and Boudaoud 2003; Gemmer and Venkataramani 2011; Klein et al. 2007; Santangelo 2009; Sharon and Efrati 2010); helical configurations (Armon et al. 2011, 2014; Wu et al. 2013); saddle-like (Kim et al. 2012; Klein et al. 2011); or dome-like surfaces (Efrati et al. 2007), and abnormal mechanical behavior of such bodies (Guest et al. 2011; Levin and Sharon 2016). The formulation shows that the nature of the particular process which leads to tissue

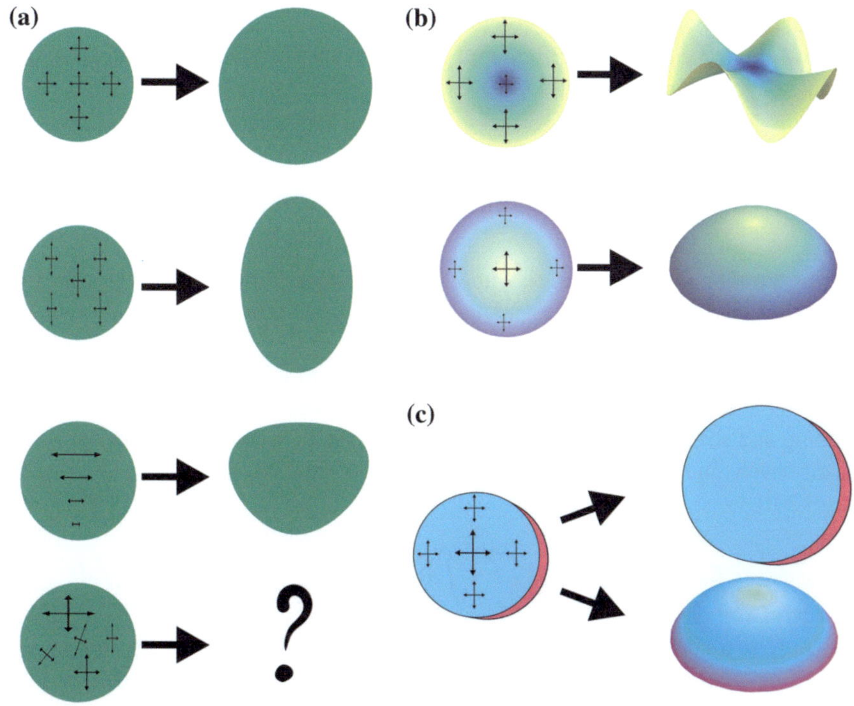

Fig. 1 **a** Simple cases in which the growth tensor dictates a Euclidean configuration, allowing the disc to remain flat. However, complicated growth patterns can lead to unpredictable shapes, which are not necessarily flat. **b** Spatial differences in growth rate can cause a flat disc to buckle. In this case, the thin sheet obeys the "target metric" given by the growth pattern. **c** Elastic frustration: if the disc has a nonzero thickness t, the same growth pattern can lead to more than one configuration, depending on the thickness. If the disc is thick it will expand while remaining flat, causing nonzero stretching energy; a thinner disc will buckle, causing nonzero bending energy. There is no stress-free configuration

growth—be it cell division, cell expansion, swelling or shrinkage of gel, or muscles contraction—does not affect the emergent shape. The shape of a thin sheet depends only on its mechanical properties and on the target geometry, dictated by the tissue expansion. Differential geometry (O'Neill 2001) tells us that in the case of 2D surfaces, there are three types of possible geometries: Euclidean (flat) geometry, Elliptic (dome-like) geometry, and hyperbolic (saddle-like) geometry. The growth distribution dictates to which of the three cases the local surface geometry belongs. Once set by growth, the surface geometry cannot be easily changed. A ball cannot be smoothly wrapped with a flat sheet of paper. Similarly, flattening a dome-like shell leads to its wrinkling or fracture. Using these notions, we realize that the growth field within a leaf which increases in size, while remaining flat, *must be consistent with Euclidean geometry*. Similarly, growth that leads to any specific 3D leaf shape must fulfill strict, well-defined constraints. Differential geometry provides us with all the exact connections between growth and shape. Though it is not always simple to follow, studying elementary differential geometry is a worthy investment for anyone who is interested in the growth of leaves, or any other thin tissue.

Leaves as Thin Sheets: Geometry and Mechanics

Leaves are thin sheets of material, and hence must obey mechanical laws, regardless of the biological processes taking place. In this section, we will present the geometrical concepts used to describe the behavior of thin elastic sheets.

The energy of thin elastic sheets can be written, to a good approximation, as a function of the geometry of a 2D surface—the midplane—and as a sum of two terms: bending and stretching. We formulate this energy, in particular, to actively growing elastic sheets, in the following way (Efrati et al. 2009; Klein et al. 2007; Sharon and Efrati 2010):

$$E_{elastic} = E_{stretching} + E_{bending} \sim t \int (a_{ij} - \bar{a}_{ij})^2 ds + t^3 \int (b_{ij} - \bar{b}_{ij})^2 ds \quad (1)$$

where ds is a surface element, the ~ relation stands for proportionality, where the proportion coefficients are constant material parameters, t is sheet thickness, and the other factors are geometric parameters defined as follows:

The **surface metric tensor**, a_{ij}, is defined at each point on a 2D surface. Using some coordinate system, it describes infinitesimal distances to the points around it. **Embedding of a metric** is a realization of a metric in space, a mapping from each point in the 2D surface to a location in 3D space in a way that fulfills the metric. It is an actual configuration of a sheet in 3D space in a way that obeys the metric. Most metrics have many embeddings.

The **surface curvature tensor**, b_{ij}, defined at each point on the surface, determines the curvatures in the coordinates' direction (and with the right transformations, in every direction).

The tensors a and b determine the configuration. Surfaces may only exist in the real three-dimensional space if they obey specific relations between metric and curvature, defined by the so-called Gauss–Mainardi–Peterson–Codazzi (GMPC) equations on which we will not elaborate here.

For physical sheets, we define the **target metric**, $\bar{a}_{ij}$, defining local lateral "rest distances" between points on the surface; these are distances that allow the sheet to abide without any internal stresses. In the context of leaves, lateral growth can be viewed as a change in the target metric. The target metric may differ from the **actual metric**, a_{ij}, which is what we measure from the sheet shape de facto. When the two are not equal, we can define the **strain tensor** as the difference between them: $u_{ij} = \frac{1}{2}(\bar{a}_{ij} - a_{ij})$. As evident from Eq. (1), the stretching energy density is proportional to $(u_{ij})^2$. Analogously, the bending energy of the sheet is a function of the difference between the target and the actual curvature tensors. A **target curvature** (also called spontaneous or rest curvature) is a result of rest lengths gradients along the thickness axis, lengths that can only be fulfilled by applying certain local curvature. A sheet obeying its target curvature will contain no bending energy.

Since physical systems are stable in the state of minimal energy, one would imagine a sheet to choose a configuration which satisfies both $a = \bar{a}$ and $b = \bar{b}$, and thus have zero elastic energy; but as target tensors are local and independent parameters of growth, they may not satisfy the constraints imposed by the abovementioned GMPC relations, and therefore there may not be a rest configuration satisfying both. In these cases, the sheet is called incompatible. It will bear residual stresses that cannot be removed even when the sheet is free in space, and the configuration is determined by a competition between bending and stretching, in order to minimize total energy. The energy depends on the sheet thickness—the stretching energy is proportional to the thickness t while the bending energy goes like t^3. As a result, a thin sheet tends to bend more easily than a thick one. For example, a thin fabric will bend (crumple/wrinkle) but would not squeeze. A thick block of sponge, however, can be squeezed, but would not crumple.

An example of a frustrated sheet would be that of a thin disc which undergoes a locally isotropic lateral growth, invariant across the disc thickness, which is larger at its central parts and smaller in the periphery (see Fig. 1c). Such growth profile prescribes zero reference curvature, due to the invariance across the thickness, but also a non-Euclidean reference metric. If the disc stays flat (no wrinkling or bending, i.e., $b = \bar{b} = 0$) its central part will be compressed, thus costing stretching energy (Fig. 1c, top). A configuration that is free of stretching energy must be an embedding of $\bar{a}$ (i.e., $a = \bar{a}$). In this case, it will be dome-like, with $b \neq \bar{b}$ (Fig. 1c, bottom). Such configurations contain bending energy. We see that there is no configuration in which both stretching and bending energies are zero. The competition between the two energy terms, depending on the thickness of the tissue, will select the equilibrium configuration.

Buckling: As mentioned before, the thinner the sheet (compared to its lateral scale)—the more bending is favored over stretching. In other words, the thinner the sheet, the smaller are the compressive strains that will lead to its buckling and the shorter are the possible buckles. In some cases, buckling can cause a desired spatial pattern to emerge (Volokh 2006). In leaves, certain pathogens (Cohen 1995), as well as genetic manipulations (Nath et al. 2003; Palatnik et al. 2003), may lead to random changes in the shape of the leaf, including its buckling into three-dimensional configurations. Shaping of a thin sheet into a desired (flat or curved) configuration implies that $\bar{a}$ and $\bar{b}$ are well tuned. Since the leaf is thin, and the buckling threshold is small, the fact that buckling is not normally observed implies that the growth of a thin leaf keeps stresses below the buckling threshold, which indicates the existence of shape regulating mechanisms.

We can define a "growth tensor" that describes how the metric changes in time (τ):

$$G_{ij} = \frac{a(\tau_2)_{ij}}{a(\tau_1)_{ij}} \tag{2}$$

This tensor contains all the data about the local growth—rates and directions. A representation of the local growth tensor is seen as double arrows in Fig. 1.

The description of growth as a tensor field, defined locally, is the ideal way to study any dynamical process. In this view, the single cell is seen as a point in a field rather than an individual, independent biological entity. As far as geometry and mechanics are concerned, its growth behavior, determined both by its own properties and the effects of its neighbors, can be described without having to go into the details of individual processes.

The Need for Growth Regulation

The leaf as a thin solid sheet is subjected to the laws of plates and shells *mechanics*. On an even more general level, the cell aggregate must fulfill the abovementioned *geometrical* constraints. At every instance, the shape of most leaves can be approximated as a two-dimensional (2D), possibly curved, surface, which is located in a three-dimensional (3D) space. This statement implies severe and specific geometrical constraints on the collection of cells, their arrangement, sizes, and possible growth. One may wonder how the cells grow in a proper way which obeys all these global constraints; how do they receive the relevant information and respond to it? Since they are not governed by a global control center, which keeps growth well regulated, they must *interact*. The activity within one region on the leaf at one instance must be mediated so that it can affect the local activity in other regions, possibly at different times.

The other obvious solution would be that growth variations are naturally very small due to some intrinsic, robust limitations on the growth rate. However, The process in which the cells expand has been shown to be very noisy, both temporally, in one-dimensional objects such as pollen tubes, in which no geometric constraints have to be considered (Winship et al. 2010); and spatiotemporally in two-dimensional organs such as leaves (Elsner et al. 2012) and sepals (Tauriello et al. 2015), which are subjected to geometric constraints. This noisy growth, with a range of growth rates rather than simple smooth expansion, is bound to create local stresses. Such stresses, if not relieved, will cause random buckling throughout the area of the leaf. However, in a healthy plant, this highly irregular growth pattern smoothes out to produce a well-defined leaf shape, which indicates the existence of growth-regulating mechanisms, possibly ones that are initiated by variations in the growth rate itself, i.e., are mechanically induced.

In this chapter, we focus on the possibility of mechanically induced interactions which affect leaf morphogenesis, as suggested by Green et al. (1970) and further developed by Shraiman (2005), who also suggested a mathematical formulation for the interaction between the growth rate and the stress caused by spatiotemporal variations.

Evidence for Mechanically Induced Shape-Conserving Mechanisms

In order to consider mechanically induced shape-conserving mechanism in leaves, evidence for the existence of shape-conserving mechanisms must be found, as well as evidence showing that mechanical signals are involved, at least to some extent. The mechanical properties of the tissue provide the crucial link between the worlds of geometry and mechanics. These properties relate strain and strain rate—which are geometrical entities—to mechanical stress. In a biological tissue, such as a leaf, they may be altered and tuned by the underlying biological activity in a way that gives a proper, geometrically consistent, growth.

Recent works in leaves indicate shape-conserving mechanisms of genetic origin. Nath and Coen (2003) demonstrated how the expression of CIN in *Antirrhinum* locally affects the way cells respond to division arrest signals, which results in excess marginal growth and a negative curvature (crinkly leaves) in CIN mutants, demonstrating how gene expression controls growth and therefore shape. Leaf morphogenesis was also found to be controlled by microRNAs, allowing a manipulation which leads to deformed leaves in *Arabidopsis* (Palatnik et al. 2003).

The idea that morphogenesis is related to mechanics is now widely accepted. The formation of leaves in phyllotactic patterns has been shown to be related to mechanical stress (Green et al. 1996; Shipman and Newell 2005). This connection is consequential: by applying mechanical stress to a meristem, the formation of ectopic leaves was induced (Green 1999). Changes in the mechanical properties of the cell have a similar affect: the cell wall loosening agent expansion was shown to trigger the formation of primordia when applied to the apical meristem (Fleming et al. 1999). Another example relating leaf morphogenesis and mechanics is that of leaf venation, which will be discussed in detail later.

Various works (Braam 2005; Coutand et al. 2000; Hellgren et al. 2004; Jaffe 1980) have shown a wide range of mechanically induced effects in plants. In particular, it has been shown (Nakayama et al. 2012) that auxin transport, which is closely related to growth, is affected by mechanical stimuli. Since plants respond to some mechanical stimuli, it is plausible that mechanisms which serve to conserve the leaf's shape may be initiated by the stress (or strain) itself. This mechanism can be either mechanical in nature, similar to the response of a viscoelastic material to stress, or originating from changes in hydraulic pressure or osmotic potential (Christmann et al. 2007; Comstock 2002; Zonia and Munnik 2007) Or it can be a biologically active mechanism, involving various functions that modify the properties of the cells in the leaf (Sampathkumar et al. 2014), mediated through processes such as ion signaling (Trewavas and Knight 1994), or other mechanosensing pathways (Vogel and Sheetz 2006). A combination of these two types is most likely.

Cortical microtubules (CMT) have been recently shown to directly relate mechanical stress and form (Hamant et al. 2008; Hervieux et al. 2016; Uyttewaal et al. 2012). CMT orientation is changed by naturally occurring mechanical stresses in the shoot apical meristem and in sepals. CMT are considered (Paredez et al. 2006) to have an effect on the orientation of cellulose microfibrils in the cell wall. Such changes in the structure of the wall may lead to changes in the mechanical properties and growth pattern of the leaf (Gibson and Gibson 2012).

The bottom line of all such processes is that growth causes mechanical stress, which in turn initiates mechanical changes in the leaf. In this section, we have reviewed works focusing mainly on relevant biological mechanisms. In the following sections, we will mention work done by our research group and others, focusing mainly on the mechanical aspects of stress and growth, which can be directly measured in leaves.

The Vascular System

For an accurate mechanical description of the leaf tissue, the vascular system has to be taken into account. It differs from its surroundings in its higher stiffness. In terms of bearing mechanical load, the veins play an important role in the leaf tissue.

The pattern in which veins are distributed on the surface of the leaf has been a focus of scientific attention, and various approaches considered the role of physical processes in the formation of this pattern. The main concept for understanding how cells differentiate and form the vascular system has been the canalization model (Sachs 1969). According to this model, the flux of auxin (Scarpella et al. 2006) can enhance and stabilize itself, thus creating stable “canals”. The canalization model predicts very realistic venation patterns (Runions et al. 2005).

The fact that the vascular system forms closed loops rather than free ends has been a challenge for the models, and can be explained as a result of a tensorial field, rather than a scalar concentration field (Couder et al. 2002). Based on this suggestion, a relatively new perspective on veins, known as the *force model* (Bohn et al. 2002), was developed. According to this model, which describes the final geometry of junctions in vasculature rather than their initial formation, each vein pulls with a force that is proportional to its diameter, and the requirement of local equilibrium at vein junctions leads to a statistical correlation between veins’ diameter and the angles between them.

More recently, a cell-based mechanical model was developed to describe the time evolution of the vascular network (Corson et al. 2009). The tissue was modeled as a network of viscoelastic cell walls, and vein cells were distinguished from ground cells by their higher rigidity. This yielded realistic venation patterns and reproduced the

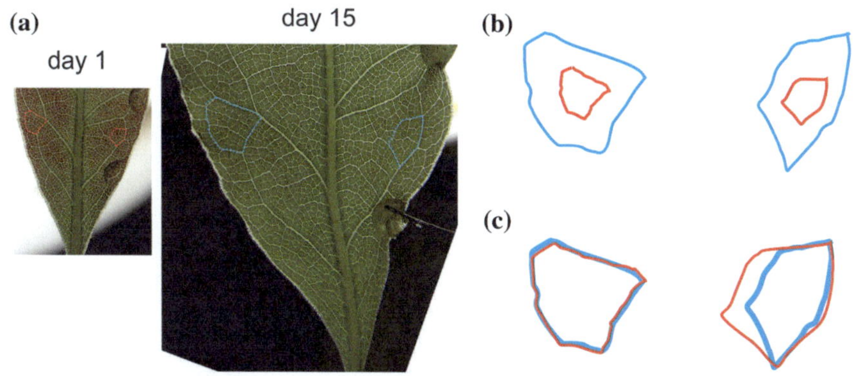

Fig. 2 **a** A bay laurel leaf was stretched on its right side while the other side serves as control. **b** A single areole as measured on day one (small, red) and day 15 (large, blue). **c** The same areole seen in b, rescaled so that changes can be easily seen. The left, unstretched side exhibits nearly isotropic expansion, while the right side, which was stretched, has grown anisotropically

experimental findings of (Bohn et al. 2002), in line with the force model. However, these studies remain correlative and do not prove that mechanical forces shape the vascular network.

In a recent experiment, we tested the force model by perturbing mechanically a growing leaf and making predictions about the effect of such a perturbation on the vascular network (Bar-Sinai et al. 2016). We chose to work with leaves in which the vein network has already formed, so as to avoid a direct coupling with differentiation mechanisms.

Each leaf of bay laurel (Laurus nobilis) was loaded by a U-shaped spring, glued to two points on its edge. The loaded leaf was allowed to grow for 15 days, during which it multiplied its area by about one order of magnitude. The vascular system of the entire leaf was repeatedly photographed. The unstretched half of the leaf was considered as a control. A clear effect on the tissue that was grown under tension was observed: while the freely grown region has grown by mere isotropic expansion, the stretched region showed a large deformation of the vascular network along the stretching direction, as can be seen in Fig. 2, thus demonstrating the validity of the force model.

The Mechanical Properties of the Leaf

Changes of mechanical properties in living tissues as a result of mechanical stimuli have been known for a while, with perhaps the most well-known example being that of bone remodeling, Wolff's law (Wolff 1986).

As mentioned above, growth of solid structures is not a purely geometric process. It greatly depends on the mechanical properties of the tissue. In the case of a leaf, the cell wall is the dominant mechanical component. The cell wall is a complex structure, and understanding its mechanical properties is challenging. Even more challenging is the application of any direct study of the wall to the behavior of the cell and the growth patterns of the intact tissue (Bidhendi and Geitmann 2016). The fact that the mechanical properties of the cell cannot be decoupled from turgor pressure makes things even more difficult (Cosgrove 2015). Even so, the cell wall is a good place to start when dealing with the mechanical properties of the leaf. Mechanically, the cell wall is commonly described (Cosgrove 1993) as a thin shell composed of a viscoelastic material, and since it is the main contributor to the mechanical strength of the cell, it would make sense to expect viscoelastic behavior of the intact leaf tissue. However, other effects, such as poro-elasticity, as well as possible biological active processes are expected to play an important role in the mechanical behavior of the leaf as a whole (Dumais and Forterre 2012).

Viscoelasticity can be described by several models, combining a spring, which represents the elastic element, and a dashpot, which represents the viscous behavior. Elasticity is characterized by an immediate response to mechanical stress, which fully and immediately recovers when the stress is removed, while conserving mechanical energy. Viscosity is characterized by a retarded response, and it does not conserve energy, which means that the response is not always fully reversed when the load is removed. When describing a viscoelastic material, the elastic and viscous elements may be combined either in parallel (Kelvin–Voigt model) or in series (Maxwell model). There are more complex models combining Maxwell and Kelvin elements, such as the generalized Maxwell model (Maxwell–Wiechert model) which includes several Maxwell elements combined in parallel, all in parallel with a lone spring. This model allows for more than one typical timescale for the viscous response (Fig. 3a–c).

Dynamic mechanical analysis (DMA) is a technique used to measure the mechanical properties of the material. We apply a periodic stress with a sinusoidal profile of frequency ω and amplitude σ_0. The leaf responds by straining with amplitude ε_0, but due to the viscous effect, there is a phase difference δ between the applied stress and the measured strain, as seen in Fig. 3b:

$$\sigma(t) = \sigma_0 \sin(\omega t)\,; \varepsilon(t) = \varepsilon_0 \sin(\omega t - \delta) \tag{3}$$

The elastic (E′) and viscous (E″) moduli (sometimes designated G′ and G″) can be calculated as

$$\mathrm{E}' = \frac{\sigma_0}{\varepsilon_0}\cos(\delta); \mathrm{E}'' = \frac{\sigma_0}{\varepsilon_0}\sin(\delta) \tag{4}$$

In general, in a complex viscoelastic material, E′ and E″ would be frequency-dependant, and may have a wide range of values depending on the frequency in which the measurement is performed.

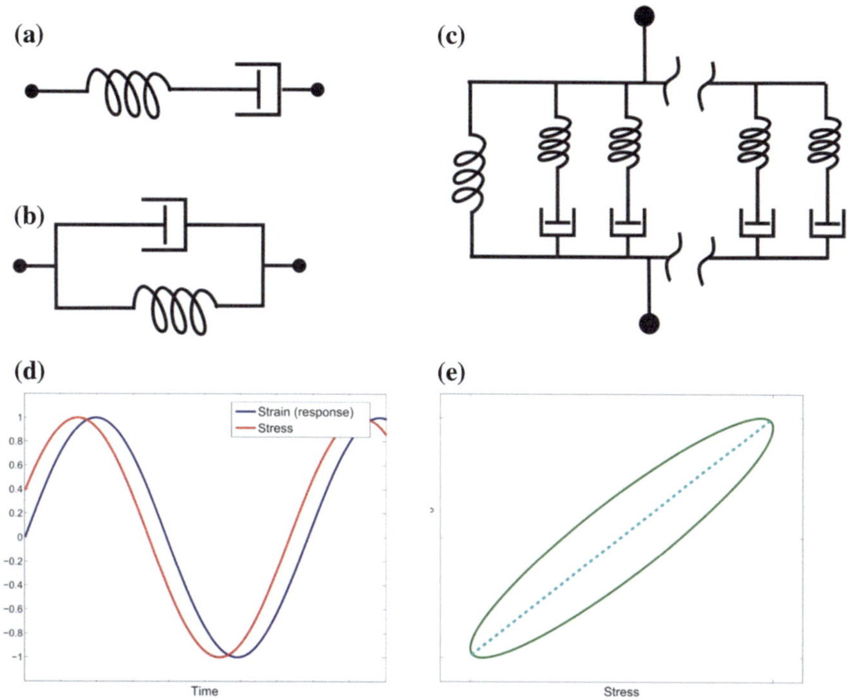

Fig. 3 **a** The Maxwell model for viscoelasticity, in which the spring, representing the elastic element, and the dashpot, representing the viscous element, are connected in series. **b** Kelvin–Voigt model, in which the spring and the dashpot are parallel. This combination fully recovers after being stretched and released. **c** The generalized Maxwell–Wiechert model, which includes several viscous and elastic components. This model allows for more than one typical timescale. **d** Stress–strain response as a function of time. In viscoelastic material, there is a lag between them—a nonzero phase difference. **e** When stress and strain are presented as a function of each other, a hysteretic loop will form. If the material is purely elastic, the curve will be linear (dashed line). If a viscous element exists, the loop will deviate from a line and have a nonzero width

When presented as a function of each other, $\sigma(\varepsilon)\, or\, \varepsilon(\sigma)$, a hysteretic loop will form. If the material is purely elastic, the curve will be linear with no hysteresis. If a nonzero viscous element exists, the loop will deviate from a line and have a nonzero width, as seen in Fig. 3e.

Changes in Mechanical Properties

In this section, we discuss the results of a recent work done by our research group, focusing directly on evidence for mechanically induced growth regulation. We

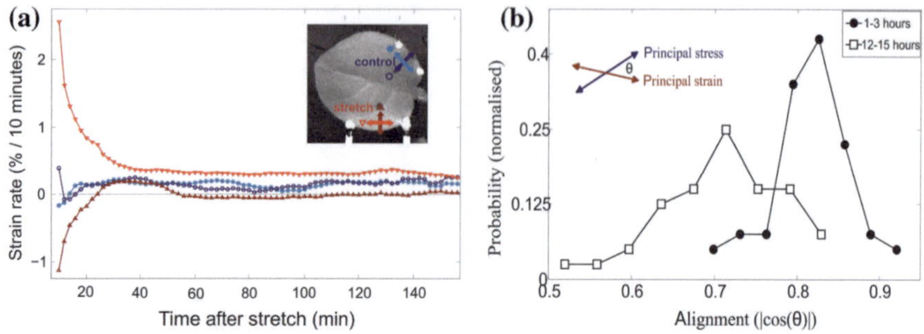

Fig. 4 **a** Strain/growth rate over 10 min in a stretched leaf, as a function of time after stretching. After an exponential decay, the rate remains relatively constant for a few hours. Growth in the stretched direction is faster than in the perpendicular direction. During this time, the strain is well aligned with the stress. **b** Changes in the stress–strain alignment over time. The initial high alignment decreases after several hours, although the leaf is still stretched with constant force. These images are extracted from (Sahaf and Sharon 2016)

observe a combination of yielding and resistance, which can help the leaf maintain a proper shape.

Hints for active regulation were found by studying the effect of mechanical loading on leaf growth patterns and on the mechanical properties of the leaf tissue (Sahaf and Sharon 2016). Inducing a dominant external, well-controlled mechanical stress provides a practical way to observe the effect of stress on growth rate and direction. We have applied constant tensile stress to one side of a growing leaf of wild-type tobacco (*Nicotiana tabacum*), and studied the growth of the loaded area compared to the other side of the leaf, which served as a control, and compared to its previous loose state.

Following an initial rapid deformation, a rather constant but anisotropic growth rate is observed. During this time we observe a high stress–strain local aligment. The relative longitudinal growth rate in the stretch direction is considerably higher than the growth rate in the control side (see Fig. 4a).

The active nature of the growing tissue becomes evident at longer times, of 12–24 h. When observed at $t > 12$ h, the stress–strain alignment decreases (Fig. 4b). This indicates that the growth becomes more isotropic, even though the leaf is subjected to the same force as before. We suggest that the leaf has changed its properties to minimize the deformation caused by the stress. Such a suggestion can be tested by directly measuring the viscoelastic properties of the leaf.

We performed dynamic mechanical analysis (DMA) which provides information on the elastic and viscous properties of the tissue before and after exposure to prolonged mechanical stress.

We have observed an increase of the elastic modulus in the stretched direction as a result of constant stretch, accompanied by a decrease in the perpendicular direction (see example in Fig. 5a). This indicates that the leaf has become stiffer in the stretched direction and more pliable in the perpendicular direction. The viscous component

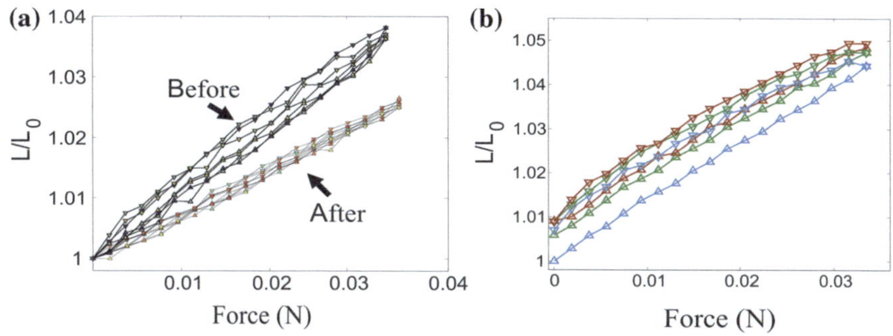

Fig. 5 **a** An example of stress-induced change in mechanical properties. The viscoelastic loops demonstrate increased rigidity after exposing the leaf to prolonged mechanical stress. **b** An example of three subsequent loops. The first (blue) loop shows irreversible, plastic deformation which is not observed in the following loops. These images are extracted from (Sahaf and Sharon 2016)

has also become anisotropic, making the leaf "flow" less in the direction of the force and more in the perpendicular direction. These changes make the leaf more resilient to stress in the direction in which the force is applied, and more responsive to stress in the perpendicular direction. As a result, growth in the direction of the force is gradually reduced while growth in the perpendicular direction increases, which may explain the reduction we see in the growth anisotropy following the stretching.

Short-term plastic deformation: As mentioned before, we have seen that yielding is the first response of the leaf when mechanical load is applied. We have observed this not only in creep tests but also in the DMA tests, in which the first loop often differs significantly from the following loops (see example in Fig. 5b). When load is first applied, the leaf deforms irreversibly, but this is not the case in the following loops in which the leaf behaves as a Kelvin–Voigt (parallel) viscoelastic material, and fully recovers its original length when the load is removed. This suggests that in some cases, the leaf does not behave as a viscoelastic material, but also has a plastic component. Such plastic deformation should result from irreversible, stress-induced processes, which are yet to be determined. One candidate could be severing of pectin links (Bidhendi and Geitmann 2016).

Statistics, Patterns, and Distributions

Measurement of leaf growth patterns is experimentally challenging. Works that have been up for the challenge (Elsner et al. 2012; Gupta and Nath 2015; Remmler and Rolland-Lagan 2012) reveal a diversity of spatial and temporal growth patterns, which can be used to formulate equations describing the growth process, and understand the underlying mechanisms. The fact that leaf growth is heterogeneous suggests that

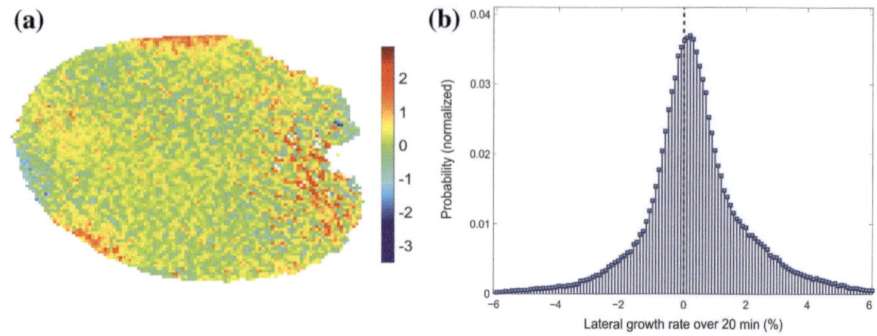

Fig. 6 **a** An example of the spatial growth field of a leaf (% over 20 min), collected over 40 min. **b** A distribution of lateral growth values (% over 20 min), collected over 24 h. The growth is noisy and includes negative values, representing shrinkage, as well as extreme local growth values

relevant information about the governing processes, or equations, is contained in the statistics of the growth field, not just in its means.

Growth measurements performed in our lab reveal this multiscale field, which involves sharp fluctuations at small time and length scales (unpublished, Fig. 6). Only at large scales in time and space does the field become smooth and well represented by its means. We observe an abundance of local shrinkage events in the tissue, i.e., a local reduction in area of small regions on the leaf, which commonly occurs during normal leaf growth (Fig. 6).

Though the surface area of large sections of a leaf increases monotonically (in a typical rate of ~4%/hour), the growth in small regions, of ~0.1 mm, oscillates between swelling and shrinking. These observations are highly relevant to the question of growth regulation, emphasizing the need for active regulation, which properly averages these large fluctuations. For example, using the distributions of growth over 2 hours, and assuming there were no spatial correlations, we estimate the stress on a flat unit area of 250×250 micron would exceed 1 MPa. Such stress would lead to significant buckling of the surface and cause the distortion of the leaf. Since our measurements provide only the actual growth of the surface, which by definition is compatible, it is likely that the fluctuations in $\bar{a}$, the "attempted growth", which generates the internal stresses, are larger.

In reality, the leaf grows flat, which indicates that some mechanism controls growth over small time and length scales. The robustness of growth regulation is probably due to the fact that unlike synthetic tissues, the leaf, as a living out-of-equilibrium "material", can manipulate its geometrical and mechanical properties.

Future Work Should Focus on Stress-Induced Changes in the Cell Wall and the Formulation of Field Equations

We see two main goals that are called for by the work we have described: (i) A more complete description of the growth process of the leaf in terms of active matter, which will include a mathematical description and account for the processes which have been observed so far, as well as others which are yet to be discovered. (ii) An understanding of the biological processes underlying these large-scale behaviors. We expect the biology underlying these processes to include changes in the cell wall, as well as processes which control the cell's water intake and turgor pressure.

Measurements on the whole leaf scale are possible and provide simultaneous information on both geometric fields (growth statistics) and on mechanical properties. As such, this type of measurements can be combined with theoretical and numerical modeling of growth, with the final goal of revealing the governing equations of motion of the growth field. Complementing measurements of biological activity in the growing/loaded leaf have the potential of completing the picture by finding the underlying biological processes that determine these effective equations.

References

Armon S, Aharoni H, Moshe M, Sharon E (2014) Shape selection in chiral ribbons: from seed pods to supramolecular assemblies. Soft Matter 10(16):2733. http://xlink.rsc.org/?DOI=c3sm52313f

Armon S, Efrati E, Kupferman R, Sharon E (2011) Geometry and mechanics in the opening of chiral seed pods. Science (New York, N.Y.) 333(6050):1726–1730. http://www.sciencemag.org/content/333/6050/1726.short

Audoly B, Boudaoud A (2003) Self-similar structures near boundaries in strained systems. Phys Rev Lett 91(8):86105. https://link.aps.org/doi/10.1103/PhysRevLett.91.086105

Bar-Sinai Y et al (2016) Mechanical stress induces remodeling of vascular networks in growing leaves. PLoS Comput Biol 12(4):e1004819. http://journals.plos.org/ploscompbiol/article?id=10.1371/journal.pcbi.1004819

Bidhendi AJ, Geitmann A (2016) Relating the mechanics of the primary plant cell wall to morphogenesis. J Exp Bot 67(2):449–461. http://jxb.oxfordjournals.org/content/67/2/449.abstract

Bohn S et al (2002) Constitutive property of the local organization of leaf venation networks. Phys Rev E 65(6):61914. http://link.aps.org/doi/10.1103/PhysRevE.65.061914

Braam J (2005) In touch: plant responses to mechanical stimuli. New Phytol 165(2):373–389. http://www.scopus.com/inward/record.url?eid=2-s2.0-13144282772&partnerID=tZOtx3y1

Christmann A, Weiler EW, Steudle E, Grill E (2007) A hydraulic signal in root-to-shoot signalling of water shortage. Plant J 52(1):167–174. http://doi.wiley.com/10.1111/j.1365-313X.2007.03234.x

Cohen J (1995) Lisianthus leaf curl a new disease of lisianthus caused by tomato yellow leaf curl virus. Plant Dis 79(4):416. http://www.cabdirect.org/abstracts/19952307537.html;jsessionid=54F0790C53EE9A2E2EE68B01F68E5AC2?freeview=true

Comstock JP (2002) Hydraulic and chemical signalling in the control of stomatal conductance and transpiration. J Exp Bot 53(367):195–200. https://academic.oup.com/jxb/article-lookup/doi/10.1093/jexbot/53.367.195

Corson F, Adda-Bedia M, Boudaoud A (2009) In silico leaf venation networks: growth and reorganization driven by mechanical forces. J Theor Biol 259(3):440–448
Cosgrove DJ (1993) How do plant cell walls extend?. Plant Physiol 102(1):1–6. http://www.pubmedcentral.nih.gov/articlerender.fcgi?artid=158739&tool=pmcentrez&rendertype=abstract
Cosgrove DJ (2015) Plant cell wall extensibility: connecting plant cell growth with cell wall structure, mechanics, and the action of wall-modifying enzymes. J Exp Bot 67(2):463–476. http://jxb.oxfordjournals.org/content/67/2/463.short
Couder Y et al (2002) The leaf venation as formed in a tensorial field. Eur Phys J B 28(2):135–138. http://www.springerlink.com/index/10.1140/epjb/e2002-00211-1
Coutand C et al (2000) Biomechanical study of the effect of a controlled bending on tomato stem elongation: global mechanical analysis. J Exp Bot 51(352):1813–1824. http://jxb.oxfordjournals.org/lookup/doi/10.1093/jexbot/51.352.1813
Dumais J, Forterre Y (2012) Vegetable dynamicks: the role of water in plant movements. Annu Rev Fluid Mech 44(1):453–478. http://www.annualreviews.org/doi/abs/10.1146/annurev-fluid-120710-101200
Efrati E, Sharon E, Kupferman R (2009) Elastic theory of unconstrained non-Euclidean plates. J Mech Phys Solids 57(4):762–775. http://www.sciencedirect.com/science/article/pii/S0022509608002160
Efrati E, Klein Y, Aharoni H, Sharon E (2007) Spontaneous buckling of elastic sheets with a prescribed non-Euclidean metric. Phys D Nonlinear Phenom 235(1–2 SPEC. ISS.):29–32
Elsner J, Michalski M, Kwiatkowska D (2012) Spatiotemporal variation of leaf epidermal cell growth: a quantitative analysis of arabidopsis thaliana wild-type and triple cyclinD3 mutant plants. Ann Bot 109(5):897–910. http://aob.oxfordjournals.org/content/early/2012/02/02/aob.mcs005.full
Fleming AJ et al (1999) Analysis of expansion-induced morphogenesis on the apical meristem of tomato. Planta 208(2):166–174. http://link.springer.com/10.1007/s004250050546
Gemmer JA, Venkataramani SC (2011) Shape selection in non-Euclidean plates. Phys D Nonlinear Phenom 240(19):1536–1552. http://linkinghub.elsevier.com/retrieve/pii/S0167278911001825
Gibson WT, Gibson MC (2012) Growing cells push back under pressure. Cell 149(2):259–261. http://www.sciencedirect.com/science/article/pii/S0092867412003959
Green PB, Erickson RO, Richmond PA (1970) On the physical basis of cell morphogenesis. Ann NY Acad Sci 175(1):712–731. http://doi.wiley.com/10.1111/j.1749-6632.1970.tb45187.x
Green PC, Steele S, Rennich SC (1996) Phyllotactic patterns: a biophysical mechanism for their origin. Ann Bot 77(5):515–528. https://academic.oup.com/aob/article-lookup/doi/10.1006/anbo.1996.0062
Green PB (1999) Invited special paper. Expression of pattern in plants: combining molecular and calculus-based biophysical paradigms. Am J Bot 86(8):1059. http://links.jstor.org/sici?sici=0002-9122%28199908%2986%3A8%3C1059%3AISPEOP%3E2.0.CO%3B2-B&origin=crossref
Guest S, Kebadze E, Pellegrino S (2011) A zero-stiffness elastic shell structure. J Mech Mater Struct 6(1–4):203–212. http://msp.org/jomms/2011/6-1/p14.xhtml
Gupta MD, Nath U (2015) Divergence in patterns of leaf growth polarity is associated with the expression divergence of miR396. Plant Cell 27(10):2785–2799. http://www.ncbi.nlm.nih.gov/pubmed/26410303
Hamant O et al (2008) Developmental patterning by mechanical signals in arabidopsis. Science 322(5908). http://science.sciencemag.org/content/322/5908/1650
Hellgren JM, Olofsson K, Sundberg B (2004) Patterns of auxin distribution during gravitational induction of reaction wood in poplar and pine. Plant Physiol 135(1):212–220. http://www.plantphysiol.org/content/135/1/212
Hervieux N et al (2016) A mechanical feedback restricts sepal growth and shape in arabidopsis. Curr Biol 26(8):1019–1028. http://linkinghub.elsevier.com/retrieve/pii/S0960982216301804
Jaffe MJ (1980) Morphogenetic responses of plants to mechanical stimuli or stress. BioScience 30(4):239–243. http://bioscience.oxfordjournals.org/content/30/4/239.abstract

Kim J et al (2012) Designing responsive buckled surfaces by halftone gel lithography. Science 335(6073)

Klein Y, Efrati E, Sharon E (2007) Shaping of elastic sheets by prescription of non-Euclidean metrics. Science 315(5815)

Klein Y, Venkataramani S, Sharon E (2011) Experimental study of shape transitions and energy scaling in thin non-Euclidean plates. Phys Rev Lett 106(11):118303. http://link.aps.org/doi/10.1103/PhysRevLett.106.118303

Levin I, Sharon E (2016) Anomalously soft non-Euclidean springs. Phys Rev Lett 116(3):35502. http://link.aps.org/doi/10.1103/PhysRevLett.116.035502

Marchetti MC et al (2013) Hydrodynamics of soft active matter. Rev Mod Phys 85(3):1143–1189. http://link.aps.org/doi/10.1103/RevModPhys.85.1143

Nakayama N et al (2012) Mechanical regulation of auxin-mediated growth. Current Biol CB 22(16):1468–1476. http://www.sciencedirect.com/science/article/pii/S0960982212007269

Nath U et al (2003) Genetic control of surface curvature. Science (New York, N.Y.) 299(5611):1404–1407. http://www.ncbi.nlm.nih.gov/pubmed/12610308 (January 2, 2016)

Nath U, Coen E (2003) Genetic control of surface curvature. Science 299(5611):1404–1407

O'Neill B (2001) Elementary differential geometry

Palatnik JF et al (2003) Control of leaf morphogenesis by microRNAs. Nature 425(6955):257–263. http://dx.doi.org/10.1038/nature01958

Paredez AR, Somerville CR, Ehrhardt DW (2006) Visualization of cellulose synthase demonstrates functional association with microtubules. Science (New York, N.Y.) 312(5779):1491–1495. http://www.sciencemag.org/content/312/5779/1491

Prost J, Jülicher F, Joanny JF (2015) Active gel physics. Nat Phys 11(2):111. http://www.nature.com/nphys/journal/v11/n2/abs/nphys3224.html

Remmler L, Rolland-Lagan AG (2012) Computational method for quantifying growth patterns at the adaxial leaf surface in three dimensions. Plant Physiol 159(1):27–39. http://www.ncbi.nlm.nih.gov/pubmed/22402928

Runions A et al (2005) Modeling and visualization of leaf venation patterns. ACM Trans Graph 24(3):702. http://portal.acm.org/citation.cfm?doid=1073204.1073251

Sachs T (1969) Polarity and the induction of organized vascular tissues. Ann Bot 33(2):263–275. https://academic.oup.com/aob/article/165334/Polarity

Sahaf M, Sharon E (2016) The rheology of a growing leaf: stress-induced changes in the mechanical properties of leaves. J Exp Bot 67(18):5509–5515. http://www.ncbi.nlm.nih.gov/pubmed/27651350

Sampathkumar A, Yan A, Krupinski P, Meyerowitz EMM (2014) Physical forces regulate plant development and morphogenesis. Curr Biol 24(10):R475–R483. http://dx.doi.org/10.1016/j.cub.2014.03.014

Santangelo CD (2009) Buckling thin disks and ribbons with non-Euclidean metrics. EPL (Europhys Lett). http://iopscience.iop.org/article/10.1209/0295-5075/86/34003/meta

Scarpella E, Marcos D, Friml J, Berleth T (2006) Control of leaf vascular patterning by polar auxin transport. Genes Dev 20(8):1015–1027. http://www.ncbi.nlm.nih.gov/pubmed/16618807

Sharon E, Efrati E (2010) The mechanics of non-Euclidean plates. Soft Matter 6(22):5693. http://xlink.rsc.org/?DOI=c0sm00479k

Shipman PD, Newell AC (2005) Polygonal planforms and phyllotaxis on plants. J Theor Biol 236(2):154–197. http://www.sciencedirect.com/science/article/pii/S0022519305000950

Shraiman BI (2005) Mechanical feedback as a possible regulator of tissue growth. Proc Natl Acad Sci USA 102(9):3318–3323. http://www.ncbi.nlm.nih.gov/pubmed/15728365

Tauriello G et al (2015) Variability and constancy in cellular growth of arabidopsis sepals. Plant Physiol 169(4):2342–2358. http://www.ncbi.nlm.nih.gov/pubmed/26432876

Thompson DW (1942) On growth and form. On Growth Form

Trewavas A, Knight M (1994) Mechanical signalling, calcium and plant form. Signals and signal transduction pathways in plants. Springer Netherlands, Dordrecht, pp 93–105. http://www.springerlink.com/index/10.1007/978-94-011-0239-1_6

Uyttewaal M et al (2012) Mechanical stress acts via katanin to amplify differences in growth rate between adjacent cells in arabidopsis. Cell 149(2):439–451. http://www.sciencedirect.com/science/article/pii/S0092867412002899

Vogel V, Sheetz M (2006) Local force and geometry sensing regulate cell functions. Nat Rev Mol Cell Biol 7(4):265–275. http://www.nature.com/doifinder/10.1038/nrm1890

Volokh KY (2006) Tissue morphogenesis: a surface buckling mechanism. Int J Dev Biol 50(2–3):359–365. http://www.ncbi.nlm.nih.gov/pubmed/16479503

Winship LJ, Obermeyer G, Geitmann A, Hepler PK (2010) Under pressure, cell walls set the pace. Trends Plant Sci 15(7):363–369. http://www.sciencedirect.com/science/article/pii/S1360138510000737

Wolff J (1986) Law bone remodelling. Springer, Berlin Heidelberg. http://link.springer.com/10.1007/978-3-642-71031-5

Wu ZL et al (2013) Three-dimensional shape transformations of hydrogel sheets induced by small-scale modulation of internal stresses. Nat Commun 4:1586. http://www.nature.com/doifinder/10.1038/ncomms2549

Zonia L, Munnik T (2007) Life under pressure: hydrostatic pressure in cell growth and function. Trends Plant Sci 12(3):90–97. http://www.sciencedirect.com/science/article/pii/S1360138507000337

Twisting Growth in Plant Roots

Hirofumi Wada and Daichi Matsumoto

Abstract One of the fundamental problems in plant morphogenesis is the molecular and cellular basis of left-right asymmetry that often leads to various chiral structures such as the coils of tendrils and twisted leaves. The twisting mutants of the *Arabidopsis* roots and hypocotyl exhibit a helical pattern of epidermal cell files with a handedness that is opposite to that of the underlying cortical microtubule arrays in the epidermis. These mutants offer the unique opportunity to investigate the genetic basis of twisting in plants, particularly in the context of cortical microtubules. In this chapter, we address the importance of large-scale mechanical forces to understand the mechanism of this hierarchical helical order, with a particular emphasis on the role of tissue tension combined with the stresses generated by differential growth. Physical processes such as elasticity and geometry might be important factors to coordinate the chirality across different length scales and to organize an entire plant body.

Introduction

Twisted shapes are ubiquitous in plants, such as vines, tendrils, flower petals, leaves, and tree trunks (Hashimoto 2002; Smyth 2016). They appear as helices, twisted ribbons, or spirals (see Fig. 1. For reference, the basic terminology on helical handedness is summarized in Fig. 2.). These structures, with the exception of a perfectly planar spiral, are often referred to as chiral, and arise from simpler slender forms such as plain rods or ribbons via deformations involving breaking of the left-right symmetry. Elucidating the biological and physical factors determining the left-right asymmetry in chiral morphologies is a fundamental issue in the biological sciences (Levin 2005; Henley 2012). For example, the helical handedness in twining plants is usually fixed and is known to be genetically regulated. In fact, in many plants,

H. Wada (✉) · D. Matsumoto
Department of Physics, Ritsumeikan University, Kusatsu, Shiga 525-8577, Japan
e-mail: hwada@fc.ritsumei.ac.jp

D. Matsumoto
e-mail: rp0006xs@ed.ritsumei.ac.jp

A. Geitmann and J. Gril (eds.), *Plant Biomechanics*,
https://doi.org/10.1007/978-3-319-79099-2_6

Fig. 1 Examples of chiral plants. (Left) A tendril of *Momordica charantia* known as bitter melon (or goya), with the left-handed helix shown (the perversion of this structure can be seen on the top of the photo). (Center) A leaf of waterweed *Vallisneria asiatica*, with a left-handed twisting ribbon shown

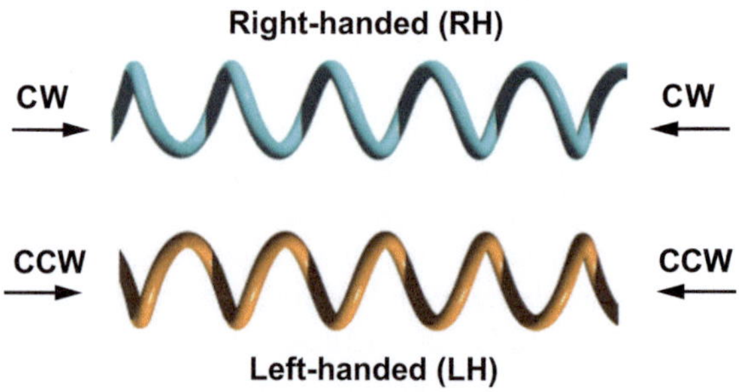

Fig. 2 Graphical definition of helical handedness. Distinction of left and right handedness is intrinsic to helical geometry, each of which is the mirror image of the other. A right-handed (RH) helix rotates clockwise (CW) while advancing from one end to the other, whereas a left-handed (LH) helix rotates counterclockwise (CCW). Note that if recession from one end to the other occurs, CCW rotation is noted in the RH helix, whereas CW rotation occurs in the LH helix. Thus, the distinction based on LH and RH, but not on CW and CCW, is generic. In this chapter, we will use the former terminology

animals, and bacteria, the symmetry-breaking factor is generally imprinted at the level of the molecular building blocks such as proteins. Although the functional advantages of chiral plants have often been addressed in terms of structure, mechanical stability, and deformations (Schulgasser and Witztum 2004; Elbaum et al. 2007; Armon et al. 2011; Gerbode et al. 2012; Zhao et al. 2015), the mechanism by which the microscopic information is translated into the macroscopic structure to robustly determine the helical morphology of plants remains unknown. Clearly, in addition to the roles of various molecular processes, including genes and morphogens (auxin), it is also essential to consider the roles of mechanical processes such as deformation, flow, and stresses in gaining a better understanding of these mechanisms (Green 1994; Boudaoud 2010; Mirabet et al. 2011; Moulia 2013).

On the one hand, plant morphogenesis is established through the three major cellular processes of cell elongation, cell division, and cell differentiation (Geitmann and Ortega 2009). On the other hand, from an architectural point of view, plants build up various complex three-dimensional shapes by combining thin (i.e., one- or two-dimensional) objects such as the roots, stems, leaves, and flower petals. These thin structures can be formed as a consequence of highly directional tissue growth, and are suitably described by geometric quantities such as curvatures (Silk 1989; Aharoni et al. 2012), which then evolve according to the laws of physical forces (Thompson 1992; Goriely and Tabor 1998; Marder et al. 2003; Sharon and Efrati 2010; Liang and Mahadevan 2011; Abraham et al. 2011).

The size and shape of a walled cell are known to be the direct consequence of the mechanical balance between the turgor pressure (i.e., hydrostatic pressure of the cytoplasm) and the tension in a load-bearing rigid cell wall enclosing the cytoplasm (Kamiya et al. 1963; Dumais and Forterre 2012). Cell elongation is realized through the chemically induced loosening of the cell wall and its consequent remodeling (Lockhart 1965; Ortega 1985; Geitmann and Ortega 2009; Burgert and Keplinger 2013). Since the turgor pressure is, by definition, isotropic, a cell always becomes a sphere if it expands isotropically with its cell wall composed of isotropic elastic materials. In fact, polarized cell expansion is attained by the built-in transverse alignment of cellulose microfibrils (Baskin 2005; Schopfer 2006; Dumais 2013), which itself is very likely to be controlled by the cortical microtubules. Many experimental lines of evidence suggest that microtubules work as cortical templates for cellulose synthase complex trajectories at the plasma membrane, and thus guide the deposition of cellulose microfibrils in cell walls (Paredez et al. 2006). Note, however, that microtubules and celluloses are not always collinear (Himmelspach et al. 2013), and growth anisotropy might not only rely on cellulose-dependent wall mechanical anisotropy given that mechanical polarities can be triggered through localized modifications of pectins (Peaucelle et al. 2015). Taken together, the cortical microtubule alignment directly regulates cell growth, and ultimately plant morphology. Thus, a possible direct link between a large-scale helical pattern and microtubule deserves detailed investigation from not only a biochemical point of view but also from physical and mathematical perspectives.

A twisting mutant in *Arabidopsis thaliana* is particularly well suited to explore the correlations between microtubule organization and helical patterns (Ishida et al. 2007b). Axial organs such as the roots and hypocotyls in wild-type plants usually maintain straight expansion (guided by gravity) during highly anisotropic growth expansion. The microtubule array of root cells appears to be a simple stack of ring-like bundles, which is aligned transverse to the primary growth axis. In contrast, in the twisting phenotype, the growth direction of epidermal cell files is systematically tilted either to the right or to the left at a certain angle (typically 10–20°). Importantly, the epidermis in twisting mutants possesses an oblique microtubule array that shows clear helical ordering. Many studies have demonstrated a striking correlation

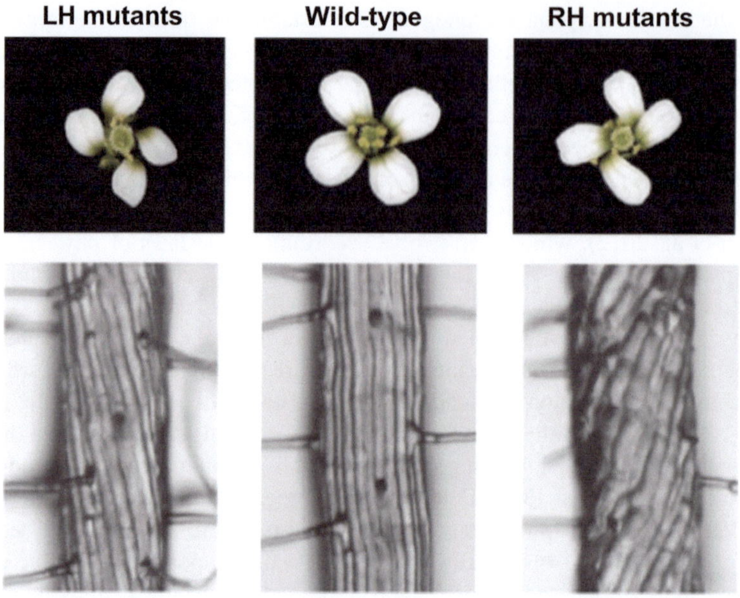

Fig. 3 Flower petals and the root epidermis of *Arabidopsis thaliana*: wild-type (center), right-handed (RH) mutant (right panel), and left-handed (LH) mutant (left panel). Image courtesy of Takashi Hashimoto

between chirality in organ growth and alignment of the helical microtubule array. More specifically, RH helical growth mutants possess LH microtubule arrays, and vice versa (Ishida et al. 2007b; Furutani et al. 2000; Thitamadee et al. 2002). This remarkably simple relationship implies an underlying mechanical basis of the macroscopic left-right asymmetry. Ishida et al. (2007b) proposed a conceptual model that the differential expansion between the inner and outer layers leads to the tilting of the epidermal cell files. Although this hypothesis might explain the tilting of cell layers essentially as a mechanical process, it cannot explain the mechanism of the selection of specific helical handedness, or its relationship with that of microtubule arrays (Figs. 3 and 4).

Model

The observed regularity and hierarchy in the helical handedness between the microtubule and epidermis provides a clue for understanding the central role of microtubules, and thus of the tubulin proteins, in plant morphogenesis (Ishida et al. 2007a). We are particularly interested in understanding how such a molecular-level event can break and regulate the macroscopic helical symmetry, when the helicity of cortical microtubules preexists.

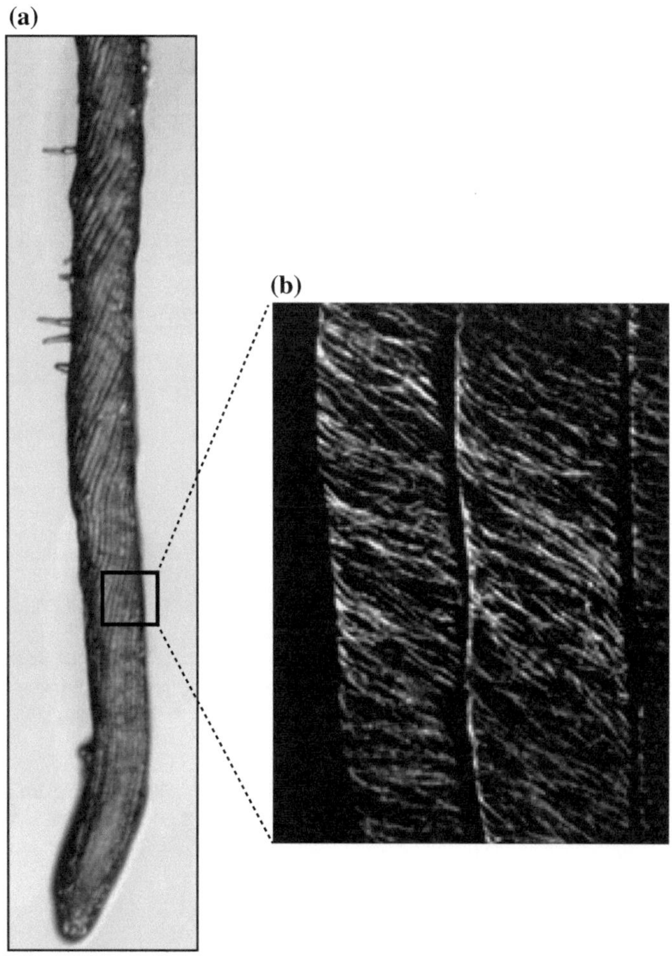

Fig. 4 **a** Root of the right-handed mutant of *Arabidopsis thaliana*. **b** Left-handed helical array of cortical microtubules in the epidermal cells of the elongation zone of the root shown in (**a**). Image courtesy of Takashi Hashimoto

In this chapter, we discuss how a biomechanical theory can explain this helical ordering of the roots of *Arabidopsis* mutants without relying on the biochemistry of the as-yet-unspecified molecules. A detailed mathematical equation based on a linear elasticity theory can be found in Wada (2012), but the main biophysical concept of the theory is completely accessible without any mathematics, as shown in Fig. 5. A fundamental assumption of this model is that plant organ morphology is governed by the large-scale mechanical stresses generated by helical growth at a single-cell level. For this, the tilted alignment of wall microfibrils directly induces helical or twisting growth of each epidermal cell. However, this requirement is incompatible at the organ scale, i.e., individual epidermal cells cannot freely twist about their own

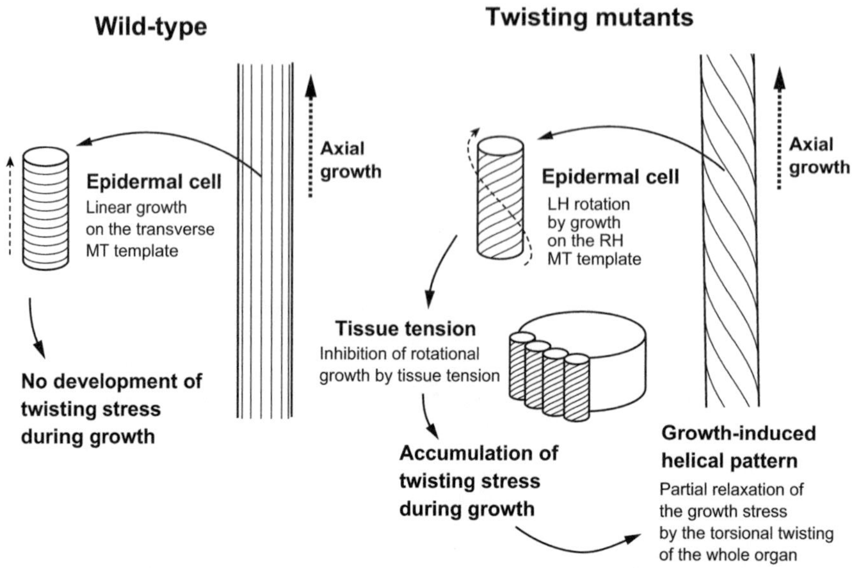

Fig. 5 Schematic illustration of the main concept of a biomechanical model that could explain the helical order in the twisting mutants of *Arabidopsis thaliana*. (Left) In the wild type, the growth direction of an epidermal cell consorts with that of the root itself, allowing for straight expansion of the root without any internal stresses. (Right) In a twisting mutant (here, the LH mutant is shown), the growth of an epidermal cell involves straight elongation as well as rotation, which can twist the cell owing to helical ordering of the underlying microtubule arrays. Thus, the exact growth orientation is incompatible with the major growth axis of the root, which develops growth-induced twisting stresses. This might lead to weak skewing of the entire organ, since it partially reduces the accumulated elastic energy driven by growth. The resulting shape can be seen as the bending (here shown as right-handed torsion) of the epidermal cell files on the root surface, as observed in experiments

axis, since they are firmly glued to each other and to the inner cortex cell layers to maintain the entire structure. This compositeness develops growth-induced stresses, which will be sufficient to enforce the entire organ to twist about the root axis so as to minimize the twisting elastic energies. Therefore, the helical growth at the single-cell level naturally organizes the helical growth of an entire organ in one coherent direction mediated only by elasticity.

Single-Cell Growth

The expansion of cells in the *Arabidopsis* roots is known to be highly anisotropic. Immediately after the cells leave the cell division zone, unidirectional expansion of an individual cell is completed in a few hours at a typical growth velocity of a few μm/min, with virtually no appreciable radial expansion. This allows for modeling

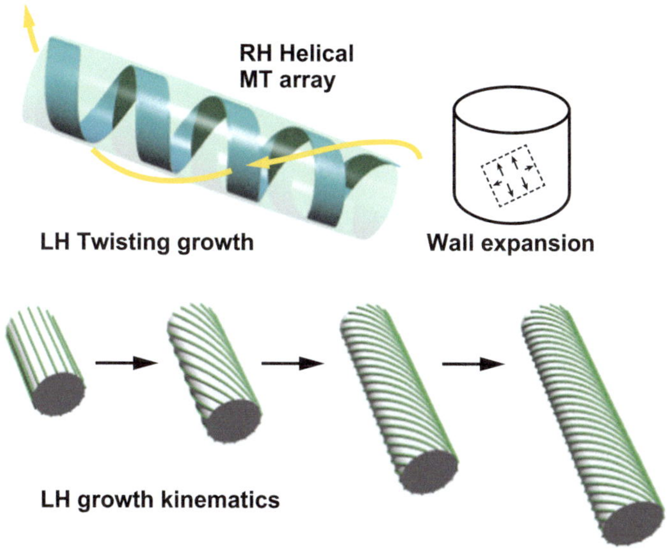

Fig. 6 Helical growth kinematics of a single cell. (Top left) A cylindrical cell with an underlying RH microtubule (MT) array (blue ribbon) undergoing LH twisted growth. (Top right) Cartoon of surface expansion of the cell wall with a tilt from the major axis. (Bottom) Kinematics of the twisted growth of a cylindrical cell with a constant radius, where the length of the cylinder increases exponentially with time. The green lines represent the material lines painted on the surface of the cylinder to show the twisting, which should not be confused with the orientation of microtubules or microfibrils

of this cell as a cylindrical rod that grows in length but maintains a constant radius. Thus, when the major growth direction is tilted away from the major axis, the cylinder twists as it elongates. See Fig. 6. This type of helical growth kinematics has long been widely recognized (Probine 1963; Roelofsen 1966) and applied to many different biological problems of walled cells that exhibit helical growth (Schulgasser and Witztum 2004; Goriely and Tabor 2011; Wolgemuth et al. 2004). Considering that the growth direction of the cell wall is transverse to the microtubule array, cells with RH microtubule alignment can be thought to exhibit LH growth and vice versa. (Further, note that the wall microfibril alignment is helical, with the same handedness as that of the microtubules.) Thus, the twisting handedness of an elongating cell is always opposite to that of the microtubule helices.

Our assumption that cells from twisting mutants might twist during growth is experimentally supported. In the study by Buschmann et al. (2009), epidermal cells from both wild-type and twisting mutants were grown in free cell suspensions, so that the cells could freely expand in fluids. They found that the wild-type cells elongated normally into a cylindrical shape, whereas the cells from twisting mutants were considerably more defective, and some of them even appeared as helical forms (Fig. 7). These findings suggest that the twist is generated within growing cells, which is the heart of organ twisting. The results shown in Fig. 6 do not consider

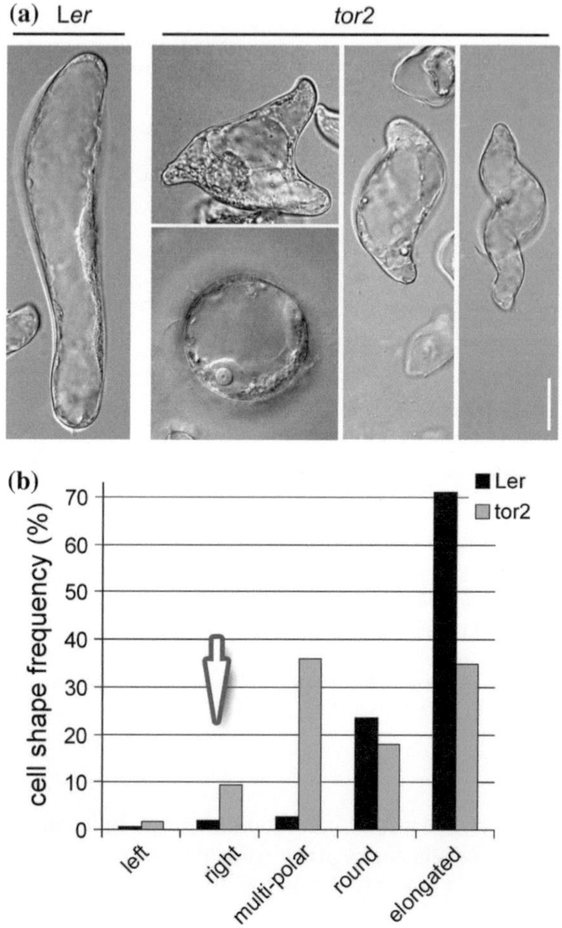

Fig. 7 Polarity defects seen in cell suspensions of *tor2*. **a** In wild-type suspension lines (Ler), cells grow in fresh medium into normal cylindrical cell shapes (one week of growth). In mutant cell (*tor2*) suspensions, cells are usually less elongated and are much more defective, including multipolar, round, or helical shapes. Scale bar: 20 μm. **b** Populations of different types of defective cell shapes observed. (Wild-type, $n = 297$; *tor2*, $n = 278$.) In *tor2* suspensions, more than 10% of the cells have helical forms, most of which are right-handed (arrow). Reproduced with permission from Buschmann et al. (2009) and is copyrighted by the American Society of Plant Biologists

any mechanics (stretching, compression, or shearing) of cell walls, but rather focus only on the idealized kinematics. Therefore, the effects of turgor pressure or residual stresses are entirely neglected. (Spiral growth by turgor pressure was previously analyzed for cell walls by using general anisotropic elastic constants (Sellen 1983).). These features should be incorporated into this simple argument in order to best understand the helical shapes of twisting mutant cells observed in the experiments.

Pattern Formation at the Tissue Scale

In plant roots, the thick and inextensible epidermal cell layer sheathes the relatively more compliant inner tissues (Baskin 2005). Thus, the entire organ is modeled as a composite elastic body consisting of the inner and outer cell layers. Therefore, the epidermal cells can elongate along the root axis, but might not twist about their own axes. This leads to the development of twisting stress as the epidermis grows. Such growth-induced stress is also transmitted to the inner cortex layer via the extracellular matrices. This load sharing is known as tissue tension (Peters and Thomos 1996). Consequently, the entire composite structure twists about the primary axis of the root in order to partially relieve the internal stress. The resulting torsional angle of the epidermal cell files is determined by the ratio of the elastic moduli between the outer and inner tissues. Importantly, the macroscopic morphological pattern is determined by the relative magnitude of the mechanical properties of the outer and inner tissues, but not by the absolute magnitude of the elastic properties. This scenario is actually considerably common and has been recognized in several recent studies of developmental processes in animals (Savin et al. 2011) as well as in plants (Yin et al. 2008).

The growth of plant organs is a very slow process; no appreciable change is noted in the physical parameters during the typical time required for the entire body to attain mechanical equilibrium. Therefore, a quasi-static assumption can be adopted where the state of the system can be determined by minimizing the elastic deformation energy of the root organ under the material parameters given by the growth process. Over a long timescale, these parameters change gradually, and the plant morphology slowly evolves with time (plastic deformation). This is a standard treatment in the biomechanics of growing tissues (Goriely et al. 2008).

Figure 8 shows a schematic representation of how the tissue tension can change the resulting appearance of the root surface pattern. This figure shows the configurations of a single twisted elastic rod wrapping the surface of a soft elastic cylinder. The slender rod mimics the cell file in the epidermis, and the cylinder corresponds to the inner body of the root (grouping all the remaining tissues). The red color on the surface of the outer rod visually indicates the magnitude of a rod's twisting degree. The equilibrium configuration shown in Fig. 8 is obtained from the numerical minimization of an appropriately constructed elastic deformation energy of this composite system (Wada 2012). Three qualitatively different cases are tested, which are explained below.

The shape shown in (a) is obtained in the absence of tissue tension between the outer rod and the cylinder. In this hypothetical case, a rod simply twists about its axis during elongation (see the helical red line), but its centerline is straight, with no helical pattern developed on the surface of the cylinder. Thus, no growth-induced stress develops in this case. In contrast, the shape shown in (b) is obtained in the presence of tissue tension, and the inner layer is considerably softer than the outer layer so that the twisting stress in the inner layer is negligible. In this case, the helical tilt angle becomes maximal. Finally, the shape shown in (c) is obtained for the inner layer that is still softer than the outer layer, but is not so soft as to allow its elasticity to resist the outer layer skewing around the cylindrical surface. The resulting helical tilt angle is reduced compared to the case shown in (b), depending on the extent of mismatch in the elastic constants of the two regions. Indeed, this situation is the closest to that of the plant root, and yields the most reasonable helical pattern

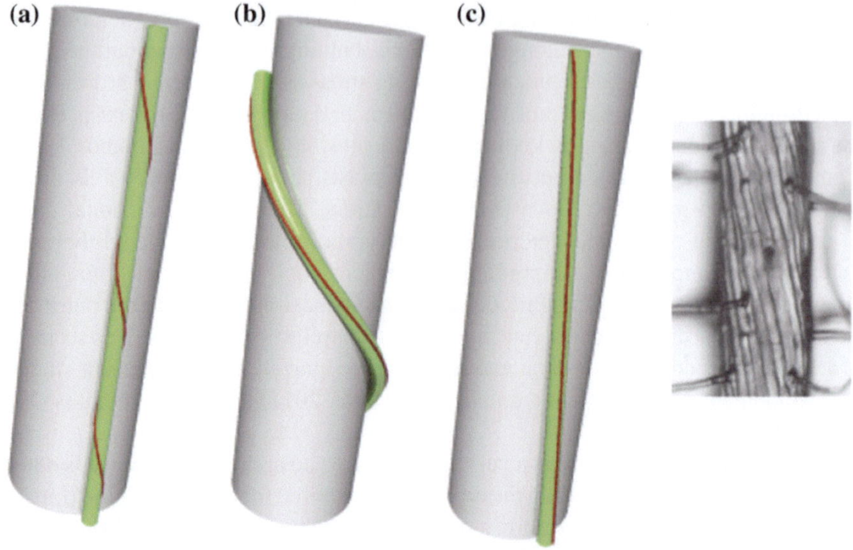

Fig. 8 Configurations of the composite body obtained from the numerical minimization of the appropriately defined elastic deformation energy. **a** Without any mechanical coupling between the outer rod and inner cylinder (i.e., no tissue tension). **b** With coupling, and the inner cylinder is considerably softer than the outer rod. **c** With coupling, and the inner cylinder is relatively softer than the outer rod. A weak helical (bending and torsional) deformation of the outer rod can be noted in (**c**), comparable to that observed in the experiments

similar to that observed in experiments. The detailed calculation and comparison with available experimental data suggest that the elastic modulus of the epidermis can be $\sim 10^2$ times larger than that of the inner cortex layer (Wada 2012). This predicted ratio is substantial and needs to be checked by conducting further experiments.

Summary

In this chapter, by examining a peculiar example of twisting plant roots, we have provided a brief overview of the importance of adopting both geometric and mechanical approaches to understand pattern formation in plants. The concept described herein is considerably simple. In order to construct a more realistic description of plant tissue growth, our argument will need to be modified to account for other physical effects such as the effects of turgor pressure, residual stress, and wall reinforcements. Most importantly, our theory does not account for the mechanism underlying the establishment of microtubule helical arrangement, the understanding of which requires physics of different lengths and time scales (Hawkins et al. 2010; Muratov and Baulin 2015). Moreover, how the large-scale forces feedback on cellular processes has not yet been discussed but is extremely important to the developmental control in plants (Hamant et al. 2008; Richter et al. 2009; Dumais 2012; Robinson et al. 2013) and animals (Shraiman 2005; Hannezo et al. 2012). The argument herein provides, based on physical concepts, a plausible explanation on how the initial symmetry-breaking of the microtubule array determines the final shape.

At this point, it is useful to briefly compare the twisting mechanism of roots with that of stems, since the tissue tension may play a different role in maintaining the uniform cylindrical structures of stems. Given the anatomical complexity of stems, it is surprising that many stems maintain a thin, uniform cylinder by regulating the highly coordinated growth of numerous constituent cells across different length scales (Baskin and Jensen 2013). In plant stems, the load-bearing outer epidermal cell layers are mechanically stiffer than the inner tissues, with the outer tissues being axially stretched while the inner tissues are compressed. Consistent with the knowledge that the microtubules align with maximal tension, the microtubules largely orient axially in the stem epidermal layers, thus preventing the oblique orientation of microtubules in individual cells, a distinctive feature from that in roots. This structure-encoded tissue tension ensures the unidirectional growth of stems. In fact, twisting stems have been observed in the csi1 mutants, in which cellulose deposition is largely uncoupled with cortical microtubules (Landrein et al. 2013). In other words, the tissue stress in stems in the wild-type may prevent twisting by enforcing the axial orientation of microtubules and cellulose in the epidermis. By contrast, in the roots, growth-induced tissue stress can promote twisting. These different roles that tissue tension plays in roots and stems may further highlight the importance of tissue tension in the morphogenesis of plants.

Our theory emphasizes that the growth-induced stress is a key factor responsible for transmitting the microtubule information to the organ shape, whose length scales can differ by 10^6 times. Indeed, at such large distances, mechanical signals could be more efficient than biochemical activities. Most importantly, tissue tension organizes a differentially strained composite structure, ensuring stress transmittance between different parts. In the developmental processes of plants (and animals), cells differentiate and assemble to form different tissues with different shapes to perform specialized biological functions. However, from a mechanical point of view, these different parts should be assembled together to work as a whole body, which share mechanical stresses. This compositeness, combined with the stresses driven by differential growth, generates the various complex morphologies observed in plants.

Acknowledgements We are grateful to Takashi Hashimoto for sharing the images shown in Figs. 1 and 2 and for valuable discussions. We also thank Tobias Baskin for suggesting relevant literature on twisted cell growth (Probine 1963; Sellen 1983).

References

Abraham Y, Tamburu C, Klein E, Dunlop JWC, Fratzl P, Raviv U, Elbaum R (2011) Tilted cellulose arrangement as a novel mechanism for hygroscopic coiling in the stork's bill awn. J R Soc Interf 9:640–647

Aharoni H, Abraham Y, Elbaum R, Sharon E, Kupferman R (2012) Emergence of spontaneous twist and curvature in non-euclidean rods: Application to erodium plant cells. Phys Rev Lett 108(238):106

Armon S, Efrati E, Kupferman R, Sharon E (2011) Geometry and mechanics in the opening of chiral seed pods. Science 333:1726–1730

Baskin TI (2005) Anisotropic expansion of the plant cell wall. Annu Rev Cell Dev Biol 21:203–222

Baskin TI, Jensen OE (2013) On the role of stress anisotropy in the growth of stems. J Exp Bot 64:4697–4707

Boudaoud A (2010) An introduction to the mechanics of morphogenesis for plant biologists. Trends Plant Sci 15:353–360

Burgert I, Keplinger T (2013) Plant micro- and nanomechanics: experimental techniques for plant cell-wall analysis. Exp Bot 64:4635–4649

Buschmann H, Hauptmann M, Niessing D, Lloyd CW, Schäffner AR (2009) Helical growth of the arabidopsis mutant tortifolia2 does not depend on cell division patterns but involves handed twisting of isolated cells. Plant Cell 21:2090–2106

Dumais J (2012) Can mechanics control pattern formation in plants? Curr Opin Plant Biol 10:58–62

Dumais J (2013) Modes of deformation of walled cells. J Exp Bot 64:4681–4695

Dumais J, Forterre Y (2012) "Vegetable Dynamicks": the role of water in plant movements. Annu Rev Fluid Mech 44:453–478

Elbaum R, Zaltzman L, Burgert I, Fratzl P (2007) The role of wheat awns in the seed dispersal unit. Science 316:884–886

Furutani I, Watanabe Y, Prieto R, Masukawa M, Suzuki K, Naoi K, Thitamadee S, Shikanai T, Hashimoto T (2000) The spiral genes are required for directional control of cell elongation in arabidopsis thaliana. Development 127:4443–4453

Geitmann A, Ortega JKE (2009) Mechanics and modeling of plant cell growth. Trends Plant Sci 14:467–478

Gerbode SJ, Puzey JR, McCormick AG, Mahadevan L (2012) How the cucumber tendril coils and overwinds. Science 337:1087–1091

Goriely A, Tabor M (1998) Spontaneous helix hand reversal and tendril perversion in climbing plants. Phys Rev Lett 80:1564

Goriely A, Tabor M (2011) Spontaneous rotational inversion in phycomyces. Phys Rev Lett 106(138):103

Goriely A, Robertson-Tessi M, Tabor M, Vandiver R (2008) Elastic growth models. In: Mondaini RP, Pardalos PM (eds) Mathematical modeling of biosystems. Springer, Berlin, Heidelberg, pp 1–44

Green PB (1994) Connecting gene and hormone action to form, pattern and organogenesis. J Exp Bot 45:1775–1788

Hamant O, Heisler MG, Joensson H, Krupinski P, Uyttewaal M, Bokov P, Corson F, Sahlin P, Boudaoud A, Meyerowitz EM, Couder Y, Traas J (2008) Developmental patterning by mechanical signals in arabidopsis. Science 322:1650–1655

Hannezo E, Prost J, Joanny JF (2012) Mechanical instability of biological tubes. Phys Rev Lett 109(018):101

Hashimoto T (2002) Molecular genetic analysis of left-right handedness in plants. Phil Trans R Soc Lond B 357:799–808

Hawkins RJ, Tindemans SH, Mulder BM (2010) Model for the orientational ordering of the plant microtubule cortical array. Phys Rev E 82(011):911

Henley CL (2012) Possible origins of macroscopic left-right asymmetry in organisms. J Stat Phys 148:741–775

Himmelspach R, Williamson RE, Wasteneys GO (2013) Cellulose microbril alignment recovers from dcb-induced disruption despite microtubule disorganization. Plant J 36:565–575
Ishida T, Kaneko Y, Iwano M, Hashimoto T (2007a) Helical microtubule arrays in a collection of twisting tubulin mutants of arabidopsis thaliana. Proc Natl Acad Sci USA 104:8544–8549
Ishida T, Thitamadee S, Hashimoto T (2007b) Twisted growth and organization of cortical microtubules. J Plant Res 120:61–70
Kamiya N, Tazawa M, Takada T (1963) The relation of tugor pressure to cell volume in nitella with special reference to mechanical properties of the cell wall. Protoplasma 57:501–521
Landrein B, Bringmann M, Lathe R, Vouillot C, Ivakov A, Boudaoud A, Persson S, Hamant O (2013) Impaired cellulose synthase guidance leads to stem torsion and twists phyllotactic patterns in arabidopsis. Curr Biol 23:895–900
Levin M (2005) Left-right asymmetry in embryonic development: a comprehensive review. Mech Dev 122:3–25
Liang H, Mahadevan L (2011) Growth, geometry, and mechanics of a blooming lily. Proc Natl Acad Sci USA 108:5516–5521
Lockhart JA (1965) An analysis of irreversible plant cell elongation. J Theor Biol 8:264–275
Marder M, Sharon E, Smith S, Roman B (2003) Theory of edges of leaves. Europhys Lett 62:498–504
Mirabet V, Das P, Boudaoud A, Hamant O (2011) The role of mechanical forces in plant morphogenesis. Ann Rev Plant Biol 62:365–385
Moulia B (2013) Plant biomechanics and mechanobiology are convergent paths to flourishing interdisciplinary research. J Exp Bot 64:4617–4633
Muratov A, Baulin VA (2015) Mechanism of dynamic reorientation of cortical microtubules due to mechanical stress. Biohys Chem 207:82–89
Ortega J (1985) Augmented growth equation for cell wall expansion. Plant Physiol 79:318–320
Paredez AR, Somerville CR, Ehrhardt DW (2006) Visualization of cellulose synthase demonstrates functional association with microtubules. Science 312:1491–1495
Peaucelle A, Wightman R, Hoefte H (2015) The control of growth symmetry breaking in the arabidopsis hypocotyl. Curr Biol 25:1746–1752
Peters WS, Thomos AD (1996) The history of tissue tension. Ann Bot 77:657–665
Probine MC (1963) Cell growth and the structure and mechanical properties of the wall in internodal cells of nitella opaca-iii. spiral growth and cell wall structure. J Exp Bot 14:101–113
Richter G, Monshausen GB, Krol A, Gilroy S (2009) Mechanical stimuli modulate lateral root organogenesis. Plant Physiol 151:1855–1866
Robinson S, Burian A, Couturier E, Landrein B, Louveaux M, Neumann ED, Peaucelle A, Weber A, Nakayama N (2013) Mechanical control of morphogenesis at the shoot apex. Exp Bot 64:4729–4744
Roelofsen PA (1966) Ultrastructure of the wall in growing cells and its relation to the direction of the growth. Adv Bot Res 2:69–149
Savin T, Kurpios NA, Shyer AE, Florescu P, Liang H, Mahadevan L, Tabin CJ (2011) On the growth and form of the gut. Nature 476:57–63
Schopfer P (2006) Biomechanics of plant growth. Am J Bot 93:1415–1425
Schulgasser K, Witztum A (2004) Spiralling upward. J Theor Biol 230:275–280
Sellen DB (1983) The response of mechanically anisotropic cylinderical cells to multiaxial stress. J Exp Bot 34:681–687
Sharon E, Efrati E (2010) The mechanics of non-euclidean plates. Soft Matter 6:5693–5704
Shraiman BI (2005) Mechanical feedback as a possible regulator of tissue growth. Proc Natl Acad Sci USA 102:3318–3323
Silk WK (1989) Growth rate patterns which maintain a helical tissue tube. J Theor Biol 138:311–327
Smyth DR (2016) Helical growth in plant organs: mechanisms and significance. Development 143:3272–3282
Thitamadee S, Tuchihara K, Hashimoto T (2002) Microtubule basis for left-handed helical growth in arabidopsis. Nature (London) 417:193–196

Thompson DW (1992) On growth and form: the complete, Revised edn. Dover, New York
Wada H (2012) Hierarchical helical order in the twisted growth of plant organs. Phys Rev Lett 109(128):104
Wolgemuth CW, Goldstein RE, Powers TR (2004) Dynamic supercoiling bifurcations of growing elastic filaments. Phys D 190:266–289
Yin J, Cao Z, Li C, Sheinman I, Chen X (2008) Stress-driven buckling patterns in spheroidal core/shell structures. Proc Natl Acad Sci USA 105:19,132–19,135
Zhao ZL, Zhao HP, Li BW, Nie BD, Feng XQ, Gao H (2015) Biomechanical tactics of chiral growth in emergent aquatic macrophytes. Sci Rep 5(12):610

Plants at Bodybuilding: Development of Plant "Muscles"

Tatyana Gorshkova, Polina Mikshina, Anna Petrova, Tatyana Chernova, Natalia Mokshina and Oleg Gorshkov

Abstract Plant fibers are the important elements to shape the mechanical properties of plant body, especially in the organs that have already ceased elongation. The major distinguishing parameters of fibers are a highly prosenchimatous cell shape and an increased cell wall thickness as compared to other types of plant cells. The increase of fiber cell length is largely achieved by intrusive growth—elongation with the increased rate as compared to the adjacent cells and squeezing between them along the middle lamellae. The highly pronounced intrusive growth is the cause of fiber bundle formation. Thickening of cell wall in fibers of many plant species is supplied by deposition of the tertiary cell wall (G-layer) of peculiar design and properties. Tension of cellulose microfibrils is developed in this cell wall layer, providing the contractile properties that permit to move plant organs. We summarize the currently available data describing the inherent to fibers mechanisms by which they attain their exclusive length (intrusive growth) and extreme cell wall thickness (tertiary cell wall deposition) and consider the results obtained by finite element modeling to realize the cause of cellulose microfibril tension. The suggested hypothesis is based on the entrapment of tissue- and stage-specific version of rhamnogalacturonan I between laterally interacting cellulose microfibrils.

T. Gorshkova (✉) · P. Mikshina · A. Petrova · T. Chernova · N. Mokshina · O. Gorshkov
The Laboratory of Plant Cell Growth Mechanisms, Kazan Institute of Biochemistry and Biophysics, FRC Kazan Scientific Center of RAS, 420111 Kazan, Russia
e-mail: gorshkova@kibb.knc.ru

P. Mikshina
e-mail: p.mikshina@gmail.com

A. Petrova
e-mail: anna.an.petrova@gmail.com

T. Chernova
e-mail: tchernova@kibb.knc.ru

N. Mokshina
e-mail: natalali@list.ru

O. Gorshkov
e-mail: o_gorshkov@mail.ru

A. Geitmann and J. Gril (eds.), *Plant Biomechanics*,
https://doi.org/10.1007/978-3-319-79099-2_7

Keywords Plant fibers · Intrusive growth · Cell wall · Tertiary cell wall G-layer · Rhamnogalacturonan I · Cellulose · Tension of cellulose microfibrils

Introduction

To withstand mechanical loads, both external and emerging during organism growth, plants develop specialized cells—fibers. Such cells are found in various plant organs, including roots, stems, petioles, tendrils, and peduncles (Esau 1965; Gorshkova et al. 2012). Fibers occur in phloem and xylem of both primary and secondary origin, and sometimes in non-vascular tissues. The major distinguishing features of fibers are the extreme length that can reach many centimeters and the thickened cell wall (Evert 2006; Gorshkova et al. 2012). Correspondingly, the especially characteristic and important stages of fiber development are cell elongation and cell wall thickening.

There are two types of plant fibers differing in design and properties. One provides static reinforcement, helping to stabilize the organism and its parts in space. The other develops contractile properties that may permit to move plant parts and as such can be considered as plant "muscles". The major difference between these two types of plant fibers is in the design of the thickened cell wall. The elongation of fibers, irrespective of their type, is largely provided by intrusive elongation that precedes cell wall thickening.

Fiber Intrusive Elongation

Intrusive growth is a type of plant cell enlargement characterized by an excessive rate of a cell elongation compared to the surrounding tissues. During intrusive elongation, a cell splits the middle lamellae of the cells on its way and squeezes in between the surrounding tissues (Esau 1965; Fahn 1990; Lev-Yadun 2001, 2015; Nezhad and Geitmann 2013). Intrusive growth is performed by several types of plant cells, like laticifers (Canaveze and Machado 2016), pollen tubes (Yang 1998), fusiform cambium initials (Mazur and Kurczynska 2012; Mazur et al. 2014), and tracheids (Jura-Morawiec 2017), with fibers being the classical example due to the extreme attained length (Esau 1943; Lev-Yadun 2001, 2015). Fiber intrusive growth is quite difficult to study since it occurs within the depth of other plant tissues and has never been reproduced in vitro (Gorshkova et al. 2012; Lev-Yadun 2015), unlike pollen tube elongation (Fan et al. 2001; Wang et al. 2008, 2015). The best studied examples of fiber intrusive elongation are primary phloem fibers of flax (Esau 1965; Ageeva et al. 2005; Snegireva et al. 2010; Gorshkova et al. 2018, in press), primary and secondary phloem fibers of hemp (Snegireva et al. 2010, 2015) and secondary xylary fibers of *Populus* (Mellerowicz et al. 2001; Gray-Mitsumune et al. 2004, 2008; Siedlecka et al. 2008). *Arabidopsis*, being a well-known model for studying cell wall thickening in fibers (Zhong et al. 2001, 2006; Zhong and Ye 2014; Cassan-Wang et al. 2013), has been barely studied in view of fiber intrusive elongation.

Basic Parameters of Fiber Intrusive Elongation

Intrusive growth is one of the early stages of fiber development. Right after initiation from the procambium, primary fibers first elongate by symplastic growth together with the surrounding tissues in the elongation zone of an organ (Esau 1943; Ageeva et al. 2005; Snegireva et al. 2010, 2015) (Fig. 1). Soon, when growth of other cells starts to slow down, fibers hold the rate of growth, or even increase it and may further elongate for several days. The onset of intrusive elongation is marked by the formation of a "knee"-like structure: the flat end of the symplastically elongating cell is transformed into a tip in order to allow penetration between other cells. In flax and hemp stems, such structures are observed at 1–2 mm from the shoot apex (Ageeva et al. 2005; Snegireva et al. 2015). "Knees" are present at both fiber ends, confirming that intrusive elongation occurs bidirectionally.

The enlargement of cell surface and volume by intrusive growth can be enormous. For example, when a flax primary phloem fiber originates from procambium, it is around 3 μm in width and at the end of symplastic growth reaches 100–200 μm in length (Ageeva et al. 2005). After accomplishing the intrusive growth, a fiber gets 20–25 μm wide with the modal length around 20,000 μm (Table 1) meaning that in average cell volume has increased about 5,000 times.

Secondary fibers, which originate from the cambium in non-elongating parts of the organ and thus are not involved in symplastic elongation, attain their increased (as compared to cambial initials) length exclusively by intrusive growth (Mellerowicz et al. 2001; Zhong et al. 2001; Snegireva et al. 2015). Secondary fibers experience especially severe mechanical constraints during intrusive elongation. Cambium activity increases organ width, so that the preformed tissues are under radial and tangential stress, especially in the xylem. Fiber intrusive elongation adds to this effect, increasing the cell number in the organ cross section. A considerable effort should be applied to provide enough space for the increase of fiber dimensions, since more mature cells located just couple of layers away from intrusively growing fibers may be already lignified and thus have fixed shapes. This can be the reason for the differences in length of phloem and xylem fibers, same as between primary and secondary fibers. In general, distribution of fibers in accordance with their modal length makes the row: primary phloem fibers >> secondary phloem fibers >> secondary xylem fibers.

During intrusive growth, the number of nuclei in primary phloem fibers is increased by asynchronous karyokinesis and can reach over hundred per cell (Ageeva et al. 2005; Snegireva et al. 2010). The mechanism that uncouples karyokinesis from cytokinesis in growing fibers is completely unknown. In primary phloem fibers of hemp, the fiber length has high positive correlation with the number of the nuclei, while in flax phloem fibers, the correlation is of moderate significance. On the other

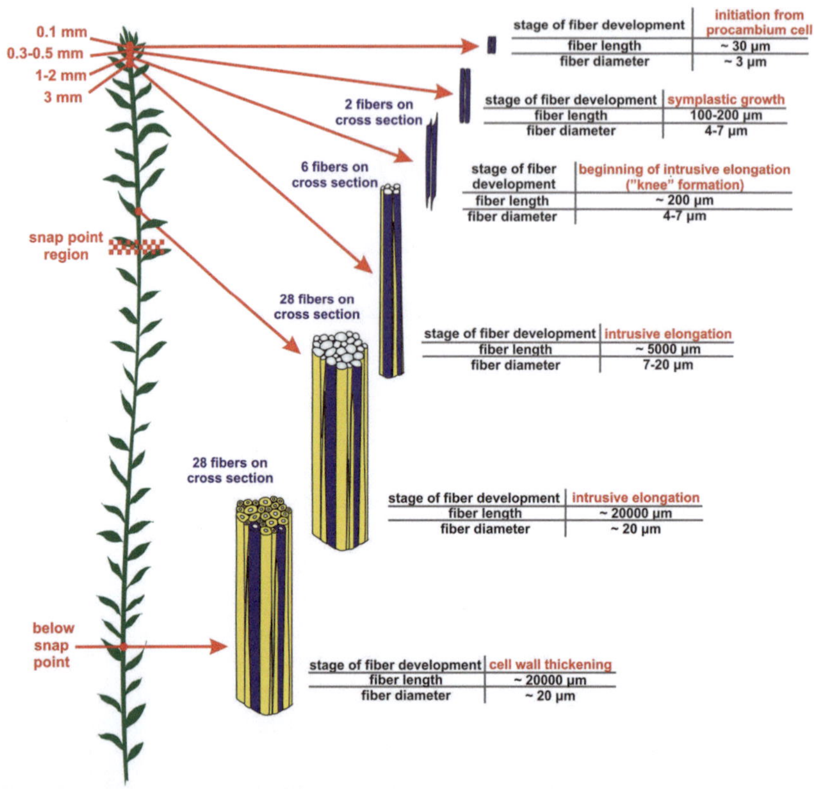

Fig. 1 Intrusive growth of flax phloem fibers: increase in cell length, localization within stem, and role in fiber bundle formation. Primary phloem fibers are initiated from procambium close to the very stem apex, then fibers grow symplastically with surrounding cells, and later, when the surrounding cells are still growing symplastically, start intrusive elongation. Manifold bidirectional intrusive elongation brings together fibers initiated from the procambium at different leaf traces leading to fiber bundle formation and increase of fiber number on the stem cross section. Fibers, which were formed by one procambium cell within one leaf trace are given in dark blue; fibers formed by procambium cells within other leaf traces—in yellow. Shown is one fiber bundle out of 20–40 bundles presented in flax stem. Intrusively growing fibers are located within several top centimeters of the stem above snap point (Gorshkova et al. 2003). Below the snap point fibers thicken their cell walls

hand, secondary phloem fibers, irrespective of their length, have only one nucleus (Snegireva et al. 2010).

Nothing is known about the trigger and stop mechanisms of intrusive growth. Moreover, there is no information about the nature of the osmoticum, which has to accumulate for providing sufficient turgor pressure during the enormous enlargement of the cell volume. Gaps in the information are largely defined by the fact that direct observation of intrusively growing fibers is practically impossible and the isolation of these cells in quantities sufficient for biochemical and molecular biological

Table 1 Major characteristics of intrusive growth of primary phloem fibers in flax and hemp

Parameter	Cell characteristic
Cell length before intrusive growth	Around 100–200 μm [1, 2]
Cell length after intrusive growth	Up to 80,000 μm; mean value—around 20,000 μm [2, 3]
Cell width before intrusive growth	3–7 μm [1, 2]
Cell width after intrusive growth	20–30 μm [1, 2]
Marker of intrusive growth initiation	Transformation of flat cell end into tapered one, formation of the "knee" [1, 2]
Evidence for intrusive growth	Increase of primary fiber number on the cross section of stem taken at the same stem height in the course of plant development [2, 3, 4]
Direction of intrusive growth	Bidirectional; "knees" from both cell ends [1, 3]
Rate of intrusive growth	Average rate—0.4 mm/h, relative rate—0.11 h^{-1}—much higher than other cells in the shoot; similar to that of cells in root tips [3]
Duration of intrusive growth	Several days [1–4]
Microtubule orientation in intrusively growing fibers	At fiber tip—transverse to the cell axis; in middle part of the cell—helical [1]
Plasmodesmata connections	Destroyed [1, 3]
Mode of elongation growth	Diffusive [1, 4]
Number of nuclei per cell	Up to 175 [3]

[1] Ageeva et al. (2005), [2] Snegireva et al. (2015), [3] Snegireva et al. 2010, [4] Gorshkova et al. (2003)

analyses is not an easy task. There are several reasons for this. First, the intrusively growing cells are located in the depth of other tissues and are difficult to observe. Second, a high ratio between fiber length and diameter (more than 1000) and the thin cell walls makes growing fibers vulnerable to damage during isolation. Third, fiber intrusive elongation is quite sensitive to injury and stops even if the fiber remains alive (Ageeva et al. 2005; Snegireva et al. 2006). Finally, as mentioned above, there were no successful attempts to reproduce fiber intrusive growth in vitro.

Tip Versus Diffusive Elongation

Plant cells can elongate either by extending the whole surface (diffusive growth) or by increasing their length only at the tips (tip growth) (Evert 2006; Geitmann and Ortega 2009; Guerriero et al. 2014). Which of these types is used by fibers is not completely clear. Primary fibers start their elongation by symplastic growth with the neighboring cells (Ageeva et al. 2005). Symplastic growth in all cell types operates through the diffusive growth, so the whole machinery for cell enlargement in primary fibers at the beginning of intrusive elongation is set up for the increase of the total cell surface. There is no evidence that the primary fibers switch toward tip growth during intrusive elongation: the so-called “clear zone” characteristic for tip growth (Bedinger et al. 1994; Hepler and Winship 2015) is absent, and cortical microtubules in fiber tip have transverse orientation, instead of longitudinal (Ageeva et al. 2005). Plasmodesmata in the primary fiber cell walls present at the stage of symplastic growth are destroyed during intrusive elongation both at the fiber/fiber and fiber/parenchyma interfaces, making a fiber a symplastically isolated domain (Ageeva et al. 2005). This points to the disturbance of the surface in the central part of fiber body, formed before intrusive elongation. The previously suggested simultaneous presence of thickened cell wall in the middle portion of a fiber and thin cell wall at fiber tips (Esau 1965) was not confirmed in detailed studies, including serial sections (Ageeva et al. 2005; Snegireva et al. 2010). Pulse-chase experiments with radioactive labeling and with staining by Calcofluor White also suggest the increase of the whole fiber surface (Gorshkova et al. 2003; Ageeva et al. 2005). All the above suggests that intrusive elongation of flax phloem fibers is carried out by diffuse growth which is characterized by expansion of entire surface of the cell (Gorshkova et al. 2003; Ageeva et al. 2005; Gorshkova et al. 2012). As pointed out in the review on the cellular mechanisms of invasive lifestyle (Nezhad and Geitmann 2013), such mechanism is quite challenging from a mechanical point of view.

The situation may be different in secondary xylem fibers that do not have the stage of symplastic elongation, because they arise in the stem part that has ceased to elongate as a whole. Intrusive growth of secondary xylem fibers occurs in the zone of xylem cell radial expansion and may be performed in a different way. Hejnowicz (1980) has suggested the specific combination of stresses in the region around cambium: due to the secondary xylem formation, cambium is pushed outwards the stem and has to increase its circumference. Such necessity may promote the intrusive growth of cambium, and probably fibers, in tangential direction. The situation is quite different from primary phloem fibers, which originate from procambium at leaf and do not get distributed along the stem circumference, but rather form the compact bundles. However, stress distribution in stem region with intrusively growing primary fibers still has to be elucidated.

A tip growth mechanism is suggested to occur in wood fibers (Wenham and Cusick 1975; Larson 1994; Gray-Mitsumune et al. 2004; Mellerowicz 2006; Siedlecka et al. 2008; Gorshkova et al. 2012). The resulting increase of fiber length may be hundreds of times less than in primary phloem fibers about 1.3–4.0 times relative to the length of cambial initials (Larson 1994). For woody plants, the intrusive growth duration is still poorly defined, and the same applies to possible combination of fiber elongation by tip growth and cell wall thickening in the middle part of the same cell. Profiling of fiber cell wall in the main cell body and at the tips by the approaches to obtain chemical and mechanical parameters is necessary to summarize growth patterns of fibers during intrusive elongation.

Bundle Formation During Fiber Intrusive Elongation

In other plant tissues, fibers can be present as individual cells (idioblasts), as radial files deposited in secondary xylem by sequential divisions of cambial cells, or in groups called bundles. The latter are formed during fiber intrusive elongation, when primary fibers initiated from the procambium at different internodes meet each other during bidirectional increase of cell length (Fig. 1). The same is true for secondary fibers initiated from cambium initials that are located at different positions relative to the longitudinal stem axis (Snegireva et al. 2015). Formation of bundles is characteristic for phloem primary and secondary fibers, which accomplish highly significant intrusive elongation. This makes intrusive growth especially important for fiber crops, such as flax and hemp.

Fibers in bundles are very tightly connected, with no intercellular spaces. This is quite important to provide large surfaces to glue fibers to each other and form strong bundles able to withstand large mechanical loads. The strength of connections between fibers gets quite obvious upon processing of fiber crops. The loosely bound surrounding cells fall apart upon retting or other types of tissue maceration, while fiber bundles stay intact (Akin 2013). The connections between fibers in bundles through the middle lamellae can be reinforced by low-molecular phenolic compounds (Gorshkova et al. 2000); at late stage of development lignin can be deposited.

Transcriptome Landscape in Intrusively Growing Fibers

The information on molecular players that perform and regulate fiber intrusive elongation is extremely scarce. Several genes, the expression of which may influence fiber length as revealed by the analysis of corresponding mutants, like *INTERFASCICULAR FIBERLESS1 (IFL1)/REVOLUTA (REV)* (Zhong and Ye 1999) or *FRAGILE FIBER (FRA2)/KATANIN (AtKTN1)* (Burk et al. 2001), do not seem to be expressed specifically at intrusive elongation stage. Additional problem to consider the results obtained on such mutants is the mistakes in identification of fibers at early stages of development, as pointed by Lev-Yadun et al. (2004). In most of performed studies, plant fibers are considered at advanced stages of development when they already have thickened cell wall.

The closest analog to fiber intrusive elongation is the growth of the pollen tube through the stigma, which has been extensively studied at molecular level (e.g., Wang et al. 2008; Qin et al. 2009). Such studies greatly benefitted from the possibility to reproduce pollen tube elongation in vitro. The only transcriptome study of intrusively growing fibers was performed on flax phloem fibers isolated by laser microdissection (Gorshkova et al. 2018, in press). The RNA-Seq analysis helped to reveal the molecular players specific for this stage of fiber development, including transcription factors, cell wall proteins, ion channels, etc. Some of them are common with the proteins, involved in pollen tube growth (Fig. 2). The list of upregulated genes

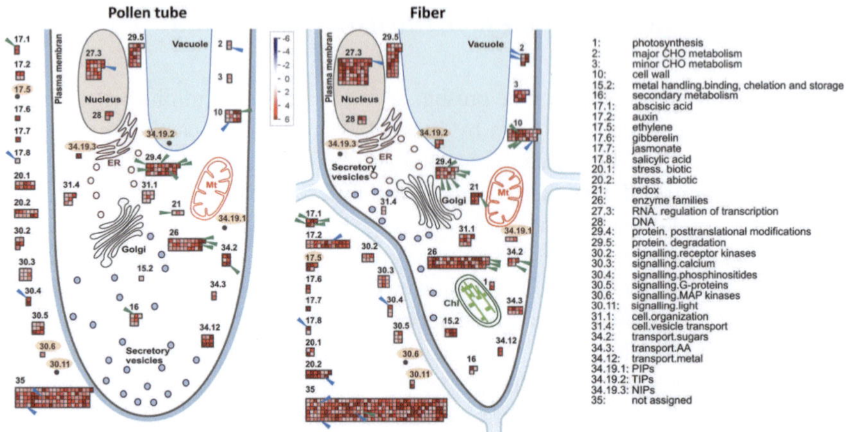

Fig. 2 Comparison of sets of genes upregulated in growing in vitro pollen tube and in intrusively elongating fibers. MAPMAN categories of genes upregulated in *Arabidopsis* pollen tube grown in vitro during 4 h versus hydrated pollen grains (to the left) and in intrusively elongating flax phloem fiber *vs* apical stem tissues (to the right). Data for pollen tube were taken from Wang et al. (2008), Qin et al. (2009) and for intrusively growing flax phloem fibers isolated by laser microdissection—from Gorshkova et al. (2018, in press). Cutoff in both cases was log2FC > 2. Blue arrowheads indicate the genes that have the counterpart with the same gene ID; green—genes for similarly annotated proteins, but with different gene IDs. Orange ovals mark categories that are upregulated only in one of analyzed plant objects

encoding similar protein is given in Table 2. Exact matching of genes upregulated in both types of growing cells is rather rare. Out of 609 genes upregulated in flax fiber and 387 genes upregulated in pollen tube of *Arabidopsis* that passed over the applied cutoff ($\log_2$FC > 2), only 10 had the counterpart with the same gene ID (flax genes were considered through the closest homologue in *Arabidopsis*). Additionally, 21 genes had similar annotation but different gene IDs (Table 2). Sometimes, one upregulated gene from *Arabidopsis* had several upregulated homologues in flax.

The list of genes that are upregulated in both considered plant systems includes the gene for CCCH protein with zinc finger domain. Such proteins that are widely spread in many types of organisms bind RNA and function in RNA processing (Wang et al. 2008). In rice, some genes for CCCH proteins with zinc finger domain were only regulated in specific tissues or under specific stresses (Li et al. 2013).

Expectedly, some genes for enzymes involved in pectin degradation are upregulated: modifications of these cell wall polymers are known to be important for growing pollen tube (Bosch and Hepler 2005; Chebli et al. 2012) and for fiber penetrating middle lamellae of the neighboring cells (Siedlecka et al. 2008). Two genes from neutral invertase family that are upregulated in flax fiber have the same homologue that is upregulated in pollen tube. Same situation is with genes encoding similar arabinogalactan proteins (AGP). Expression of numerous representatives of AGP multigene family is finely tuned, so that genes of specific AGPs are up- and down-regulated in the course of various developmental transitions. Similar AGP expressed

Table 2 List of genes with upregulated expression both in intrusively growing fibers and pollen tubes

Description (MAPMAN BinCode)	Lus ID	At homolog
Genes that have the counterpart with the same gene ID		
CCCH-type zinc fingerfamily protein with RNA-binding domain (27)	Lus10006379	AT2G05160
Plasma membrane associated cation-binding protein 1 (20, 30)	Lus10036243	AT4G20260
	Lus10038383	AT4G20260
O-fucosyltransferase family protein (17)	Lus10037475	AT2G44500
Carbohydrate-binding protein (35)	Lus10011718	AT1G10150
Arabinogalactan protein 20 (35)	Lus10032355	AT3G61640
	Lus10033939	AT3G61640
Plant neutral invertase family protein (2)	Lus10017095	AT3G06500
	Lus10037817	AT3G06500
S-adenosyl-L-methionine-dependent methyltransferases superfamily protein (17)	Lus10036547	AT5G66430
Genes for similarly annotated proteins, but with different gene ID		
GDSL-like Lipase/Acylhydrolase superfamily protein (26)	Lus10008784	AT3G27950
	Lus10024690	AT1G29670
	Lus10032316	AT1G29660
GRAM domain-containing protein/ ABA-responsive protein-related (17)	Lus10010167	AT5G08350
	Lus10013445	AT5G08350
	Lus10040999	AT5G08350
	Lus10010166	AT5G08350
Pectin lyase-like superfamily protein (10, 35)	Lus10011885	AT3G07010
	Lus10016837	AT3G16850
	Lus10022817	AT3G24670
	Lus10036231	AT4G20050
Protein kinase superfamily protein (29)	Lus10006643	AT5G15080

(continued)

Table 2 (continued)

Description (MAPMAN BinCode)	Lus ID	At homolog
	Lus10016382	AT1G33260
	Lus10036534	AT2G45590
	Lus10003155	AT4G13020
	Lus10002518	AT3G27580
	Lus10022272	AT3G27580
Sugar transporter (34)	Lus10009714	AT1G77210
	Lus10009428	AT1G22710
Thioredoxin superfamily protein (21)	Lus10040898	AT3G62950
UDP-glucoronosyl and UDP-glucosyl transferase	Lus10040725	AT5G59590

Several genes in flax genome may be homologues to the same gene from *Arabidopsis*

both in pollen tube and in intrusively growing fiber indicate some general similarities in the considered processes.

At the same time, there are major differences. Some groups of genes are upregulated specifically in intrusively growing fibers (Fig. 2). These include numerous genes of aquaporins, both located in plasma membrane (PIP) and in tonoplast (TIP). Indeed, the enormous cell enlargement is coupled to the increase of vacuole that occupies the major portion of the "grown-up" fiber. Aquporins are necessary to provide the adequate flux of water. In pollen tube, the situation is different, since the growing tip is separated from the preformed tube part by callose plugs (Chebli et al. 2012).

Using the applied cutoff, upregulated genes for proteins involved in hormone metabolism and signaling have different sets in pollen tube and fiber. This is especially true for ethylene (several upregulated genes in fiber and none in pollen tube) and auxin (the list of upregulated genes is considerably extended in fibers as compared to pollen tube).

Further analysis of data obtained by such powerful techniques as RNA-Seq and laser microdissection may strongly advance understanding of the mechanism of fiber intrusive growth.

Cell Wall Thickening

Two Types of Thick Cell Walls Can Be Present in Fibers

After accomplishing intrusive elongation, fibers start to thicken cell wall. Fibers that provide static reinforcement deposit secondary cell wall (SCW), which may consist of several layers with different orientations of cellulose microfibrils, are named S1–S3 according to the sequence of deposition. The major constituents of secondary cell walls in fibers are cellulose, xylan, and lignin (Fig. 3). Fibers that can be considered as plant "muscles" form tertiary cell wall that is deposited after at least one layer of secondary cell wall and is often named G-layer. There is no lignin and xylan in tertiary cell wall, cellulose amounts up to 90% of cell wall. The major matrix polysaccharide of tertiary cell wall is tissue—and stage-specific version of rhamnogalacturonan I (RG-I); some glucomannans and arabinogalactan proteins are also present (Fig. 3). The whole machinery of cell wall formation, including the set of regulating transcription factors, differs significantly between secondary and tertiary cell walls, as comes from the transcriptome analysis of cells at corresponding stages of development (Gorshkov et al. 2017).

Tertiary cell wall is fiber-specific, since it was described only in this cell type, while secondary cell wall is present in many cell types: vessels, fibers, parenchyma, etc. Tertiary cell wall can be formed constitutively, in the course of normal plant development, like in fibers of many crops (flax, hemp, ramie, nettle, etc.) that experience severe mechanical stresses due to high ratio of plant height to stem width. Deposition of tertiary cell wall can be induced, as it happens in tension wood fibers after plant inclination. The occurrence and functions of fibers with tertiary cell wall in plant kingdom have been recently reviewed (Gorshkova et al. 2018), same as the major ideas on the mechanisms that provide contractile properties to plant "muscles" (Alméras and Clair 2016). Here, we focus on the modeling of creation of cellulose microfibril tension.

Fiber-Specific Tertiary Cell Wall is a Dynamic Structure

Tertiary cell wall undergoes significant modification in the course of fiber development. This is well illustrated by the ultrastructural analysis of fiber development in flax (Gorshkova et al. 2003; Gorshkova and Morvan 2006) and hemp (Chernova et al. 2018). The newly deposited portions of tertiary cell wall (Gn) look "loosened", with some irregular "dark" strips between electron-transparent material. Such strips originate as the result of secretion of "dark" material from Golgi vesicles (Salnikov et al. 2008). The stripped appearance of the newly deposited cell wall portions is transient; later in fiber development it turns into the relatively uniform one (G), without pronounced "dark" strips (Fig. 4). Concomitantly, new portions of cell wall material

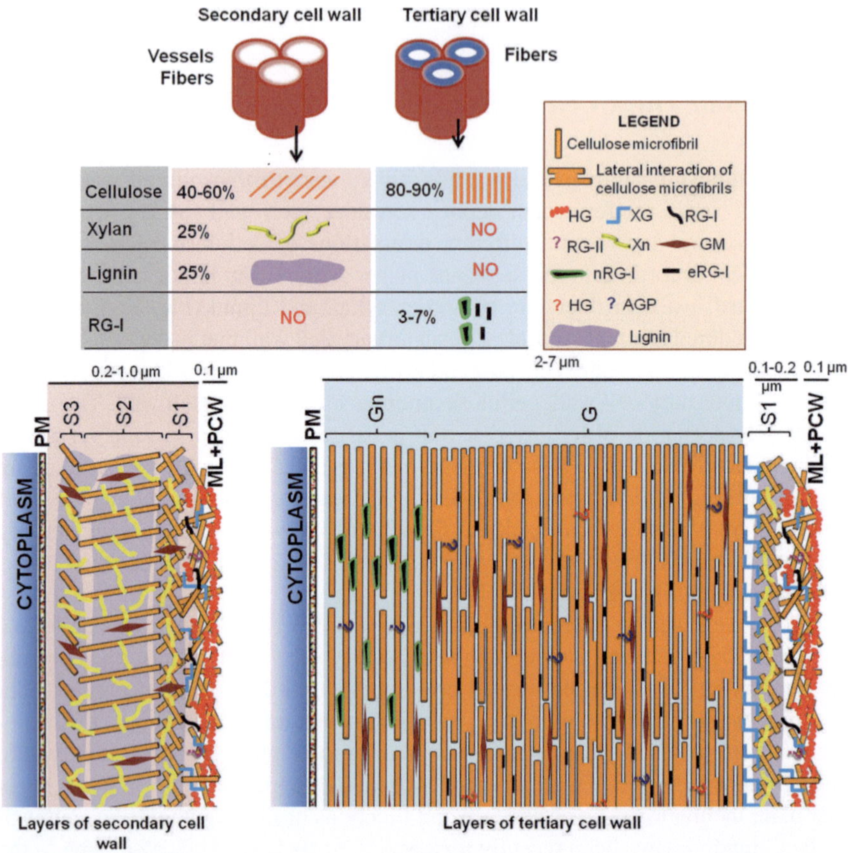

Fig. 3 Composition and structure of secondary and tertiary cell wall are quite different. PM—plasma membrane, ML—middle lamella, PCW—primary cell wall, S1–S3—layers of secondary cell wall, G—dense layer of tertiary cell wall, Gn—newly deposed layer of tertiary cell wall, HG—homogalacturonan, XG—xyloglucan, RG-I—rhamnogalacturonan I, RG-II—rhamnogalacturonan II, Xn—xylan, GM—glucomannan, nRG-I—nascent rhamnogalacturonan I, eRG-I—rhamnogalacturonan I entrapped by cellulose microfibrils (after in muro modification of nascent RG-I by galactosidase), AGP—arabinogalactan proteins. Bottom part—scheme of longitudinal sections through a cell with secondary cell wall and through a fiber with tertiary cell wall

are deposited that follow the same sequence of events. Finally, at fiber maturity, the whole tertiary cell wall looks quite “solid” and uniform (Fig. 4).

Immunolabelling with LM5 (Gorshkova et al. 2004), biochemical analysis of cell wall polymers (Gorshkova et al. 2010) and analysis of transgenic plants with downregulation of the β-galactosidase specifically expressed in fibers at this stage of development (Roach et al. 2011) have confirmed that such changes in cell wall structure are due to the partial trimming off the β-1 → 4-galactan chains of RG-I. Rhamnogalacturonan I initially synthesized with long galactan side chains is secreted

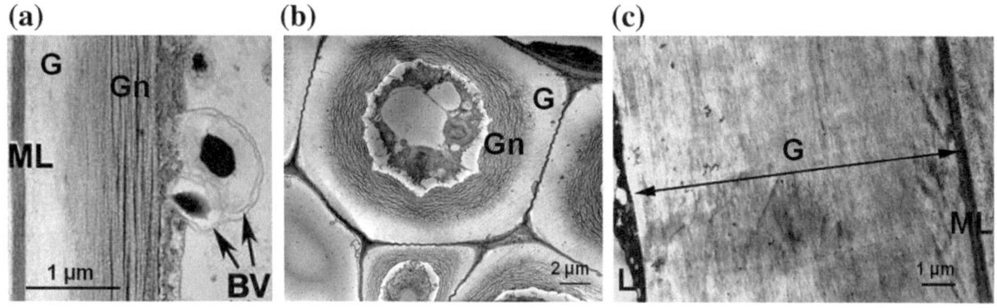

Fig. 4 Tertiary cell wall undergoes post-deposition modification. **a–c** Successive stages of tertiary cell wall deposition in flax phloem fibers. **a** longitudinal section of fiber at the beginning of tertiary cell wall deposition, (Salnikov et al. 2008), figure reprinted with kind permission from Springer **b** cross section of fibers with non-mature tertiary cell walls (Gorshkova and Morvan 2006), figure reprinted with kind permission from Springer **c** longitudinal section of fiber with mature cell wall (Gorshkova et al. 2003) figure reprinted with kind permission from Elsevier. ML—middle lamella together with primary cell wall and S1 of secondary cell wall; BV—bicolor vesicles secreting nascent nRG-I (black), Gn—newly deposited loosened layer of tertiary cell wall, G—dense layer of tertiary cell wall, L—lumen

into the periplasmic space and covers a large part of cellulose microfibril surfaces (Salnikov et al. 2008), thereby preventing their lateral interactions. This corresponds to the formation of loosened Gn-layer (Fig. 4). Subsequent removal of a part of galactan side chains from rhamnogalacturonan I backbone allows the lateral interactions of cellulose microfibrils and turns Gn into G-layer. In addition, such in muro post-synthetic modification significantly alters the elastic-plastic properties of RG-I (Mikshina et al. 2015b), so that the nascent RG-I differs from that one entrapped by cellulose microfibrils (Table 3).

Similar processes may occur in fibers of different origin. Transformation of Gn into G-layer is well resolved in primary and secondary phloem fibers of hemp by microscopy; biochemical analysis reveals the presence of tissue—and stage-specific RG-I with β-1 $\rightarrow$ 4-galactan chains Chernova et al. (2018). Similar RG-I and high β-galactosidase activity demonstrated in developing tension wood fibers (Gorshkova et al. 2015), as well as the detection of epitope for antibody raised against flax fiber-specific galactosidase in tertiary cell wall of aspen tension wood (Mokshina et al. 2012), suggest the involvement of this enzyme in β-(1 $\rightarrow$ 4)-galactan processing in tension wood. However, Gn-layer is poorly observed by microscopy during tertiary cell wall deposition in tension wood. This may be caused by rapid transition of newly deposed Gn-layer into the dense G-layer due to permanent and severe mechanical loads in the case of tension wood.

Table 3 Comparison of physicochemical parameters of nascent RG-I and RG-I entrapped by cellulose microfibrils in flax fibers

Parameter	Nascent RG-I	Entrapped RG-I
Backbone [1–6]	[→4)-α-D-Gal*p*A-(1→2)-α-L-Rha*p*(1→]	
GalA/Rha	1.0–2.2	0.9–1.4
α-L-Rha*p* branched with side chains	More than 90%	72%
Side chains [1–7]	Main chains: β-(1→4)-Gal*p*, attached to the O-4 of Rha*p*	
	(a) Short chains of one or two Gal; (b) Long (non-branched) chains of at least 26 Gal; (c) Chains of 3–12 Gal branched at O-2 and O-3 positions with single Gal; (d) Chains of at least 17 Gal residues, decorated with single Ara	(a) Short chains of one Gal (47% Rha is branched with only terminal Gal); (b) Oligomeric Gal chains (14 Gal in average); (c) Gal chains containing 1→2-Gal and 1→6-Gal; (d) Gal chains decorated with Ara
Content of RG-I [4, 6, 8]	1–2% of the cell wall mass of fiber-enriched peels or ~0.1 mg per growing stem	1.3–1.6% from mature fiber dry weight or 2.7–4.3% from the residue obtained after KOH treatment (cellulose+RG-I)
GPC elution region [1–9]	700–2000 kDa	100–400 kDa
Spatial organization	Associate of molecules [5, 6, 10]	
	Part of neutral galactan chains, interacting with each other, forms the core zone and holds the molecules in associates; other neutral chains are directed outside from backbone and are removed by galactosidase during modification *in muro*	Neutral galactan chains, interacting with each other, form the core zone and hold the molecules in associates; charged RG-I backbone is located at the periphery of associate
Hydrodynamic properties	(a) Tend to self-association in solution [6, 10]; b) Retain hydrodynamic volumes after molecular weight decrease by galactanase treatment [1, 2]; c) Contain strongly retained water molecules [11]	
Hydrodynamic radius	~25–35 nm [3]	~17.5–20 nm [6]
Availability of OH groups for solvent [11]	In 75% of OH groups protons are very slowly exchanged on deuterium	In 55% of OH groups protons are very slowly exchanged on deuterium; 10% of OH groups are sterically inaccessible for the solvent

(continued)

Table 3 (continued)

Parameter	Nascent RG-I	Entrapped RG-I
Rheological properties [11, 12]	(a) Form strong hydrogels at room temperature after microwave irradiation; the minimal concentration required for gel formation—4% (b) Gels from RGs-I are characterized by hyperelastic properties (c) Gelation of RG-I is based on galactan side chain interaction and does not involve chemical modifications	
Intrinsic viscosity	690 mg/g	380 mg/g
Young's modulus	9.3 kPa	13.7 kPa
Poisson's ratio	0.495	0.483
Stress–strain curves	Concave shapes	

[1] Gurjanov et al. (2007), [2] Mikshina et al. (2012), [3] Gorshkova et al. (2004), [4] Gurjanov et al. (2008), [5] Gorshkova et al. (2009), [6] Mikshina et al. (2015a), [7] Gorshkova et al. (1996), [8] Gorshkova and Morvan (2006), [9] Chernova et al. (2007), [10] Mikshina et al. (2013), [11] Mikshina et al. (2015b), [12] Mikshina et al. (2017)

Tertiary Cell Wall as the Basis of Plant "Muscles"

Due to high cellulose content in tertiary cell wall, microfibrils are not covered by layers of matrix polymers, making possible lateral interaction. Turgor pressure is necessary to push the microfibrils toward each other close enough to interact laterally. The consequences of lateral interactions are the increased degree of crystallinity and size of crystallites characteristic for G-layer (Dadswell and Wardrop 1955; Viëtor et al. 2002; Mellerowicz and Gorshkova 2012). Once formed, lateral interactions of cellulose microfibrils get strong enough to keep the whole system, so that G-layer is able to provide contractile properties in the absence of turgor pressure, as it may happen in tension wood fibers, when the protoplast dies off during the long life of a tree.

Due to the absence of hydrophobic lignin and presence of hydrophilic RG-I, tertiary cell walls are especially enriched in water (Schreiber et al. 2010). Mechanical properties of tertiary cell wall to a certain extent are dependent on high water content (Goswamy et al. 2008). However, the mechanism of tension creation is not fully based on different swelling capacity of cell wall layers with different orientation of microfibrils, as it happens in pine cones or wheat awns (Dawson et al. 1997; Elbaum et al. 2007). The action of plant "muscles" is not as dependent on the humidity as the abovementioned mechanism due to additional player—tissue—and stage-specific version of RG-I. Its characteristic feature is the ability to form supramolecular complexes; water is an important element of such complexes (Mikshina et al. 2015a, b).

The hypothesis on the involvement of RG-I in creation of cellulose microfibril tension is based on the original idea of Mellerowicz et al. (2008), suggesting that the entrapment of matrix polysaccharide between laterally interacting cellulose microfibrils may cause their tension. To validate the hypothesis and to check if the

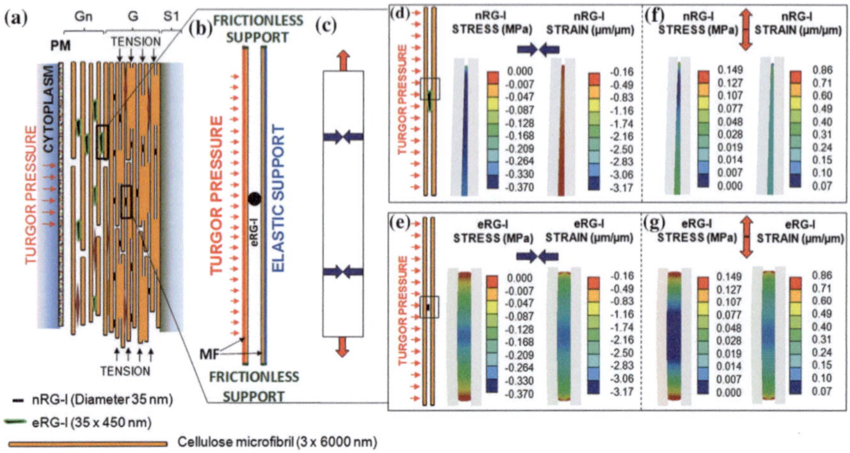

Fig. 5 Stress and strain of hyperelastic rhamnogalacturonans I at entrapment between two cellulose microfibrils, as determined by finite element model. **a** Schematic representation of tertiary cell wall with Gn and G layers; **b** initial model that includes RG-I from G-layer (eRG-I) captured by cellulose microfibrils (MF), one of which is under the pressure corresponding to the turgor pressure (0.3 MPa (Zimmermann et al. 1980), red); the other microfibril is elastically fixed (blue); ends of microfibrils are frictionless fixed to prevent slipping (dark green); **c** direction of material deformation (blue arrows—compression (negative values), red arrows—extension (positive values)); **d**, **f** stress and strain of nascent RG-I with long chains of β-1,4-galactan (nRG-I) from Gn-layer; **e**, **g** stress and strain of the RG-I with shorter β-1,4-galactan chains (eRG-I) that is entrapped by cellulose microfibrils in G-layer (based on data from Mikshina et al. 2015b); **d**, **e** measurements in the direction of turgor action (blue arrows pointing toward each other), **f**, **g** perpendicularly to turgor action (red double-headed arrow). Young modulus and Poisson's ratio were determined experimentally for nascent and entrapped RGs-I of flax fibers (Mikshina et al. 2015b) and taken from Sturcova et al. (2004) and Bos (2004) for cellulose microfibrils. Black rectangles in **d** and **e** mark the regions enlarged for demonstration of RG-I stress and strain

fiber-specific version of RG-I is effective in such system, we applied finite element modeling to figure out the stress and strain of captured RG-I (Fig. 5) and of cellulose microfibrils (Fig. 6). The model was based on the experimentally determined for RGs-I of flax fibers physicochemical parameters and particle sizes (Table 3, Mikshina et al. 2015a, b).

Finite Element Modeling of Tension Creation in Tertiary Cell Wall

The simplest version of the model included single rhamnogalacturonan I particle located between two cellulose microfibrils (Fig. 5b). One of the cellulose microfibrils was under the pressure corresponding to turgor pressure (0.3 MPa) (Zimmermann et al. 1980). The ends of this microfibril were frictionless fixed to prevent slipping of

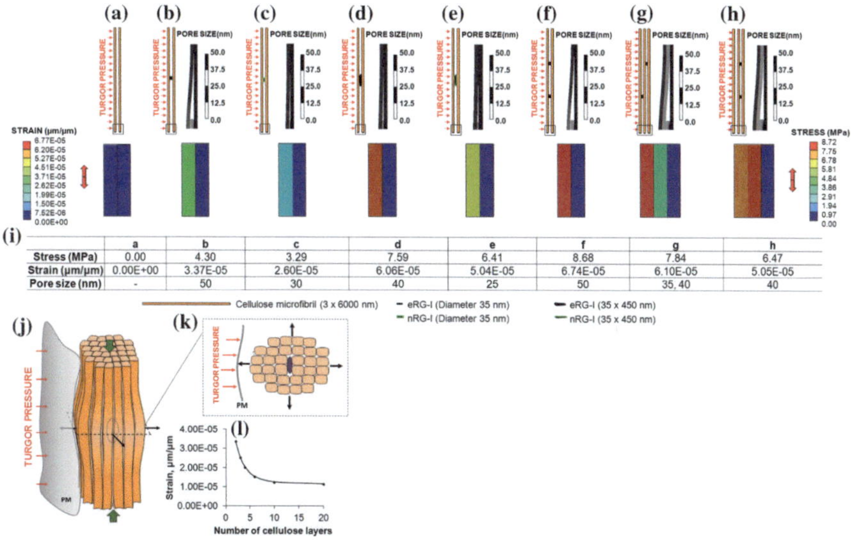

	a	b	c	d	e	f	g	h
Stress (MPa)	0.00	4.30	3.29	7.59	6.41	8.68	7.84	6.47
Strain (µm/µm)	0.00E+00	3.37E-05	2.60E-05	6.06E-05	5.04E-05	6.74E-05	6.10E-05	5.05E-05
Pore size (nm)	-	50	30	40	25	50	35, 40	40

Fig. 6 Stress and strain of cellulose microfibrils and size of pores between them at RG-I entrapment, as determined by finite element model. **a–h** Results of calculations for different properties, sizes, and distribution of captured RG-I: **a** without entrapped RG-I; **b** eRG-I (Ø 35 nm), **c** nRG-I (Ø 35 nm), **d** eRG-I (35 × 450 nm), **e** nRG-I (35 × 450 nm), **f** two eRG-I (Ø 35 nm); **g** two eRG-I (Ø 35 nm) between three different microfibrils, **h** two eRG-I (Ø 35 nm) between two microfibrils in the model with three microfibrils, with no RGs-I between second and third microfibrils. General design of models is similar to Fig. 5b. Top row—length of pores formed in the vicinity of entrapped RG-I. Black rectangles in **a–h** mark the regions enlarged in second row for demonstration of microfibril stress and strain (red double-headed arrows—direction of stress/strain calculation). **i** table of stress, strain of microfibril ends and pore size values obtained from **a–h** models. **j**, **k** schematic representation of 32 non-fixed cellulose microfibrils (beige) and their deformation around one eRG-I (purple and dotted ovals), PM—plasma membrane, red arrows—turgor pressure, black arrows—direction of microfibril lateral strain, green arrows—direction of microfibril longitudinal strain: **j** 3D representation, **k** cross section; **i** curve describing the attenuation of microfibril strain at increasing number of cellulose layers relative to eRG-I position. eRG-I—entrapped RG-I, nRG-I—nascent RG-I; sizes of eRG-I were obtained by DLS measurement (Mikshina et al. 2015a), sizes of nRG-I were calculated from photo (Fig. 4a, Salnikov et al. 2008)

microfibrils under pressure. Another microfibril was elastically fixed; the stiffness that is necessary to determine cell wall fixing value was calculated using Young's modulus for tertiary cell wall of oak (Yamamoto et al. 2010). Characteristics of flax cellulose were taken from the literature: Young's modulus—130 GPa (Sturcova et al. 2004), Poisson's ratio—0.4 (Bos 2004), microfibril length—6 µm, and diameter—3 nm (Perez and Mazeau 2004; Ioelovich, 2008). At further steps, additional elements were added to the model (Fig. 6).

The calculations were made in two versions—for nascent and for entrapped rhamnogalacturonans I. All necessary parameters for modeling (Young's modulus, Poisson's ratio, the Mooney–Rivlin constants describing large deformation (Egholm et al. 2006)) were obtained from the uniaxial compression experiments of RG-I gels.

Physicochemical parameters of nRG-I and eRG-I differ (Table 3), but both of them have hyperelastic properties (Mikshina et al. 2015b). Diameter of eRG-I associate was determined by light scattering and constituted 35 nm (Mikshina et al. 2015a), nRG-I size was measured on tertiary cell wall photo (Fig. 3a, Salnikov et al. 2008) and was taken as 35×450 nm.

According to the calculation, RGs-I have anisotropic stress/strain, since their values significantly differ in perpendicular directions (Fig. 5d–g). Based on the model, there is no complete spreading of RGs-I over microfibrils. Thus, hyperelastic properties of flax fiber RG-I allow it to withstand pressure corresponding to the cell turgor, remaining rather compact. eRG-I has higher stiffness and remains more compact as compared to nRG-I.

Entrapment of RG-I causes deformation of cellulose microfibrils (Fig. 6). In provided model, same as within cell wall, cellulose microfibrils are fixed, therefore total microfibril stress and strain in the presence of entrapped RG-I are directed to tension. In the absence of RG-I, microfibrils are not deformed under applied turgor pressure (0.3 MPa) (Fig. 6a). To compare the effectiveness of nRG-I and eRG-I to cause tension within cell wall, calculations were made for both of them at two different sizes of particles (Fig. 6b–e). Stress and strain of microfibrils are different depending on the properties of the captured RG-I: capture of eRG-I causes higher stresses than nRG-I of the same size.

The characteristic feature of tertiary cell wall is its mesoporosity. The pores have ink-bottle shape (narrow beginning with further expansion); their size increases during cell wall maturation (Clair et al. 2008; Chang et al. 2009, 2015). As comes from the proposed model (Fig. 6), such pores can be formed at RG-I capture by laterally interacting microfibrils. To calculate the size of the formed pores, the distance of 0.27–0.3 nm needed for hydrogen bond formation (Hubbe 2006) was used as a boundary condition. The size of pores formed at RG-I capture varies from 25 to 50 nm (Fig. 6), being the same level as the experimentally determined values for tension wood fibers (Clair et al. 2008; Chang et al. 2009, 2015). According to the model, the size of the pore is higher at entrapment of eRG-I, than nRG-I. It means that the in muro modification of nascent galactan by galactosidase, in addition to “cleaning” of microfibril surface covered by long chains of β-($1 \rightarrow 4$)-galactan to increase lateral interaction of cellulose, changes the properties of RG-I. RG-I modification in muro may be the cause of pore size increase upon tertiary cell wall maturation. This is consistent with the results obtained for poplar tension wood (Chang et al. 2015).

The content of eRG-I in tertiary cell wall is quite low: as determined by sugar content, the proportion between cellulose and eRG-I is 97:3 (Gurjanov et al. 2008; Mikshina et al. 2015a). Low content of the entrapped matrix polysaccharide is necessary to provide conditions for lateral interaction of cellulose microfibrils and to escape considerable swelling that may ruin the whole system, as pointed by Alméras and Clair (2016). The knowledge of approximate cellulose microfibril mass (volume $\times$ density) and of cellulose/eRG-I proportion allows to calculate quantitative ratio of cellulose microfibrils to eRG-I “balls”—about 32/1 (Fig. 6j, k). Increasing of cellulose layer number relative to eRG-I position leads to attenuation of microfibril strain (Fig. 6i). Based on the increased cellulose crystallite size of tertiary cell wall

(Dadswell and Wardrop 1955; Viëtor et al. 2002; Muller et al. 2006), it was concluded that eRG-I was entrapped between four cellulose microfibrils (Mellerowicz et al. 2008). Consequently, at least nine cellulose layers take part in tension creation at one eRG-I entrapping. Taking into account the length of mature flax fiber (2 cm in average, Table 1), thickness of tertiary cell wall, and values of microfibril strains for nine layers (Fig. 6i), the calculated strain value is 0.25% of total fiber length. This value is in accord with magnitude of released longitudinal maturation strain of *Eucalyptus nitens* (Clair et al. 2013).

Altogether, finite element modeling validates the hypothesis that the entrapment of tertiary cell wall-specific version of RG-I with hyperelastic properties between laterally interacting cellulose microfibrils provides tension that is the basis for contractile properties of plant "muscles".

Conclusions

Three major processes were previously suggested as the basis to accomplish the movements of plant organs: irreversible tissue growth, reversible turgor-regulated changes in tissue tension, and differential shrinkage through tissue drying (Geitmann 2016). Fibers with tertiary cell wall demonstrate the fourth mechanism. While using some elements of the listed above, the fourth mechanism is developed within tertiary cell wall and is based on its specific design and post-synthetic modification of matrix polymer. The major players are cellulose microfibrils, all of which have axial orientation, and tissue—and stage-specific version of rhamnogalacturonan I. The entrapment of RG-I between laterally interacting cellulose microfibrils causes their tension that is the basis of fiber contractile properties. Hyperelastic properties, specific spatial organization with charged regular backbone located at the surface of associate of the polysaccharide molecules, ability to strongly retain water molecules and to form strong hydrogel distinguish rhamnogalacturonan I of tertiary cell wall from other known polysaccharides of this type.

Fibers with tertiary cell wall are widely spread in plant kingdom; they develop the mechanism for the movement of plant organs that have already ceased elongation growth. While supporting plant body they provide flexibility in addition to mechanical strength, helping, for example, long narrow stems to withstand severe mechanical stresses. However, the quantitative differences in mechanical properties between the fibers that deposit only secondary cell wall and the fibers that additionally develop tertiary cell wall still have to be systematically analyzed.

Cell wall thickening in plant fibers is preceded by intrusive elongation, which is especially pronounced in phloem fibers. The importance of fiber intrusive growth is often overlooked; understanding of its mechanisms is at its infancy. At that, it is the intrusive growth that provides the enlarged surface for subsequent cell wall thickening and leads to fiber bundle formation, considerably reshaping the cell contacts of fibers. Thus, the fiber intrusive elongation forms the basis to develop mechanical properties of plant body and largely determines the yield and quality of biomass, especially in fiber crops. The comparison of the mechanisms of symplastic, protrusive, and intrusive elongation may help to generalize the notion of plant cell growth—one of the basic processes that distinguishes plant organism. Thus, plant fibers are the useful models to study several processes important for plant biomechanics.

Acknowledgements The study was supported by Russian Science Foundation (project 16-14-10256—GT, CT, MN, GO; comparison of sets of genes up-regulated in intrusively elongating fibers and growing in vitro pollen tube; analysis of post-deposition modification of tertiary cell wall), and Program of the President of Russian Federation for Young Scientists (project MK-8393.2016.4—MP; analysis of the rhamnogalacturonan I ability to gelation and finite element modeling of tension creation in tertiary cell wall; project MK-8014.2016.4—PA, MN; study of the tissue-specific galactosidase role in the gelatinous cell wall formation).

References

Ageeva M, Petrovska B, Kieft H et al (2005) Intrusive growth of flax phloem fibres is of intercalary type. Planta 222:565–574

Akin DE (2013) Linen most useful: perspectives on structure, chemistry, and enzymes for retting flax. ISRN Biotechnol 2013:186534

Alméras T, Clair B (2016) Critical review on the mechanisms of maturation stress generation in trees. J R Soc Interface 13(122):20160550

Bedinger PA, Hardeman KJ, Loukides CA (1994) Travelling in style: the cell biology of pollen. Trends Cell Biol 4:132–138

Bos HL (2004) The potential of flax fibres as reinforcement for composite materials. Dissertation, Technische Universiteit Eindhoven

Bosch M, Hepler PK (2005) Pectin methylesterases and pectin dynamics in pollen tubes. Plant Cell 17:3219–3226

Burk DH, Liu B, Zhong R, Morrison WH, Ye Z-H (2001) A katanin-like protein regulates normal cell wall biosynthesis and cell elongation. Plant Cell 13(4):807–827

Canaveze Y, Machado SR (2016) The occurrence of intrusive growth associated with articulated laticifers in *Tabernaemontana catharinensis* A.DC., a New Record for Apocynaceae. Int J Plant Sci 177(5):458–467

Cassan-Wang H, Goué N, Saidi MN et al (2013) Identification of novel transcription factors regulating secondary cell wall formation in Arabidopsis. Front Plant Sci 4:189

Chang SS, Clair B, Ruelle J et al (2009) Mesoporosity as a new parameter for understanding tension stress generation in trees. J Exp Bot 60(11):3023–3030

Chang SS, Quignard F, Almeras T, Clair B (2015) Mesoporosity changes from cambium to mature tension wood: a new step toward the understanding of maturation stress generation in trees. New Phytol 205(3):1277–1287

Chebli Y, Kaneda M, Zerzour R, Geitmann A (2012) The cell wall of the Arabidopsis pollen tube—spatial distribution, recycling, and network formation of polysaccharides. Plant Physiol 160(4):1940–1955

Chernova TE, Gurjanov OP, Brach NB et al (2007) Variability in the composition of tissue-specific galactan from flax fibers. Russ J Plant Physiol 54(6):876–884

Chernova TE, Mikshina PV, Salnikov VV et al (2018) Development of distinct cell wall layers both in primary and secondary phloem fibers of hemp (*Cannabis sativa* L.). Ind Crop Prod 117:97–109

Clair B, Gril J, Di Renzo F et al (2008) Characterization of a gel in the cell wall to elucidate the paradoxical shrinkage of tension wood. Biomacromolecules 9:494–498

Clair B, Alteyrac J, Gronvold A et al (2013) Patterns of longitudinal and tangential maturation stresses in Eucalyptus nitens plantation trees. Ann For Sci 70:801–811

Dadswell HE, Wardrop AB (1955) The structure and properties of tension wood. Holzforschung 9(4):97–104

Dawson C, Vincent JFV, Rocca A-M (1997) How pine cones open. Nature 390:668

Egholm RD, Christensen SF, Szabo P (2006) Stress-strain behavior in uniaxial compression of polymer gel beads. J Appl Polym Sci 102:3037–3047

Elbaum R, Zaltzman L, Burgert I, Fratzl P (2007) The role of wheat awns in the seed dispersal unit. Science 316(5826):884–886

Esau K (1943) Vascular differentiation in the vegetative shoot of linum. III. The origin of the bast fibers. Am J Bot 30(8):579–586

Esau K (1965) Plant anatomy, 2nd edn. Wiley, New York

Evert RF (2006) Esau's plant anatomy: meristems, cells, and tissues of the plant body: their structure, function, and development, 3rd edn. Wiley, New York

Fahn A (1990) Plant anatomy, 4th edn. Pergamon Press, Oxford

Fan L-M, Wang Y-F, Wang H, Wu W-H (2001) In vitro Arabidopsis pollen germination and characterisation of the inward potassium currents in Arabidopsis pollen grain protoplasts. J Exp Bot 52:1603–1614

Geitmann A, Ortega JKE (2009) Mechanics and modeling of plant cell growth. Trends Plant Sci 14(9):467–478

Geitmann A (2016) Actuators acting without actin. Cell 166(1):15–17

Gorshkov O, Mokshina N, Gorshkov V et al (2017) Transcriptome portrait of cellulose-enriched flax fibers at advanced stage of specialization. Plant Mol Biol 93:431–449

Gorshkova TA, Wyatt SE, Salnikov VV et al (1996) Cell-wall polysaccharides of developing flax plants. Plant Physiol 110(2):721–729

Gorshkova TA, Salnikov VV, Pogodina NM et al (2000) Composition and distribution of cell wall phenolic compounds in flax (Linum usitatissimum L.) stem tissues. Ann Bot 85(4):477–486

Gorshkova TA, Sal'nikov VV, Chemikosova SB et al (2003) The snap point: transition point in *Linum usitatissimum* L. bast fiber development. Ind Crops Prod 18:213–221

Gorshkova TA, Chemikosova SB, Salnikov VV et al (2004) Occurrence of cell-specific galactan is coinciding with bast fibre developmental transition in flax. Ind Crops Prod 19:217–224

Gorshkova T, Morvan C (2006) Secondary cell-wall assembly in flax phloem fibres: Role of galactans. Planta 223:149–158

Gorshkova TA, Mikshina PV, Ibragimova NN et al (2009) Pectins in secondary cell walls: modifications during cell wall assembly and maturation. In: Schols HA, Visser RGF, Voragen AGJ (eds) Pectins and pectinases. Wageningen Academic Publishers, The Netherlands, pp 149–165

Gorshkova TA, Gurjanov OP, Mikshina PV et al (2010) A special type of secondary cell wall, formed by plant fibers. Russ J Plant Physiol 57:346–361

Gorshkova T, Brutch N, Chabbert B et al (2012) Plant fibre formation: state of the art, recent and expected progress, and open questions. CRC Crit Rev Plant Sci 31:201–228

Gorshkova T, Mokshina N, Chernova T et al (2015) Aspen tension wood fibers contain β-(1 → 4)-galactans and acidic arabinogalactans retained by cellulose microfibrils in gelatinous walls. Plant Physiol 169(3):2048–2063

Gorshkova TA, Chernova TE, Gorshkov VY et al (2018) Intrusive growth of flax fibers: major players revealed by transcriptome analysis (in press)

Gorshkova T, Chernova T, Mokshina N et al (2018) Plant "muscles": fibers with a tertiary cell wall. New Phytol 218(1):66–72

Goswami L, Dunlop JWC, Jungnikl K et al (2008) Stress generation in the tension wood of poplar is based on the lateral swelling power of the G-layer. Plant J 56:531–538

Gray-Mitsumune M, Mellerowicz EJ, Abe H et al (2004) Expansins abundant in secondary xylem belong to subgroup A of the α-expansin gene family. Plant Physiol 135:1552–1564

Gray-Mitsumune M, Blomquist K, McQueen-Mason S et al (2008) Ectopic expression of a wood-abundant expansin PttEXPA1 promotes cell expansion in primary and secondary tissues in aspen. Plant Biotechnol J 6:62–72

Guerriero G, Hausman J-F, Cai G (2014) No stress! Relax! Mechanisms governing growth and shape in plant cells. Int J Mol Sci 15:5094–5114
Gurjanov OP, Gorshkova TA, Kabel MA et al (2007) MALDI-TOF MS evidence for the linking of flax bast fibre galactan to rhamnogalacturonan backbone. Carbohydr Polym 67:86–96
Gurjanov OP, Ibragimova NN, Gnezdilov OI, Gorshkova TA (2008) Polysaccharides, tightly bound to cellulose in the cell wall of flax bast fibre: Isolation and identification. Carbohydr Res 72:719–729
Hejnowicz Z (1980) Tensional stress in the cambium and its developmental significance. Am J Bot 67:1–5
Hepler PK, Winship LJ (2015) The pollen tube clear zone: clues to the mechanism of polarized growth. J Integr Plant Biol 57(1):79–92
Hubbe MA (2006) Bonding between cellulosic fibers in the absence and presence of dry-strength agents. BioResources 1(2):281–318
Ioelovich M (2008) Cellulose as a nanostructured polymer: a short. BioResources 3(4):1403–1418
Jura-Morawiec J (2017) Atypical origin, structure and arrangement of secondary tracheary elements in the stem of the monocotyledonous dragon tree, *Dracaena draco*. Planta 245(1):93–99
Larson PR (1994) The vascular cambium. Development and structure. Springer, Berlin
Lev-Yadun S (2001) Intrusive growth—the plant analog of dendrite and axon growth in animals. New Phytol 150:508–512
Lev-Yadun S, Dafni A, Flaishman MA et al (2004) Plant coloration undermines herbivorous insect camouflage. BioEssays 26:1126–1130
Lev-Yadun S (2015) Plant development: cell movement relative to each other is both common and very important. Plant Sig Behav 10(3):e991566–1
Li WT, He M, Wang J, Wang YP (2013) Zinc finger protein (ZFP) in plants-a review. POJ 6(6):474–480
Mazur E, Kurczynska EU (2012) Rays, intrusive growth, and storied cambium in the inflorescence stems of *Arabidopsis thaliana* (L.) Heynh. Protoplasma 249:217–220
Mazur E, Kurczyńska EU, Friml J (2014) Cellular events during interfascicular cambium ontogenesis in inflorescence stems of Arabidopsis. Protoplasma 251:1125–1139
Mellerowicz EJ, Baucher M, Sundberg B, Boerjan W (2001) Unravelling cell wall formation in the woody dicot stem. Plant Mol Biol 47:274–329
Mellerowicz EJ (2006) Xylem cell expansion—Lessons from poplar. In: Hayashi T (ed) The science and lore of the plant cell wall. Universal Publishers, Brown Walker Press, Boca Raton
Mellerowicz EJ, Immerzeel P, Hayashi T (2008) Xyloglucan: the molecular muscle of trees. Ann Bot 102:659–665
Mellerowicz EJ, Gorshkova TA (2012) Tensional stress generation in gelatinous fibres: a review and possible mechanism based on cell-wall structure and composition. J Exp Bot 63:551–565
Mikshina PV, Gurjanov OP, Mukhitova FK et al (2012) Structural details of pectic galactan from the secondary cell walls of flax (*Linum usitatissimum* L.) phloem fibres. Carbohydr Polym 87:853–861
Mikshina PV, Chernova TE, Chemikosova SB et al (2013) Cellulosic fibers: role of matrix polysaccharides in structure and function. In: van de Ven T, Godbout L (eds) Cellulose—fundamental aspects. InTech, Rijeka, pp 91–113
Mikshina PV, Idiyatullin BZ, Petrova AA et al (2015a) Physicochemical properties of complex rhamnogalacturonan I from gelatinous cell walls of flax fibers. Carbohydr Polym 117:853–861
Mikshina PV, Petrova AA, Idiyatullin BZ et al (2015b) Tissue-specific rhamnogalacturonan I forms the gel with hyperelastic properties. Biochem (Mosc) 80:915–924
Mikshina PV, Makshakova ON, Petrova AA et al (2017) Gelation of rhamnogalacturonan I is based on galactan side chain interaction and does not involve chemical modifications. Carbohydr Polym 171:143–151
Mokshina NE, Ibragimova NN, Salnikov VV et al (2012) Galactosidase of plant fibers with gelatinous cell wall: identification and localization. Russ J Plant Physiol 59(2):246–254

Müller M, Burghammer M, Sugiyama J (2006) Direct investigation of the structural properties of tension wood cellulose microfibrils using microbeam X–ray fibre diffraction. Holzforschung 60:474–479

Nezhad AS, Geitmann A (2013) The cellular mechanics of an invasive lifestyle. J Exp Bot 64(15):4709–4728

Perez S, Mazeau K (2004) Conformations, structures, and morphologies of celluloses. In: Dumitriu S (ed) Polysaccharides, structure and functional versatility, 2nd edn. CRC Press, New York, pp 41–68

Qin Y, Leydon AR, Manziello A et al (2009) Penetration of the stigma and style elicits a novel transcriptome in pollen tubes, pointing to genes critical for growth in a pistil. PLoS Genet 5:e1000621

Roach MJ, Mokshina NY, Badhan A et al (2011) Development of cellulosic secondary walls in flax fibers requires β-galactosidase. Plant Physiol 156(3):1351–1363

Salnikov VV, Ageeva MV, Gorshkova TA (2008) Homofusion of Golgi secretory vesicles in flax phloem fibers during formation of the gelatinous secondary cell wall. Protoplasma 233:269–273

Schreiber N, Gierlinger N, Pütz N et al (2010) G-fibres in storage roots of *Trifolium pratense* (Fabaceae): tensile stress generators for contraction. Plant J 61:854–861

Siedlecka A, Wiklund S, Péronne M-A et al (2008) Pectin methyl esterase inhibits intrusive and symplastic cell growth in developing wood cells of Populus. Plant Physiol 146:554–565

Snegireva AV, Ageeva MV, Vorob'ev VN et al (2006) Plant fiber intrusive growth characterized by NMR method. Russ J Plant Physiol 53:163–168

Snegireva A, Ageeva M, Amenitskii S et al (2010) Intrusive growth of sclerenchyma fbers. Russ J Plant Physiol 57:342–355

Snegireva A, Chernova T, Ageeva M et al (2015) Intrusive growth of primary and secondary phloem fibres in hemp stem determines fibre-bundle formation and structure. AoB Plants 7:plv061

Sturcova A, His I, Apperley DC et al (2004) Structural details of crystalline cellulose from higher plants. Biomacromol 5:1333–1339

Viëtor RJ, Newman RH, Ha MA et al (2002) Conformational features of crystal-surface cellulose from higher plants. Plant J 30:721–731

Wang Y, Zhang WZ, Song LF et al (2008) Transcriptome analyses show changes in gene expression to accompany pollen germination and tube growth in *Arabidopsis*. Plant Physiol 148:1201–1211

Wang SS, Diao WZ, Yang X et al (2015) Arabidopsis thaliana CML25 mediates the Ca(2+) regulation of K(+) transmembrane trafficking during pollen germination and tube elongation. Plant Cell Environ 38(11):2372–2386

Wenham MW, Cusick F (1975) The growth of secondary wood fibres. New Phytol 74:247–261

Yamamoto H, Ruelle J, Arakawa Y et al (2010) Origin of the characteristic hygro-mechanical properties of the gelatinous layer in tension wood from Kunugi oak (*Quercus acutissima*). Wood Sci Technol 44:149–163

Yang Zh (1998) Signaling tip growth in plants. Curr Opin Plant Biol 1:525–530

Zhong R, Ye Zh-H (1999) IFL1, a gene regulating interfascicular fiber differentiation in Arabidopsis, encodes a homeodomainleucine zipper protein. Plant Cell 11:2139–2152

Zhong R, Burk DH, Ye ZH (2001) Fibers. A model for studying cell differentiation, cell elongation, and cell wall biosynthesis. Plant Physiol 126(2):477–479

Zhong R, Demura T, Ye Zh-H (2006) SND1, a NAC domain transcription factor, is a key regulator of secondary wall synthesis in fibers of Arabidopsis. Plant Cell 18(11):3158–3170

Zhong R, Ye Zh-H (2014) Complexity of the transcriptional network controlling secondary wall biosynthesis. Plant Sci 229:193–207

Zimmermann U, Husken D, Schulze ED (1980) Direct turgor pressure measurements in individual leaf cells of Tradescantia virginiana. Planta 149:445–453

Modeling Plant Morphogenesis: An Introduction

Anne-Lise Routier-Kierzkowska and Adam Runions

Abstract In this chapter, we provide an overview of some of the computational models used to understand morphogenesis in plants. In particular, we focus on models of growth and patterning processes in primary tissues, prior to the onset of lignification. We explain the assumptions behind these models and how they relate to biological evidence. Our aim is to provide some basic intuitions regarding the construction, operation, and interpretation of such models.

Keywords Computational modeling · Physically based models
Auxin patterning · Phyllotaxis · Vein formation · Leaf development · FEM
Buckling · Growth

Introduction

Computational modeling of morphogenesis is highly interdisciplinary, falling at the interface of computer science, physics, mathematics, biochemistry, and developmental biology. Evaluating the relevance, usefulness, and dynamics of computer simulations requires some prior knowledge in these different fields, which can be difficult to find in a condensed manner. In this chapter, we aim at giving some key insights into different types of models of morphogenesis, as an aid in assessing their usefulness and understanding how they relate to biological data. For the sake of clarity, we do not provide a comprehensive review of the relevant modeling literature, but instead focus on representative models that help to illustrate key points. In particular, the models we discuss capture shape emergence and patterning in fast-growing primary tissues. For a broader overview of the various types of models and related concepts,

A.-L. Routier-Kierzkowska (✉) · A. Runions
Department of Comparative Development and Genetics,
Max Planck Institute for Plant Breeding Research, Cologne, Germany
e-mail: al.routier@umontreal.ca

A.-L. Routier-Kierzkowska
Département de Sciences Biologiques, Institut de Recherche En Biologie Végétale, Montréal, Canada

A. Geitmann and J. Gril (eds.), *Plant Biomechanics*,
https://doi.org/10.1007/978-3-319-79099-2_8

see (Chickarmane et al. 2010; Prusinkiewicz and Runions 2012; Runions et al. 2014; Ali et al. 2014).

Due to the self-organizing nature of biological systems, models of morphogenesis often exhibit complex emergent behaviors that cannot be predicted a priori. Thus, in some sense, computer simulations constitute "in silico experiments", where the output of a simulation can be compared to the outcome of an experiment. Evaluating the relevance of an experiment or a model therefore requires answering very similar questions:

(a) What do we want to test?
(b) What are the assumptions? Limits of validity?
(c) How can the model be validated?
(d) What conclusions can be drawn from the results?

In the following, we will attempt to answer each question for different types of models. The question of "what is the hypothesis tested" is the most important, and answering it is often nontrivial. Morphogenesis results from the combination of different biological processes which often interact with each other. Disentangling one aspect from another and testing it separately requires careful model design. It is somewhat similar in spirit to controlling environmental conditions during an experiment, so that the effect of each variable can be tested in isolation. In a model, processes that are not the primary focus of the simulation can be described in advance by the modeler, fixing the "environmental conditions" so to speak. In contrast, key aspects of the simulation are those that emerge from simple rules that follow the working hypotheses guiding model construction. We call *predictive capacity* the model's ability to give an emerging property as an output. The examples we consider here illustrate how predictive capacity is more subtle than an on/off switch, instead resembling a progressive rotary dial from fully descriptive to completely emergent.

From a modeling perspective, the processes shaping an organ can be broadly divided into three categories: pattern formation, organ expansion—including cell expansion—and cell division. Models often combine all three. In practice, however, typically only some aspects of models will be emergent, while the remaining aspects are handled in a more descriptive fashion. Here, we use concrete examples to dissect the models and understand their predictive capacity either for patterning or growth. In each category, the modeler must make design choices affecting the model's execution, e.g., the model dimensionality or level of detail. Such choices are driven by the hypothesis being tested, but also by the availability of pertinent biological information and technical limitations.

Models of Pattern Formation

Almost since the advent of digital computing, computational modeling has been used to explain how patterns of gene expression, hormone transport, or groups of differentiated cells emerge within seemingly homogeneous tissues. Unlike animal tissues, which can use cell migration or intercalation as patterning mechanisms, plant tissues cannot rely on cellular movement, since plant cells are permanently attached to their neighbors via their shared cell walls. Consequently, plant patterning and development relies entirely on information stored within the cell (*cell-autonomous* processes) and communication with its immediate neighbors (*cell nonautonomous* processes) via chemical processes like diffusion and active transport, or mechanics. Plants take advantage of inter-cellular communication to define stable regions of cell identity or gene expression in spite of organ expansion. For example, roots (Grieneisen et al. 2007) and shoots (Fujita et al. 2011) maintain a pool of stem cells of a relatively constant size that follow their tips as they grow, despite the fact that the individual cells comprising this pool slowly change over time.

From a modeling perspective, patterning mechanisms can often be separated from growth. Many patterning feedbacks do not explicitly involve growth and, conversely, growth can occur without the emergence of new patterns. Some organs, such as hypocotyls and cotyledons, are already well organized into different tissue types during embryogenesis, so their adult form can be simply considered as larger version of their embryonic counterparts. In contrast to animal development though, the architecture of a plant changes substantially over its lifespan as new branches or lateral roots are continuously produced. Differences in growth rates between organs can lead to seemingly very different shapes at the scale of the whole plant, making it more difficult to immediately spot common underlying patterns. For example, the leaf rosette of *Arabidopsis* looks very different from its mature inflorescence, even though they share the same pattern of organ emergence (i.e., the same phyllotaxy). A certain level of abstraction is therefore necessary for describing and modeling patterns. Finally, growth also provides a key input to patterning by increasing the space in which spatial patterning processes operate. For example, the spaces between leaf veins and trichomes increase as a leaf grows, allowing for the patterning of higher order veins (Marcos and Berleth 2014) or additional trichomes. Here again growth can obscure common patterning mechanisms, since the final pattern may depend on how the organ expanded over time.

Levels of Abstraction

Similar general principles can explain pattern formation in very different systems, from hair follicle initiation in mice to stripes on a seashell (reviewed by Kondo and Miura 2010; Meinhardt 2009). Models constructed at different levels of abstraction can thus reveal similarities that are obscured by the complexity of biological or molecular details. More abstract models can also highlight the essence of a patterning mechanism and help reveal the underlying logic of a biological process. Common levels of abstraction for patterning include geometric models, which focus on spatiotemporal relations, physical analogies, which exploit analogies with well-understood physical process and cellular-level models (the most directly relatable to biological observations). We outline these levels of abstraction, some of their applications, and related considerations in the remainder of this section.

Geometric Models

Geometric models have a long history in studies of plant development, with early examples including rules for cell division (Hofmeister 1863; Errera 1886 and Sachs 1878) and phyllotaxis (Hofmeister Hofmeister 1868). The primary strength of these models is their ability to represent and explore spatiotemporal relations and their consequences. For instance, new organs typically appear at the tip of growing shoots in the largest empty space between preexisting organs (Fig. 1a). This observation is attributed to Hoffmeister (Hofmeister 1868) and can be formalized as *Hoffmeister's rule*, which proposes that primordia locally inhibit the initiation of new organs, and a new primordium appears where this inhibition is weakest. Models built on this simple rule (Mitchison 1977) relate phyllotactic patterns to the relative size of primordia compared to the meristem: small primordia reproduce the conspicuous spiral arrangement of organs at the golden angle (137.5°), while large primordia cause organs to initiate on alternating sides of the shoot (180°). These models can also reproduce a broader range of phyllotactic patterns (Smith et al. 2006a, b) by accounting for a gradual decrease of the inhibition away from the primordium or over time (Fig. 1a3). Notably, Hoffmeister's rule can also be applied to leaf shape (Fig. 1b) (Runions et al. 2017). Here, additional protrusions are introduced along the leaf margin as space becomes available due to growth, generating diverse leaf forms, from simple leaves to prominently lobed or even compound (i.e., consisting of numerous leaflets). Such a simple model can therefore highlight deep similarities between seemingly distant developmental processes. As it turned out, molecular processes initiating organs are closely related to those shaping leaf form (Alvarez et al. 2016; Bilsborough et al. 2011; Floyd and Bowman 2010; Prusinkiewicz and Runions 2012).

A notable aspect of geometric models is their relative independence of the underlying molecular mechanism. For instance, phyllotaxis models based on Hoffmeister's rule assume that each primordium introduces a local inhibition. Whether this inhibition results from an inhibitor produced at primordia (Meinhardt et al. 1998), the collection of an activator from surrounding tissues (Jönsson et al. 2006; Smith et al. 2006a, b), or even mechanical signals (Heisler et al. 2010) is secondary. This highlights the capability of geometric models to capture the essence of a morphogenetic process in a conceptually crisp manner. At the same time, this can lead to cases where the precise molecular implementation of a model may be unclear. This is currently the case for geometric models of vein formation (Rodkaew et al. 2003; Runions et al. 2005, 2017; Owens et al. 2016), where connections to polar auxin transport have been proposed, but have yet to be explored in cellular-level models.

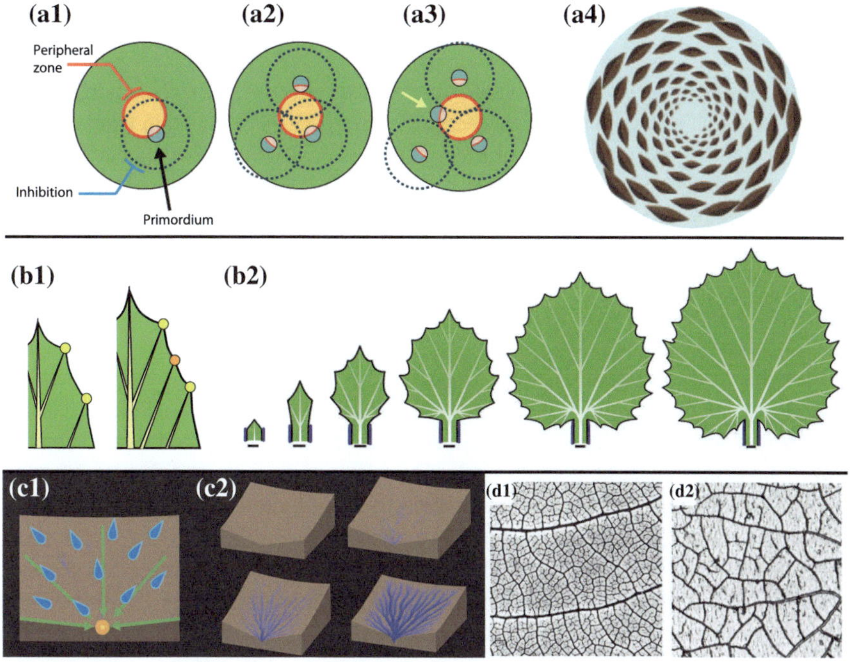

Fig. 1 Models of patterning processes using geometric rules and physical analogies. **a** Phyllotactic patterning based on Hoffmeister's rule. **a1** The shoot apical meristem is divided into central zone (orange) and peripheral zone (red circle). Primordia are introduced in the peripheral zone, but existing primordia inhibit their introduction (dashed circle). **a2** An arrangement of three primordia inhibits the entire peripheral zone. **a3** Growth increases the distance of the three primordia from the shoot tip, allowing for the introduction of a new primordium (marked by an arrow). **a4** A spiral phyllotactic arrangement (137.5 degree divergence angle) generated by an inhibition model of phyllotaxis (adapted from Smith et al. 2006a, b). **b** A geometric model of leaf shape based on Hoffmeister's rule (adapted from Runions et al. 2017). Patterning is restricted to the leaf margin and excluded from the petiole (marked in blue). Veins determine growth directions. **b1** Leaf growth increases the distance between protrusions along the leaf margin, when this distance exceeds a threshold a new vein is introduced. **b2** Starting from a small primordium with a single vein, the leaf grows, extending along the axis defined by the vein. This increases the distance of points on the margin from the tip of the midvein, leading to the introduction of two new veins. These veins introduce new growth directions, causing the leaf lamina to broaden. As growth continues, further veins are introduced, terminating at the spaces along the margin generated by growth. The gradual introduction of veins dynamically changes the growth of the leaf margin, leading to the self-organization of a serrated simple leaf. **c** A simulation of river canalization (adapted from Runions 2014). **c1** The terrain slopes (green arrows) toward a sink (yellow disk). Water droplets are randomly placed and move down the slope. As it moves, each droplet erodes the terrain, locally changing the slope. **c2** At first, droplets move directly toward the sink due to the slope in the terrain. However, the path left by earlier droplets attracts later droplets, leading to the emergence of discrete canals forming a river network emptying at the sink. **d** Comparison of networked veins in *Polygonum Polystachium* (**d1**) to a crack pattern produced by drying silicon gel (**d2**), adapted from Couder et al. (2002) with kind permission of The European Physical Journal

Physical Analogies

Closely related to geometric models are those relying on physical analogies. An example is Sach's canalization hypothesis (Sachs 2003, as modeled in Runions et al. 2014), which proposes that leaf vein formation is analogous to rivers carving their path in soft terrain (Fig. 1c). As water flows over the terrain, it erodes the soil. This creates a deeper path, which attracts more water and further enhances the size and capacity of the channel. Over time, this produces a feedback where local enhancement of flux (water flow) leads to the emergence of an efficient transport network. The canalization hypothesis posits a mechanism explaining the formation of branched vascular patterns, as seen in the major veins of many leaves (such as oaks and maples) and the veins in the stems of many plants. In this model, water is analogous to the phytohormone auxin, and the channels to the developing veins. Conceptually, it implies that vascular strands are sustained through auxin flux from nearby regions toward the vein. Consequently, like river networks, canalization partitions the domain into discrete drainage basins. Nearby veins thus compete for auxin and deplete it nearby, inhibiting the advancement of neighboring veins toward them. As such, vein tips naturally repel each other, leading to the emergence of a branched network topology. Thus, this analogy does not readily explain the closed loops that are present in various vascular patterns. Instead, this feature is captured by the pattern of cracks emerging in stressed materials (e.g., glazed ceramics). This analogy was explored by Couder et al. (2002), who found that by drying gel films to induce stresses it was possible to generate remarkably plausible networked vein patterns (Fig. 1d). In a stressed medium, cracks propagate perpendicular to the direction of greatest stress. Once formed, the widening of a crack relaxes stresses perpendicular to its path. This shifts the direction of greatest stress close to the crack by 90°, causing it to align with the crack's path. Thus, when a new crack emerges, it propagates away from the earlier crack supporting it. However, when the crack tip enters the vicinity of another crack, progressing perpendicular to the direction of maximal stress causes them to collide at a right angle, comparable to how high-order veins connect to lower order veins to generate closed loops (Marcos and Berleth 2014). Subsequently, Laguna et al. (2008) found that a similar mechanical instability can be modeled by assuming that mesophyll cells in a growing tissue are compressed by the epidermis, and that the stresses induced cause mesophyll cells to collapse. The pattern of collapsed mesophyll cells produces networked patterns visually and statistically similar to networked vein patterns.

An advantage of these approaches is that they relate complicated patterning phenomena to well-understood physical processes. Thus, these analogies provide a clear conceptualization of the patterning processes they capture, as well as tools for their mathematical description and analysis. As with geometric models, the precise relation to the underlying molecular processes may be unclear. For example, the actual mechanism of vein formation could be very different from crack formation and involve only molecular mechanisms. A relevant example is the analogy between heat and chemical diffusion, which have an identical mathematical description but consist of distinct physical phenomena (transfer of thermal energy vs. movement of particles). However, analogies can help identify the key ingredients in pattern emergence and provide a stepping stone to understanding the underlying biological mechanism. For example, the canalization hypothesis permits a cellular implementation (Mitchison 1980; Mitchison et al. 1981; discussed below), which accounts for many biological observations (Sachs 2003; Rolland-Lagan and Prusinkiewicz 2005; Runions et al. 2014).

Cellular Level

Cellular-level models explicitly capture the geometry and topology of cells (i.e., neighborhood relations). This level of abstraction permits the direct representation of cellular processes and their spatial–temporal evolution during development. Consequently, it is typically the most immediately relatable to biological observations of developing tissues. This has made it a popular choice for computational studies of patterning and morphogenesis. However, detailed cellular simulations can be computationally intensive. Furthermore, the construction of such models requires many choices of representation related to the implementation of growth, cell division, and molecular processes. These implementation details are important, as they can have a large impact on simulation results and their interpretation.

Molecular processes can be captured by storing the concentration or amount of a substance (such as hormones or proteins) in the various compartments represented by the model (Fig. 2a; e.g., the cellular concentration of auxin in a cell or amount of PIN on a cell membrane) and simulating the changes of these quantities over time. Many spatial patterns require communication between cells (cell nonautonomous processes). Plant cells are separated by their respective cell walls and an extracellular space (the *apoplast*). Additionally, neighboring cells are connected by *plasmodesmata*, small channels in the cell wall that directly connect the interior of each cell. Chemical signals move between cells either directly, via the plasmodesmata, or by passing sequentially through the plasma membrane of each cell and the extracellular space separating them. In either case, additional processes regulate the diffusion or transport of substances.

Diffusion via either pathway provides a simple means for cells to communicate with each other. Tissue-level gradients can be established by diffusion of a compound from a source within the tissue (Wartlick et al. 2009). Once a gradient is established, the cellular concentration of the diffusing substance provides information regarding the cells' position relative to the source. Such positional information (Wolpert 1969) can determine the fate or differentiation status of the cell (Mähönen et al. 2014; Kazama et al. 2010). In growing tissues, the form of the gradient emerges from the combined effects of diffusion, decay, and growth-driven dilution. For instance, cell division arrest in the leaf can be explained by a gradient established by the diffusion and decay of a substance emanating from the leaf base (Kazama et al. 2010). In contrast, the division of the root into distinct meristematic, elongation, and differen-

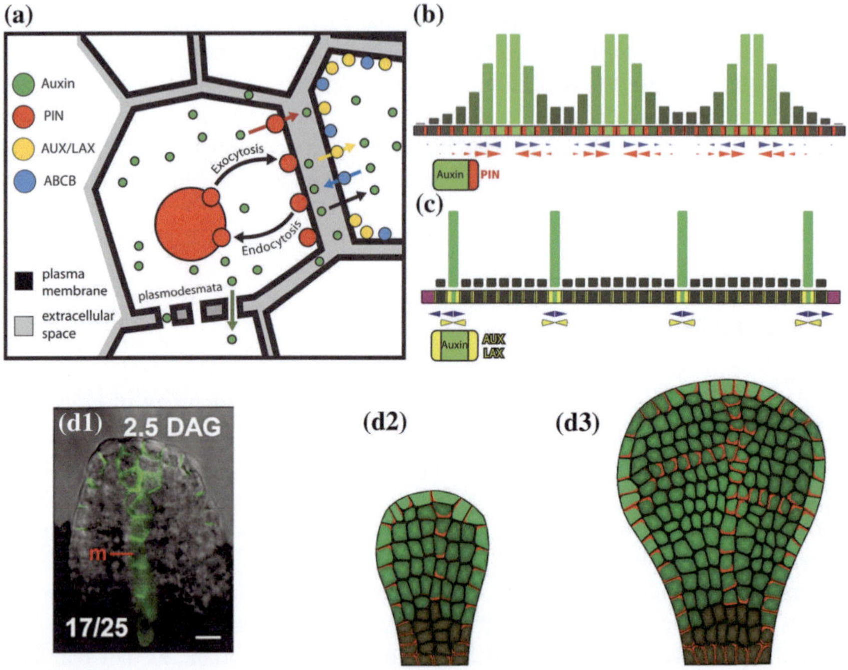

Fig. 2 Cellular-level patterning models. **a** Processes underlying patterning models based on auxin transport. Auxin exporters from the PIN and ABCB families (red and blue arrows respectively) transport auxin from the cell into the extracellular space. Auxin moves from the extracellular space to the cell via diffusion (black arrow) and import by AUX/LAX importers (yellow arrow). Auxin can move directly between cells via plasmodesmata (green arrow). The localization of PIN at the plasma membrane is nonuniform and changes dynamically due to allocation (exocytosis) and deallocation (endocytosis). ABCB and AUX/LAX proteins have a uniform membrane localization. **b–c** Emergence of equidistant auxin maxima in a 1D file of cells (adapted from Runions et al. 2014). **b** Maxima are organized by up-the-gradient PIN polarization, and represent a steady state between PIN transport (red arrows) and diffusion (blue arrows). **c** Maxima emerge from an importer-based feedback, where auxin promotes AUX/LAX expression. Pink cells indicate auxin sinks. In this case, patterning is driven by competition between neighboring cells for auxin in the extracellular space, and represents a steady state between apolar export by ABCBs (blue arrows) and apolar import by AUX/LAX proteins (yellow arrows). **d** Combined emergence of auxin maxima and vascular strands in a growing leaf (similar to Smith and Bayer 2009). **d1** PIN1:GFP expression in a young leaf of *Arabidopsis thaliana* (adapted from Scarpella et al. 2006). PIN1 polarities converge toward the leaf tip in the margin. In the interior, a strand of PIN1 expression marks the site of midvein initiation. These cells are polarized away from the convergence point at the tip. **d2–3** The model represents a cross section through the leaf. Dark cells at the leaf base are auxin sinks. PIN in cells on the leaf margin polarize up-the-gradient, whereas those in sub-epidermal layers polarize with-the-flux. **d2** An auxin maxima emerges at the leaf tip. Auxin at this site leaks into sub-epidermal layers, driving the formation of a polarized strand patterning the midvein. **d3** Growth of the leaf creates more space along the margin, leading to the emergence of additional maxima, and corresponding second-order veins (adapted from Smith 2007)

tiation zones (root zonation) has been modeled based on a gradient of PLETHORA established through its diffusion and gradual dilution by growth (Mähönen et al. 2014). In these models, biological plausibility can be determined via comparison of the diffusion, turnover, and growth rates to those measured experimentally. Here, live microscopy of proteins tagged with fluorescent markers offers a variety of techniques for such measurements at the cellular scale. For instance, Mähönen et al. (2014) estimated the degradation rate of PLETHORA2 by measuring the time needed to abolish fluorescent signal from a tagged version of the protein following inducible RNA interference (RNAi). Inducing the expression of the tagged protein in only a few cells can be used to observe the diffusion of a compound away from its production site (Mähönen et al. 2014; Wartlick et al. 2009). Microscopy techniques such as fluorescence recovery after photobleaching (FRAP) can even be used to measure protein diffusion rates (e.g., Kicheva et al. 2007).

Diffusion alone becomes insufficient to explain more complex dynamic patterns, like the appearance of trichomes in the leaf (Bouyer et al. 2008; Smith 2011) or periodic initiation of primordia in the shoot (Jönsson et al. 2006; Smith et al. 2006b). These patterns involve the establishment of spatially separated activation maxima in a growing tissue. Reaction–diffusion processes have been proposed to explain the emergence of local peaks in gene activity in both animals and plants (Kondo and Miura 2010; Meinhardt et al. 1998), from the formation of zebra stripes to the regular spacing of trichomes in *Arabidopsis* (Bouyer et al. 2008; Smith 2011). In plants, the active transport of the hormone auxin provides an additional mechanism to explain the establishment of tissue-level gradients, strands pre-patterning vasculature, and local maxima (Runions et al. 2014).

Auxin is transported in and out of the cell by dedicated exporters and importers localized to the cell membrane (Fig. 2a). Furthermore, it regulates its own transport and controls the expression and subcellular localization of its transporters. Here, we focus on patterning feedbacks involving the auxin exporter PIN-FORMED (PIN), the currently most investigated theme in computational models. Such feedbacks underly diverse phenomena, including the positioning of organs (Jönsson et al. 2006; Smith et al. 2006a, b), regulation of leaf form (Bilsborough et al. 2011), patterning of vasculature (Mitchison 1980, 1981; Rolland-Lagan and Prusinkiewicz 2005; Feugier et al. 2005), maintenance and growth of the root meristem (Grieneisen et al. 2007), and apical dominance (Prusinkiewicz et al. 2009). Auxin importers (AUX/LAX proteins) also play a role in auxin transport, and patterning feedbacks are also possible (Fig. 2c, Laskowski et al. 2008; Smith and Bayer 2009), but are not discussed in detail here.

PIN proteins have a nonuniform localization, preferentially localizing to subdomains of the plasma membrane. In the shoot, PIN polarities change dynamically during morphogenesis, and both PINs and their dynamic localization are important. Consequently, a key question is how this localization is determined. Studies observing PIN polarization point to two seemingly distinct patterns of polarization in the epidermis and sub-epidermal layers. In the epidermis, PINs appear to polarize *up-the-gradient* of auxin concentration, forming peaks. In internal layers, highly polarized strands of PIN expression are seen, where PINs polarize *with-the-flux* of auxin transport (Fig. 2d). To illustrate basic aspects of patterning driven by polar auxin transport, we focus on the dual-polarization model (Bayer et al. 2009). This model combines two mechanisms to provide a unified explanation of auxin-based patterning in the shoot. It should be noted, however, that alternative polarization models have been proposed (reviewed in Runions et al. 2014), based on extracellular auxin (Roussel and Slingerland 2012; Abley et al. 2013; Wabnik et al. 2010), subcellular gradients (Kramer 2009), or mechanical stresses (Heisler et al. 2010).

The first mechanism accounts for PIN polarization in the epidermis, where PINs polarize toward points of auxin accumulation (Reinhardt et al. 2003; Smith et al. 2006a, b). These points correspond to the sites of organ initiation in the shoot (Jönsson et al. 2006; Smith et al. 2006a, b), and in the leaf margin pattern the protrusions later developing into serrations, lobes, or leaflets (Hay et al. 2006; Bilsborough et al. 2011; Barkoulas et al. 2008; Koenig et al. 2009). Such patterns can be reproduced by assuming that PINs are allocated to each cell wall based on the auxin concentration in the adjoining cell (Jönsson et al. 2006; Smith et al. 2006a, b). This produces a feedback where slight differences in auxin concentration are reinforced. As PINs polarize toward cells with higher auxin concentrations, transport toward these cells is increased. This increases the auxin concentration in these cells, causing neighboring cells to polarize more strongly toward it. This causes a periodic pattern of auxin maxima to emerge with PIN-polarized up-the-gradient (Fig. 2b). In the growing tissues of the leaf margin (Bilsborough et al. 2011) and shoot epidermis (Jönsson et al. 2006; Smith et al. 2006a, b), this leads to the gradual organization of maxima (Fig. 2d) in the space generated by growth.

Once formed, auxin at the maxima in the epidermis moves to internal tissues. In sub-epidermal tissues, PINs polarize away from the epidermis, and are eventually restricted to narrow, high-flux strands pre-patterning veins (Scarpella et al. 2006; Bayer et al. 2009), connecting the auxin *source* in the epidermis to preexisting veins (which act as an auxin *sink*). The second mechanism, accounting for the patterning of vascular strands, is inspired by Sachs canalization hypothesis (Sachs 2003, as modeled in Fig. 1c). Its cellular implementation (Mitchison 1980, 1981; Rolland-Lagan and Prusinkiewicz 2005) assumes that PINs are allocated to the cell wall based on the net flux of auxin through the wall (total outflux—total influx). This causes initial directions of auxin flow to be reinforced, leading to a coherent polarization *with-the-flux* of auxin transport. Different patterns emerge based on the strength of this feedback. A strong, nonlinear feedback of flux on polarization leads to the emergence of preferred flux pathways, pre-patterning vascular strands (the canals in the canalization hypothesis; Fig. 2d). Nonetheless, the fluxes generated by the strong feedback can overpower weaker fluxes guiding developing strands to auxin sinks, making it difficult to efficiently connect auxin sources and sinks (Bayer et al. 2009; Smith and Bayer 2009). In contrast, a weak feedback (linear) will not produce strands, but instead generates a broad tissue polarization (Feugier et al. 2005; Stoma et al. 2008) robustly connecting sources and sinks (analogous to the initial slope of the terrain in Fig. 1c). Consequently, a gradual transition from weak to strong polarization provides a more reliable mechanism to pattern vascular strands (O'Connor et al. 2014).

The broad biological relevance of cellular-level models has lead to the introduction of many approaches for representing cells and the molecular processes therein. Consequently, two aspects of model evaluation are the plausibility of molecular processes and the suitability of the chosen cellular representation to the problem at hand. Formulating molecular process in terms of chemical reactions (c.f. Shapiro et al. 2013) and relying on computational gene regulatory networks (see Bolouri 2008) helps ensure their plausibility. Such considerations underlay initial critiques of up-the-gradient and with-the-flux polarization models, regarding molecularly plausible mechanisms for, respectively, measuring auxin in a neighboring cell (i.e., across the apoplast) or net flux through the cell membrane; an issue only recently resolved (Cieslak et al. 2015). Moving to cellular representations, it is common to represent the cell interior, apoplast, and cell walls as well mixed compartments, and sometimes omit the apoplast completely. However, finer discretizations are required to explicitly represent subcellular gradients or detailed PIN localizations at the cell wall (e.g., Kramer 2009; Abley et al. 2013). Similarly, omission of the extracellular space (causing transport to occur directly between cells) may be questionable when AUX/LAX proteins are included in models, as in this case import would occur by directly removing auxin from the neighboring cell. As can be seen from these considerations, the relatively low level of abstraction in cellular models places a greater emphasis on model details. It is therefore more difficult to crisply relate low-level model dynamics to the collective behavior of tissues.

Organ Growth

How does an organ control its growth? This seemingly simple question involves a whole field of scientific research. The complex nature of growth mechanisms has led to the development of various computational approaches, examining growth regulation from the molecular scale to that of entire organs. The final shape of an organ is not the only useful output of a growth model, as the key question is also *how* we got there. Even at the level of single cells, completely different local growth patterns can result in similar shapes. For example, long and thin cells such as pollen tubes or trichome branches can grow while maintaining their shape either by expanding only at their tip (Fayant et al. 2010; Rojas et al. 2011) or all along their shank (Yanagisawa et al. 2015). These two expansion patterns reflect distinct biological mechanisms, namely local deposition of pectin-rich wall material at the pollen tube tip versus loosening of the preexisting cellulose matrix along the trichome flanks. For a model to be biologically relevant, it is therefore essential to account for patterns of local growth, provided they are known (Geitmann and Ortega 2009).

Models of organ growth and pattern formation are often combined together, as growth can drive patterning and patterning processes can direct growth. Simulations can also help us understand how molecular factors modify the shape of an organ. Although many genes have been linked to organ shape and growth in plants, their action is necessarily indirect as it must be mediated by mechanical processes. At the cellular level, growth must occur via the deformation of the remarkably rigid wall surrounding the cells (Cosgrove 2005). Owing to the cell wall, plant cells can typically withstand an internal pressure about twice as high as the pressure of a car tire, or even higher in specialized cells like stomata (Geitmann and Ortega 2009). Cell walls exhibit a viscoelastic behavior, meaning that they deform under mechanical load in a time-dependent manner. By definition, in non-growing cells, the cell wall is stretched by turgor pressure in a *reversible* (or *anelastic*) way, i.e., the wall can return to its original shape if pressure vanishes. Due to the viscous behavior of the cell wall, this return to the original configuration can be slow (Nolte and Schopfer 1997). In a growing cell, a part of the cell wall deformation is *irreversible* (or plastic), meaning it would not completely return to its original length if pressure was to drop to zero (Routier-Kierzkowska and Smith 2012). Molecular growth factors (e.g., wall-loosening enzymes and growth hormones) facilitate irreversible deformation by modifying the mechanical properties of the cell wall (Cosgrove 2016).

Comparing cells with small balloons, it would seem that the increase of cell volume resulting from growth should generate a decrease in internal pressure. Plant cells, however, actively regulate their pressure within minutes (Cosgrove 1986). In contrast, irreversible cell wall deformation usually occurs over a much longer timescale (hours) (Nolte and Schopfer 1997), so the internal pressure can be considered constant during cell growth in most cases (Schopfer 2006). Variation in turgor pressure due to growth has not been detected so far, even in very fast-growing cells such as pollen tubes (Winship et al. 2010), although the effect of rapid growth on turgor deserves further investigation using noninvasive techniques (Vogler et al. 2013). Some of the tension generated by turgor in the cell wall is "transferred" into irreversible deformation, similar to a chewing gum that becomes floppy when stretched. For this reason, cell wall growth constitutes a *stress relaxation* mechanism (Schopfer 2006).

Ultimately, organ growth arises from the integrated growth of individual cells. Plant tissues and organs have a few properties that make them particularly good subjects for biomechanical modeling. In plants, morphogenesis requires the coordinated deformation of individual cells and can occur even in the absence of cell division (Kwiatkowska 2004). Apart from a few notable exceptions, plant cells are stuck to each other by a strong pectin layer (the middle lamella). This means that plant cells cannot migrate, slide relative to each other or intercalate. Consequently, plant tissues as a whole are well approximated as solids, in contrast to the more fluid-like behavior of some animal tissues during their development, since animal cells can crawl past each other and exchange neighbors. Cell walls within plant tissues form a continuum. Therefore, mechanical forces acting on individual cells are also transmitted to their neighbors and build up at the organ level, resulting in *tissue stresses* (also called *tissue tensions*) (Hejnowicz et al. 2000; Peters and Tomos 1996). These forces are responsible for some plant movements and are also thought to play an important role in driving growth (Baskin and Jensen 2013).

Predictive Capacity of Growth Models

At first glance, it may seem that modeling the growth process entirely—from gene expression to final organ shape—would always be most desirable. However, in practice, we often do not have enough data to model all the levels of growth control. Instead of attempting to simulate growth as a completely emergent property, modelers can introduce some descriptive components to the model. The descriptive aspects of the model are considered as a given, and there is often no attempt to elucidate the biological mechanisms controlling them. For example, some cells can be assumed to possess an intrinsic polarization. From a modeling perspective, the cause of this polarization does not need to be known. Fixing a preferred direction opens, however, the possibility to test the effects of some parameters, e.g., assign different growth rates along the perpendicular and parallel orientations to the main axis (Kennaway et al. 2011). Such a strategy is extremely useful in testing only one hypothesis at a time, while keeping all other conditions constant.

In a purely descriptive growth model, both the organ shape and deformation are given as inputs to the model. This is particularly useful to test the effects of growth on patterning mechanisms. For example, patterns of auxin transport establishing leaf venation can be modeled on a growing leaf (Smith and Bayer 2009; Smith 2011). The modeler can specify leaf shape at different steps of development and the deformation of different regions of the leaf between those steps (as in Fig. 2d). This way, we can examine how the emerging part of the model (auxin patterning) depends on geometry and growth, which are given here as an input. Various patterning mechanisms can then be implemented using the same growth template, e.g., to test how local auxin maxima emerge and fluxes evolve by expanding leaves. Such descriptive growth models could be readily adapted to take actual growth data (e.g., from Rolland-Lagan et al. 2014) as input.

In contrast, a purely emergent growth model (Fig. 3) would predict organ shape from the collective mechanical behavior of pressurized cells. Predicting growth at the level of individual cells already presents challenges, as shown by models of pollen tube growth (Fig. 3a–b4, Fayant et al. 2010). Cell shape and the gradient of local cell wall properties are given as inputs to the model, while final cell shape and patterns of local expansion (i.e., strain) are obtained as outputs (Fig. 3a). The initial shape of the pollen tube tip results in a gradient of mechanical tension, which increases from the summit to the flank. A very strong gradient of cell wall stiffness is needed to counteract the stress gradient and obtain the correct pollen tube shape (Fig. 3b4). A similar balance between cell wall elasticity and mechanical stress controls organ deformation, as shown in growth simulations of the mature Arabidopsis embryo (Bassel et al. 2014) (Fig. 3c). In this model, the shape and size of individual cells vary within the embryo, determining a gradient of mechanical stress at the organ level (Fig. 3c1). This model can explain why the tip of the root, in which the expression and activity of growth-inducing factors is maximal (Fig. 3c3), is not the region of maximal expansion (Fig. 3c2). The concentration of growth factors represents the patterning component of the model, which in this case is fully specified by the modeler based on experimental data (here, expression and activity of gibberellic acid and expansions). Specifying one aspect of the simulation (gene expression) makes it possible to clearly identify cell geometry as another key ingredient of organ growth regulation.

The preceding examples illustrate the concept of descriptive vs emergent growth models with very contrasting cases. However, most growth models fall somewhere in between, as we will see in the next section.

What is the Level of Abstraction Needed to Model Mechanics?

Since plant cells are constrained by their walls, organ growth can be described entirely as the deformation of cell walls under turgor pressure (Green 1962). Mechanically, the cell wall is a composite material, which contains very stiff cellulose fibers embedded in a pectin matrix (Cosgrove 2005). Cellulose fibers tend to have a preferred orientation within the cell wall, causing it to be mechanically anisotropic, i.e., stiffer along one direction (Baskin 2005). Because cell wall components are produced within the cell, new layers are deposited from the inside upon already tensed walls. This creates a structure in which mechanical stresses are not distributed evenly throughout the wall's depth (Hejnowicz and Borowska-Wykret 2005; Dyson and Jensen 2010), raising the question of which layer is most relevant for growth (Baskin and Jensen 2013).

Owing to its fascinating complexity, the role of cell wall structure and mechanics in controlling growth has received much attention (see review Bidhendi and Geitmann 2016). As discussed in the previous section, however, growth patterns emerging from mechanical models can be counterintuitive, since they do not necessarily follow the expression patterns of wall-softening proteins (Bassel et al. 2014). Mechanical stresses at the tissue level have also been proposed to explain the discrepancy sometimes observed between growth directions and cellulose orientation (Baskin and Jensen 2013). Modeling morphogenesis therefore requires us to account for the many intermediaries between genetic processes, cell expansion, and final organ shape (Fig. 4). Here, we list some levels of abstractions used for modeling growth.

Highest level of Abstraction: The Organ as a Continuum

At the scale of entire organs, it is convenient to consider that some tissues or regions instruct the growth of the whole. For example, we could assume that leaf growth is directly controlled at the level of its contour (or margin) (Bilsborough et al. 2011). This model ignores the blade and veins, and simulates the emergence of serrations by locally pushing the contour outward, based on auxin concentration along the leaf margin. Such a hypothesis is plausible because auxin-driven patterning is known to play a crucial role in shaping the leaf, and auxin maxima form along the leaf margin.

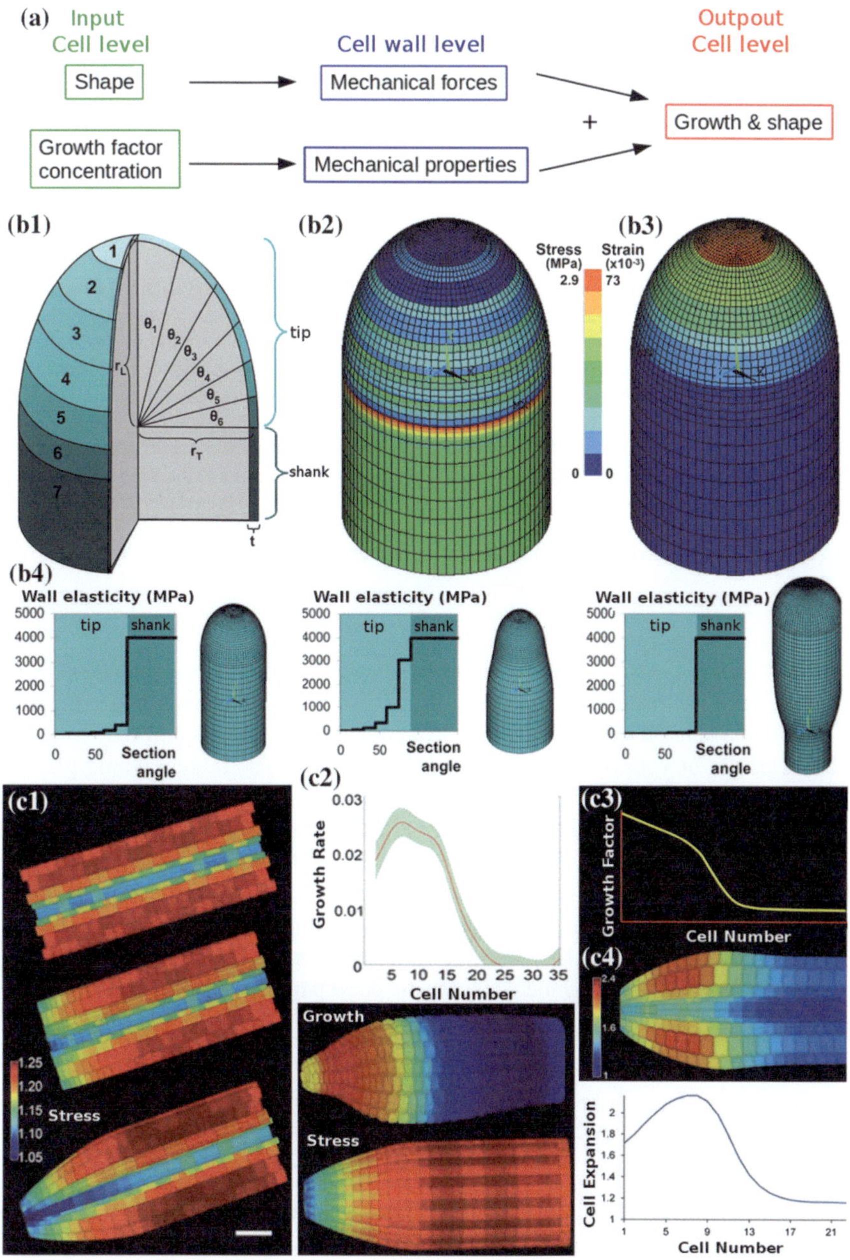
(a) Input
Cell level
Shape
Growth factor concentration
Cell wall level
Mechanical forces
Mechanical properties
+
Outpout
Cell level
Growth & shape
(b1)
tip
shank
(b2)
Stress (MPa)
Strain (x10-3)
(b3)
(b4)
Wall elasticity (MPa)
Section angle
(c1)
Stress
(c2)
Growth Rate
Cell Number
Growth
Stress
(c3)
Growth Factor
Cell Number
(c4)
Cell Expansion
Cell Number

◂**Fig. 3** Growth as an emergent property of mechanical models. **a** From a physical perspective, cell growth is driven by mechanical forces (i.e., stress) in the cell wall and restricted by the mechanical properties of the wall. In a cellular model, mechanical stresses are predicted based on cell geometry and turgor pressure. Cell wall mechanical properties are assumed to depend on the local concentration of a growth factor (e.g., wall-loosening enzyme, growth hormone). **b** To model pollen tube growth, the cell wall is divided into small rings from tip to shank and assigned different mechanical properties (**b1**). Due to the cell's shape, stress generated by turgor is minimal at the very tip of the tube and increases toward the shank (**b2**). However, local deformation (strain) observed during growth is maximal at the tip (**b3**). To compensate for the stress gradient, the cell wall is assumed to be very soft at the tip and stiffer at shank. The stiffness gradient near the tip has to follow a specific pattern in order to precisely reproduce the form of a growing pollen tube (**b4**) (adapted from Fayant et al. 2010). **c** Mechanical models can also predict organ growth, here the shoot and root of a mature embryo in *Arabidopsis*. Pressurizing 3D cellular templates with different geometries shows that the stress patterns depend on geometry both at the cell and organ level. Assuming a realistic geometry, i.e., tapering of the root and smaller cells toward the tip, result in smaller stresses at the root tip (**c1**). The distribution of stresses cannot, by itself, explain the measured cellular growth rates, which are maximal in a region close to the tip (**c2**). The distribution of growth factors (here, gibberellic acid) also does not correspond exactly to local growth, since they are maximal at the very tip of the root (**c3**). A realistic pattern of cell growth is instead obtained by combining the effects of geometry and the gradient of growth factor (**c4**) (adapted from Bassel et al. 2014). Note that the average stress per cell was computed by Bassel et al. using the elastic deformation of cells with uniform, isotropic walls

On the other hand, we could assume that the shape of a leaf is entirely dictated by its stiff vascular system and that the leaf lamina only follows passively, filling in the gaps between the veins. In such a model, the leaf contour would emerge from connecting the end points of vein branches (Runions et al. 2017). The lamina would play no role in controlling leaf growth and would only be used for visualization purposes (Fig. 1b). We can picture the leaf margin in this case as a sort of rubber membrane surrounding stiff branches representing the veins. As the veins grow, they push on the rubber, causing it to deform. Different shapes of leaf serrations can be obtained by modulating the membrane resistance to bending and stretching (i.e., how stiff it is). Note, however, that such models are not used to make quantitative mechanical predictions, and they employ mechanically inspired parameters as a means to ensure a smooth, realistic contour.

Mechanical Compatibility Between Growing Regions

In the models discussed above, no assumption is made concerning the mechanical state of the lamina. To picture leaf mechanics more realistically, we could imagine the rigid veins attached all along their length to the lamina, represented by a soft sheet of plastic. A piece of lamina attached from both sides to elongating veins would stretch due to vein deformation. The lamina would therefore pull on the veins, restricting their growth in return. Considering a plant organ as a continuum ensures that the growth of each of its components is mechanically constrained by the others.

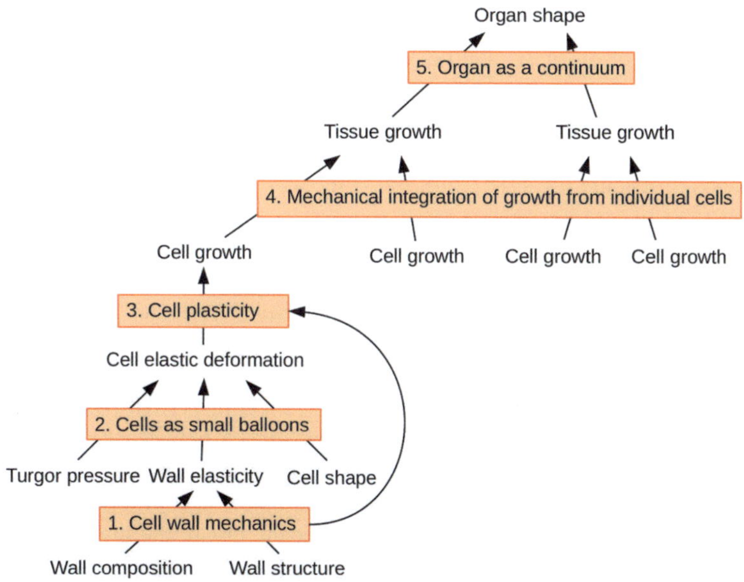

Fig. 4 Different levels of abstraction used in growth models. Although organ growth is ultimately instructed by genes, it is regulated at many different scales. Using a simplified view of the lower scales helps to focus growth models on specific aspects of growth. The cell wall is a crucial factor in determining organ growth and shape; its mechanics are fundamental to our understanding of growth regulation. While cell wall composition and structure can be assessed experimentally, mechanical models (e.g., Dyson and Jensen 2010) are needed to understand how such a complex structure behaves mechanically (level 1). In addition to cell wall mechanics, the cell shape determines how the cell deforms under internal pressure (level 2). The elastic deformation of pressurized cells can be computed in 3D using FEM modeling (e.g., Mosca et al. 2017; Weber et al. 2015). Determining how the turgor-driven, reversible cell deformation relates to plastic deformation (growth) requires yet another level of assumptions concerning wall mechanics (level 3). Often plastic deformation is simply considered as proportional to the elastic one (e.g., Bassel et al. 2014; Boudon et al. 2015; Fayant et al. 2010). Such models can predict cell growth as an emergent property of mechanics and geometry. Nonetheless, it is also possible to consider the growth of individual cells or small regions of the tissue as given (i.e., a descriptive input to simulations). In this case, tissue growth and shape emerges from the mechanical integration of the regional expansion rates assigned by the modeler to different portions of the organ (e.g., Dumais and Steele 2000; Kennaway et al. 2011) (level 4). At yet a higher level of abstraction (level 5), the growth of tissues itself is considered as an input. Growth of the whole is constrained by geometry (e.g., Bilsborough et al. 2011), given the basic assumption that organs are continuous

Models that consider tissues as solids are called physically based simulations and are often solved using the Finite Element Method (FEM) (e.g., Dumais and Steele 2000; Liang and Mahadevan 2009; Kenneway and al. 2011; Kuchen et al. 2012). In such models, a patch of fast-growing cells pushes on their slow-growing neighbors, creating local compressions. Conversely, slower growing cells pull on their fast-growing surroundings and create internal tensions. Paul Green pioneered the idea that the internal forces resulting from the differences in growth between regions could explain patterns commonly observed in plants (Green et al. 1996, reviewed in Dumais 2007). For example, that the regular waves often observed on grass leaves could emerge from a mismatch in growth between the center of the blade and leaf margin. The compressive forces created by the extra growth in the middle would cause the thin blade to buckle, similar to how a knitted piece of fabric ruffles when adding more stitches in some rows (Fig. 5a). Buckling provides an elegant explanation for the emergence of complex shapes, since even very simple initial conditions—such as attaching together two regions of different growths—can result in intricate periodic patterns of bulges. These ideas have been further developed to reproduce the regular arrangement of florets emerging on a sunflower head (Hernandez and Green 1993; Dumais and Steele 2000) and various ruffling patterns in leaves or petals (Fig. 5b) (Sharon et al. 2007; Liang and Mahadevan 2009; Liang and Mahadevan 2011).

The out-of-plane deformation occurring during buckling partially releases internal compressions and tensions resulting from the mismatch between growing tissues. However, ruffling will not occur if the surface is too thick. Deforming a solid is always associated with an energetic cost; the shape of an object at equilibrium is given by minimizing the sum of all the costs. The thicker the organ, the higher the energy cost associated with bending the surface, until it becomes more favorable to stay flat and conserve the internal forces, called *residual stresses*. As a result, the slower growing regions will be under tension, causing them to extend more than if they had been isolated from the rest of the tissue. Faster growing regions, on the other hand, will be prevented from expanding to their full capacity due to local compression. Kennaway et al. (2011) call this a "conflict between *specified* and *resultant* growth". *Specified growth* is the growth that would happen to a region of the tissue if it grew in isolation. *Resultant growth* is the growth that can be directly observed by time-lapse imaging. It is a compromise resulting from the specified growth and the residual stresses at mechanical equilibrium (Fig. 5c). Patterns of specified growth can be very simple (slow vs. fast regions), as discussed above in the case of buckling mechanisms (Fig. 5a1). However, we can expect specified growth to be instructed by gene expression or hormone concentration and be smoothly distributed along the organ (e.g., Kuchen et al. 2012). The shape of resultant growth gradients emerges from both the pattern of specified growth, mechanical properties of tissues, and the organ's initial shape (Fig. 5c).

Plant organs often grow at different rates in different directions, i.e., the growth is anisotropic (from the greek an—"not", isos—"equal", tropos—"way"). Anisotropic tissue growth is exploited by plants to produce very diverse forms despite the limitations imposed by the attachments between cell walls. Simulating growth anisotropy also opens new possibilities for models. For example, the anisotropy of resultant and specified growth does not necessarily match each other. Specified growth patterns that do not present any preferred direction (isotropic) can even, in some instances, generate anisotropic resultant growth (Kennaway et al. 2011).

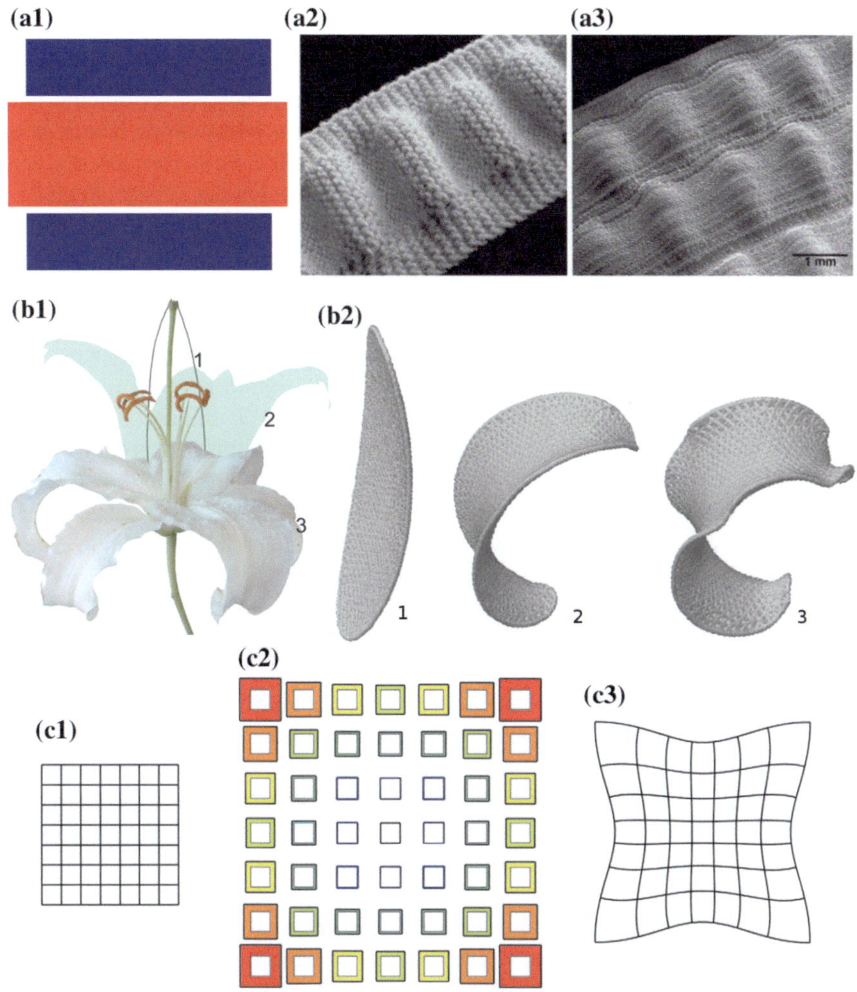

Fig. 5 Physically based models of tissue growth. Patterning and organ shape can be simulated by considering tissues as homogeneous solids. For example, gluing two bands of slow-growing tissues (in blue) on each side of a fast-growing one (in red) (**a1**) can cause the middle band to wrinkle (or buckle). Paul Green illustrated the concept of buckling by increasing the number of stitches in the central rows of a knitted fabric (**a2**) and proposed it as a mechanism to explain the formation of periodic patterns in different systems, for example, the rippling of grass leaf blades (**a3**) (adapted from Dumais 2007). **b** Buckling processes have been used to model the formation of different kinds of wavy patterns, e.g., wrinkles of lily petals (**b1**). In this case, the petal margin expands faster than its midvein, causing the thin petal to buckle as it grows during flower opening (**b2**) (adapted from Liang and Mahadevan 2011). **c** In thicker organs, the mechanical stresses caused by incompatibility in growth between regions cannot be relieved by buckling. If we divide the tissue in its initial state into a regular grid (**c1**) and assign a different value of *specified growth* to each square (**c2**, increasing growth rate from blue to red), the tissue in its final configuration (**c3**) will still bear some *residual stresses* due to local differences in specified growth. In this case, the *resultant growth* between initial and final state will have a smoother pattern than the specified growth (adapted from Kennaway et al. 2011)

Cells as Small Balloons

The physically based models considered above describe tissues as a continuum. This assumption is very useful if we want to know how tissues containing hundreds of cells behave as a whole, i.e., how gradients of mechanical properties affect tissue shape. However, continuous models do not take into account the effects of cell size and shape on tissue mechanics. In spherical cells, for example, the tension is directly proportional to cell size. Tension in cylindrical cells is also proportional to the cell's radius, but is asymmetric, being twice as large in one direction as the other (Baskin and Jensen 2013). Mechanical stresses (i.e., forces within the cell wall) in pressurized spheres and cylinders can be described with simple, well-known formulas. However, predicting the complex stress patterns in pressurized tissues requires mechanical simulations (Fig. 6a) (Bassel et al. 2014; Mosca et al. 2017; Boudon et al. 2015). Recent modeling of 3D pressurized cells revealed their sometimes surprising behavior. Intuitively, we would expect cells to expand in all directions while their internal pressure rises, similar to a rubber balloon. Simulations show, however, that assuming a common square cell shape with walls stiffer in one direction is enough to obtain cells that shrink instead of expanding along one direction (Fig. 6b), creating a pulling force at the tissue level (Hofhuis et al. 2016). Thus, cell and organ geometry appears to be a crucial element in linking the mechanical behavior at the cellular and tissue levels.

Stresses within cell walls depend *only* on turgor pressure and the shape of the cell once it is pressurized (i.e., shape at mechanical equilibrium). This means that if we can extract the exact 3D shape of an isolated turgid cell, we can predict local stresses within the cell wall without knowing its mechanical properties (e.g., Bassel et al. 2014). However, if we want to compute local deformations (or strains) occurring when the cell is pressurized, we need to assign elastic properties to the cell wall. In cylindrical cells, for example, cell wall tension is two times higher in the hoop direction than in the longitudinal direction. If the cell wall elasticity was equal in both directions, the cell diameter would increase twice as much as its length when turgor pressure goes up. On the other hand, the cell wall would have to be twice as stiff along the hoop just to obtain an equal deformation in all directions. This simple example shows how behavior at the cell level (isotropic expansion) does not match the assigned cell wall properties (anisotropic elasticity) due to cell geometry. Physically based models of pressurized cells therefore provide an essential tool to understand how subcellular properties such as turgor pressure and the mechanical properties of cell walls affect cell mechanics and, ultimately, growth.

Elastic Versus Plastic Deformations in Cells

We have seen above how cells deform reversibly due to turgor pressure. Simulating growth requires assumptions to be made regarding how reversible (elastic) and irre-

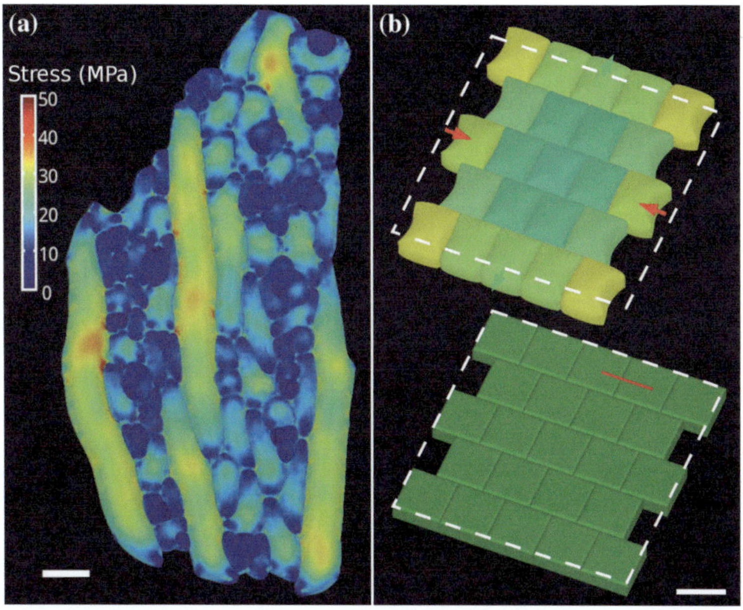

Fig. 6 Effects of geometry on mechanical behavior of pressurized cells in 3D. At equal pressure, larger cells will generally bear more stress than small ones, as demonstrated in a pressurized 3D template of a sepal (**a**, adapted from Mosca et al. 2017). In most cases, cells are expected to expand in all directions as the pressure increases. However, assuming a simple square shape and a stiffer cell wall in one direction (**b**, bottom, red line) is enough to cause a pressurized tissue to shrink in one direction (**b**, top, red arrows) and expand in the other (yellow arrows) (adapted from Hofhuis et al. 2016). Scale bars: 20 μm (**a**), 50 μm (**b**)

versible (plastic) deformations are related. In most cellular models (Hamant et al. 2008; Corson et al. 2009; Smith 2011; Merks et al. 2011; Bassel et al. 2014; Boudon et al. 2015; Prusinkiewicz and Runions 2012), growth is a modification of the reference configuration, i.e., the shape of the object if it was free from any mechanical force. A common assumption is that growth occurs if the stress generated by turgor exceeds a certain threshold, as proposed by Lockhardt (1965). In volumetric models of pressurized cells, growth of the cell walls follows the three-dimensional elastic deformation (i.e., the elastic strain) resulting from internal pressure, cell geometry, and wall elasticity (Bassel et al. 2014; Boudon et al. 2015; Fayant et al. 2010). To implement Lockhart's hypothesis at the local cell wall level, it is assumed that the reference configuration of a cell wall changes in proportion to its elastic strain (Goriely et al. 2008). We can illustrate this concept with a simple spring: in this case, the reference configuration is also called rest length, and it is the length of the spring when nothing pushes or pulls on it. Pulling on the spring causes it to elongate, and this deformation is usually reversible or elastic. However, if we heat the spring to a high temperature while it is stretched, the spring will not be able to fully return to its original length after being released. Part of the elastic (reversible) deformation has been converted into an irreversible one by softening of the metal, changing its crystalline structure with heat. Similarly, the cell wall structure is known to be modified during growth, e.g., by breaking and re-establishing links in the cellulose–hemicellulose network (Cosgrove 2005; Schopfer 2006). In simulations of growing cells above cited, the amount of reversible deformation converted into irreversible is determined by growth factors and the magnitude of the elastic deformation. This agrees with experimental evidence showing that cells undergoing larger elastic deformations upon changes in pressure generally grow faster (Kierzkowski et al. 2012, reviewed by Routier-Kierzkowska and Smith 2012).

Conclusions

The utility of computational models arises from their ability to explicitly test causal relations and their implications. In some cases, simulations can be seen as a practical test of "mental models" inferred from biological experiments. Implementing models can help to identify missing elements from initial hypothesis, while examining simulation output can reveal unexpected emergent properties. Conversely, experiments can then be designed to confirm or falsify model predictions and discriminate between alternative model assumptions.

To take full advantage of the potential of modeling, it is thus essential for biologists and modelers to work in close collaboration. Establishing a model might require much back-and-forth between simulations and experiments throughout a research project. Such integrated approaches allow the modeler to understand as much as possible of the biology and make informed decisions about the relevant level of abstraction for models, or the refinement of hypotheses. For biologists, receiving feedback from modelers at an early stage of the project makes it possible to design specific experiments aimed at measuring quantities that are essential inputs to the model, and to more rigorously test model predictions.

Nonetheless, comparing model output with experimental data can be more subtle than it may first appear. For instance, the effect of patterning processes occurring at an early stage of development can be masked by subsequent growth. An example is given by phyllotactic patterns, a classical subject of modeling in plant development. The arrangement of primordia around the meristem is often reflected by the position of lateral organs (flowers, fruits, or leaves) around the stem. Lateral organs are much easier to observe than tiny primordia; however, differential growth of the stem between organs can modify their relative positions. Irregular (Burian et al. 2015) or even opposite (Landrein et al. 2013) positioning of mature organs along the stem can occur in mutants, despite a normal spiral phyllotactic pattern at the meristem. Similarly, the complexity of leaf forms can change substantially during the expansive phase of leaf growth. For example, leaves expressing genes associated with leaflet initiation can develop first as a compound primordia but present with a single, simplified blade at maturity (Bharathan et al. 2002). Such examples highlight the importance of choosing the appropriate stage of development for comparing forms predicted by the model and actual organ shape.

Another crucial aspect of model validation is assessing the dynamics of processes in space and time. Models of growth should reproduce not only the final shape of the organ or cell but also patterns of local expansion generating the shape (Fayant et al. 2010). Technical advances in 3D imaging and growth analysis at the cellular level (e.g., Barbier de Reuille et al. 2015; Vlad et al. 2014) will further contribute to the refinement of growth models. Time-lapse data can also help disentangle different aspects of morphogenesis. For example, cell expansion and division can be uncoupled in plants via a process called *compensation* (reviewed in Hisanaga et al. 2015). Cell size in a mature organ can therefore reflect how fast the cells grew as well as how many times they divided. Here, modeling the dynamics of cell division and expansion will help us understand how plants regulate their cell size (Roeder 2012).

More broadly, many aspects of morphogenesis are still under-explored. Among others, the feedbacks between cell division planes and molecular processes (De Rybel et al. 2014; Yoshida et al. 2014) or mechanical forces and cell polarity (Heisler et al. 2010; Bozorg et al. 2014) seem particularly promising. Further pursuing these themes requires ever more integrated models combining feedbacks between growth, patterning, cell division, and mechanics. The increasing complexity of processes explored by such models likewise calls for an ever tighter integration of modeling and biology.

Acknowledgements We thank Richard Smith for helpful discussions and providing the simulation of auxin patterning in a developing leaf. We also would like the thank Przemyslaw Prusinkiewicz for discussions that helped to formulate some of the ideas appearing in this chapter. Support for this work was provided by the Bundesministerium für Bildung und Forschung grant 031A492, the Human Frontier Science Program grant RGP0008/2013 and the Max Planck Society. Funding from the European Commission from a Marie Skłodowska-Curie individual fellowship (Horizon 2020, 703886) is also gratefully acknowledged by AR.

References

Abley K, De Reuille PB, Strutt D et al (2013) An intracellular partitioning-based framework for tissue cell polarity in plants and animals. Development 140:2061–2074. https://doi.org/10.1242/dev.062984

Ali O, Mirabet V, Godin C, Traas J (2014) Physical models of plant development. Annu Rev Cell Dev Biol 30:59–78. https://doi.org/10.1146/annurev-cellbio-101512-122410

Alvarez JP, Furumizu C, Efroni I et al (2016) Active suppression of a leaf meristem orchestrates determinate leaf growth. Elife 5:1–17. https://doi.org/10.7554/elife.15023

Barbier de Reuille P, Routier-Kierzkowska A-L, Kierzkowski D et al (2015) MorphoGraphX: a platform for quantifying morphogenesis in 4D. Elife 4:1–20. https://doi.org/10.7554/elife.05864

Barkoulas M, Hay A, Kougioumoutzi E, Tsiantis M (2008) A developmental framework for dissected leaf formation in the *Arabidopsis* relative *Cardamine hirsuta*. Nat Genet 40:1136–41. https://doi.org/10.1038/ng.189

Baskin TI (2005) Anisotropic expansion of the plant cell wall. Annu Rev Cell Dev Biol 21:203–222. https://doi.org/10.1146/annurev.cellbio.20.082503.103053

Baskin TI, Jensen OE (2013) On the role of stress anisotropy in the growth of stems. J Exp Bot. https://doi.org/10.1093/jxb/ert176
Bassel GW, Stamm P, Mosca G et al (2014) Mechanical constraints imposed by 3D cellular geometry and arrangement modulate growth patterns in the Arabidopsis embryo. Proc Natl Acad Sci USA 111:8685–8690. https://doi.org/10.1073/pnas.1404616111
Bayer EM, Smith RS, Mandel T et al (2009) Integration of transport-based models for phyllotaxis and midvein formation. Genes Dev 5:373–384. https://doi.org/10.1101/gad.497009.5
Bharathan G, Goliber TE, Moore C et al (2002) Homologies in leaf form inferred from KNOXI gene expression during development. Science 296:1858–1860. https://doi.org/10.1126/science.1070343
Bidhendi AJ, Geitmann A (2016) Relating the mechanics of the primary plant cell wall to morphogenesis. J Exp Bot 67:449–461. https://doi.org/10.1093/jxb/erv535
Bilsborough GD, Runions A, Barkoulas M et al (2011) Model for the regulation of Arabidopsis thaliana leaf margin development. Proc Natl Acad Sci 108:3424–3429. https://doi.org/10.1073/pnas.1015162108
Bolouri H (2008) Computational modeling of gene regulatory networks: a primer. World Scientific Publishing Co Inc
Boudon F, Chopard J, Ali O et al (2015) A computational framework for 3D mechanical modeling of plant morphogenesis with cellular resolution. PLoS Comput Biol 11:e1003950. https://doi.org/10.1371/journal.pcbi.1003950
Bouyer D, Geier F, Kragler F et al (2008) Two-dimensional patterning by a trapping/depletion mechanism: The role of TTG1 and GL3 in Arabidopsis trichome formation. PLoS Biol 6:1166–1177. https://doi.org/10.1371/journal.pbio.0060141
Bozorg B, Krupinski P, Jönsson H (2014) Stress and strain provide positional and directional cues in development. PLoS Comput Biol. https://doi.org/10.1371/journal.pcbi.1003410
Burian A, Raczyńska-Szajgin M, Borowska-Wykręt D et al (2015) The CUP-SHAPED COTYLEDON2 and 3 genes have a post-meristematic effect on Arabidopsis thaliana phyllotaxis. Ann Bot 115:807–820. https://doi.org/10.1093/aob/mcv013
Chickarmane V, Roeder AHK, Tarr PT et al (2010) Computational morphodynamics: a modeling framework to understand plant growth. Annu Rev Plant Biol 61:65–87. https://doi.org/10.1146/annurev-arplant-042809-112213
Cieslak M, Runions A, Prusinkiewicz P (2015) Auxin-driven patterning with unidirectional fluxes. J Exp Bot 66:5083–5102. https://doi.org/10.1093/jxb/erv262
Corson F, Hamant O, Bohn S (2009) Turning a plant tissue into a living cell froth through isotropic growth. Proc Natl Acad Sci 106:8453–8458. https://doi.org/10.1073/pnas.0812493106
Cosgrove D (1986) Biophysical control of plant cell growth. Annu Rev Plant Physiol 37:377–405. https://doi.org/10.1146/annurev.pp.37.060186.002113
Cosgrove DJ (2005) Growth of the plant cell wall. Nat Rev Mol Cell Biol 6:850–861. https://doi.org/10.1038/nrm1746
Cosgrove DJ (2016) Plant cell wall extensibility: connecting plant cell growth with cell wall structure, mechanics, and the action of wall-modifying enzymes. J Exp Bot 67:463–476. https://doi.org/10.1093/jxb/erv511
Couder Y, Pauchard L, Allain C et al (2002) The leaf venation as formed in a tensorial field. Eur Phys J B 28:135–138. https://doi.org/10.1140/epjb/e2002-00211-1
De Rybel B, Adibi M, Breda AS et al (2014) Integration of growth and patterning during vascular tissue formation in Arabidopsis. Science 345(80):1255215–1255215. https://doi.org/10.1126/science.1255215
Dumais J (2007) Can mechanics control pattern formation in plants? Curr Opin Plant Biol 10:58–62. https://doi.org/10.1016/j.pbi.2006.11.014
Dumais J, Steele C (2000) New evidence for the role of mechanical forces in the shoot apical meristem. J Plant Growth Regul 7–18. https://doi.org/10.1007/s003440000003
Dyson RJ, Jensen OE (2010) A fibre-reinforced fluid model of anisotropic plant cell growth. J Fluid Mech 655:472–503. https://doi.org/10.1017/s002211201000100x

Errera L (1886) Sur une condition fondamentale d'équilibre des cellules vivantes. Comptes Rendus Hebdomadaires des Seances de l'Academie des Sciences 103:822–824

Fayant P, Girlanda O, Chebli Y et al (2010) Finite element model of polar growth in pollen tubes. Plant Cell 22:2579–2593. https://doi.org/10.1105/tpc.110.075754

Feugier FG, Mochizuki A, Iwasa Y (2005) Self-organization of the vascular system in plant leaves: inter-dependent dynamics of auxin flux and carrier proteins. J Theor Biol 236:366–375. https://doi.org/10.1016/j.jtbi.2005.03.017

Floyd SK, Bowman JL (2010) Gene expression patterns in seed plant shoot meristems and leaves: homoplasy or homology? J Plant Res 123:43–55. https://doi.org/10.1007/s10265-009-0256-2

Fujita H, Toyokura K, Okada K, Kawaguchi M (2011) Reaction-diffusion pattern in shoot apical meristem of plants. PLoS ONE. https://doi.org/10.1371/journal.pone.0018243

Geitmann A, Ortega JKE (2009) Mechanics and modeling of plant cell growth. Trends Plant Sci 14:467–478. https://doi.org/10.1016/j.tplants.2009.07.006

Goriely A, Robertson-Tessi M, Tabor M, Vandiver R (2008) Elastic growth models. In: Mathematical modelling of biosystems. Applied optimization, vol 102. Springer, Berlin, Heidelberg. https://doi.org/10.1007/978-3-540-76784-8_1

Green P (1962) Mechanism for plant cellular morphogenesis

Green PB, Steele CS, Rennich SC (1996) Phyllotactic patterns: a biophysical mechanism for their origin. Ann Bot 77:515–527. https://doi.org/10.1006/anbo.1996.0062

Grieneisen VA, Xu J, Marée AFM et al (2007) Auxin transport is sufficient to generate a maximum and gradient guiding root growth. Nature 449:1008–1013. https://doi.org/10.1038/nature06215

Hamant O, Heisler MG, Jönsson H et al (2008) Developmental patterning by mechanical signals in Arabidopsis. Science 322:1650–1655. https://doi.org/10.1126/science.1165594

Hay A, Barkoulas M, Tsiantis M (2006) ASYMMETRIC LEAVES1 and auxin activities converge to repress BREVIPEDICELLUS expression and promote leaf development in Arabidopsis. Development 133:3955–3961. https://doi.org/10.1242/dev.02545

Heisler MG, Hamant O, Krupinski P et al (2010) Alignment between PIN1 polarity and microtubule orientation in the shoot apical meristem reveals a tight coupling between morphogenesis and auxin transport. PLoS Biol 8:e1000516. https://doi.org/10.1371/journal.pbio.1000516

Hejnowicz Z, Borowska-Wykręt D (2005) Buckling of inner cell wall layers after manipulations to reduce tensile stress: observations and interpretations for stress transmission. Planta 220:465–473. https://doi.org/10.1007/s00425-004-1353-z

Hejnowicz Z, Rusin A, Rusin T (2000) Tensile tissue stress affects the orientation of cortical microtubules in the epidermis of sunflower hypocotyl. J Plant Growth Regul 19:31–44. https://doi.org/10.1007/s003440000005

Hernandez L, Green P (1993) Transductions for the expression of structural pattern: analysis in sunflower. Plant Cell 5:1725–1738. https://doi.org/10.1105/tpc.5.12.1725

Hisanaga T, Kawade K, Tsukaya H (2015) Compensation: a key to clarifying the organ-level regulation of lateral organ size in plants. J Exp Bot 66:1055–1063. https://doi.org/10.1093/jxb/erv028

Hofmeister W (1863) Zusatze und berichtigungen zu den 1851 veröffentlichen untersuchungen der entwicklung höherer kryptogamen. Jahrbucher für Wissenschaft und Botanik 3:259–293

Hofmeister W (1868) *Allgemeine Morphologie der Gewächse*. W. Engelmann

Hofhuis H, Moulton D, Lessinnes T et al (2016) Morphomechanical innovation drives explosive seed dispersal. Cell 166:222–233. https://doi.org/10.1016/j.cell.2016.05.002

Jönsson H, Heisler MG, Shapiro BE et al (2006) An auxin-driven polarized transport model for phyllotaxis. Proc Natl Acad Sci USA 103:1633–1638. https://doi.org/10.1073/pnas.0509839103

Kazama T, Ichihashi Y, Murata S, Tsukaya H (2010) The mechanism of cell cycle arrest front progression explained by a KLUH/CYP78A5-dependent mobile growth factor in developing leaves of arabidopsis thaliana. Plant Cell Physiol 51:1046–1054. https://doi.org/10.1093/pcp/pcq051

Kennaway R, Coen E, Green A, Bangham A (2011) Generation of diverse biological forms through combinatorial interactions between tissue polarity and growth. PLoS Comput Biol 7:e1002071. https://doi.org/10.1371/journal.pcbi.1002071

Kierzkowski D, Nakayama N, Routier-Kierzkowska A-L et al (2012) Elastic domains regulate growth and organogenesis in the plant shoot apical meristem. Science 335(80):1096. https://doi.org/10.1126/science.1213100

Kicheva A, Pantazis P, Bollenbach T, Kalaidzidis Y, Bittig T, Jülicher F, González-Gaitán M (2007) Kinetics of morphogen gradient formation. Science 315:521–525. https://doi.org/10.1126/science.1135774

Koenig D, Bayer E, Kang J, Kuhlemeier C, Sinha N (2009) Auxin patterns *Solanum lycopersicum* leaf morphogenesis. Development 136:2997–3006. https://doi.org/10.1242/dev.033811

Kondo S, Miura T (2010) Reaction-diffusion model as a framework for understanding biological pattern formation. Science 329(80):1616–20. https://doi.org/10.1126/science.1179047

Kramer EM (2009) Auxin-regulated cell polarity: an inside job? Trends Plant Sci 14:242–247. https://doi.org/10.1016/j.tplants.2009.02.005

Kuchen EE, Fox S, de Reuille PB et al (2012) Generation of leaf shape through early patterns of growth and tissue polarity. Science 335:1092–1096. https://doi.org/10.1126/science.1214678

Kwiatkowska D (2004) Structural integration at the shoot apical meristem: models, measurements, and experiments. Am J Bot 91:1277–1293. https://doi.org/10.3732/ajb.91.9.1277

Laguna MF, Bohn S, Jagla EA (2008) The role of elastic stresses on leaf venation morphogenesis. PLoS Comput Biol. https://doi.org/10.1371/journal.pcbi.1000055

Landrein B, Lathe R, Bringmann M et al (2013) Impaired cellulose synthase guidance leads to stem torsion and twists phyllotactic patterns in arabidopsis. Curr Biol 23:895–900. https://doi.org/10.1016/j.cub.2013.04.013

Laskowski M, Grieneisen VA, Hofhuis H et al (2008) Root system architecture from coupling cell shape to auxin transport. PLoS Biol 6:2721–2735. https://doi.org/10.1371/journal.pbio.0060307

Liang H, Mahadevan L (2009) The shape of a long leaf. Proc Natl Acad Sci USA 106:22049–22054. https://doi.org/10.1073/pnas.0911954106

Liang H, Mahadevan L (2011) Growth, geometry, and mechanics of a blooming lily. Proc Natl Acad Sci USA 108:5516–5521. https://doi.org/10.1073/pnas.1007808108

Lockhart JA (1965) An analysis of irreversible plant cell elongation. J Theor Biol 8:264–275. https://doi.org/10.1016/0022-5193(65)90077-9

Mähönen AP, Ten Tusscher K, Siligato R et al (2014) PLETHORA gradient formation mechanism separates auxin responses. Nature 515:125–129. https://doi.org/10.1038/nature13663

Marcos D, Berleth T (2014) Dynamic auxin transport patterns preceding vein formation revealed by live-imaging of *Arabidopsis* leaf primordia. Front Plant Sci 5:235. https://doi.org/10.3389/fpls.2014.00235

Meinhardt H (2009) The algorithmic beauty of sea shells. Springer Science & Business Media. https://doi.org/10.1007/978-3-662-03617-4

Meinhardt H, Koch A-J, Bernasconi G (1998) Models of pattern formation applied to plant development. In: Barabe D, Jean RV (eds) Symmetry in plants. World Scientific, Singapore, pp 723–758. https://doi.org/10.1142/9789814261074_0027

Merks RMH, Guravage M, Inze D, Beemster GTS (2011) VirtualLeaf: an open-source framework for cell-based modeling of plant tissue growth and development. Plant Physiol 155:656–666. https://doi.org/10.1104/pp.110.167619

Mitchison GJ (1980) The dynamics of auxin transport. Proc R Soc London Ser B Biol Sci 209:489–511. https://doi.org/10.1098/rspb.1980.0109

Mitchison GJ (1977) Phyllotaxis and the fibonacci series. Science 196:270–275. https://doi.org/10.1126/science.196.4287.270

Mitchison GJ, Hanke DE, Sheldrake AR (1981) The polar transport of auxin and vein patterns in plants [and discussion]. Philos Trans R Soc B Biol Sci 295:461–471. https://doi.org/10.1098/rstb.1981.0154

Mosca G, Sapala A, Strauss S et al (2017) On the micro-indentation of plant cells in a tissue context. Phys Biol 14:15003. https://doi.org/10.1088/1478-3975/aa5698

Nolte T, Schopfer P (1997) Viscoelastic versus plastic cell wall extensibility in growing seedling organs: a contribution to avoid some misconceptions. J Exp Bot 48:2103–2107. https://doi.org/10.1093/jxb/48.12.2103

O'Connor DL, Runions A, Sluis A et al (2014) A division in PIN-Mediated auxin patterning during organ initiation in grasses. PLoS Comput Biol 10:21–24. https://doi.org/10.1371/journal.pcbi.1003447

Owens A, Cieslak M, Hart J et al (2016) Modeling dense inflorescences. ACM Trans Graph 35:1–14. https://doi.org/10.1145/2897824.2925982

Peters WS, Tomos AD (1996) The history of tissue tension. Ann Bot 77:657–65. https://doi.org/10.1006/anbo.1996.0082

Prusinkiewicz P, Crawford S, Smith RS et al (2009) Control of bud activation by an auxin transport switch. Proc Natl Acad Sci USA 106:17431–17436. https://doi.org/10.1073/pnas.0906696106

Prusinkiewicz P, Runions A (2012) Computational models of plant development and form. New Phytol 193:549–569. https://doi.org/10.1111/j.1469-8137.2011.04009.x

Reinhardt D, Pesce E-R, Stieger P et al (2003) Regulation of phyllotaxis by polar auxin transport. Nature 426:255–260. https://doi.org/10.1038/nature02081

Rodkaew Y, Chongstitvatana P, Siripant S, Lursinsap C (2003) Particle systems for plant modeling. In: B-G Hu, M Jaeger (eds) Plant growth modeling and applications. Proceedings of PMA03, Tsinghua University Press and Springer, Beijing, pp 210–217

Roeder AH (2012) When and where plant cells divide: a perspective from computational modeling. Curr Opin Plant Biol 15:638–644. https://doi.org/10.1016/j.pbi.2012.08.002

Rojas ER, Hotton S, Dumais J (2011) Chemically mediated mechanical expansion of the pollen tube cell wall. Biophys J 101:1844–1853. https://doi.org/10.1016/j.bpj.2011.08.016

Rolland-Lagan A-G, Remmler L, Girard-Bock C (2014) Quantifying shape changes and tissue deformation in leaf development. Plant Physiol 165:496–505. https://doi.org/10.1104/pp.113.231258

Rolland-Lagan AG, Prusinkiewicz P (2005) Reviewing models of auxin canalization in the context of leaf vein pattern formation in Arabidopsis. Plant J 44:854–865. https://doi.org/10.1111/j.1365-313x.2005.02581.x

Roussel MR, Slingerland MJ (2012) A biochemically semi-detailed model of auxin-mediated vein formation in plant leaves. BioSystems 109:475–487. https://doi.org/10.1016/j.biosystems.2012.05.010

Routier-Kierzkowska A-L, Smith RS (2012) Measuring the mechanics of morphogenesis. Curr Op Plant Biol 16:25–32. https://doi.org/10.1016/j.pbi.2012.11.002

Runions A, Fuhrer M, Lane B et al (2005) Modeling and visualization of leaf venation patterns. ACM Trans Graph 24:702. https://doi.org/10.1145/1073204.1073251

Runions A (2014) Computational modeling of leaf development and form (Doctoral dissertation, University of Calgary)

Runions A, Smith RS, Prusinkiewicz P et al (2014) Computational models of auxin-driven development. Auxin Its Role Plant Dev 9783709115:315–357. https://doi.org/10.1007/978-3-7091-1526-8

Runions A, Tsiantis M, Prusinkiewicz P (2017) A common developmental programme can produce diverse leaf shapes. New Phytol. https://doi.org/10.1111/nph.14449

Sachs J (1878) Über die anordnung der zellen in jüngsten pflanzentheilen. Arb bot Inst Wurzburg 2:46

Sachs T (2003) Collective specification of cellular development. BioEssays 25:897–903. https://doi.org/10.1002/bies.10328

Scarpella E, Marcos D, Friml J, Berleth T (2006) Control of leaf vascular patterning by polar auxin transport. Genes Dev 20:1015–1027. https://doi.org/10.1101/gad.1402406

Schopfer P (2006) Biomechanics of plant growth. Am J Bot 93:1415–1425. https://doi.org/10.3732/ajb.93.10.1415

Shapiro BE, Meyerowitz EM, Mjolsness E (2013) Using cellzilla for plant growth simulations at the cellular level. Front Plant Sci 4:408. https://doi.org/10.3389/fpls.2013.00408

Sharon E, Roman B, Swinney HL (2007) Geometrically driven wrinkling observed in free plastic sheets and leaves. Phys Rev E Stat Nonlinear Soft Matter Phys 75:1–7. https://doi.org/10.1103/physreve.75.046211

Smith RS (2007) Simulation models of phyllotaxis and morphogenesis in plants. (Doctoral dissertation, University of Calgary)

Smith RS (2011) Modeling plant morphogenesis and growth. In: New trends in the physics and mechanics of biological systems. Oxford University Press, pp 301–338

Smith RS, Bayer EM (2009) Auxin transport-feedback models of patterning in plants. Plant Cell Environ 32:1258–1271. https://doi.org/10.1111/j.1365-3040.2009.01997.x

Smith RS, Guyomarc'h S, Mandel T et al (2006a) A plausible model of phyllotaxis. Proc Natl Acad Sci USA 103:1301–6. https://doi.org/10.1073/pnas.0510457103

Smith RS, Kuhlemeier C, Prusinkiewicz P (2006b) Inhibition fields for phyllotactic pattern formation: a simulation study. This article is one of a selection of papers published on the Special Theme of Shoot Apical Meristems. Can J Bot 84:1635–1649. https://doi.org/10.1139/b06-133

Stoma S, Lucas M, Chopard J et al (2008) Flux-based transport enhancement as a plausible unifying mechanism for auxin transport in meristem development. PLoS Comput Biol 4:e1000207. https://doi.org/10.1371/journal.pcbi.1000207

Vlad D, Kierzkowski D, Rast MI et al (2014) Leaf shape evolution through duplication, regulatory diversification, and loss of a homeobox gene. Science 343(80):780–3. https://doi.org/10.1126/science.1248384

Vogler H, Draeger C, Weber A, Felekis D et al (2013) The pollen tube: a soft shell with a hard core. Plant J 73:617–627. https://doi.org/10.1111/tpj.12061

Wabnik K, Kleine-Vehn J, Balla J et al (2010) Emergence of tissue polarization from synergy of intracellular and extracellular auxin signaling. Mol Syst Biol 6:447. https://doi.org/10.1038/msb.2010.103

Wartlick O, Kicheva A, González-Gaitán M (2009) Morphogen gradient formation. Cold Spring Harb Perspect Biol 1:1–23. https://doi.org/10.1101/cshperspect.a001255

Weber A, Braybrook S, Huflejt M, Mosca G et al (2015) Measuring the mechanical properties of plant cells by combining micro-indentation with osmotic treatments. J Exp Bot 66:3229–3241. https://doi.org/10.1093/jxb/erv135

Winship LJ, Obermeyer G, Geitmann A, Hepler PK (2010) Under pressure, cell walls set the pace. Trends Plant Sci 15:363–369. https://doi.org/10.1016/j.tplants.2010.04.005

Wolpert L (1969) Positional information and the spatial pattern of cellular differentiation. J Theor Biol 25:1–47. https://doi.org/10.1016/s0022-5193(69)80016-0

Yanagisawa M, Desyatova AS, Belteton SA et al (2015) Patterning mechanisms of cytoskeletal and cell wall systems during leaf trichome morphogenesis. Nat Plants 1:15014. https://doi.org/10.1038/nplants.2015.14

Yoshida S, Barbier de Reuille P, Lane B et al (2014) Genetic control of plant development by overriding a geometric division rule. Dev Cell 29:75–87. https://doi.org/10.1016/j.devcel.2014.02.002

Mechanical Conflicts in Growth Heterogeneity

Nathan Hervieux and Olivier Hamant

Abstract Morphogenesis involves coordinated cell division and cellular growth. Beyond average growth rate and direction, multicellular growth can also be characterized by its variance, i.e., the level of heterogeneity between individual cells or regions. Because final shapes are usually very reproducible, this raises the question of the contribution of local variability in growth in morphogenesis. Here, we focus on the mechanical conflicts that are associated with differential growth and discuss how mechanical forces, emerging from growth heterogeneity, can serve as cues to channel morphogenesis.

Keywords Mechanical signals · Stochasticity · Robustness · Feedback
Quantitative imaging · Proprioception

Introduction

Mechanical conflicts in development shape living organisms: Most biological structures are pre-stressed (also called auto-stressed), meaning that their shapes emerge and are maintained through a balance between tension and compression, as in an inflated balloon or a suspension bridge for instance. It follows that shapes prescribe a pattern of stress; as nicely stated by D'Arcy Wentworth Thompson: "*In short, the form of an object is a diagram of forces*" (Thompson 1917). Tissues thus experience multiple mechanical conflicts, either through shape-derived stress when growth is homogeneous, through growth-derived stress when growth rate and/or growth direction differ between regions, or through a combination of both. Although our understanding of the role of such mechanical signals in morphogenesis is still in its infancy, it finds a new echo today with the development of quantitative imaging coupled with genetics (Box1, see also Coen and Rebocho 2016). Because plant shapes

N. Hervieux · O. Hamant (✉)
Laboratoire Reproduction et Développement des Plantes, Université de Lyon,
ENS de Lyon, UCB Lyon 1, CNRS, INRA, 69342 Lyon, France
e-mail: Olivier.hamant@ens-lyon.fr

A. Geitmann and J. Gril (eds.), *Plant Biomechanics*,
https://doi.org/10.1007/978-3-319-79099-2_9

and associated mechanics mainly depend on the cell wall, plants are ideal systems to investigate the contribution of mechanical heterogeneities and associated mechanical conflicts, in development.

Box 1. Progress in quantitative cell growth assessment of multicellular organisms

- Live imaging protocols are now available for most tissues. Progress in microscopy makes image acquisition faster, while retaining high resolution. This is, for instance, illustrated with the development of spinning disk confocal microscopy and resonant scanner confocal microscopy. Although it is not adaptable to all samples, notable thick and opaque samples, light sheet microscopy is also becoming more and more widespread (e.g., Keller et al. 2008; Maizel et al. 2011).
- Over the past few years, tremendous progress has been achieved regarding micromechanical tools. In particular, microindenters have been used to monitor the stiffness of pollen tube (Zerzour et al. 2009) and tissues (Beauzamy et al. 2015; Routier-Kierzkowska et al. 2012). Micro-rheometry has been developed to measure whole cell stiffness and viscosity (Durand-Smet et al. 2014). Local stiffness measurements have also been possible, thanks to atomic force microscopy (Peaucelle et al. 2011; Milani et al. 2011, 2014; Peaucelle 2014).
- Image analysis has become more and more resolutive and integrative. The growth of individual cells and the tracking of cell lineages through time in 3D is now routinely performed (e.g., Fernandez et al. 2010; Bassel et al. 2014). Note, however, that 3D segmentation is computationally expensive and technically challenging. MorphoGraphX bridges the gap between 3D and 2D quantification by working directly with curved surface images extracted from 3D data (Barbier de Reuille et al. 2015). Briefly, from an image stack, the software detects the edges of the object and then creates a mesh on this surface. Then, fluorescence below the surface is projected onto the surface, providing signal on a surface in 3D. This is perfectly suited when focusing on growth in the epidermis. For example, it was used to analyze leaflet growth (Vlad et al. 2014), to reveal mechanical patterns at tomato vegetative shoot apices (Kierzkowski et al. 2012) or to correlate curvature, and stress pattern, to gene expression (Landrein et al. 2015).
- Modeling: The finite element method is now widespread to represent the different faces of plant cells, in 3D tissues. This still involves a number of technical challenges but has been key to explore the link between growth and cell shape in the hypocotyl (Bassel et al. 2014), to investigate the contribution of wall stiffness and anisotropy in organogenesis at the shoot apical meristem (Sassi et al. 2014; Boudon et al. 2015) and to explore the roles of strain and stress in development (Hamant et al. 2008; Bozorg et al. 2014).

Mechanical Conflicts at the Single Cell Level

Growing plant cells continuously experience and respond to mechanical stress in their walls: turgor pressure puts the cell wall under tension, and, in turn, the cell wall reacts to such tension through biosynthesis and remodeling (Cosgrove 2016). This is the typical case of a pre-stressed system with an envelope under tension and a content under compression, as in a balloon. However, this averaged picture masks some hidden complexity. In particular, the shape of the cell impacts the local pattern of stress in the wall, at the individual cell level: using a pressure vessel analogy, one can easily calculate the stress anisotropy in plant cell walls. If a cell were purely spherical, then tensile stress would be isotropic at any spot at the cell wall surface (Fig. 1a). However, most cells are slightly elongated, as in an ellipsoid, and therefore a maximum of tensile stress appears along the transverse direction of the cell (Fig. 1a). One of the best macroscopic analogies for this is the observation that sausages grilled on a barbecue (thus, with an envelope under tension as in a pressure vessel) break open lengthwise, because maximal tensile stress is transverse (Beer and Johnston 1992). Depending on the shape of the cell, the stress pattern will be more or less uniform and certain regions of the cell will experience more or less stress, or similar stress magnitude, but in different directions. For instance, in the jigsaw puzzle shaped pavement cell, the pattern of stress in the cell wall can be quite complex (Sampathkumar et al. 2014, Fig. 1a).

This local conflict in stress direction at the cell wall can impact the biochemistry of the cell wall. Cellulose microfibrils are considered the main load-bearing components and act as a corset around the cell channeling growth direction (Castle 1937). Each cellulose microfibril is formed from the “bundling” and crystallization of dozens of (1, 4)-linked β-d-glucan chains, each made by a plant cellulose synthase protein (CESA, Cosgrove 2005). Microtubules were first discovered in plants (Ledbetter and Porter 1963) and cortical microtubules (CMTs) were soon found to be aligned at the cell cortex parallel to cellulose fibrils in the wall (Baskin 2005). Their role in wall synthesis was predicted even before they were formally identified, as disruption of microtubules with the spindle depolymerizing drug Colchicine led to isotropic growth (Green 1962). Further work using mutants with disorganized CMTs (Buschmann and Lloyd 2008) or other microtubule depolymerizing drugs like oryzalin (Corson et al. 2009) further confirmed the role of CMTs in controlling growth anisotropy. These effects were formally related to defects in cellulose microfibril orientation (Baskin 2005). Those observations led to the proposal that CMTs guide the trajectory of cellulose synthase complexes (CSCs) as they synthesize microfibrils in the cell wall (Heath 1974). Using tagged CESA and live imaging using spinning disk confocal microscopy, Paredez et al. showed that CSC trajectories in the plasma membrane are parallel to CMTs (Paredez et al. 2006). Note that the CMT–cellulose colinearity is not universal. For instance, transverse cellulose microfibrils deposition could occur in the absence of transverse microtubules in the thermosensitive *mor1* mutant (Himmelspach et al. 2003). While this suggests that the colinearity between cellulose deposition and microtubule alignment is not strict, recent evidence supports a dominant role of microtubules in guiding cellulose synthase. In particular, the functional link between CSCs and CMTs was consolidated with the identification of the cellulose synthase interactive protein 1 (CSI1), which acts as a physical linker between CSCs and CMTs in *Arabidopsis* (Bringmann et al. 2012). Models suggest that CSI1 might actively contribute to the guidance of CSCs, by sliding or actively

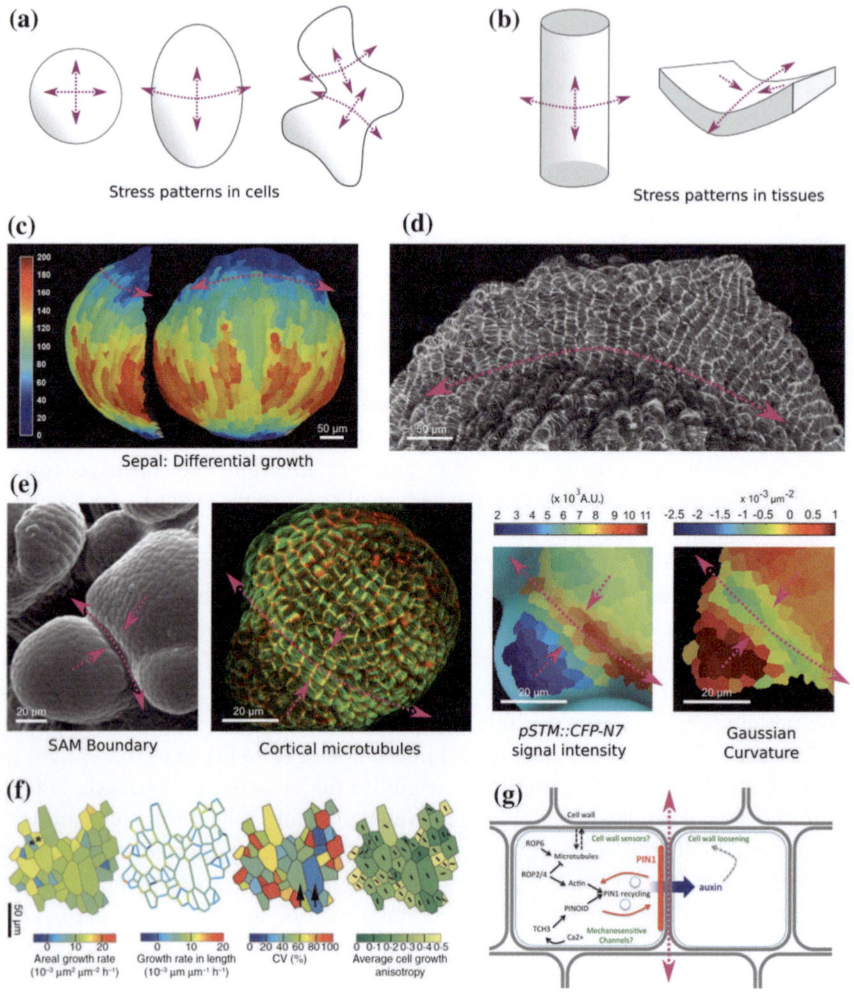

migrating along microtubules during cellulose synthesis by CSCs. A rotary model also suggests that CSI1 may be important for the activity of CESA complexes: in this model, several CSI1 proteins would be attached to a CSC but only one or two would be attached to the neighboring microtubule. The synthesis of cellulose microfibrils would push the complex forward and would break the interaction between microtubules and CSI1 protein allowing in turn the neighboring CSI1 present on the CESA complexes to interact with cortical microtubules (Bringmann et al. 2012).

For a long time, it was proposed that mechanical stress can orient the fibrillar structure of the cell wall (Castle 1937). The relationship between microtubules and strain was also analyzed (Green 1962) and it was proposed that stress can control

◂**Fig. 1** Multiscale mechanical conflicts in growth heterogeneity. **a** Pattern of stress (pink) for various cell shapes (pressure vessel analogy: wall under tension): spherical (e.g., protoplast), ellipsoidal (e.g., epidermal cell), and jigsaw puzzle shaped cells (e.g., pavement cell). **b** Pattern of stress (pink) for various tissue shapes (pressure vessel analogy: assuming epidermis under tension): cylindrical (e.g., stem) and folded (e.g., meristem boundary). **c** Predicted stress pattern (pink) emerging from growth heterogeneities between regions (e.g., Arabidopsis sepal). Growth rate is colored as % of area extension. Adapted with authorization from (Hervieux et al. 2016). **d** Confocal image of the tip of an Arabidopsis sepal, after a period of microtubule depolymerization with oryzalin and isotropic growth, and imaged when microtubules repolymerized. Note the supracellular alignment of CMTs at the tip, matching the stress pattern (pink) emerging from differential growth between the tip and center of the sepal at that stage. Adapted with authorization from (Hervieux et al. 2016). **e** Scanning electron micrograph of the Arabidopsis shoot apical meristem. In all panels, the meristem is at the center, a young organ is emerging near the bottom left corner and a saddle-shaped boundary is formed between meristem and organ, where tension (pink) is highly anisotropic. Confocal imaging of CMTs at the shoot apex shows a supracellular alignment of CMTs at the boundary matching the shape- and growth-derived maximal stress direction in that domain. Using MorphoGraphX on confocal images of *pSTM::CFP-N7* expressing meristems, the promoter activity of *STM* was quantified, and shown to be increased at the boundary. Adapted with authorization from (Laufs et al. 1998; Hamant et al. 2008; Landrein et al. 2015). **f** Growth heterogeneity in triple Cyclin D3 mutant leaves, as revealed using the replica method (SEM images of epoxy resin replicas of living tissues): relative growth rate in length refers to straight line segments between two vertices joined by the same anticlinal cell wall; coefficient of variation (CV) refers to the growth rates in length of segments belonging to individual cells. Note the high level of heterogeneity in wall extension, growth rate, and growth anisotropy. Asterisks: adjacent cells of similar size and distinct growth rates in area; arrows: adjacent cells of similar size but with different values of the coefficient of variation. Adapted with authorization from (Elsner et al. 2012). **g** Predicted stress pattern (pink) emerging from growth heterogeneities between adjacent cells. Differential auxin concentrations between adjacent cells would contribute to growth heterogeneities by differentially loosening cell walls, and in turn, mechanical conflicts between adjacent cells would increase membrane tension, inhibiting endocytosis of auxin efflux carrier PIN1, thus maintaining PIN1 polarities and differential auxin concentrations between adjacent cells. Adapted with authorization from (Hamant et al. 2010)

the orientation of the microtubule arrays (Green and King 1966; Williamson 1990) and in turn control the deposition of cellulose microfibrils. This theory was tested using protoplasts (i.e., wall-less plant cells): a reorientation of microtubules parallel to the centrifugal force vector was observed after centrifuging protoplasts and lead to growth perpendicular to this axis in recovered protoplast, corresponding to the stage when the cell wall is re-synthesized (Wymer et al. 1996). Similarly, the stretching of protoplast could impact the orientation of microtubule arrays (Fisher and Cyr 2000). More recently, using modeling and micromechanical perturbations, cortical microtubules were found to align along maximal tensile stress in hypocotyls (Hejnowicz et al. 2000), meristematic cells (Hamant et al. 2008), sepals (Hervieux et al. 2016), and growing pavement cells (Sampathkumar et al. 2014). Although such reports are more rare in underground tissues, CMTs have been found to reorient after wounding in roots (Geitmann et al. 1997).

The cortical microtubule alignment with maximal tension has several interesting implications at the individual cell level. In particular, since Sachs (1878) and Errera (1886), it was proposed that cells divide along the shortest path, i.e., generate the shortest possible division plane, albeit following the constraints of surface tension, as in soap bubbles. This deterministic rule was recently extended to a probabilistic rule, since several minima may exist for a given cell morphology (Besson and Dumais 2011; Thompson 1917). Interestingly, if a cell is elongated, then maximal tension also prescribes the shortest cell division plane direction, this time in a deterministic way (Louveaux et al. 2016). The impact of mechanical perturbations on cell division plane orientation in various tissues further confirms the role of mechanical stress in cell division plane orientation (Lintilhac and Vesecky 1984; Yeoman and Brown 1971; Louveaux et al. 2016). As cell division plane orientation largely relies on CMT orientation, it is thus likely that, to divide, cells sense their own geometry at least partly through the perception of maximal tension in their walls.

Mechanical Conflicts Between Internal Layers and the Epidermis

As in single cell, patterns of mechanical stress are also highly heterogeneous in a multicellular context. To what extent does this pattern depend on tissue shape?

For a long time, the outer cell layers of plant tissues have been thought to mechanically restrict the growth of inner tissues. This puts the outer cell layers under tension and the inner tissues under compression (Hofmeister 1859; Peters and Tomos 1996). This can be shown experimentally by peeling the outer teguments of etiolated sunflower hypocotyls. After peeling, the outer tegument retracts instantly whereas the inner tissues expand (Kutschera and Niklas 2007). This simple experiment supports the notion that the epidermis in plants is under tension and surrounds internal tissues that are under compression; it suggests that the epidermis restricts and thus channels growth. This is also consistent with the observation that epidermal cells usually have thicker outer cell walls than cells of inner tissues. Interestingly, an implication of this mechanical balance is that plant tissues should behave like pressurized shells rather than "cell agglomerates". In other words, plant tissues may behave like a "big plant cell" (Fig. 1b, Hofmeister 1859; Kutschera and Niklas 2007). The elastic nature of plant tissues can be revealed through compression with a micro-vice (Hamant et al. 2008) or a micro-indenter (Beauzamy et al. 2015), which further supports the pressurized shell hypothesis. Therefore, as in individual cells, the stress pattern (emerging from the mechanical conflict between internal and external cell layers) should become highly dependent on the shape of the tissue.

Using CMTs as readout of maximal tensile stress in tissues, this question was explored at the shoot apical meristem. In the apical dome, approximated as a spherical dome where tensile stress is supposed to be isotropic, CMTs do not display preferential orientations. In contrast, the geometry of the boundary between the meristem and a young primordium prescribes strongly anisotropic tensile stresses. CMT arrays in that domain are highly anisotropic and aligned along the axis of the boundary, matching maximal tensile stress direction (Fig. 1e). The correlation between maximal tension and CMT orientation was further consolidated with local perturbations of the stress pattern either via compression or ablation (Hamant et al.

2008). Beyond the shoot apical meristem, the alignment of CMTs with maximal stress was also observed in the epidermis of cotyledons (Sampathkumar et al. 2014) and sepals (Fig. 1c, d; Hervieux et al. 2016).

Note that while microtubules and cellulose microfibrils are classically pictured as the main drivers of growth anisotropy (thus closing the feedback loop between shape-derived stress and CMT response to stress), mechanical tests suggest that CMTs react to stress, and thus may not be the initial trigger of growth anisotropy. In particular, a mechanical polarity of cell walls was measured with AFM on hypocotyl epidermis, before anisotropic cell expansion and this was associated with pectin demethylesterification instead (Peaucelle et al. 2015). Yet, the role of CMTs and cellulose orientation would play a dominant role later on, as the cell starts to expand in a preferential direction.

Differential growth can be a strong source of mechanical stress between internal tissue layers. For instance, lateral roots literally push on the root cortex to make their way out from the pericycle. This involves a complete remodeling of the root cortex mechanics, notably through a regulation of aquaporin expression (Péret et al. 2012) and the perception of stress in the endodermis from the growing pericycle (Vermeer et al. 2014). The growth of the albumen is also pushing on one of the outer layers of immature seeds, and this has been related to wall thickening, microtubule bundling and alignment, as well as *ELA1* gene induction in that stretched layer (Creff et al. 2015).

Differential Growth at the Epidermis: Mechanical Conflicts Coordinate Cell Behavior

If the mechanical conflict between internal and external cell layers is essential to understand morphogenesis, notably through the pressure vessel analogy, differential growth within the epidermis can also affect the local stress pattern. In particular, the contiguity between a slow-growing region and a fast-growing region generates a mechanical conflict in the form of a directional mechanical stress at the interface between those two zones. Several studies, mainly arising from research in Drosophila, point at a key role of epithelial growth-derived stress in morphogenesis.

Growth-derived stresses have been proposed to play a central role in the control of organ size and shape. In particular, the accumulation of mechanical stresses generated by differential growth was proposed to trigger growth arrest at the level of the whole organ (Shraiman 2005). While this proposal is still debated, it may apply to the wing imaginal disk in Drosophila, where faster growth in the outer part of the disk compresses internal cells and would provoke an arrest in cell division (Aegerter-Wilmsen et al. 2007). Mechanical conflicts also polarize cell divisions in that tissue (Legoff et al. 2013) and the planar polarity of the wing blade has been related to the mechanical conflict generated by the contraction of the hinge of the developing wing (Aigouy et al. 2010). Mechanical conflicts were also shown to display a cadastral role: straight tension lines between two wing disk subdomains in Drosophila prevent cell movement from one domain to the other (Landsberg et al. 2009; Aliee et al. 2012). More recently, a mechanical conflict emerging from differential growth was shown to increase the activity of the Hippo signaling pathway leading to a decrease in growth rate (Pan et al. 2016).

In plants, a mechanical conflict was identified between the fast-growing center and slow-growing tip of the sepal during its early development; the resulting transverse tensile stress at the tip channels sepal shape, through an impact of stress on microtubule and cellulose deposition (Fig. 1c, d). Such mechanical feedback may thus act as an organ shape-sensing mechanism: the perception of tension at the sepal tip triggers a directional and multicellular cell wall reinforcement, which, if maintained and amplified, could lead to growth arrest from the tip to the bottom of the sepal (Hervieux et al. 2016). Note that the alignment of CMTs at the boundary between shoot meristem and organs also falls into that category: CMTs not only align along shape-derived stress, they also align along growth-derived stress as the emerging organ is growing much faster than the boundary and the meristem (Fig. 1e, Burian et al. 2013; Hamant et al. 2008). The synergy between growth-derived and shape-derived stress may actually explain why the CMT alignment is so strong at the meristem boundary and tissue folding so stereotypical in that tissue.

Beyond CMTs and cellulose, mechanical conflicts at the epidermis likely contribute to the robustness of gene expression patterns too. In the shoot apical meristem, the promoter activity of the homeobox gene *SHOOT MERISTEMLESS (STM)* is quantitatively correlated to the curvature in the saddle-shaped boundary domain of the shoot apical meristem in *Arabidopsis* (Fig. 1e, Landrein et al. 2015). As this boundary reflects the presence of a highly anisotropic mechanical stress prescribed by tissues shape and differential growth (Hamant et al. 2008; Burian et al. 2013), *STM* expression was proposed to be promoted by mechanical stress in that domain (Fig. 1e). As STM activity rather promotes a reduction in growth rate, this may be viewed as another feedback loop in which differential growth triggers a mechanical conflict that in turn reduces growth (Landrein et al. 2015). Other boundary genes, like *CUC3*, but not *CUC1* or *CUC2*, are also induced by such mechanical stress in the meristem boundary (Landrein et al. 2015; Fal et al. 2015). This result echoes other findings in animals, and notably in Drosophila (Farge 2003) or zebrafish (Brunet et al. 2013) where mesoderm patterning genes are induced by growth-induced mechanical deformations.

Growth Heterogeneity: How to Generate Reproducible Shapes Despite Developmental Noise?

While mechanical conflicts emerge at the epidermis between fast- and slow-growing regions, careful quantification of growth also highlights the surprising level of local growth heterogeneity, i.e., between adjacent individual cells (Fig. 1f, Elsner et al. 2012). For instance, at the shoot apical meristem, a clear correlation could be confirmed between shape-derived stress direction and CMT orientation at the crease-shaped boundary of the meristem, but the correlation was weak in the peripheral zone of the shoot apical meristem, even though the somewhat cylindrical shape of this domain would prescribe a bias in the stress direction. This local discrepancy could be resolved when taking into account an additional factor: the presence of growth heterogeneities may add noise to the stress pattern and thus to CMT behavior (Burian et al. 2013). In fact, the coefficient of variability of growth rate between adjacent cells was measured in that tissue and could reach as much as 600% (Uyttewaal et al. 2012)! If mechanical conflicts arise in the presence of differential growth, such level of growth heterogeneity must also generate stress. And if so, how does the tissue cope (or use) such local heterogeneities?

Growth-derived stress between adjacent cells may help coordinate the local behavior of key cell effectors. Based on careful imaging of the auxin efflux carrier PIN1 at the shoot apical meristem, PIN1 had been proposed to accumulate on the membrane facing the cell containing the highest concentration of auxin, in a positive feedback loop (Heisler et al. 2005; Smith et al. 2006). This observation raises the question of the mechanism allowing a cell to "measure" the auxin content in an adjacent cell. Several cues may be involved, including mechanical stress. Indeed, assuming that the accumulation of auxin would locally loosen the cell wall, knowing that membrane tension prevents endocytosis and that PIN1 polarity relies on such vesicle trafficking, PIN1 was found to accumulate on membranes exhibiting the highest level of tension, through modeling, correlative imaging, and micromechanical perturbations (Fig. 1g, Heisler et al. 2010; Hamant et al. 2010; Nakayama et al. 2012). Although the exact contribution of mechanical stress in PIN1 polarity remains to be investigated, this demonstrates that growth-derived mechanical conflicts between adjacent cells can affect cell behavior and indirectly contribute to large-scale patterns, like phyllotaxis.

Conversely, growth heterogeneity may negatively impact organ shape reproducibility: local differences in growth rate between adjacent cells may distort the tissue locally. How does the organ cope with such heterogeneity (and associated variability)? An accumulation of reactive oxygen species (ROS) in *ftsh4* mutant was shown to increase variability in sepal size and shape (Hong et al. 2016). Surprisingly, this increased shape variability is associated with a decrease of growth variability at the cell level. It was thus proposed that reproducible organ shape in the wild-type results from spatiotemporal averaging of cellular variability, i.e., compensation between fast- and slow-growing regions of the sepal. More locally, fast growing trichomes generate a local pattern of stress, that is thought to channel the radial expansion of socket cells in leaves and sepals; in turn, the microtubule response to stress prevents the distorsion of the whole tissues when trichome emerge ("mechanical shielding", Hervieux et al. 2017). It seems therefore that mechanical conflicts are perceived before they are filtered out, to generate reproducible shapes in the end.

How is growth heterogeneity actively maintained or monitored? While this question remains largely unexplored, there is evidence that mechanical signals are involved. Let us revisit the mechanical conflicts arising from local growth heterogeneities between adjacent cells at the shoot apical meristem. Differences in cell growth rates, and associated mechanical conflicts, should lead to local microtubule reorganizations. Interestingly, using modeling and experiments, it was shown that the microtubule response to such growth-derived stress does not promote growth homogeneity, but instead fuels growth heterogeneity. To reach that conclusion, it was necessary to obtain plants with different levels of response to microtubule response to mechanical stress. The katanin mutant exhibits reduced microtubule severing, and thus reduced microtubule self-organization dynamics also implying delayed microtubule response to mechanical perturbations, when compared to the wild type. When analyzing growth in the katanin mutant, the level of growth heterogeneity was also found to be reduced, consistent with a model predicting that a strong microtubule response to stress, through the impact on the mechanical anisotropy of cell walls, would actually fuel growth heterogeneity (Uyttewaal et al. 2012). In other words, this work suggests that wild-type cells over-react to mechanical stress, and by doing so, maintain growth heterogeneity. The maintenance of growth heterogeneity through such mechanical feedback (in synergy with other pathways) may prime cells for differential growth and thus potentiate organogenesis: if, instead, growth was actively maintained homogeneous, the tissue would have a harder time generating a local differential growth event, the initial step for organ emergence (Uyttewaal et al. 2012). In that scenario, the mechanical feedback on CMTs would not act as buffering mechanism, but would instead increase noise at the tissue level in order to produce reproducible organs or achieve proper morphogenesis at the organism level.

Conclusion

To give rise to reproducible shapes, while undergoing fluctuations in most processes relating to physiology and development, one may propose, for sake of argument, that most if not all parameters are tightly regulated. However, there is accumulating evidence that stochasticity is widespread in cellular and molecular mechanisms. In the most extreme scenario, in which reproducible shapes are only the fruit of tight regulations, this would suggest that randomness corresponds to developmental white noise and is unimportant. This obviously raises a paradox: how can regulation be strict and at the same time ignores small fluctuations? The idea that a "program" controls developmental robustness is obviously too one sided and it is more likely that developmental robustness emerges both from the gene network (its topology and regulation), but also on multiple feedback, some of them arising from growth-derived mechanical conflicts.

While this "stochastic view" of development is more and more consensual, it only starts to be addressed experimentally. For instance, dorsal closure in Drosophila embryogenesis is not the result of a well-choreographed and synchronized cell contraction, but instead is the result of stochastic contractions (Martin et al. 2009; Pouille et al. 2009). Such stochasticity is also observed at all scales in plants. Interestingly, plastic plant phenotypes that vary between environments or variable phenotypes that vary stochastically within an environment might be advantageous in some cases. In other words, stochasticity occurs during development and can have a significant impact on promoting phenotypic diversity and survival in some environmental conditions at the scale of the individual (Abley et al. 2016).

As shown above, thanks to rapid progress in live imaging and growth quantification methods, variability in the growth rate of individual cells was revealed in most developing plant tissues (Box 1, Fig. 1f, Elsner et al. 2012; Uyttewaal et al. 2012; Hong et al. 2016). In addition, no robust correlation between growth rate and cell size, nuclear size, or anisotropy has been established, suggesting that plant cell growth is also largely stochastic. Cell size variability in the sepal epidermis was actually related to the stochasticity in the time at which cells stop dividing and enter the specialized endoreduplication cell cycle (Roeder et al. 2010). As mechanical conflicts are associated with such growth and shape stochasticity, this opens exciting prospects in the biophysics and development field (Hong et al. 2018).

To conclude, mechanical conflicts can be interpreted by cells and tissues as proprioceptive cues, amenable to inform on their current growth and form status (Hamant and Moulia 2016). The presence of growth fluctuations in space and time is widespread and likely play an essential role in development, notably through the generation and perception of mechanical signals. Recent progress in modeling, live imaging, and image analysis now allow us to address that question, not only in relation to differential growth but also including the stochasticity of growth.

References

Abley K, Locke JCW, Leyser HMO (2016) Developmental mechanisms underlying variable, invariant and plastic phenotypes. Ann Bot 117:733–748

Aegerter-Wilmsen T, Aegerter CM, Hafen E, Basler K (2007) Model for the regulation of size in the wing imaginal disc of Drosophila. Mech Dev 124:318–326

Aigouy B, Farhadifar R, Staple DB, Sagner A, Röper J-C, Jülicher F, Eaton S (2010) Cell flow reorients the axis of planar polarity in the wing epithelium of Drosophila. Cell 142:773–786

Aliee M, Röper J-C, Landsberg KP, Pentzold C, Widmann TJ, Jülicher F, Dahmann C (2012) Physical mechanisms shaping the Drosophila dorsoventral compartment boundary. Curr Biol CB 22:967–976

Barbier de Reuille P, Routier-Kierzkowska A-L, Kierzkowski D, Bassel GW, Schüpbach T, Tauriello G, Bajpai N, Strauss S, Weber A, Kiss A et al (2015) MorphoGraphX: a platform for quantifying morphogenesis in 4D. eLife 4:05864

Baskin TI (2005) Anisotropic expansion of the plant cell wall. Annu Rev Cell Dev Biol 21:203–222

Bassel GW, Stamm P, Mosca G, Barbier de Reuille P, Gibbs DJ, Winter R, Janka A, Holdsworth MJ, Smith RS (2014) Mechanical constraints imposed by 3D cellular geometry and arrangement modulate growth patterns in the Arabidopsis embryo. Proc Natl Acad Sci USA 111:8685–8690

Beauzamy L, Louveaux M, Hamant O, Boudaoud A (2015) Mechanically, the shoot apical meristem of arabidopsis behaves like a shell inflated by a pressure of about 1 MPa. Front Plant Sci 6:1038

Beer FP, Johnston ER (1992) Mechanics of material. McGraw-Hill

Besson S, Dumais J (2011) Universal rule for the symmetric division of plant cells. Proc Natl Acad Sci US A 108:6294–6299

Boudon F, Chopard J, Ali O, Gilles B, Hamant O, Boudaoud A, Traas J, Godin C (2015) A computational framework for 3D mechanical modeling of plant morphogenesis with cellular resolution. PLoS Comput, Biol, p 11

Bozorg B, Krupinski P, Jönsson H (2014) Stress and strain provide positional and directional cues in development. PLoS Comput Biol 10:e1003410

Bringmann M, Landrein B, Schudoma C, Hamant O, Hauser M-T, Persson S (2012) Cracking the elusive alignment hypothesis: the microtubule-cellulose synthase nexus unraveled. Trends Plant Sci 17:666–674

Brunet T, Bouclet A, Ahmadi P, Mitrossilis D, Driquez B, Brunet A-C, Henry L, Serman F, Béalle G, Ménager C et al (2013) Evolutionary conservation of early mesoderm specification by mechanotransduction in Bilateria. Nat Commun 4:2821

Burian A, Ludynia M, Uyttewaal M, Traas J, Boudaoud A, Hamant O, Kwiatkowska D (2013) A correlative microscopy approach relates microtubule behaviour, local organ geometry, and cell growth at the Arabidopsis shoot apical meristem. J Exp Bot 64:5753–5767

Buschmann H, Lloyd CW (2008) Arabidopsis mutants and the network of microtubule-associated functions. Mol Plant 1:888–898

Castle ES (1937) Membrane tension and orientation of structure in the plant cell wall. J Cell Comp Physiol 10:113–121

Coen E, Rebocho AB (2016) Resolving conflicts: modeling genetic control of plant morphogenesis. Dev Cell 38:579–583

Corson F, Hamant O, Bohn S, Traas J, Boudaoud A, Couder Y (2009) Turning a plant tissue into a living cell froth through isotropic growth. Proc Natl Acad Sci USA 106:8453–8458

Cosgrove DJ (2005) Growth of the plant cell wall. Nat Rev Mol Cell Biol 6:850–861

Cosgrove DJ (2016) Plant cell wall extensibility: connecting plant cell growth with cell wall structure, mechanics, and the action of wall-modifying enzymes. J Exp Bot 67:463–476

Creff A, Brocard L, Ingram G (2015) A mechanically sensitive cell layer regulates the physical properties of the Arabidopsis seed coat. Nat Commun 6:6382

Durand-Smet P, Chastrette N, Guiroy A, Richert A, Berne-Dedieu A, Szecsi J, Boudaoud A, Frachisse J-M, Bendahmane M, Hamant O et al (2014) A comparative mechanical analysis of plant and animal cells reveals convergence across kingdoms. Biophys J 107:2237–2244

Elsner J, Michalski M, Kwiatkowska D (2012) Spatiotemporal variation of leaf epidermal cell growth: a quantitative analysis of Arabidopsis thaliana wild-type and triple cyclinD3 mutant plants. Ann Bot 109:897–910

Errera L (1886) Sur une condition fondamentale d'e´ quilibre des cellules vivantes. C R Hebd Seances Acad Sci 822–824

Fal K, Landrein B, Hamant O (2015) Interplay between miRNA regulation and mechanical stress for CUC gene expression at the shoot apical meristem. Plant Signal, Behav

Farge E (2003) Mechanical induction of Twist in the Drosophila foregut/stomodeal primordium. Curr Biol CB 13:1365–1377

Fernandez R, Das P, Mirabet V, Moscardi E, Traas J, Verdeil J-L, Malandain G, Godin C (2010) Imaging plant growth in 4D: robust tissue reconstruction and lineaging at cell resolution. Nat Methods 7:547–553

Fisher DD, Cyr RJ (2000) Mechanical forces in plant growth and development. Bull Publ Am Soc Gravit Space Biol. 13:67–73

Geitmann A, Hush JM, Overall RL (1997) Inhibition of ethylene biosynthesis does not block microtubule re-orientation in wounded pea roots. Protoplasma 198:135–142

Green PB (1962) Mechanism for plant cellular morphogenesis. Science 138:1404–1405

Green P, King A (1966) A mechanism for the origin of specifically oriented textures in development with special reference to Nitella wall texture. Aust J Biol Sci 421–437

Hamant O, Moulia B (2016) How do plants read their own shapes? New Phytol 212:333–337

Hamant O, Heisler MG, Jonsson H, Krupinski P, Uyttewaal M, Bokov P, Corson F, Sahlin P, Boudaoud A, Meyerowitz EM et al (2008) Developmental patterning by mechanical signals in Arabidopsis. Science 322:1650–1655

Hamant O, Traas J, Boudaoud A (2010) Regulation of shape and patterning in plant development. Curr Opin Genet Dev 20:454–459

Heath IB (1974) A unified hypothesis for the role of membrane bound enzyme complexes and microtubules in plant cell wall synthesis. J Theor Biol 48:445–449

Heisler MG, Ohno C, Das P, Sieber P, Reddy GV, Long JA, Meyerowitz EM (2005) Patterns of auxin transport and gene expression during primordium development revealed by live imaging of the Arabidopsis inflorescence meristem. Curr Biol CB 15:1899–1911

Heisler MG, Hamant O, Krupinski P, Uyttewaal M, Ohno C, Jonsson H, Traas J, Meyerowitz EM (2010) Alignment between PIN1 polarity and microtubule orientation in the shoot apical meristem reveals a tight coupling between morphogenesis and auxin transport. PLoS Biol 8

Hejnowicz Z, Rusin A, Rusin T (2000) Tensile tissue stress affects the orientation of cortical microtubules in the epidermis of sunflower hypocotyl. J Plant Growth Regul 19:31–44

Hervieux N, Dumond M, Sapala A, Routier-Kierzkowska A-L, Kierzkowski D, Roeder AHK, Smith RS, Boudaoud A, Hamant O (2016) A mechanical feedback restricts sepal growth and shape in arabidopsis. Curr Biol CB

Hervieux N, Tsugawa S, Fruleux A, Dumond M, Routier-Kierzkowska AL, Komatsuzaki T, Boudaoud A, Larkin JC, Smith RS, Li CB, Hamant O (2017) Mechanical shielding of rapidly growing cells buffers growth heterogeneity and contributes to organ shape reproducibility. Curr Biol 27(22):3468–3479.e4

Himmelspach R, Williamson RE, Wasteneys GO (2003) Cellulose microfibril alignment recovers from DCB-induced disruption despite microtubule disorganization. Plant J Cell Mol Biol 36:565–575

Hofmeister W (1859) Über die Beugungen saftreicher Pflanzenteile nach Erschütterung. Ber Verh Ges Wiss Leipz 175–204

Hong L, Dumond M, Tsugawa S, Sapala A, Routier-Kierzkowska A-L, Zhou Y, Chen C, Kiss A, Zhu M, Hamant O et al (2016) Variable cell growth yields reproducible organdevelopment through spatiotemporal averaging. Dev Cell 38:15–32

Hong L, Dumond M, Zhu M, Tsugawa S, Li CB, Boudaoud A, Hamant O, Roeder AHK (2018) Heterogeneity and robustness in plant morphogenesis: from cells to organs. Annu Rev Plant Biol. https://doi.org/10.1146/annurev-arplant-042817-040517

Keller PJ, Schmidt AD, Wittbrodt J, Stelzer EHK (2008) Reconstruction of zebrafish early embryonic development by scanned light sheet microscopy. Science 322:1065–1069

Kierzkowski D, Nakayama N, Routier-Kierzkowska A-L, Weber A, Bayer E, Schorderet M, Reinhardt D, Kuhlemeier C, Smith RS (2012) Elastic domains regulate growth and organogenesis in the plant shoot apical meristem. Science 335:1096–1099

Kutschera U, Niklas KJ (2007) The epidermal-growth-control theory of stem elongation: an old and a new perspective. J Plant Physiol 164:1395–1409

Landrein B, Kiss A, Sassi M, Chauvet A, Das P, Cortizo M, Laufs P, Takeda S, Aida M, Traas J et al (2015) Mechanical stress contributes to the expression of the STM homeobox gene in Arabidopsis shoot meristems. eLife 4:e07811

Landsberg KP, Farhadifar R, Ranft J, Umetsu D, Widmann TJ, Bittig T, Said A, Jülicher F, Dahmann C (2009) Increased cell bond tension governs cell sorting at the Drosophila anteroposterior compartment boundary. Curr Biol CB 19:1950–1955

Laufs P, Grandjean O, Jonak C, Kiêu K, Traas J (1998) Cellular parameters of the shoot apical meristem in Arabidopsis. Plant Cell 10:1375–1390

Ledbetter MC, Porter KR (1963) A "microtubule" in plant cell fine structure. J Cell Biol 19:239–250

Legoff L, Rouault H, Lecuit T (2013) A global pattern of mechanical stress polarizes cell divisions and cell shape in the growing Drosophila wing disc. Dev Camb Engl. 140:4051–4059

Lintilhac PM, Vesecky TB (1984) Stress-induced alignment of division plane in plant tissues grown in vitro. Nature 307:363–364

Louveaux M, Julien J-D, Mirabet V, Boudaoud A, Hamant O (2016) Cell division plane orientation based on tensile stress in Arabidopsis thaliana. Proc Natl Acad Sci USA 113:E4294–E4303

Maizel A, von Wangenheim D, Federici F, Haseloff J, Stelzer EHK (2011) High-resolution live imaging of plant growth in near physiological bright conditions using light sheet fluorescence microscopy. Plant J Cell Mol Biol 68:377–385

Martin AC, Kaschube M, Wieschaus EF (2009) Pulsed contractions of an actin-myosin network drive apical constriction. Nature 457:495–499

Milani P, Gholamirad M, Traas J, Arneodo A, Boudaoud A, Argoul F, Hamant O (2011) In vivo analysis of local wall stiffness at the shoot apical meristem in Arabidopsis using atomic force microscopy. Plant J Cell Mol Biol 67:1116–1123

Milani P, Mirabet V, Cellier C, Rozier F, Hamant O, Das P, Boudaoud A (2014) Matching patterns of gene expression to mechanical stiffness at cell resolution through quantitative tandem epifluorescence and nanoindentation. Plant Physiol 165

Nakayama N, Smith RS, Mandel T, Robinson S, Kimura S, Boudaoud A, Kuhlemeier C (2012) Mechanical regulation of auxin-mediated growth. Curr Biol CB 22:1468–1476

Pan Y, Heemskerk I, Ibar C, Shraiman BI, Irvine KD (2016) Differential growth triggers mechanical feedback that elevates Hippo signaling. Proc Natl Acad Sci USA

Paredez AR, Somerville CR, Ehrhardt DW (2006) Visualization of cellulose synthase demonstrates functional association with microtubules. Science 312:1491–1495

Peaucelle A (2014) AFM-based mapping of the elastic properties of cell walls: at tissue, cellular, and subcellular resolutions. J Vis Exp JoVE

Peaucelle A, Braybrook SA, Le Guillou L, Bron E, Kuhlemeier C, Höfte H (2011) Pectin-induced changes in cell wall mechanics underlie organ initiation in Arabidopsis. Curr Biol CB 21:1720–1726

Peaucelle A, Wightman R, Höfte H (2015) The control of growth symmetry breaking in the Arabidopsis Hypocotyl. Curr Biol CB 25:1746–1752

Péret B, Li G, Zhao J, Band LR, Voß U, Postaire O, Luu D-T, Da Ines O, Casimiro I, Lucas M et al (2012) Auxin regulates aquaporin function to facilitate lateral root emergence. Nat Cell Biol 14:991–998

Peters WS, Tomos AD (1996) The history of tissue tension. Ann Bot 77:657–665

Pouille PA et al (2009) Mechanical signals trigger Myosin II redistribution and mesoderm invagination in Drosophila embryos. Sci Signal 2:ra16

Roeder AHK, Chickarmane V, Cunha A, Obara B, Manjunath BS, Meyerowitz EM (2010) Variability in the control of cell division underlies sepal epidermal patterning in Arabidopsis thaliana. PLoS Biol 8:e1000367

Routier-Kierzkowska A-L, Weber A, Kochova P, Felekis D, Nelson BJ, Kuhlemeier C, Smith RS (2012) Cellular force microscopy for in vivo measurements of plant tissue mechanics. Plant Physiol 158:1514–1522

Sachs J (1878) Über die Anordnung der Zellen in jüngsten Pflanzentheilen. Arb Bot Inst 46–104

Sampathkumar A, Krupinski P, Wightman R, Milani P, Berquand A, Boudaoud A, Hamant O, Jonsson H, Meyerowitz EM (2014) Subcellular and supracellular mechanical stress prescribes cytoskeleton behavior in Arabidopsis cotyledon pavement cells. eLife 3

Sassi M, Ali O, Boudon F, Cloarec G, Abad U, Cellier C, Chen X, Gilles B, Milani P, Friml J et al (2014) An auxin-mediated shift toward growth isotropy promotes organ formation at the shoot meristem in Arabidopsis. Curr Biol CB 24:2335–2342

Shraiman BI (2005) Mechanical feedback as a possible regulator of tissue growth. Proc Natl Acad Sci USA 102:3318–3323

Smith RS, Guyomarc'h S, Mandel T, Reinhardt D, Kuhlemeier C, Prusinkiewicz P (2006) A plausible model of phyllotaxis. Proc Natl Acad Sci USA 103:1301–1306

Thompson DW (1917) On growth and form. Cambridge University Press, UK

Uyttewaal M, Burian A, Alim K, Landrein B, Borowska-Wykret D, Dedieu A, Peaucelle A, Ludynia M, Traas J, Boudaoud A et al (2012) Mechanical stress acts via katanin to amplify differences in growth rate between adjacent cells in Arabidopsis. Cell 149:439–451

Vermeer JEM, von Wangenheim D, Barberon M, Lee Y, Stelzer EHK, Maizel A, Geldner N (2014) A spatial accommodation by neighboring cells is required for organ initiation in Arabidopsis. Science 343:178–183

Vlad D, Kierzkowski D, Rast MI, Vuolo F, Dello Ioio R, Galinha C, Gan X, Hajheidari M, Hay A, Smith RS et al (2014) Leaf shape evolution through duplication, regulatory diversification, and loss of a homeobox gene. Science 343:780–783
Williamson R (1990) Alignment of cortical microtubules by anisotropic wall stresses. Aust J Plant Physiol 601–613
Wymer CL, Wymer SA, Cosgrove DJ, Cyr RJ (1996) Plant cell growth responds to external forces and the response requires intact microtubules. Plant Physiol 110:425–430
Yeoman PM, Brown R (1971) Effects of mechanical stress on the plane of cell division in developing callus cultures. Ann Bot 1102–1112
Zerzour R, Kroeger J, Geitmann A (2009) Polar growth in pollen tubes is associated with spatially confined dynamic changes in cell mechanical properties. Dev Biol 334:437–446

Folding, Wrinkling, and Buckling in Plant Cell Walls

Dorota Borowska-Wykręt and Dorota Kwiatkowska

Abstract In this chapter, we discuss various cases of cell and tissue wrinkling or folding from the perspective of a putative mechanism of their formation—tissue folding in the contractile roots; cell or meristem surface folding in phyllotaxis generation; the formation of the stomata pore and various types of gas spaces; the development of jigsaw puzzle-shaped epidermal cells; and the wrinkling of cell wall layers after the removal of tensile stress. We also address the biological role of such shaped cells or tissues and the mechanical property or state of the cell wall or tissue that is manifested by its folding or wrinkling. Buckling and differential growth are likely ways to generate folds or wrinkles. The former is an intuitive mechanism from the mechanical perspective, while the latter derives from biology. Some cases of cell or tissue morphogenesis suggest that locally the two mechanisms may simultaneously contribute to the formation of a wavy shape.

Keywords Aerenchyma · Cell wall buckling · Contractile roots · Differential growth · Intercalary gas spaces · Leaf and petal epidermis · Phyllotaxis

Introduction

The regular patterns of wrinkles or folds that appear on a plant cell or tissue surface have drawn the attention of both biologists and physicists for long time. Though the regularity of wrinkling or folding as such is fascinating, the main issue is what we can learn from these patterns. Depending on how this question is addressed, the following aspects of plant development, structure, or function may be studied: (i) the mechanism of the formation of a given shape; (ii) the biological function of such shaped cells or tissues; or (iii) the mechanical property or state of the cell wall or tissue that is manifested by its shape. It is noteworthy that the cell wall mechanics

D. Borowska-Wykręt · D. Kwiatkowska (✉)
Faculty of Biology and Environment Protection, Department of Biophysics and Morphogenesis of Plants, University of Silesia in Katowice, Jagiellońska 28, 40-032 Katowice, Poland
e-mail: dorota.kwiatkowska@us.edu.pl

A. Geitmann and J. Gril (eds.), *Plant Biomechanics*,
https://doi.org/10.1007/978-3-319-79099-2_10

is important for all three of these topics. In this chapter, we will discuss various cases of cell and tissue wrinkling or folding from the perspective of a putative mechanism of their formation, while at the same time addressing their function.

Buckling (Green 1999; Hejnowicz 2011) and differential growth (Coen et al. 2004; Kennaway et al. 2011) are likely the mechanisms that are responsible for the generation of the wrinkles or folds of cell walls or tissue surfaces. The former is often an intuitive mechanism from the mechanical perspective, while the latter derives from biology. There are two major differences between these putative mechanisms. First, plant growth is an irreversible expansion of the cell wall surface while in the course of buckling, the surface does not expand irreversibly. Second, growth requires the tensile in-plane stress of the cell wall while compressive in-plane stress is a prerequisite for buckling. Nevertheless, some cases of the formation of folded or wrinkled cell walls or tissues suggest that locally the two mechanisms may contribute to the formation of the shape. This could be possible when the tensile stress necessary for cell wall growth was not derived from turgor pressure but from swelling of the matrix in some layers of the wall (Hejnowicz 2011).

Buckling of Cell Wall or Tissue Surface

A plate or a smooth shell that is initially flat may obtain a folded or wrinkled shape due to buckling. Such deformation occurs when a critical value of the in-plane compressive force is surpassed and, as a result, the plate or shell becomes mechanically unstable (Ugural 1999). The pattern of folds or wrinkles that appear due to buckling depends on the geometry, thickness, and mechanical properties of the plate or shell and on the value of the applied compressive force. The examples of folded or wrinkled cell walls and tissue surfaces discussed below differ in these traits and in the origin of the compressive force.

Compression Results from the Tension Generated in Adjacent Cells

A rather obvious example of the buckling-based formation of folds or wrinkles is the deformation of the tissues in contractile roots. The substantial pulling forces generated by contractile roots (measured, e.g., by Pütz 1992; Cresswell et al. 1999; Abasolo et al. 2009) displace bulbs, corms, or other shoot organs to a location deeper into the soil as an adaptation to vegetative propagation or for survival in unfavorable seasons. Another example is the taproot contraction of young seedlings that leads to the displacement of the cotyledons closer to the soil surface, which is important for the future development of the seedling and mature plant (Pütz 1992). Pulling forces are generated by the contraction of some but not all of the root tissues. As a result, the remaining tissues are exposed to compression and buckle. Thus, the root surface in the contractile zone becomes wrinkled like in the roots of a hyacinth *Hyacinthus orientalis* bulb (Cyr et al. 1988) or the seedling taproots of plants from the pea family (Fabaceae) and cycads (Fisher 2008) in which the outer cortex and epidermis buckle. At the same time, some inner root tissues may become folded, i.e., attain a characteristic sinuous shape in a longitudinal root section like the older tracheary elements of cycad seedling taproots (Fisher 2008; Tomlinson et al. 2014). In the annual desert plant *Gymnarrhena micrantha* from the sunflower family (Asteraceae), wrinkling of the radial anticlinal (i.e., perpendicular to the root surface) cell walls of the parenchyma were also reported (Zamski et al. 1983).

There is still a debate on the mechanism of root contraction that generates compressive forces in some of the root tissues. This mechanism is most likely not the same in different plant groups. In the hyacinth and other monocots, contraction is associated with the shortening of the inner parenchyma cells of the root cortex along the root axis and their simultaneous elongation along the root radius (Cyr et al. 1988). Since this process is accompanied by the deposition of new cell wall material and a fourfold increase in the cell volume, it is a true growth phenomenon. It is noteworthy that the same cortex cells elongate along the root axis prior to the contraction. This means that their growth anisotropy changes—prior to the contraction, the direction of maximal growth is along the root axis and during contraction it is along the root radius. This change is attributed to a reorganization of the cortical microtubule (cMT) arrays and the related change in the orientation of the newly deposited cellulose microfibrils (CMFs) (Smith-Huerta and Jernstead 1989, 1990). The strongly anisotropic growth of the cell wall is accompanied by the orientation of both arrays, which are generally kept perpendicular to the direction of maximal growth. In a number of species, packets of collapsed dead parenchyma cells appear between the packets of radially elongated cortex cells (e.g., Fisher 2008). On this basis, it was proposed that root contraction is an example of cohesive plant organ movements (reviewed in Romberger et al. 1993). Cortex contraction would then be generated by cohesive forces between water molecules and their adhesion to the walls of dead cells. Root contraction can also be driven by G-fibers, which are similar to those in the tension wood of broad-leaved trees (Fisher 2008; Schreiber et al. 2010; Tomlinson et al. 2014). They are present in the xylem and phloem of the contractile taproots and hypocotyls of many Fabaceae species (Fisher 2008). For example, in clover (*Trifolium pratense*) contractile taproots, G-fibers have a structure and chemical composition of the cell wall that is similar to the G-fibers of the tension wood of trees (Schreiber et al. 2010). However, in roots, G-fibers are present all over the root cross section rather than eccentrically as in the tension wood (Schreiber et al. 2010; Tomlinson et al. 2014). Thus, the forces that are generated by the G-fibers lead to root contraction instead of bending. In roots where the G-fibers are present, such a mechanism of root contraction seems to be more probable than the parenchyma-based mechanism. In particular, it has been questioned whether the cortex parenchyma cells can generate forces that are sufficient to deform the xylem tissues (Schreiber et al. 2010) and the collapse of the cortex parenchyma cells was postulated as being the result rather than the cause of root contraction (Tomlinson et al. 2014).

Compression Is a Result of Cell or Tissue Geometry

Cell wall buckling resulting from a specific geometry, e.g., a saddle-like shape, is one of the mechanisms that have been postulated to trigger appendage formation in the green alga *Acetabularia acetabulum* (synonym of *A. mediterranea*) (Dumais and Harisson 2000). During the diplophase stage of its life cycle at the tip of the thallus, which is a giant single cell, whorls of lateral appendages, i.e., hairs or gametophores, are periodically formed (Dumais et al. 2000). The growing cell tip has a tapered shape that is similar to pollen tubes or root hairs, which also extend via tip growth. However, prior to the whorl formation, the tip flattens and growth is redistributed from its apex to a subapical annulus. While the tip still broadens, appendages are initiated at its periphery. According to a mechanical model of the whorl morphogenesis proposed by Martynov (1975), buckling of the tip cell wall is responsible for the appendage initiation. Although turgor pressure generally leads to the tensile in-plane stress of the cell wall, the flattened tip is mechanically unstable and local compression is predicted at its periphery. This compression may lead to buckling that results in regularly spaced inconspicuous folds of the cell wall. Since plant cells are postulated to recognize their shape (Hamant and Moulia 2016), such an emerging pre-pattern of cell wall geometry can become a template for the appendages. It should be noted, however, that the empirical data concerning *Acetabularia* morphogenesis are more consistent with the chemical model of whorl morphogenesis proposed by Harrison et al. (1981), and Dumais and Harrison (2000). This model is based on experiments showing that the temperature in which *Acetabularia* plants grow affects the spacing of the appendages. The observed relationship resembles the temperature dependence of the rate parameters of a chemical reaction. Nevertheless, the contribution of a buckling mechanism cannot be excluded.

Buckling was also proposed as the mechanism that is responsible for the initiation of primordia in the multicellular shoot apical meristems, which leads to the generation of regular phyllotactic patterns (Green 1999). In a model proposed by Green et al. (1996), the tunica, i.e., the cell layers on the meristem surface and the underlying corpus, are regarded as two contacting material layers. Their properties imply the specific spacing of the meristem surface folds, i.e., the wavelength of undulation. The compressive in-plane force that is assumed to exist in the meristem surface is balanced by the tunica's reluctance to bend and the corpus's reluctance to displace in the direction that is normal to the tunica surface. The buckling model based on such assumptions showed the de novo formation and propagation of a whorled primordia pattern on an annulus-shaped meristem periphery and a spiral pattern formation if the spirals were included as boundary conditions. The existence of in-plane compression in the meristem surface, which is a critical component of the buckling mechanism, has only been demonstrated for the capitulum of sunflower *Helianthus annuus* (Dumais and Steele 2000). The pressurized shell model where the tunica, especially its superficial cell walls, is regarded as a shell under the pressure that is generated by the corpus predicts the local compressive forces in the circumferential direction for some capitulum shapes. It is noteworthy that when the young capitulum attains such shapes during development, primordia are indeed initiated in the regions of the predicted compressive stress in a circumferential direction. The central meristem zone is under tension in all directions, which may explain the absence of primordia initiation. The applicability of the pressurized shell model to the shoot apical meristem is supported by empirical data (Beauzamy et al. 2015) and the model predictions were further confirmed by cutting experiments (Dumais and Steele 2000). The buckling model can also explain the variation of the cotyledon number in the somatic embryos of larch *Larix* x *leptoeuropaea* (Harrison and von Aderkas 2004). Consistently spaced cotyledons are initiated in the whorls on the shoot apices of multicellular somatic embryos, which are analogous to the giant cell tip of *Acetabularia*. Moreover, in this case, the apex flattens before the primordia appear, and therefore compressive in-plane forces can be expected. The contribution of buckling to primordia initiation is, however, not supported for the shoot apical meristems in general. The meristem shape is usually a dome or paraboloid that is unlike the young sunflower capitulum or larch embryos and the boundaries between the primordia and the meristem are the only regions where compression is expected in the meristem surface (Burian et al. 2013).

Cell Wall Layers Buckle After the Removal of the Stress

The removal of tensile stress from the primary cell walls, especially the relatively thick ones, may lead to the formation of wrinkles on the innermost, i.e., the recently formed, cell wall layer (Fig. 1). This has been shown for the epidermal cells, e.g., in the sunflower hypocotyl (Hejnowicz and Borowska-Wykręt 2005) or the coleoptile of the barley *Hordeum vulgare* (Fig. 1d, e) as well as for collenchyma cells like in the peduncle of the dandelion *Taraxacum officinale* (Fig. 1f, g). The epidermal or collenchyma cell walls are under the tensile in-plane stress directly driven by turgor. Since these are the outer tissues of cylindrically shaped organs, in situ the tensile tissue stress is superimposed on the directly turgor-driven stress and the direction of the maximal stress is longitudinal (Hejnowicz and Sievers 1996; Niklas and Paolillo 1998; reviewed in Baskin and Jensen 2013). When the tensile in-plane stress of the walls is removed (the tissue stress by tissue isolation from the organ and the turgor-driven stress by plasmolysis), the innermost wall layers wrinkle. The wrinkles are then transverse, i.e., perpendicular to the direction of maximal tensile stress in situ (Fig. 1d, f). The postulated mechanism of wrinkle formation is buckling (Hejnowicz and Borowska-Wykręt 2005). Models explaining the buckling of layered structures, which are applied to plant organs like fruits, assume that the structure is composed of a thin stiff film that is attached to a thick "substrate" that has a much lower Young's modulus (Gough et al. 1940; Yin et al. 2008; Sharon and Efrati 2010; Chen and Yin 2010). However, it seems unlikely that in the still expanding primary cell walls of epidermis or collenchyma of the hypocotyls, coleoptiles or peduncles, the youngest cell wall layer is much stiffer than the remaining wall portion. Moreover, the buckling cell wall layer, which has recently been formed, is not expected to be under tension. Rather the innermost layer is the least stressed while the stress in other layers increases with their age, i.e., with the distance from the cell protoplast (Hejnowicz and Borowska-Wykręt 2005). Therefore, in this case, wrinkling of the cell wall surface due to the removal of the stress is most likely a manifestation of differences between the wall layers in the tensile in-plane stress in the longitudinal direction.

Differential Growth and Local Buckling of the Cell Wall

In tissues such as the aerenchyma, which has stellate cells, or a leaf epidermis with jigsaw puzzle-shaped cells, the morphogenesis that leads to a complex cell shape is accompanied by extensive tissue growth. In such cases, the differential growth of individual cell walls is predicted to assist in the formation of the cell shape as has been supported empirically for epidermal cells (Armour et al. 2015). The question arises whether the tensile in-plane stress in the wall in such shaped cells, which is a prerequisite for the wall growth, results exclusively from turgor. Similar to the case of a pressurized shell discussed above, tensile turgor-driven stress is expected in the plane of the walls that are bent outward from the cell lumen or that are aligned along the shortest path between the cell wall junctions. This condition is not observed in folded cell walls. An extreme case in which the tensile in-plane stress could not be driven by turgor would be a wall fragment that is folded into the cell (like a wall invagination), which would grow more and more into the cell, while the entire cell was not growing. In such cases, an alternative mechanism for generating stress was postulated (Hejnowicz 2011), i.e., the tensile stress in a wall layer can be generated by swelling in the adjoining layer. In a theoretical situation, if the two layers are not attached, the one that was swelling would simply expand more than the other (Fig. 2a-a'). However, if the layers were tightly joined like in the cell wall, the static equilibrium condition would require that the sum of the forces is zero, and therefore the compression in the swelling layer would bring the other layer under tension. In the case of the cell wall, the swelling of a wall matrix that fills the spaces between stiff CMFs would result in compressive stress in the swelling wall layer (e.g., B1 in Fig. 2b, c). If the CMFs in the swelling layer were aligned, i.e., their arrangement was anisotropic, the compressive in-plane stress in this layer would also be anisotropic with the greatest compression in the direction perpendicular to the orientation of the CMFs. Accordingly, the tensile stress generated in the adjoining layer (A1 in Fig. 2b, c) would also be anisotropic with the maximal tension in the same direction. If the distribution of the swelling layers within the compound wall (comprising two walls of neighboring cells that are tightly joined by middle lamella) was symmetric and all of the layers remained attached, the tensile stress generated in such a manner could explain the wall growth (Fig. 2b-b'). However, in some cases, such a mode of stress generation could lead to local cell wall buckling. An example is a straight wall fragment, the two ends of which are fixed by the remaining cell walls, where the distribution of the swelling layers is not symmetrical (Fig. 2c-c'). In such a case,

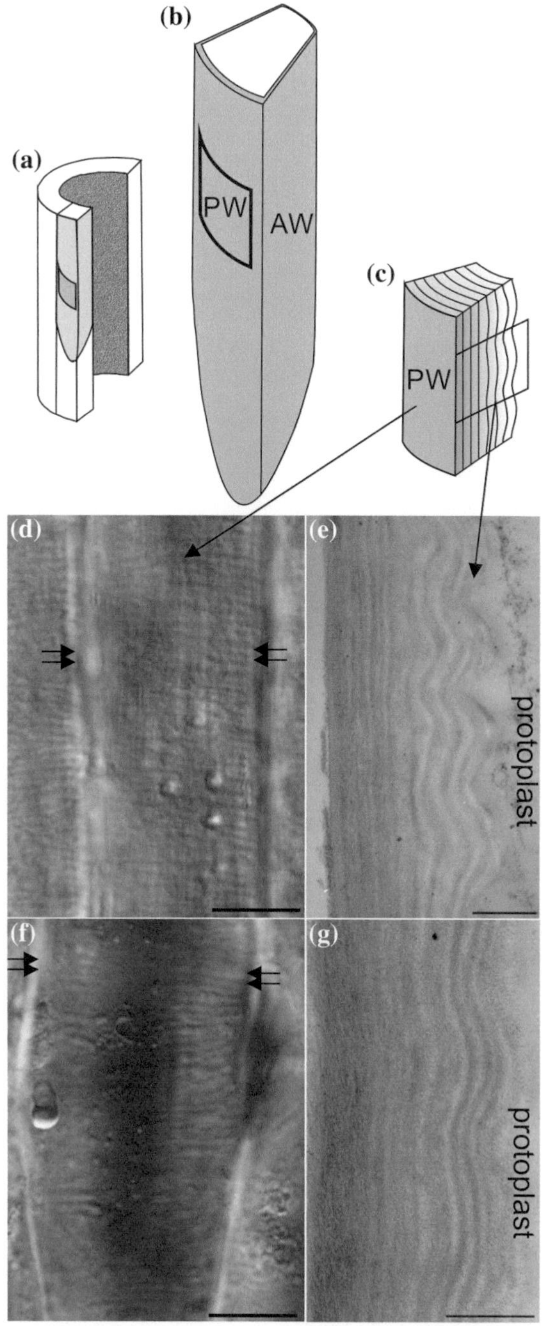

Fig. 1 Buckling of cell wall layers after the removal of the stress. **a–c** Schematic representation of an outer tissue of a cylindrically shaped organ (*a*), its exemplary cell (*b*), and a fragment of the outer periclinal wall (*c*) shown in an optical section obtained using Nomarsky microscopy (*d, f*) and in a longitudinal radial section observed using transmission electron microscopy (*e, g*). *PW* labels periclinal walls; *AW*—anticlinal walls. **d–g** Micrographs of wall fragments of the barley coleoptile epidermis (*d, e*) and dandelion peduncle collenchyma (*f, g*). Wrinkles visible in the wall plane (*d, f*) appear as undulated layers in the sections (*e, g*). *Arrows* in *d, f* point to anticlinal walls, which are perpendicular to the image plane. Scale bars 10 μm (*d, f*), 1 μm (*e, g*)

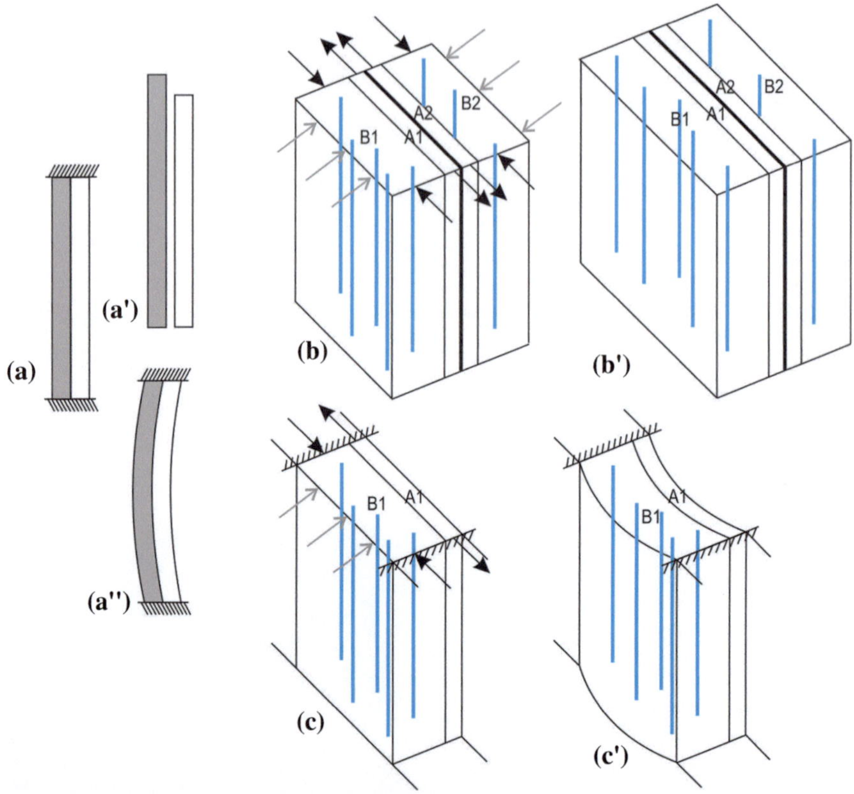

Fig. 2 Schematic representation of the putative changes in the shape of a cell wall fragment due to matrix swelling. **a** If two unbound layers (*a*) extend differently, no buckling occurs (*a'*) but if they are bound and their ends are fixed, when the compression threshold is surpassed, they lose stability and bend due to buckling (*a"*). **b** and **c** Schemes of a cell wall fragment comprising layers with different swelling capacities. Layers are labeled with capital letters (*A, B*) and a number, *1* or *2*, referring to different cells. *B* indicates layers with an increased capacity for matrix swelling. The orientation of CMFs is represented by *blue (gray) line segments*; *black arrows* represent forces; *gray arrows*—water penetration. A boundary between the walls of two cells is indicated by a thicker line. **b** Symmetric swelling on both sides of the fragment of a compound wall (the wall comprising the adhering walls of two cells bound by the middle lamella) results in the compression of the *B1, B2* layers and tension of layers *A1, A2* (*b*), which is a prerequisite for growth (*b'*). **c** If the distribution of the swelling layers is asymmetric, like in a wall fragment of a single cell (*c*), buckling occurs (*c'*) analogous to *a"*

when a critical value of the compressive in-plane force is surpassed, the wall fragment loses its stability and the two joined layers buckle. This could explain the initiation of a cell wall folding into the cell lumen, which is an initial stage in the generation of a complex cell shape.

Below we discuss examples of cell morphogenesis in which differential growth and local buckling are likely involved.

Bending of a Single Wall into the Cell Lumen

Stomata Pore Formation

During stomata morphogenesis, the two guard cells, which exhibit a structural symmetry, undergo profound shape changes that are accompanied by growth. In particular, they are transformed from generally elongated cells into kidney-shaped cells (Fig. 3a, b) like in the dicot pea *Pisum sativum* (Singh and Srivastava 1973), monocot *Allium* (Bünning and Biegert 1953; Stebbins and Jain 1960), and the fern *Asplenium nidus* (Apostolakos and Galatis 1998) or into dumbbell-shaped cells in the grasses *Poaceae* (Stebbins and Shan 1960; Kaufman et al. 1970; Srivastava and Singh 1972). The final stage of stomata maturation and the most interesting from the mechanical point of view is the formation of the stomata pore (aperture) that occurs after the guard cells stop growing (Fig. 3a). The pores are usually the only places in the leaf epidermis where the anticlinal walls (walls that are perpendicular to the organ surface) of the adjacent cells are separated. Although in most plants the stomata comprise two guard cells, a fully functional stoma of the moss *Funaria* is a single binucleate cell with a central pore (Sack and Paolillo 1983a, b). Similar unicellular and binucleate aberrant stomata have been reported for leaf epidermis of maize *Zea mays* that had been treated with caffeine, which inhibits cytokinesis. It is noteworthy that even in such aberrant stomata, the anticlinal walls lining the "pore" become separated (Galatis and Apostolakos 1991).

During pore formation, the two single anticlinal walls of the guard cells, which are initially glued by the middle lamella and thus comprise a compound anticlinal wall, become separated (Fig. 3c-c'). The separation occurs within the thickened portion of the compound wall. The anticlinal walls lining the pore are called ventral walls, while the other anticlinal walls are called dorsal walls (Fig. 3b). The thickening of the ventral walls is not uniform; they are thickest at the junctions with the periclinal (i.e., parallel to the organ surface) walls (Singh and Srivastava 1973; Palevitz and Hepler 1976; Sack and Paolillo 1983a, b). The orientation of the CMFs in the anticlinal walls is perpendicular to the epidermis surface, while in the periclinal walls, the CMFs form radiating arrays that run from the ventral to dorsal walls (Fig. 3b), which is also called radial micellations (Singh and Srivastava 1973; Palevitz and Hepler 1976; Sack 1987). In all of the cases that were examined including the unicellular stomata of *Funaria*, wall separation is a gradual process. It starts locally, usually at junctions between the ventral and two periclinal walls, from where it proceeds to the adjacent parts of the compound ventral wall (Sack and Paolillo 1983a, b; Sack 1987; Galatis and Apostolakos 2004; Apostolakos et al. 2009). Only in some species, such as the ferns *Asplenium nidus* and *Adiantum capillus-veneris* (Apostolakos and Galatis 1998, 1999) or the pea (Singh and Srivastava 1973), does the separation start at the center of the compound ventral wall.

The presence of nonuniform symmetric thickenings of the compound ventral wall suggests that the forces that pull apart the two single ventral walls that lead to their separation could result from buckling, due to which each wall would bend into

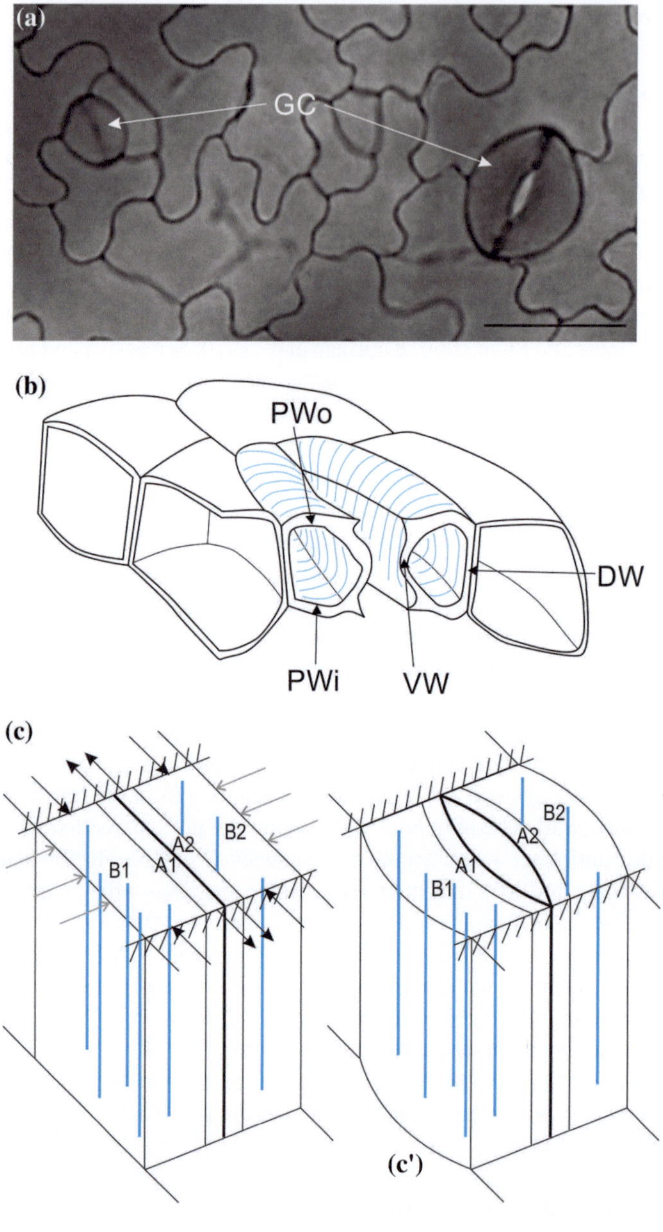
(a)
GC
(b)
PWo
DW
PWi
VW
(c)
B1
A1
A2
B2
(c')

◀**Fig. 3** Pore opening during stomata development. **a** Two stages of stomata development in the micrograph of the *Arabidopsis* leaf epidermis—in the stomata on left, the two guard cells (*GC*) are not yet separated; in the stomata on right, the pore is already formed. Scale bars 20 μm. **b** Schematic representation of an epidermis fragment with a stomata cut at its center across the pore. *Blue (gray) lines* represent CMFs orientation; *PWi* and *PWo* indicate the inner and outer periclinal walls, respectively; *DW*—the dorsal anticlinal wall; *VW*—the ventral anticlinal wall. **c** Scheme of a fragment of a compound cell wall comprising layers with different swelling capacities. Labeling as in Fig. 2b, c. Symmetric swelling on both sides of the compound wall results in the compression of the *B1, B2* layers and tension in the *A1, A2* layers (*c*). If the fragment ends are fixed by remaining cell walls and the middle lamella is weakened, the two single walls buckle and the space between them opens forming a pore (*c'*)

the cell lumen. The necessary compressive stress may arise in the inner layers of the thickened wall portions, i.e., those facing the protoplast (B1 and B2 in Fig. 3c), where it results from an increase in the matrix swelling capacity during the maturation of the wall. This tendency of the matrix to swell may be due to the deposition of callose, the presence of which in the ventral walls seems to be correlated with stomatal pore formation (Apostolakos and Galatis 1999; Galatis and Apostolakos 2004; Apostolakos et al. 2009). Because the orientation of the CMFs in the ventral walls is perpendicular to the epidermis surface, the maximal compressive stress would be in the direction running parallel to the surface, i.e., along the future pore. The two ends of the future pore are fixed by the remaining cell walls. Thus, as a result of the increasing swelling, the compressive stress in the inner wall layers would attain a threshold value and the release of the stress would lead to wall buckling, i.e., its bending toward the cell lumen (Fig. 3c'). This spontaneous process could be facilitated by a weakening of the middle lamella at the location of the future pore. In some plants, the middle lamella of a young compound ventral wall seems to be lacking pectins, which glue the two individual walls together (Galatis 1980; Galatis and Mitrakos 1980). In the unicellular stomata of *Funaria*, in turn, the loci of the wall weakness could be the sites where the cuticular precursors are cumulated (Sack and Paolillo 1983a, b).

The aligned CMFs and cell wall thickenings in the guard cells of the kidney-like stomata are also crucial for the functioning of the stomata at maturity, i.e., their ability to open and close (Srivastava and Singh 1972; Singh and Srivastava 1973; Palevitz and Hepler 1976; Galatis 1980; Galatis and Apostolakos 2004). When cell turgor increases, the radial arrangement of the CMFs in periclinal walls facilitates the reversible expansion of these walls mainly along the pore, thereby forcing the deformation of the anticlinal walls. Since the dorsal walls are relatively thin, they expand to a larger extent and bend outward from the pore. The thickenings of ventral walls, in turn, prevent their expansion but allow bending, which results in the pore opening.

Schizogenous, Expansiogenous, and Intercalary Gas Spaces in Parenchyma

Intercellular spaces form a complex system of usually gas-filled canals and chambers within various types of parenchyma. Apart from gas transport, the spaces may be involved in light scattering, provide buoyancy in water, or form domatia for plant-associated organisms (Raven 1996). It is noteworthy that even aerenchyma can contribute to a mechanical organ function—in the giant leaf petioles in *Amorphophallus*, a turgid aerenchyma inside the cylindrically shaped organ improves its mechanical stability while reducing the specific density (Hejnowicz and Barthlott and 2005). Gas spaces are classified according to their origin into lysigenous, schizogenous, expansigenous, and intercalary. Lysigenous gas spaces are formed in places at which cells undergo lysis. The formation of the other two space types is important from the mechanical perspective. Schizogenous spaces develop at the corners of cells, i.e., three-way or less often four-way cell wall junctions, via cell wall separation (Raven 1996). Spaces of the recently introduced expansigenous type develop by the expansion of schizogenous intercellular spaces into large air chambers by the enlargement and/or divisions of the space lining cells without any further cell separation or collapse (Sego et al. 2005). However, gas spaces may form not only at the cell corners but also between the walls of only two cells, i.e., within a compound cell wall (Fig. 4a). Such spaces, which are called intercalary gas spaces, are lined by walls of two rather than three or more cells. They are rather common in the diaphragms of water-submerged organs that have aerenchyma (Sifton 1945) and also occur in the mesophyll, e.g., in the soybean *Glycine max* (Weston and Cass 1973), *Nigella damascena* (Wernicke et al. 1993), the wheat *Triticum aestivum* (Jung and Wernicke 1990), or the fern *Adiantum capillus-veneris* (Panteris et al. 1993a).

Schizogenous space formation at the cell corners begins when the cells are still growing and dividing. For example, in the common bean *Phaseolus vulgaris*, the spaces open in the palisade and spongy mesophyll of developing leaves that are only 1–3% of their final size (Jeffree et al. 1986). In some tissues, i.e., the parenchyma of the pea root and the hypocotyl of the bean *Phaseolus aureus*, a structurally distinguished splitting layer was identified at three-way cell wall junctions. The cell wall cleavage is restricted to this layer, which is deposited prior to the space opening and exhibits an enzymatic activity (Roland 1978). The wall segments that border the future intercellular space are thicker than the remaining cell walls (Roland 1978; Kallöffel and Linssen 1984) and are reinforced by being enriched with weakly or non-esterified pectin epitopes (Parker et al. 2001; Sotiriou et al. 2016). A theoretical consideration of the cell separation forces at variously shaped three-way cell wall junctions showed that the in-plane forces that were induced by turgor in the walls of contacting cells pulled the cells apart (Jarvis 1998). Due to these same forces, the released cell wall fragments are expected to be bent into the newly opened space (Fig. 4b). However, starting from the early space development, the cell walls lining the spaces sometimes exhibit the opposite convexity—they are bent into the cell lumen (Hejnowicz 2011). The reason for this may be that the in-plane stress originating from the swelling of the cell wall matrix contributes to the growth of the

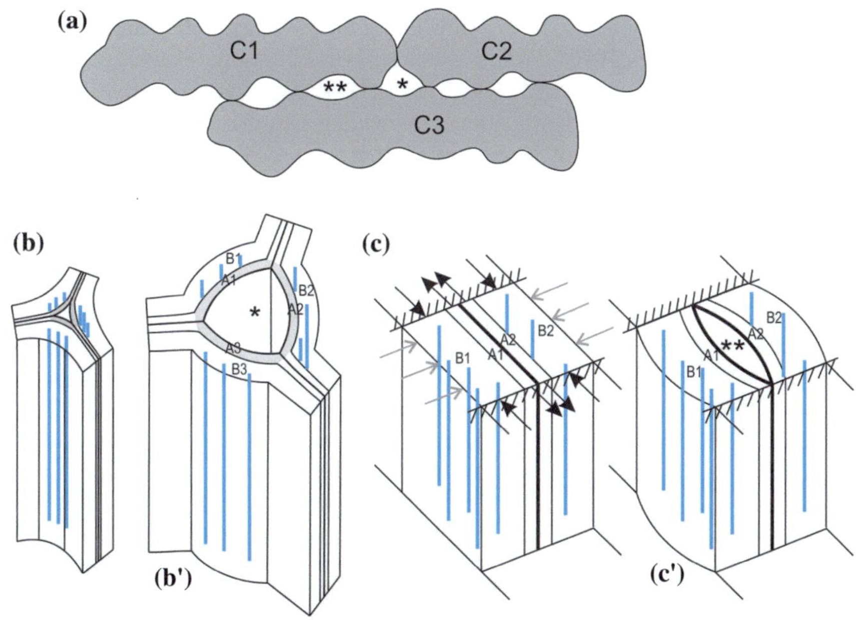

Fig. 4 Schizogenous and intercalary gas spaces between mesophyll cells. **a** Three cells (*C1-3*), with shapes resembling the wheat mesophyll with a schizogenous space (*) at the three-way cell wall junction (between cells *C1, 2, 3*) and several intercalary spaces (**) between two adjacent cells. **b** Opening of a schizogenous space at the three-way junction of the cell walls (*b*) facilitated by turgor-driven forces and the consecutive growth of walls lining the space (shaded in *b'*) due to the presence of layers with different swelling capacities. Labeling as in Fig. 2b, c. **c** Formation of intercalary space within a fragment of a compound cell wall comprising layers with different swelling capacities. Symmetric swelling on both sides of the compound wall results in the compression of the *B1, B2* layers and tension in the *A1, A2* layers (*c*). If the fragment ends are fixed by the remaining cell walls and the middle lamella is weakened, the two single walls buckle and the space between them opens as the gas space (*c'*)

wall lining the space that occurs during its opening. This postulate is supported by observations on the alignment of the cMTs under the walls lining the future space and the corresponding alignment of the CMFs in the walls prior to the space opening. For example, the cMTs are aligned along the axis of the future space canal in the *Sagittaria* diaphragm (Liang et al. 2008) and in the spongy mesophyll of the *Vigna* leaf (Sotiriou et al. 2016; here, the corresponding orientation of the CMFs was also shown). Thus, the swelling of the wall matrix between the CMFs of the inner (protoplasm facing) wall layer would bring the adjoining outer layer under the tensile stress in the direction perpendicular to the canal axis (Fig. 4b'). The walls lining the gas spaces are also often bent into the cell lumen during their expansion, which leads to an increase in the space diameter. For example, the micro-casting of the intercellular gas space system in the *Vigna radiata* hypocotyl shows that the spaces have tubular shape (Prat et al. 1997). Moreover, in the freeze-fractured palisade mesophyll of a young bean leaf, the cell walls lining the gas spaces are concave toward the space (Jeffree et al. 1986). Similar space shapes occur in the diaphragms of the shoot internodes in the *Scirpus validus* (Kaul 1971) or *Sagittaria trifolia* leaf petiole (Liang et al. 2008). Such cell walls that are bending into the cell lumen cannot be brought under tensile stress by turgor and their growth is likely to be driven by the in-plane wall tension originating from swelling stress (Hejnowicz 2011).

A very likely mechanism for the formation of intercalary gas spaces is the buckling that results from swelling stress in the inner cell wall layers (Fig. 4a, c-c'), where cells are not pulled apart by the turgor-originating forces. The walls lining the intercalary spaces bend into the cell from the moment a space is initiated. The initiation of the intercalary space is likely driven by the enzymatic dissolution of the middle lamella accompanied by the deposition of new wall material at the future space lining walls (Weston and Cass 1973) or the local thickening of the cell wall (Liang et al. 2008). Prior to the initiation of the intercalary space, the cell walls form distinct subdomains—in the mesophyll cells of diverse species such as maize and the fern *Asplenium nidus*, the future space lining walls and the cell contacts that remain after the space opening differ in their polysaccharide composition (Giannoutsou et al. 2013; Sotiriou et al. 2016). In all of the species examined, the orientation of the cMTs and CMFs is perpendicular to the direction of the cell wall expansion and the cMT and CMF patterns are symmetric in the adjacent cells between which the spaces are formed (Jung and Wernicke 1990; Wernicke et al. 1993; Panteris et al. 1993a; Sotiriou et al. 2016). All of these observations support the postulate that the intercalary gas spaces are initiated due to the buckling that is caused by the swelling of matrix in the wall layers adjacent to the cell lumen, which is somewhat similar to the stomata pore formation (Fig. 4c-c'). Additional support comes from experiments with the colchicine treatment of developing *Adiantum* leaflets where the disruption of the cMTs disabled the intercalary space formation (Panteris et al. 1993a).

Later during development, when elaborate cell shapes are formed, like in stellate aerenchyma, the differential growth of the cell walls plays a crucial role. For example, in the course of radial expansigeny in water-submerged roots, cell expansion dominates along the radial walls lining the gas spaces, which were formed at the three-way wall junctions. As a result, the spaces enlarge and become elongated in the radial direction (Sego et al. 2005). The arms of the developing stellate aerenchyma cells of the stem internodes in *Scirpus* are most likely stretched by vertically oriented plates of cells (partitions) in which there are unmatched cell elongations and divisions (Kaul 1971).

Wall Ingrowths in Mesophyll of Pine Needles

The mesophyll cells of xeromorphic leaves (needles) of pines (*Pinus*) and some other gymnosperms have specific cell wall ingrowths (e.g., *In* in Fig. 5a) that are sometimes referred to as infolds or invaginations (Romberger et al. 1993). During the mesophyll development, the cells become arranged in plates, which are one cell thick (*MPl* in Fig. 5b). The plates are perpendicular to the needle axis and are separated from

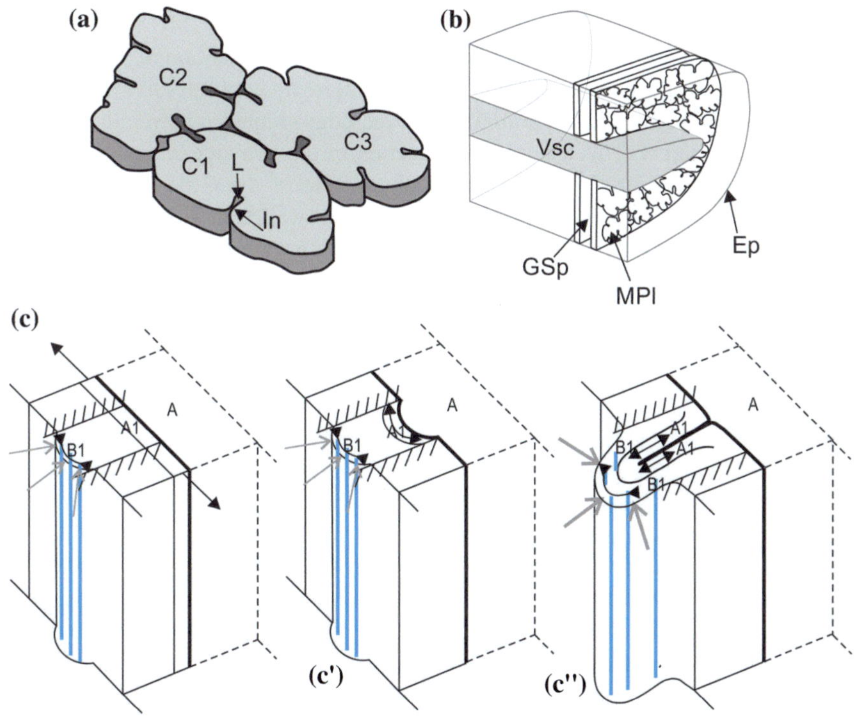

Fig. 5 Cell morphogenesis in a pine needle mesophyll. **a** Three adjacent mesophyll cells (*C1-3*) forming a mesophyll plate fragment. An exemplary wall invagination (*In*) terminated by a loop (*L*) is labeled in *C1*. **b** Fragment of a needle cut along the vascular bundle (*Vsc*) with two exemplary mesophyll plates (*MPl*) spreading between the vascular bundle and the epidermis (*Ep*). A large gas space (*GSp*) is present between the plates. **c** Ingrowth initiation in one of two contacting cell walls due to the swelling of the local wall thickening (*B1* in *c*) is followed by the expansion of the ingrowth assisted by the stresses generated by *B1* layer (*c'*, *c"*)

one another by large intercellular gas spaces (Campbell 1972; Wiebe and Al-Saadi 1976). The ingrowths, which protrude quite far into the cell lumen, appear on the walls that are perpendicular to the plate surface (Fig. 5a, b). They comprise a folded, appressed fragment of the wall belonging to one cell, which is joined by the middle lamella, and are often terminated by a loop that is formed by this single wall (*L* in Fig. 5a) and that extends from one plate surface to another (Harris 1971; Campbell 1972; Wiebe and Al-Saadi 1976; Gambles and Dengler 1982a, b). It is postulated that the gas spaces within the ingrowth loops and the spaces between the mesophyll plates (*GSp* in Fig. 5b) form a continuous air system from the needle base to its apex (Wiebe and Al-Saadi 1976). Apart from the gas space connection, the basic function of these ingrowths is to increase the inner cytoplasmic surface and to mechanically stabilize the cell walls (Hoss and Wernicke 1995).

The formation of wall ingrowths is interesting in the context of the perspective of this chapter since its mechanism involves differential growth and possibly also buckling. The process begins with the appearance of cMT and CMF bundles along the edge of the future ingrowth, which is followed by the formation of a minor fold of the wall (Hoss and Wernicke 1995). Next, the ingrowth continues to extend while the cMTs and CMFs remain parallel to its free edge. Finally, the middle lamella between the folded wall weakens, the separation of the two side walls begins, and a loop is formed (Harris 1971). The postulated role of the CMF bundles is the local reinforcement of the wall so that only the thin-walled cell portions between the reinforced fragments grow and bend outward from the cell lumen (Hoss and Wernicke 1995). This is a differential growth mechanism. However, the process of ingrowth initiation could also be assisted by buckling. The swelling of the wall matrix between the CMFs that are present where the ingrowth is initiated may generate compressive stress in the inner wall layers (B1 in Fig. 5c). The release of this stress perpendicular to the CMFs orientation could lead to the local wall buckling, i.e., folding into the cell lumen that is initiating the ingrowth (Fig. 5c'). Since the ingrowth formation starts while the mesophyll cells are still growing (Hoss and Wernicke 1995), the discrimination between these two mechanisms (differential growth and swelling-driven buckling) of ingrowth initiation would require a detailed in vivo growth analysis. Later on, during cell morphogenesis after the ingrowth was initiated, the deposition of new material and the swelling in inner wall layers (B1 in Fig. 5c") probably brought the non-swelling layers of the folded wall under tensile stress (A1 in Fig. 5c"), thus facilitating its growth into the cell lumen.

Intercalary Gas Spaces in the Petal Epidermis

The wide variety of epidermal cell shapes that are characteristic for the petals of angiosperm flowers is related to the petal's function of attracting pollinators (Kay et al. 1981). One of the special features of the petal epidermis in a number of species is the presence of intercalary gas spaces between the anticlinal walls of cells (Hiller 1872), which affect the optical properties of the petal surface. This does not mean that the petal surface is discontinuous since the epidermis is covered by a cuticle that is also continuous above the gas spaces (spaces are apparent in an optical section through the epidermal cells shown in Fig. 6a" but the epidermis surface shown in Fig. 6b is continuous). In the petals of the pimpernel *Anagallis grandiflora*, both the anticlinal (Fig. 6a-a") and periclinal (Fig. 6b) cell walls are undulated (the formation of undulations is addressed in the next section) and numerous intercalary gas spaces form between anticlinal walls (Kay et al. 1981; Quintana et al. 2007). The gas space is initiated as a local thickening of the undulated anticlinal wall (Fig. 6a). First, a space with a drop-like shape is formed bending into the cell lumen (Fig. 6a'). Its further expansion is most likely accompanied by further cell wall separation (Fig. 6a"). Our preliminary results showed that prior to the space opening, a large amount of cellulose was present at the site of the space formation. Since cMTs form bundles perpendicular to the petal surface, a corresponding arrangement of the CMFs is expected in the cellulose-enriched bands (Raczyńska-Szajgin and Kwiatkowska, unpublished data). Such a wall structure suggests that buckling may contribute to the

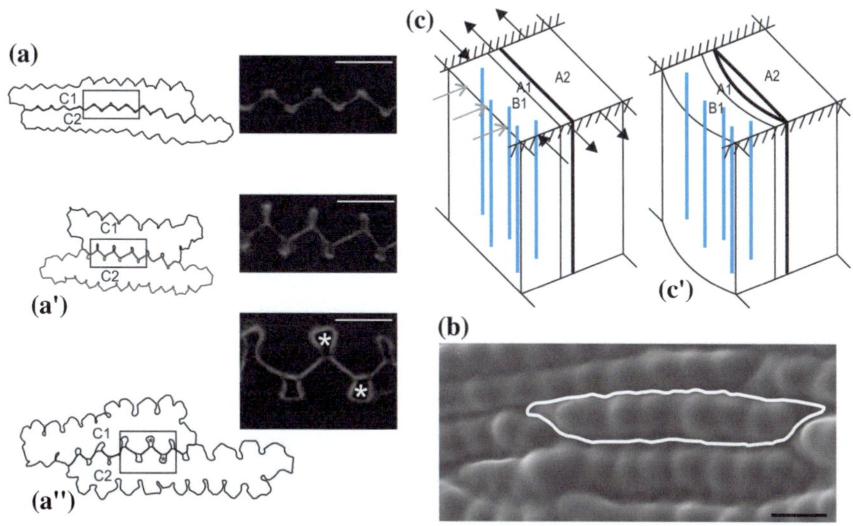

Fig. 6 Formation of the intercalary gas spaces in the petal epidermis of the pimpernel *Anagallis grandiflora*. **a** Epidermal cell outlines and magnified fragments of their anticlinal walls shown in optical sections through the walls obtained using confocal laser scanning microscopy. Consecutive developmental stages are shown. First, anticlinal wall thickenings appear at the tips of the lobes in young epidermal cells (*a*). Next, drop-shaped gas spaces (*a'*) are formed. Finally, in the mature petal (*a"*), the spaces (*) have expanded due to the growth of the space lining walls and further wall separation. **b** Micrograph of the mature petal surface obtained using scanning electron microscopy. An exemplary cell is outlined. The undulated surface of the outer periclinal walls is covered with a continuous cuticle layer. Scale bars 10 μm (*a-a"*), 40 μm (*b*). **c** Asymmetric swelling in one of the single walls comprising a wall fragment with fixed ends (*c*) leads to the buckling of this wall (*B1, A1*) if the middle lamella is weakened (*c'*). As a result, the intercalary gas space opens

space opening (Fig. 6c-c') similar to that of the intercalary spaces in the mesophyll. The same direction of the cMTs is preserved during the further expansion of the anticlinal wall lining the gas space, which means that the direction of the maximal expansion of this wall is perpendicular to the direction of the CMFs. Interestingly, the fastest expansion of the spaces occurs when the growth of the epidermal cells is already very slow, thus suggesting that the tensile in-plane stress that is necessary for wall growth is due to the matrix swelling in the inner cell layers.

Formation of Folds by the Compound Anticlinal Walls of the Epidermal Pavement Cells

The epidermal pavement cells of the leaves in vascular plants often have wavy (undulated) outlines due to the formation of lobes by the anticlinal cell walls (Fig. 7a, b). The lobes that are formed by the epidermal cells are interdigitating and both of the neighboring cells participate in the formation of the undulations of their common compound anticlinal wall. Thus, the same fragment of the compound wall is concave toward one cell but convex toward the other. The jigsaw puzzle-shaped epidermal cells of *Arabidopsis thaliana* leaves and cotyledons have anticlinal walls that exhibit a waviness of a similar amplitude throughout their entire depth. In the maize epidermis, the anticlinal walls of juvenile leaves are lobed similar to *Arabidopsis* (though the overall cell shape is an elongated rectangle unlike the irregular shape of the *Arabidopsis* pavement cells), while an extreme waviness develops in adult leaves that form a zipper-like pattern (Moose and Sisco 1994). These anticlinal walls are less wavy in the deeper portions of the wall and straight at the point of contact with the inner periclinal walls. A similar epidermal cell shape also occurs in some other grasses and in the sedge family Cyperaceae. Wavy anticlinal walls are supposed to enhance intercellular adhesion by increasing the anticlinal wall contact area (Sylvester and Smith 2009). This also increases the cell wall fraction in the epidermis thus presumably also improving its mechanical properties. In *Erianthus arundinaceus*, a drought-tolerant relative of sugarcane *Saccharum officinarum*, wavy anticlinal walls develop in the epidermis of leaves under drought stress. Their postulated biological role is in enhancing sucrose accumulation in the apoplast, which would improve drought tolerance (Augustine et al. 2015).

Although much research has been performed on the development of puzzle-shaped cells, especially in *Arabidopsis*, the data on the growth accompanying their morphogenesis are not complete as yet. It has, nevertheless, been shown that (i) lobe initiation involves the local breaking of growth symmetry (Zhang et al. 2011); (ii) most of the new lobes are initiated during a period of fast cell growth (Elsner et al. 2012; Armour et al. 2015); (iii) lobe formation is gradual, i.e., new lobes are added to already wavy walls (Zhang et al. 2011; Elsner et al. 2012; Staff et al. 2012); (iv) growth of the lobed cells is differential, in particular the periclinal cell walls grow preferentially at the sites of lobe formation; and (v) the growth of the periclinal wall regions near the anticlinal wall that is bending into the cell lumen is slower and anisotropic with a maximum along the anticlinal walls, while the regions near walls bending outward from the cell grow faster and less anisotropically (Armour et al. 2015). In the case of anticlinal walls, theoretically, since fast wall expansion at the lobe tip would generate friction with the adjacent concave bend of neighboring cell, the highest expansion is probably expected at the side walls of lobes (Geitmann and Ortega 2009). Nevertheless, more measurements on the growth of the anticlinal walls during lobe formation are needed.

The main targets of the molecular signals regulating puzzle-shaped cell morphogenesis, which have been extensively studied in *Arabidopsis* and maize, are the cytoskeleton and cell wall (reviewed, e.g., in Kotzer and Wasteneys 2006). Szymanski (2014) hypothesized that a lobe is initiated by the local nucleation of cMTs in one cell that is caused by a locally increased mechanical stress at the junction between the periclinal and anticlinal walls (such a relationship between stress distribution, which was assessed by a finite element modeling and cMT arrangement, was shown for expanding but already lobed epidermal cells by Sampathkumar et al. 2014). These cMTs form aligned bundles that are perpendicular to the leaf surface under the anti-

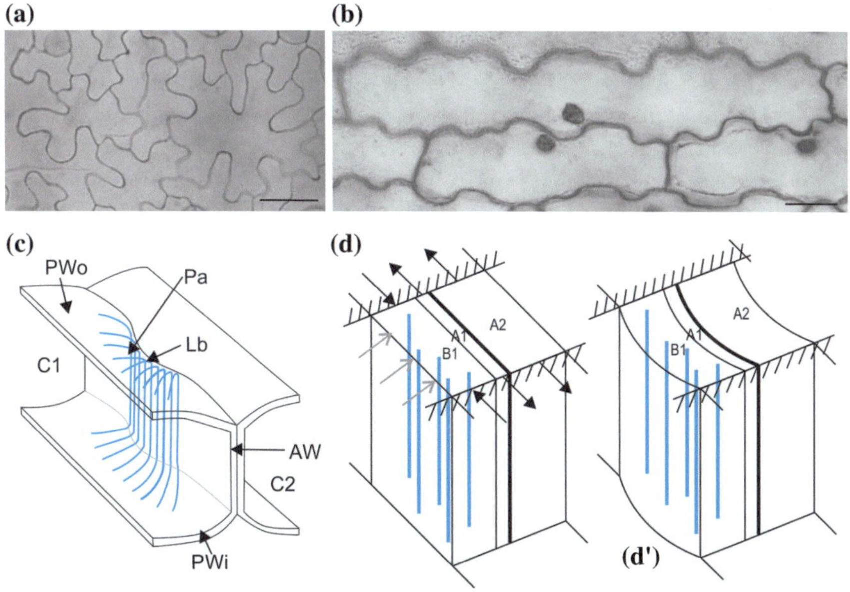

Fig. 7 Jigsaw puzzle-shaped epidermal cells. **a** and **b** Micrographs showing optical paradermal sections of the leaf epidermis in *Arabidopsis* (*a*) and maize (*b*). Scale bars 20 μm. **c** At the site of lobe (*Lb*) formation, CMFs (*blue/gray lines*) form bundles in the anticlinal wall (*AW*) of *C1* into which lumen the wall is bending. The CMF bundles extend radiating into the outer and inner periclinal walls (*PWo* and *PWi*, respectively) of *C1*. Pads (*Pa*) are formed at the junctions between the anticlinal and periclinal walls. **d** Asymmetric swelling in one of the single walls comprising a compound wall fragment with fixed ends leads to the buckling of the wall (*B1, A1, A2* in *d*). As a result, a lobe is initiated (*d'*)

clinal wall (Fig. 7c). The bundles extend radially under the outer periclinal wall. Since the CMFs are deposited parallel to the cMTs, such a reinforced portion of the periclinal and anticlinal cell walls grows anisotropically with its maximal growth along the periclinal–anticlinal wall junction. This generates strain in the periclinal wall of the neighboring cell with a maximal direction that is perpendicular to their common anticlinal wall. As a result, the compound anticlinal wall bends into the cell with the cMT bundles. Such a cMT arrangement and similarly aligned CMFs have been reported for a number of species that form puzzle-shaped epidermal cells by Panteris et al. (1993b, 1994). It has also been confirmed for *Arabidopsis* by an in vivo study of growth and cMTs, although the spatiotemporal relation between the cMTs and lobe formation turned out to be not so straightforward (Zhang et al. 2011; Armour et al. 2015). Panteris et al. (1993b, 1994) showed that the cMT bundles are accompanied by cell wall pads that are located at the junction between the anticlinal and outer, sometimes also inner, periclinal walls in all of the examined species. It has been postulated that the cell wall pads are the places where the local loosening between the radial CMFs occurs that leads to the local expansion of the periclinal cell walls mainly along the junction with the anticlinal wall, i.e., perpendicular to the CMFs, which supports the Szymanski (2014) model. Moreover, in the maize leaf epidermis, bundles of cMTs that are perpendicular to the leaf surface are present at the future anticlinal cell wall indentations and were proposed to direct periodic cell wall thickenings with intervening thinner regions spreading to periclinal cell walls (Frank and Smith 2002). The role of the CMFs in lobe formation is supported by the *anisotropy1* mutant phenotype. This mutation in cellulose synthase of primary cell walls, which reduces cellulose crystallinity, also causes impaired lobe formation in *Arabidopsis* pavement cells (Fujita et al. 2013). Moreover, exogenous cellulase treatment leads to smooth-shaped pavement cells in *Arabidopsis* (Higaki et al. 2016).

Such a scenario of lobe initiation could also be interpreted as the local buckling of the compound anticlinal cell wall. Namely, the swelling of cell wall matrix between the CMFs in the pad of one cell that is not accompanied by similar changes in the adjoining cell wall (B1 in Fig. 7d) would generate a compression that leads to the buckling of the compound wall (Fig. 7d'). Quite a different buckling mechanism of anticlinal wall undulation has been proposed for the *ric1* knockout mutant of *Arabidopsis* (Higaki et al. 2016). RIC1 together with ROP6 promote the ordering of cMTs and, as a result, they locally suppress the wall growth of one cell while in a complementing domain of the adjacent cell, ROP2 together with RIC4 induces the formation of the diffuse actin filaments that promote lobe formation. Rather surprisingly, in the cotyledon epidermis of the *ric1* mutant, the cells are lobed. Moreover, exogenous cellulase treatment does not affect lobe formation, unlike in the wild-type plants, although it promotes cell elongation. It has been postulated and supported by modeling that the lobed (undulated) anticlinal walls in the *ric1* mutant result from an excess of the anticlinal wall growth, thus leading to compression due to geometric constraints and finally to wall buckling. In the mutant, lobe formation would not be affected by cellulase treatment because cells respond to such treatment by secreting extra material into their walls, which would result in even longer walls and buckling. It is noteworthy that in such a scenario, the buckling applies to the entire segment of the compound anticlinal wall between the neighboring three-way junctions of the epidermal cell walls.

It is instructive to compare the putative contribution of buckling in the initiation of lobes of the compound anticlinal walls and in the opening of intercalary gas spaces. Both concern the anticlinal walls, but in the case of lobe initiation, the buckling would be at the compound wall scale, while in the case of space opening, it would be at the single wall scale (compare Fig. 7d-d' with Fig. 4c-c'). Thus, during lobe initiation, a single cell wall, which will bend outward from the cell due to buckling (A2 in Fig. 7d), would resist the tension resulting from the swelling in the other single wall (B1 in Fig. 7d) that is brought under compression. In turn, in the case of gas space formation, in both neighboring cells, the layers facing the cell lumen (B1 and B2 in Fig. 4c) would swell and be brought under compression, while the outer layers would resist the tension (A1 and A2 in Fig. 4c). The analogy between these two cases could be illustrated by the epidermis in *Adiantum* leaflets. Both lobes and gas spaces are formed in these epidermal cells (Panteris et al. 1993b). The compound anticlinal walls of the epidermal cells are undulated, while between the inner periclinal walls of the epidermis and the outer periclinal walls of subepidermis (mesophyll), intercalary gas spaces are formed. Along both the anticlinal and inner periclinal walls of the epidermal cells, cMTs form bundles that extend radially under the adjoining walls. However, the bundles under the inner periclinal walls have the opposite arrangement with respect to those under the adjoining walls of the subepidermal cells, while the cMTs under the anticlinal walls alternate with the neighboring epidermal cells. In both cases, the cMT bundles are accompanied by cell wall thickenings with aligned CMFs that are also opposite at the gas spaces but that alternate at the lobes.

Concluding Remarks and Future Perspectives

The various examples discussed above show that the folds or wrinkles of a cell or tissue surface may have different origins. We have focused on the cases that can be explained by buckling or the differential growth of cell walls. The clear cases of a buckling-based mechanism are tissue folding in contractile roots and during the initial stages of the formation of the intercalary gas spaces in the leaf mesophyll or petal epidermis. In the remaining cases presented above, differential growth can play a dominating role in the process. However, in both the mechanism based on buckling and in the one based on differential growth, the cell wall mechanics plays a crucial role. The mechanical properties and state of the cell wall in situ are also fundamental to explain the wrinkling of the cell wall surface after the removal of the stress. What definitely requires attention in the future and needs empirical support is the hypothesis about swelling-based stresses in cell walls. Such empirical studies would verify the postulates on the buckling-based mechanisms of cell wall wrinkling and folding as well as the non-turgor-driven growth presented in this chapter.

We have not addressed some cases of wall folding and wrinkling, mainly because of the elusive explanations of their formation mechanism. One of the very interesting examples is the wrinkling of cell walls that have been reported for the roots of *Arabidopsis* plants with a decreased level of CEL1 protein (Tsabary et al. 2003). This protein is a secreted cell wall enzyme that is most likely responsible for relaxation of the xyloglucan cross-links between the CMFs (Urbanowicz et al. 2007; Cosgrove 2016). The longitudinal walls of root epidermis in these plants, both anticlinal and periclinal, are wrinkled and the wrinkles are perpendicular to the long cell axis (Tsabary et al. 2003). This is puzzling because these root portions are still growing though to a lesser extent than in wild-type plants. In particular, the wrinkled cell walls elongate, i.e., expand in the direction across the wrinkles. It has been postulated that a reduced enzyme level results in only a partial cell wall relaxation while the CMFs that remain bound are responsible for the cell wall wrinkling. However, how the cell walls are brought under tensile stress so that growth is facilitated remains an open question.

Acknowledgements Work in D.K. research team is supported by the National Science Centre, Poland, research grant MAESTRO no. 2011/02/A/NZ3/00079. We thank Dr. Agata Burian for the discussions and valuable comments on this manuscript and Dr. Magdalena Raczyńska-Szajgin for the micrographs of the *A. grandiflora* petal epidermis. The drawings presented in the figures were prepared using Adobe Design Premium CS4 (Adobe Systems Inc. USA) and CorelDRAW X6 (Corel Corp.).

References

Abasolo WP, Yoshida M, Yamamoto H, Okuyama T (2009) Stress generation in aerial roots of *Ficus elastica* (Moraceae). IAWA J 30:216–224

Apostolakos P, Galatis B (1998) Probable involvement of cytoskeleton in stomatal-pore formation in *Asplenium nidus* L. Protoplasma 203:48–57

Apostolakos P, Galatis B (1999) Microtubule and actin filament organization during stomatal morphogenesis in the fern *Asplenium nidus*. II. Guard cells. New Phytol 141:209–223

Apostolakos P, Livanos P, Galatis B (2009) Microtubule involvement in the deposition of radial fibrillar callose arrays in stomata of the fern *Asplenium nidus* L. Cytoskeleton 66:342–349

Armour WJ, Barton DA, Law AMK, Overall RL (2015) Differential growth in periclinal and anticlinal walls during lobe formation in *Arabidopsis* cotyledon pavement cells. Plant Cell 27:2484–2500

Augustine SM, Cherian AV, Syamaladevi DP, Subramonian N (2015) *Erianthus arundinaceus* HSP70 (*EaHSP70*) acts as a key regulator in the formation of anisotropic interdigitation in sugarcane (*Saccharum* spp. hybrid) in response to drought stress. Plant Cell Physiol 56:2368–2380

Baskin TI, Jensen OE (2013) On the role of stress anisotropy in the growth of stems. J Exp Bot 64:4697–4707

Beauzamy L, Louveaux M, Hamant O, Boudaoud A (2015) Mechanically, the shoot apical meristem of arabidopsis behaves like a shell inflated by a pressure of about 1 MPa. Front Plant Sci 6:1038

Burian A, Ludynia M, Uyttewaal M, Traas J, Boudaoud A, Hamant O, Kwiatkowska D (2013) A correlative microscopy approach relates microtubule behaviour, local organ geometry and cell growth at the *Arabidopsis* shoot apical meristem. J Exp Bot 64:5753–5767

Bünning E, Biegert F (1953) Die Bildung der Spaltöffnungsinitialen bei *Allium cepa*. Z Bot 41:17–39

Campbell R (1972) Electron microscopy of the development of needles of *Pinus nigra* var. *maritima*. Ann Bot 36:711–720

Chen X, Yin J (2010) Buckling patterns of thin films on curved compliant substrates with applications to morphogenesis and three-dimensional micro-fabrication. Soft Matter 6:5667–5680

Coen E, Rolland-Lagan A-G, Matthews M, Bangham JA, Prusinkiewicz P (2004) The genetics of geometry. Proc Natl Acad Sci USA 101:4728–4735

Cosgrove DJ (2016) Catalysts of plant cell wall loosening [version1; referees: 2 approved]. F1000Research, 5:F1000 Faculty Rev-119

Cresswell A, Sackville Hamilton NR, Thomas H, Charnock RB, Cookson AR, Thomas BJ (1999) Evidence for root contraction in white clover (*Trifolium repens* L.). Ann Bot 84:359–369

Cyr RJ, Lin B-L, Jernstedt JA (1988) Root contraction in hyacinth. II. Changes in tubulin levels, microtubule number and orientation associated with differential cell expansion. Planta 174:446–452

Dumais J, Serikawa K, Mandoli DF (2000) *Acetabularia*: a unicellular model for understanding subcellular localization and morphogenesis during development. J Plant Growth Regul 19:253–264

Dumais J, Harrison L (2000) Whorl morphogenesis in the dasycladalean algae: the pattern formation viewpoint. Phil Trans R Soc Lond B 355:281–305

Dumais J, Steele ChR (2000) New evidence for role of mechanical forces in the shoot apical meristem. J Plant Growth Regul 19:7–18

Elsner J, Michalski M, Kwiatkowska D (2012) Spatiotemporal variation of leaf epidermal cell growth: a quantitative analysis of *Arabidopsis thaliana* wild-type and triple *cyclinD3* mutant plants. Ann Bot 109:897–910

Fisher JB (2008) Anatomy of axis contraction in seedlings from a fire prone habitat. Am J Bot 95:1337–1348

Frank MJ, Smith LG (2002) A small, novel protein highly conserved in plants and animals promotes the polarized growth and division of maize leaf epidermal cells. Curr Biol 12:849–853

Fujita M, Himmelspach R, Ward J, Whittington A, Hasenbein N, Liu Ch, Truong TT, Galway ME, Mansfield SD, Hocart ChH, Wasteneys GO (2013) The *anisotropy1* D604 N mutation in the *Arabidopsis* cellulose synthase1 catalytic domain reduces cell wall crystallinity and the velocity of cellulose synthase complexes. Plant Physiol 162:74–85

Galatis B (1980) Microtubules and guard-cell morphogenesis in *Zea mays* L. J Cell Sci 45:211–244

Galatis B, Mitrakos K (1980) The ultrastructural cytology of the differentiating guard cells of *Vigna sinensis*. Am J Bot 67:1243–1261

Galatis B, Apostolakos P (1991) Microtubule organization and morphogenesis of stomata in caffeine-affected seedlings of *Zea mays*. Protoplasma 165:11–26

Galatis B, Apostolakos P (2004) The role of the cytoskeleton in the morphogenesis and function of stomatal complexes. New Phytol 161:613–639

Gambles RL, Dengler RE (1982a) The anatomy of the leaf of red pine, *Pinus resinosa*. I. Nonvascular tissues. Can J Bot 60:2788–2803

Gambles RL, Dengler RE (1982b) The anatomy of the leaf of red pine, *Pinus resinosa*. II. Vascular tissues. Can J Bot 60:2804–2824

Geitmann A, Ortega JKE (2009) Mechanics and modelling of plant cell growth. Trends Plant Sci 14:467–478

Giannoutsou E, Sotiriou P, Apostolakos P, Galatis B (2013) Early local differentiation of cell wall matrix defines the contact sites in lobed mesophyll cells of *Zea mays*. Ann Bot 112:1067–1081

Gough HJ, Elam CF, de Bruyne NA (1940) The stabilization of a thin sheet by a continuous supporting medium. J R Aeronaut Soc 44:12–43

Green PB (1999) Expression of pattern in plants: combining molecular and calculus-based biophysical paradigms. Am J Bot 86:1059–1076

Green PB, Steele CS, Rennich SC (1996) Phyllotactic patterns: a biophysical mechanism for their origin. Ann Bot 77:515–527

Hamant O, Moulia B (2016) How do plants read their own shapes? New Phytol 212:333–337

Harris WM (1971) Ultrastructural observations on the mesophyll cells of pine leaves. Can J Bot 49:1107–1109

Harrison LG, Snell J, Verdi R, Vogt DE, Zeiss GD, Green BR (1981) Hair morphogenesis in *Acetabularia mediterranea*: temperature-dependent spacing and models of morphogen waves. Protoplasma 106:211–221

Harrison LG, von Aderkas P (2004) Spatially quantitative control of the number of cotyledons in a clonal population of somatic embryos of hybrid larch *Larix* x *leptoeuropaea*. Ann Bot 93:423–434

Hejnowicz Z (2011) Plants as mechano-osmotic transducers. In: Wojtaszek P (ed) Mechanical integration of plant cells and plants. Springer, Berlin, Heidelberg, pp p241–p267

Hejnowicz Z, Barthlott W (2005) Structural and mechanical peculiarities of the petioles of giant leaves of *Amorphophallus* (Araceae). Am J Bot 92:391–403

Hejnowicz Z, Borowska-Wykręt D (2005) Buckling of inner cell wall layers after manipulations to reduce tensile stress: observations and interpretations for stress transmission. Planta 220:465–473

Hejnowicz Z, Sievers A (1996) Tissue stresses in organs of herbaceous plants. III. Elastic properties of the tissues of sunflower hypocotyl and origin of tissue stresses. J Exp Bot 47:519–528

Higaki T, Takigawa-Imamura H, Akita K, Kutsuna N, Kobayashi R, Hasezawa S, Miura T (2016) Exogenous cellulose switches cell interdigitation to cell elongation in an RIC1-dependent manner in *Arabidopsis thaliana* cotyledon pavement cells. Plant Cell Physiol 58:106–119

Hiller GH (1872) Untersuchungen über die Epidermis der Blüthenblätter. Jahrb f wiss Bot 15:411–452

Hoss S, Wernicke W (1995) Microtubules and the establishment of apparent cell wall invaginations in mesophyll cells of *Pinus silvestris* L. J Plant Physiol 147:474–476

Jarvis MC (1998) Intercellular separation forces generated by intracellular pressure. Plant Cell Environ 21:1307–1310

Jeffree CE, Dale JE, Fry SC (1986) The genesis of intercellular spaces in developing leaves of *Phaseolus vulgaris* L. Protoplasma 132:90–98

Jung G, Wernicke W (1990) Cell shaping and microtubules in developing mesophyll of wheat (*Triticum aestivum* L). Protoplasma 153:141–148

Kennaway R, Coen E, Green A, Bangham A (2011) Generation of diverse biological forms through combinatorial interactions between tissue polarity and growth. PLoS Comput Biol 7:e1002071

Kollöffel C, Linssen PW (1984) The formation of intercellular spaces in the cotyledons of developing and germinating pea seeds. Protoplasma 120:12–19

Kaufman PB, Petering LB, Yocum CS, Baic D (1970) Ultrastructural studies on stomata development in internodes of *Avena sativa*. Am J Bot 57:33–49

Kaul RB (1971) Diaphragms and aerenchyma in *Scirpus validus*. Am J Bot 58:808–816

Kay QON, Daoud HS, Stirton CH (1981) Pigment distribution, light reflection and cell structure in petals. Bot J Lin Soc 83:57–84

Kotzer AM, Wasteneys GO (2006) Mechanisms behind the puzzle: microtubule-microfilament cross-talk in pavement cell formation. Can J Bot 84:594–603

Liang F, Shen L-Z, Chen M, Yang Q (2008) Formation of intercellular gas space in the diaphragm during the development of aerenchyma in the leaf petiole of *Sagittaria trifolia*. Aqu Bot 88:185–195

Martynov LA (1975) A morphogenetic mechanism involving instability of initial form. J Theor Biol 52:471–480

Moose SP, Sisco PH (1994) *Glossy15* controls the epidermal juvenile-to-adult phase transition in maize. Plant Cell 6:1343–1355

Niklas KJ, Paolillo DJ (1998) Preferential states of longitudinal tension in the outer tissues of *Taraxacum officinale* (Asteraceae) peduncles. Am J Bot 85:1068–1081

Palevitz BA, Hepler PK (1976) Cellulose microfibril orientation and cell shaping in developing guard cells of *Allium*: the role of microtubules and ion accumulation. Planta 132:71–93

Panteris E, Apostolakos P, Galatis B (1993a) Microtubule organization, mesophyll cell morphogenesis and intercellular space formation in *Adiantum capillus-veneris* leaflets. Protoplasma 172:97–110

Panteris E, Apostolakos P, Galatis B (1993b) Microtubule organization and cell morphogenesis in two semi-lobed cell types of *Adiantum capillus-veneris* L. leaflets. New Phytol 125:509–520

Panteris E, Apostolakos P, Galatis B (1994) Sinuous ordinary epidermal cells: behind several patterns of waviness, a common morphogenetic mechanism. New Phytol 127:771–780

Parker ChC, Parker ML, Smith AC, Waldron KW (2001) Pectin distribution at the surface of potato parenchyma cells in relation to cell-cell adhesion. J Agric Food Chem 49:4364–4371

Prat R, André JP, Mutaftschiev S, Catesson AM (1997) Three-dimensional study of the intercellular gas space in *Vigna radiate* hypocotyl. Protoplasma 196:69–77

Pütz N (1992) Measurement of the pulling force of a single contractile root. Can J Bot 70:1433–1439

Quintana A, Albrechtová J, Griesbach RJ, Freyre R (2007) Anatomical and biochemical studies of anthocyanidins in flowers of *Anagallis monelli* L. (Primulaceae) hybrids. Sci Hort 112:413–421

Raven JA (1996) Into the voids: the distribution, function, development and maintenance of gas spaces in plants. Ann Bot 78:137–142

Roland JC (1978) Cell wall differentiation and stages involved with intercellular gas space opening. J Cell Sci 32:325–336

Romberger JA, Hejnowicz Z, Hill JF (1993) Plant structure: function and development. Springer, Berlin

Sack FD, Paolillo DJ Jr (1983a) Stomatal pore and cuticle formation in *Funaria*. Protoplasma 116:1–13

Sack FD, Paolillo DJ Jr (1983b) Structure and development of walls in *Funaria* stomata. Am J Bot 70:1019–1030

Sack FD (1987) The development and structure of stomata. In: Zeiger E, Farquhar GD, Cowan IR (eds) Stomatal function. Stanford University Press, Stanford, pp p59–p89

Sampathkumar A, Krupiński P, Wightman R, Milani P, Berquand A, Boudaoud A, Hamant O, Jönsson H, Meyerowitz EM (2014) Subcellular and supracellular mechanical stress prescribes cytoskeleton behavior in *Arabidopsis* cotyledon pavement cells. eLife 3:e01967

Schreiber N, Gierlinger N, Putz N, Fratzl P, Neinhuis Ch, Burgert I (2010) G-fibres in storage roots of *Trifolium pratense* (Fabaceae): tensile stress generators for contraction. Plant J 61:854–861

Sego JL Jr, Marsh LC, Stevens KJ, Soukup A, Votrubová O, Enstone DE (2005) A re-examination of the root cortex on wetland flowering plants with respect to aerenchyma. Ann Bot 96:565–579

Sharon E, Efrati E (2010) The mechanics of non-Euclidean plates. Soft Matter 6:5693–5704

Sifton HB (1945) Air-space tissue in plants. Bot Rev 11:108–143

Singh AP, Srivastava LM (1973) The fine structure of pea stomata. Protoplasma 76:61–82

Smith-Huerta NL, Jernstedt JA (1989) Root contraction in hyacinth III. Orientation of cortical microtubules visualized by immunofluorescence microscopy. Protoplasma 151:1–10

Smith-Huerta NL, Jernstedt JA (1990) Root contraction in hyacinth IV. Orientation of cellulose microfibrils in radial longitudinal and transverse cell walls. Protoplasma 154:161–171

Sotiriou P, Giannoutsou E, Panteris E, Apostolakos P, Galatis B (2016) Cell wall matrix polysaccharide distribution and cortical microtubule organization: two factors controlling mesophyll cell morphogenesis in land plants. Ann Bot 117:401–419

Srivastava LM, Singh AP (1972) Stomatal structure in corn leaves. J Ultrastruct Res 39:345–363

Staff L, Hurd P, Reale L, Seoighe C, Rockwood A, Gehring C (2012) The hidden geometries of the *Arabidopsis thaliana* epidermis. PLoS ONE 7:e43546

Stebbins GL, Jain SK (1960) Developmental studies of cell differentiation in the epidermis of monocotyledons. I. *Allium*, *Rhoeo* and *Commelina*. Dev Biol 2:409–426

Stebbins GL, Shan SS (1960) Development studies of cell differentiation in the epidermis of monocotyledons. II. Cytological features of stomatal development in the Graminae. Dev Biol 2:477–500

Sylvester AW, Smith LG (2009) Cell biology of maize leaf development. In: Bennetzen JL, Hake SC (eds) Handbook of maize: its biology. Springer, New York, pp p179–p203

Szymanski DB (2014) The kinematics and mechanics of leaf expansion: new pieces to the *Arabidopsis* puzzle. Curr Opin Plant Biol 22:141–148

Tomlinson PB, Magellan TM, Griffith MP (2014) Root contraction in *Cycas* and *Zamia* (Cycadales) determined by gelatinous fibers. Am J Bot 101:1275–1285

Tsabary G, Shani Z, Roiz L, Levy I, Riov J, Shoseyov O (2003) Abnormal 'wrinkled' cell walls and retarded development of transgenic *Arabidopsis thaliana* plants expressing endo-1, 4-β-glucanase (*cel*1) antisense. Plant Mol Biol 51:213–224

Ugural AC (1999) Stresses in plates and shells. WCD McGraw-Hill, Boston-Toronto

Urbanowicz BR, Bennett AB, del Campillo E, Catalá C, Hayashi T, Henrissat B, Höfte H, McQueen-Mason SJ, Patterson SE, Shoseyov O, Teeri TT, Rose JKC (2007) Structural organization and a standardized nomenclature for plant endo-1, 4-β-glucanases (cellulases) of glycosyl hydrolase family 9. Plant Physiol 144:1693–1696

Wernicke W, Günther P, Jung G (1993) Microtubules and cell shaping in the mesophyll of *Nigella damascena* L. Protoplasma 173:8–12

Weston GD, Cass DD (1973) Observations on the development of the paraveinal mesophyll of soybean leaves. Bot Gaz 134:332–335

Wiebe HH, Al-Saadi HA (1976) The role of invaginations in armed mesophyll cells of pine needles. New Phytol 77:773–775

Yin J, Cao Z, Li C, Sheinman I, Chen X (2008) Stress-driven buckling patterns in spheroidal core/shell structures. P Natl Acad Sci USA 105:19132–19135

Zamski E, Ucko O, Koller D (1983) The mechanism of root contraction in *Gymnarrhena micranatha*, a desert plant. New Phytol 95:29–35

Zhang C, Halsey LE, Szymanski DB (2011) The development and geometry of shape change in *Arabidopsis thaliana* cotyledon pavement cells. BMC Plant Biol 11:27

Structural Principles in the Design of Hygroscopically Moving Plant Cells

Rivka Elbaum

Abstract Plants do not have mineralized skeletons. Instead, each of the plant's cells has an envelope of a cellulose-based wall, which provides a mechanical support to the organism. This stiff wall enables plants to assume flexible body shapes. However, the wall interferes with proteinous muscle-like movements of cells and organs because it is too stiff to yield to forces generated by motor proteins. Nevertheless, plants move constantly. The movements rely on water translocations, which result in the swelling (or growth) of cells located strategically. Water may swell protoplasts in movements that require live cells, like tip growth, tropism, and gas exchange. Other movements are initiated by the swelling of cell walls. These occur in dead tissues that can afford drying. The hygroscopically based movement is very common in seed dispersal mechanisms. The seed that detaches from the mother plant is carried by a cellulosic device. This device was synthesized by the plant and programmed to do some mechanical work, like jumping, crawling, and sowing, in order to deliver the seed to a germination location. This nonliving device provides the seed with means to move away from its mother and siblings. The movement may utilize several types of cells, which differ in the arrangement of cell wall cellulose microfibrils. I present here three types of contracting cells that, together with stiff fiber cells resisting any contraction, create a variety of hygroscopic movements.

Keywords Secondary cell wall · Hygroscopic movement · Cellulose microfibril angles · Seed dispersal · Coiling cells · Contraction · Fiber cells

Movements in Plants

Plants have a plastic shape modulated in reaction to external conditions, with no mineralized skeleton. The structural basic element that affords plants their mechanical capabilities is a stiff wall enveloping each cell. This element enables the plants

R. Elbaum (✉)
RH Smith Institute for Plant Sciences and Genetics in Agriculture, Hebrew University of Jerusalem, Rehovot 7010001, Israel
e-mail: rivka.elbaum@mail.huji.ac.il

A. Geitmann and J. Gril (eds.), *Plant Biomechanics*,
https://doi.org/10.1007/978-3-319-79099-2_11

to assume varied architectures, as each part of their body carries its weight (with the exception of climbers that get some support from their surroundings). While granting plants the freedom to design their bodies in accordance with their environment, the stiff walls impair motor protein-driven movements of cells and tissues. This is because the proteins are capable of generating forces that are too weak to strain the cells' walls (Geitmann 2016). Movements inside the plant cell by motor protein (myosin, actin, tubulin, and dynein) are described in chapter "Modelling, Evaluation and Biomechanical Consequences of Growth Stress Profiles Inside Tree Stems" in the book by Koller and Van Volkenburgh (2011). Nevertheless, plants exhibit significant large-scale movements all the time. Such movements are required for gas exchange through stomata, for following resources such as light and water, and avoiding hostile environments, and for reproduction. Instead of motor proteins, motion is driven by changes in the volume of cells as water moves in or out of them. Since the movement of water is relatively slow, most of the movements are too slow to grasp by our eyes (Geitmann 2016).

Some movements involve permanent cell growth. For example, the root and shoot tips are propelled by a growth-mediated motor tissue, which is situated behind the apex. Meristematic cells at the apex divide and feed into this motor. Within the motor, the cells elongate, pushing the growing apex into a new environment. As soon as the cells reach their final length, they leave the motor region and become a support to the moving tip. Thus, the growth-mediated motors are in a constant state of flux, following the movement that they create. Differential rates of growth in opposite areas of the growing tissue change the orientation of the supported growing tip. By these means, roots may forage for essential nutrients and water, and shoots direct themselves to light (Koller and Van Volkenburgh 2011).

Tendril tips too utilize growth-mediated motors to survey their surroundings in search of support. Once in touch with a potential support, contracting cells pull the plant onto it (Bowling and Vaughn 2009). The mechanism utilized to contract the cells is similar to that found in reaction wood tissues that pull tree branches upward against gravity. The cells, called gelatinous or G-fibers, contract during their development in parallel to the deposition of a gelatinous cell wall, which eventually fills most of their lumen (Clair et al. 2010). The tension persists in hydrated non-viable cells and is released after removal of the G-layer (Goswami et al. 2008). We thus infer that the walls, and specifically the G-layer, produce the contracting forces.

In mature tissues, movement cannot rely on cell's growth. Instead, we detect a change in the turgidity of cells that are located at strategic points in the tissue. In grasses, two rows of big bulliform (bubble-shaped) cells flank the major veins. When grass leaves experience drought, those bulliform cells contract and cause the leaf blade to roll along its length. This movement reduces the leaf exposed surface, which results in decreased sun exposure and transpiration. As water becomes available, the movement is reversed by the inflation of the bulliform cells, which flatten the leaf (Koller and Van Volkenburgh 2011). Other turgor-driven movement is the closing of mimosa leaves in response to touch. There, specialized motor organs called pulvini are located at the bases of the leaflets and petioles. The movement is initiated by a triggered unloading of sucrose from the phloem, resulting in the lowering of apoplastic water potential. Increase in the membrane permeability of cells at the lower sector of the pulvinus results in the loss of K^+ and Cl^- ions together with water, causing a turgor drop (Braam 2005). The movement is thought to protect the leaves from herbivory and mechanical damage. Recovery of turgor and spreading of the leaves will occur with energy-dependent import of K^+ and Cl^- ions. In contrast to the fluxing growth-mediated motor tissue, the turgor-mediated motors are static within the plant, creating reversible movements (Koller and Van Volkenburgh 2011).

Seed Dispersal

Seed dispersal is an extreme case, in which movement plays a very obvious and important role. Seeds normally detach from the mother plant, utilizing dispersal mechanisms that evolved over millions of years to optimize the survival odds of the new generation. Dispersal involves scattering away from the mother plant, establishing new populations, and escaping competition from mother and siblings. An opposing force of selection restricts dispersal distance, as the habitat that supported a successful mother plant may also support the development of a fertile new generation. We thus find intricate and versatile dispersal mechanisms that reflect the balance between these two conflicting selection forces (Fahn and Zohary 1955; Fahn and Werker 1972).

The explosive dispersion of cardamine seeds represents a case where metabolism of lignin could explain the evolution of a movement mechanism promoting distant seed dispersal (Hofhuis et al. 2016). The cardamine fruit is made of two long and narrow boat-shaped valves. Mature valves coil outward quickly and shoot the seeds that are loosely attached to them. Partial lignification of the inner cell layer of the valves stiffens the inner layer of the fruit, while allowing the valve to buckle transversally. The motor for the movement is the expansion of the valve's epidermis cells. During their final development, the cells swell transversely and contract longitudinally, pulling on the valve lengthwise. In parallel, the tissue connecting the valve‘s margins weakens by the activity of cell wall enzymatic modifications (Vaughn et al. 2011). This dehiscence tissue that is designed to crack allows the tensed valves to buckle, coil quickly, and shoot the loosely attached seeds to some distance. *Cardamine* is the only genus within the Brassicaceae family where shooting of seeds evolved, marking an evolutionary change in seed dispersal from a close to distant range (Hofhuis et al. 2016).

Plant Movement in Dead Tissues

In many cases, a module that carries the seeds is made of nonliving plant tissue. Movements can still take place in such structures. The cellulosic tissue may crawl and jump, delivering the seed to a safe germination locus. The motor for these

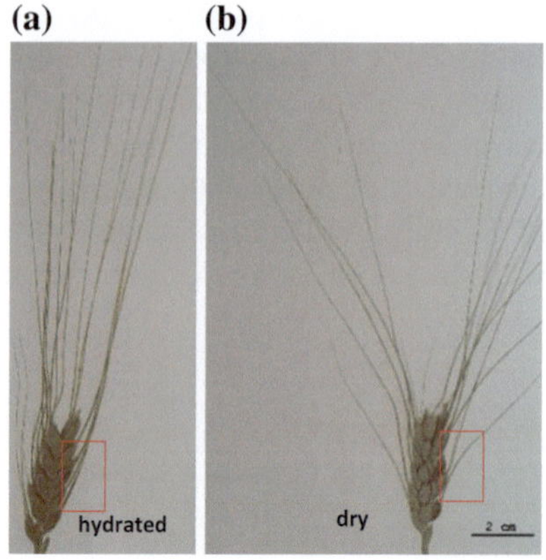

Fig. 1 Wheat awns bend as they dry. **a** Wet spike of domestic bread wheat (*Triticum aestivum*) presents straight awns (red rectangle). **b** With drying, the awns bend at their bases (red rectangle). Figures 2 and 3 show sections from a similar bending region. Scale bar is common to both panels. Figure adapted from Elbaum and Abraham (2014)

movements is hygroscopic, based on changes in the dimensions of cell walls as a result of changes in its hydration, (Koller and van Volkenburgh 2011). Hygroscopic movement may have a significant adaptive value because of its sensitivity to the momentary humidity. For example, in wind-dispersed pine seeds, dry conditions are likely to promote distant dispersal because the dry seed dispersal units will be lighter and less sticky, and may spread farther than wet units. Pinecone scales bend open as they dry, releasing seeds in weather more likely to carry the dispersal units on vertical winds (Nathan et al. 1999). The spontaneous movement of the dead pinecone tissue has inspired scientists to build self-moving windows that respond to meteorological conditions (Reichert et al. 2015). Another example of hygroscopic movement is presented in Fig. 1, showing that when a wheat spike dries, the long awns, which are attached to the seeds, bend at their bases. This property evolved in wild wheat plants, as a means for the seed dispersal units to push their seeds along and into the soil (Elbaum et al. 2007).

Structures of Cell Walls that Enable Motion in Dead Tissues

Both bending tissues in wheat awns and in pinecone scales are constructed of a bilayer: one layer is soft and contracts lengthwise as it dries, while the other layer is stiff and resists the contraction. The resulting dry structure is bent (Fig. 1b). Since the tissue is dead, the controlled movement must be imprinted in the structure of the cell walls. The walls can be viewed as a composite material made of partially crystalline cellulose bundles, called microfibrils. The microfibrils aggregate to bigger units of macrofibers by bonding to a less structured matrix of hemicellulose, pectin, proteins, lignin, and other aromatic molecules (Fig. 2). The orientation of the cellulose microfibrils, their cross-linkage through the matrix, and the composition

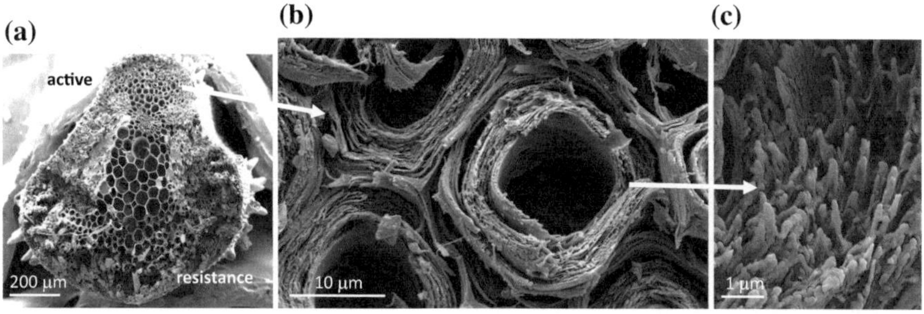

Fig. 2 Cellulose micro- and macrofibrils. a A cross section in a hygroscopically active wheat awn taken from pasta wheat (*Triticum turgidum*), imaged under scanning electron microscope. The region marked as "active" contracts as it dries (similar to the region marked by rectangles in Fig. 1). **b** A close-up of the active region, showing cell walls enveloping empty (dead) lumens. The walls are arranged in layers. **c** Cryo Scanning electron micrograph showing macrofibrils made of cellulose microfibrils that are glued by cell wall matrix materials into bundles (image taken by Stanislav Gorb). Panels a and b are adapted from Abraham and Elbaum (2013b)

of the matrix determine the mechanical properties of individual cells. For example, in cell walls of wood tissues, cellulose contributes to strength under tension, while lignin contributes to strength under compression (Salmén 2015 and Chap. 10). The microfibril orientation effect can be demonstrated in hairs of cottonseeds, where single cells form the fibers that we spin into threads. Cotton fibers are made of almost pure cellulose microfibrils, which are laid at an angle of 20–30° to the fiber cell's long axis (Sfiligoj Smole et al. 2013). This angle is called the microfibril angle (MFA), which is an important concept when discussing hygroscopic contraction. The deviation of the crystalline microfibers from the fiber axis makes this material more elastic as compared to flax, which has an MFA of about 10° (Baley 2002). Higher organization levels include the distribution pattern of these load-bearing cells in a tissue and the general shape of the organ, which will determine the ability of a specific plant organ to fulfill its mechanical role (Fig. 3).

We studied dispersal units in the grass and geranium families, which contain tissues that react to the daily cycle of air humidity. The dispersal units are capable of reversibly bending, twisting, or coiling, yielding intriguing structures (Elbaum and Abraham 2014). Seeds attached to such tissues may be shot off the mother plant, pushed along soil surfaces until trapped in porous soil, or even sown in the ground (Elbaum et al. 2007; Abraham and Elbaum 2013a).

Axially Contracting Cells

Many plant hygroscopic actuators contain long, stiff cells. The movement is created by cells that we call "active" that dry and contract in a predicted way. In such cells, the MFA is large (the cellulose microfibrils are aligned at a large angle to the cell's

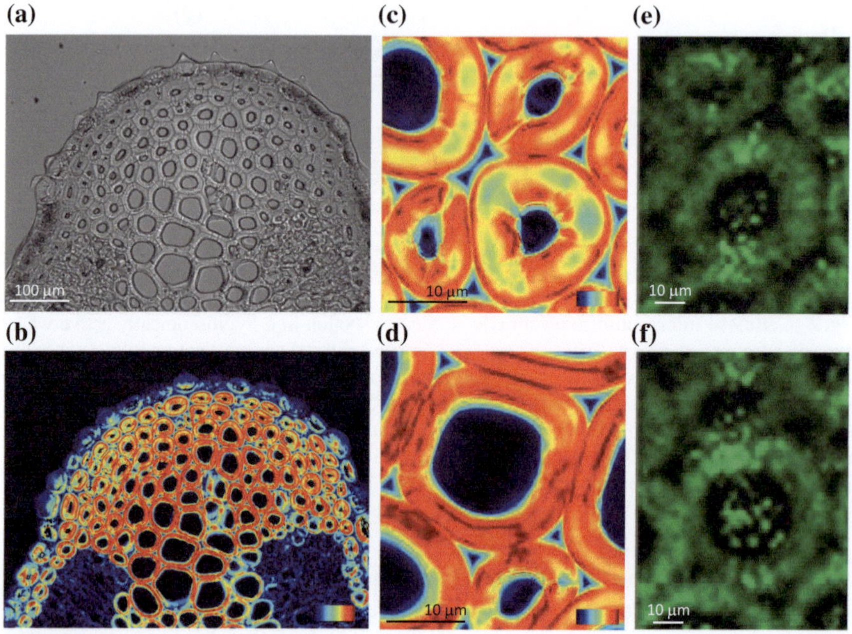

Fig. 3 Stiffness gradient in the active tissue of the wheat awn taken from pasta wheat (*Triticum turgidum*). **a** Bright-field and **b** polarized light images of the active tissue in a cross section of the awn bending region (similar to the region marked by rectangles in Fig. 1) (scale bar is common to both panels). Close-up (polarized light microscopy) of cells from the periphery **c** and the center **d** of the active region show variation in light retardance between the regions, which indicates a variation in cellulose microfibril angles (MFAs). The lower light retardance values at the periphery of the awn **c** indicate higher MFAs, making a stiffer shell that may stabilize the structure. **e** and **f** Raman images of the peak at 370 cm^{-1} (cellulose, orientation-insensitive) in peripheral **e** and central **f** cells, demonstrating that there is no variation in the cellulose concentration between the edge and center of the awn. Similar color codes were used in both panels, spanning 50–290 arbitrary Raman intensity units. Figure was adapted from Abraham and Elbaum (2013b)

long axis), and when they dry, the cell wall matrix contracts and pulls the microfibrils closer to each other (Fig. 4a). This results in a lengthwise contraction of single cells. The degree of contraction increases with the MFA. Isolated cells may exhibit a twist in addition to the contraction. Twist seems to be a result of the unchanging length of the microfibrils that limits some lengthwise contraction during the cell's lateral contraction (Burgert et al. 2005). Cells with rhomboid cross section that are fixed in a tissue are predicted to exhibit much smaller contraction as compared to tubular cells that may twist to some extent even when embedded in a tissue (Burgert et al. 2007).

In many cases, hygroscopically active tissues include two distinct types of aligned cells that are arranged in a bilayer: one layer contains active cells, while a second layer contains cells with a very dense cell wall, where the microfibrils are laid at a small MFA. These cells have a high tensile modulus, and hardly shorten when they desiccate. The force balance between the contracting and resisting layers results in the bending of the bilayer (Fig. 4), similar to a bi-metallic thermometer strip, though it is based on differential contraction with drying instead of cooling.

An example of such movement is shown in Figs. 1 and 2: in wheat awns, the active, softer layer contracts when dry, while the stiffer resisting layer does not shorten. The resulting shape is curved. The reversible change in shape with the cycling daily humidity enables the wild wheat dispersal unit to crawl along surfaces (video clip of crawling wheat seed dispersal in Elbaum et al. 2007). Bending motion is commonly applied in seed dispersal systems, including pinecone scales (Dawson et al. 1997), primary rays of wild carrot (Lacey et al. 1983), and some members of the geranium family (Abraham and Elbaum 2013a).

Isotropically Contracting Cells

Enforcing isotropically contracting cells by stiff fiber cells with very low MFA may also create a directed movement. In the legume family, twisting valves are constructed of a parenchyma layer that is sandwiched between two stiff fiber cells layers; each is aligned at a distinct angle to the pod's long axis (Fahn and Zohary 1955). Movement is initiated by the isotropic contraction of the parenchyma cells. The cellulose alignment

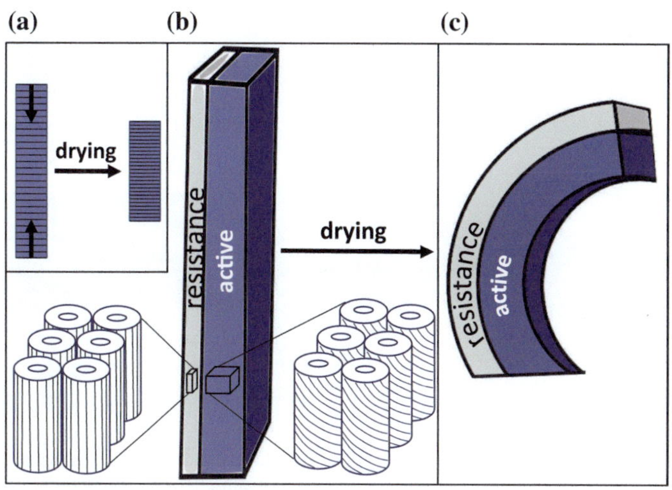

Fig. 4 General bilayered structure of a hygroscopic device made of cell walls. **a** A scheme of a 2D sheet consisting of stiff long elements (fibrils or fibers) (black) embedded in an amorphous contractible matrix (purple). The contraction of the sheet occurs perpendicular to the stiff elements. **b** A hydrated straight structure is constructed of a resistance tissue (left) made of stiff cells with a very low MFA. On the right, the active tissue is constructed of cells with large MFA. **c** While drying, the active cells contract and pull the bilayer into a bent position. Adapted from Elbaum and Abraham (2014)

in the parenchyma cells is not detected via X-ray diffraction (Armon et al. 2011), suggesting that the microfibrils do not conform to a specific alignment. Variations in the width of the pod and the angle between the stiff cells and the long axis of the pod may lead to changes in the type of movement, from bending and twisting to coiling (Armon et al. 2011).

A variation on this motif is found in the opening mechanism of sesame capsules, where the parenchyma cells form an active layer that pulls on stiff sclerenchymatous fiber cells. The fiber cells are arranged in two layers, parallel and perpendicular to the long axis of the capsule (Fig. 5). The relative thickness of the longitudinal fiber cell layer in relation to the parenchyma moderates the extent of movement, creating a twist in the capsule (Shtein et al. 2016).

Coiling Cells

The geranium family members rely on hygroscopic bending and coiling to disperse their seeds. We uncovered differences in the structure of awns, relating them to the diverse seed dispersal strategies that are exhibited within this family (Abraham and

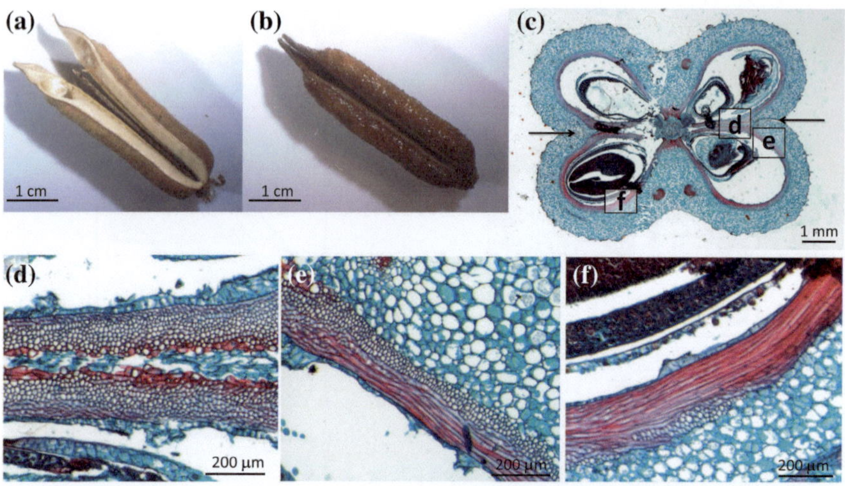

Fig. 5 Twist in sesame (*Sesamum indicum*) capsules. Capsule opening when dry **a** and closing when wet (**b**). **c** A cross section from the middle of the sesame capsule stained with safranin-fast green. The lignified endocarp (red) resisting the isotropic contraction of the parenchyma (stained in blue). Black arrows indicate on the false septum making up the disconnecting dehiscence tissue, which bisects the capsule. Black squares mark the regions enlarged in panels (**d**)–(**f**), presenting variation in the thicknesses of longitudinal and circumferential cell layers, which build a fiber bilayer around the endocarp. The variation in the thickness of the longitudinal fiber layer and the parenchyma control the longitudinal curvature of the capsule around its circumference (Shtein et al. 2016). Figure adapted from Elbaum and Abraham (2014)

Elbaum 2013a). In addition to bilayer structures, we observed coiling structures (video clip of a coiling awn, in Abraham et al. 2012). In contrast to bending motion, a bilayer is not necessary for the movement, coiling of the tissue stems from intrinsically coiling cells (video clip of coiling cells, in Abraham et al. 2012). The coil tightness increases while drying, creating motion (Fig. 6). Thus, in this case, the hygroscopic movement stems from a single cell type (Abraham et al. 2012).

The coiling of cells is a direct result of the cellulose microfibril arrangement. The stiff microfibrils are laid at a changing MFA around the circumference of the cell (Aharoni et al. 2012; Abraham et al. 2012). Such change in angle is demonstrated by the variation in polarized light retardance along a single cell wall layer (Fig. 7). This is in contrast to non-coiling cells, where the MFA is fixed around the circumference of each cell wall layer.

Summary

The arrangement of cellulose microfibrils within the cell wall determines the direction in which the cell wall material would be strongest under tension. The perpendicular direction is prone to contraction during drying. Matrix polymers (mainly hemicellulose, pectin, and lignin) cross-link the cellulose microfibrils. The matrix increases the resistance of the composite material to compression and dictates the way it interacts

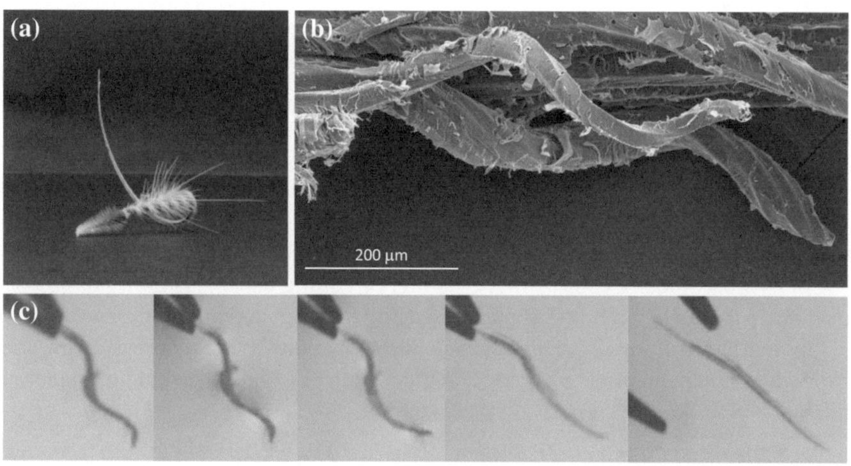

Fig. 6 Reaction of a coiling cell to desiccation. **a** Awn of stork's bill (*Erodium gruinum*) after its dissemination. The total length of the uncoiled awn is about 10 cm, the coiling part being about 3.8 cm when straight. **b** Scanning electron image of fractured dry awn, showing coiling of single cells and a group of cells (adapted from Abraham et al. 2012). **c** A sequence of pictures of a dry cell of about 1 mm (left) isolated from the awn, which straightens upon wetting. The time spanned by the image sequence is 1.5 s

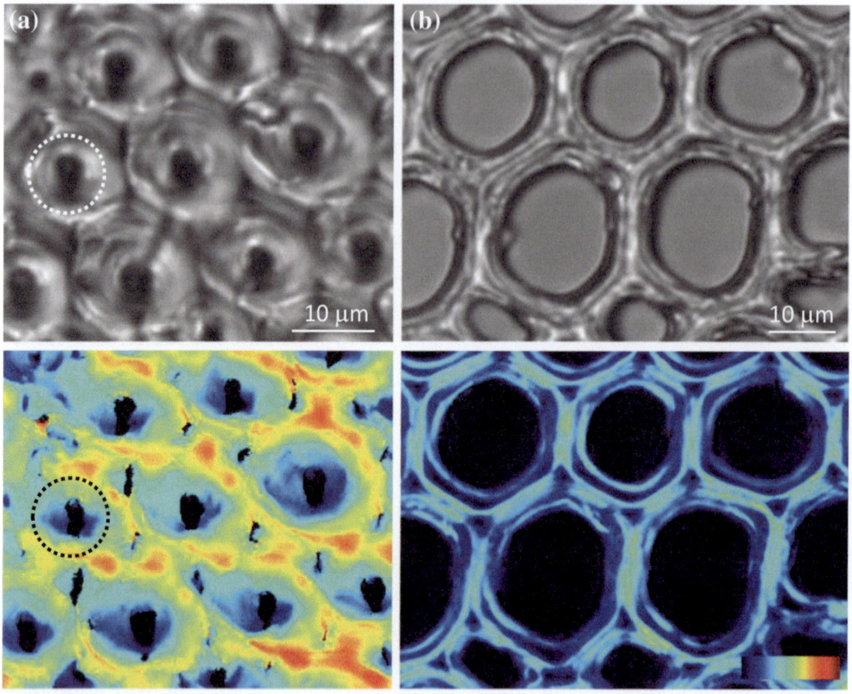

Fig. 7 Cross sections of plant tissues demonstrating variation in light retardance patterns. **a** Coiling cells of *Erodium gruinum* bright-field (top) and light retardance image (bottom). The retardance, which reflects the microfibril tilt angle, varies along the cell wall layers. A circle is marked on both images to help following one cell wall layer. **b** Bright-field (top) and light retardance (bottom) images of a cross section of an *Araucaria excelsa* branch. The microfibril angle in these cells is fixed along the cell wall layers, as demonstrated by the uniform retardance. The retardance scale, spanning 0–273 nm, is common to bottom panels. Figure adapted from Abraham and Elbaum (2013b)

with water. Versatility in cellulose deposition by the protoplast enables the formation of hygroscopically contracting, twisting, or coiling cells. It is even possible to imagine a single fiber cell that contains twisting, coiling, and inert regions along its length. Further exploration of hygroscopic dispersal units is likely to reveal new and surprising movement strategies, enriching the scope of our understanding material designs.

Acknowledgements I wish to thank Yael Abraham for 6 years of productive work. Thanks to Michael Elbaum and Jaime Kigel for critically reading the manuscript, and to Stanislav Gorb for the Cryo-SEM image.

References

Abraham Y, Elbaum R (2013a) Hygroscopic movements in Geraniaceae: the structural variations that are responsible for coiling or bending. New Phytol 199:584–594. https://doi.org/10.1111/nph.12254

Abraham Y, Elbaum R (2013b) Quantification of microfibril angle in secondary cell walls at subcellular resolution by means of polarized light microscopy. New Phytol 197:1012–1019. https://doi.org/10.1111/nph.12070

Abraham Y, Tamburu C, Klein E, Dunlop JWC, Fratzl P, Raviv U, Elbaum R (2012) Tilted cellulose arrangement as a novel mechanism for hygroscopic coiling in the stork's bill awn. J R Soc Interface 9:640–647. https://doi.org/10.1098/rsif.2011.0395

Aharoni H, Abraham Y, Elbaum R, Sharon E, Kupferman R (2012) Emergence of spontaneous twist and curvature in non-euclidean rods: application to *Erodium* plant cells. Phys Rev Lett 108:238106. https://doi.org/10.1103/physrevlett.108.238106

Armon S, Efrati E, Kupferman R, Sharon E (2011) Geometry and mechanics in the opening of chiral seed pods. Science 333:1726–1730. https://doi.org/10.1126/science.1203874

Baley C (2002) Analysis of the flax fibres tensile behaviour and analysis of the tensile stiffness increase. Compos Part Appl Sci Manuf 33:939–948. https://doi.org/10.1016/S1359-835X(02)00040-4

Bowling AJ, Vaughn KC (2009) Gelatinous fibers are widespread in coiling tendrils and twining vines. Am J Bot 96:719–727. https://doi.org/10.3732/ajb.0800373

Braam J (2005) In touch: plant responses to mechanical stimuli. New Phytol 165:373–389. https://doi.org/10.1111/j.1469-8137.2004.01263.x

Burgert I, Eder M, Gierlinger N, Fratzl P (2007) Tensile and compressive stresses in tracheids are induced by swelling based on geometrical constraints of the wood cell. Planta 226:981–987. https://doi.org/10.1007/s00425-007-0544-9

Burgert I, Frühmann K, Keckes J, Fratzl P, Stanzl-Tschegg S (2005) Properties of chemically and mechanically isolated fibres of spruce (Picea abies [L] Karst). Part 2: Twisting phenomena. Holzforschung 59:247–251

Clair B, Almeras T, Pilate G, Jullien D, Sugiyama J, Riekel C (2010) Maturation stress generation in poplar tension wood studied by synchrotron radiation microdiffraction. Plant Physiol 152:1650–1658. https://doi.org/10.1104/pp.109.149542

Dawson C, Vincent JFV, Rocca A-M (1997) How pine cones open. Nature 390:668

Elbaum R, Abraham Y (2014) Insights into the microstructures of hygroscopic movement in plant seed dispersal. Plant Sci 223:124–133. https://doi.org/10.1016/j.plantsci.2014.03.014

Elbaum R, Zaltzman L, Burgert I, Fratzl P (2007) The role of wheat awns in the seed dispersal unit. Science 316:884–886. https://doi.org/10.1126/science.1140097

Fahn A, Werker E (1972) Anatomical mechanisms of seed dispersal. Seed biology, Kozlowsky TT. Academic Press, New York, pp 151–221

Fahn A, Zohary M (1955) On the pericarpial structure of the legumen, its evolution and relation to dehiscence. Phytomorphology 5:99–111

Geitmann A (2016) Actuators acting without actin. Cell 166:15–17. https://doi.org/10.1016/j.cell.2016.06.030

Goswami L, Dunlop JWC, Jungnikl K, Eder M, Gierlinger N, Coutand C, Jeronimidis G, Fratzl P, Burgert I (2008) Stress generation in the tension wood of poplar is based on the lateral swelling power of the G-layer. Plant J 56:531–538. https://doi.org/10.1111/j.1365-313X.2008.03617.x

Hofhuis H, Moulton D, Lessinnes T, Routier-Kierzkowska A-L, Bomphrey RJ, Mosca G, Reinhardt H, Sarchet P, Gan X, Tsiantis M, Ventikos Y, Walker S, Goriely A, Smith R, Hay A (2016) Morphomechanical innovation drives explosive seed dispersal. Cell 166:222–233. https://doi.org/10.1016/j.cell.2016.05.002

Koller D, Van Volkenburgh E (2011) The Restless Plant. Harvard University Press, Cambridge, Mass

Lacey EP, Kaufman PB, Dayanandan P (1983) The anatomical basis for hygroscopic movement in primary rays of Daucus carota ssp. carota (Apiaceae). Bot Gaz 144:371–375

Nathan R, Safriel UN, Noy-Meir I, Schiller G (1999) Seed release without fire in Pinus halepensis, a Mediterranean serotinous wind-dispersed tree. J Ecol 87:659–669. https://doi.org/10.1046/j.1365-2745.1999.00382.x

Reichert S, Menges A, Correa D (2015) Meteorosensitive architecture: Biomimetic building skins based on materially embedded and hygroscopically enabled responsiveness. Comput-Aided Des 60:50–69. https://doi.org/10.1016/j.cad.2014.02.010

Salmén L (2015) Wood morphology and properties from molecular perspectives. Ann For Sci 72:679–684. https://doi.org/10.1007/s13595-014-0403-3

Sfiligoj Smole M, Hribernik S, Stana Kleinschek K, Kreže T (2013) Plant fibres for textile and technical applications. In: Advances in agrophysical research

Shtein I, Elbaum R, Bar-On B (2016) The hygroscopic opening of sesame fruits is induced by a functionally graded pericarp architecture. Plant Biophys Model 1501. https://doi.org/10.3389/fpls.2016.01501

Vaughn KC, Bowling AJ, Ruel KJ (2011) The mechanism for explosive seed dispersal in Cardamine hirsuta (Brassicaceae). Am J Bot 98:1276–1285. https://doi.org/10.3732/ajb.1000374

Using Modeling to Understand the Hygromechanical and Hysteretic Behavior of the S2 Cell Wall Layer of Wood

Dominique Derome, Karol Kulasinski, Chi Zhang, Mingyang Chen and Jan Carmeliet

Abstract To understand moisture sorption and swelling of wood requires revealing the behavior at the S2 cell wall layer, one of the layers of the secondary cell wall, at the atomistic scale. Difficulties in experimentally determining the organization and properties of S2 layer at such a small scale are a stumbling block for understanding of swelling and sorption in their full complexity. Recent works using atomistic modeling (Molecular Dynamics (MD) and Grand Canonical Monte Carlo) provide complementary insights. The linear dependence between moisture content, swelling, and porosity change is found to be correlated with the number and location of water–polymer hydrogen bonds within the system. Such information is upscaled for general use within a poromechanical framework. This chapter summarizes recent new physical insights in the sorption and swelling behavior of the S2 cell wall layer, stemming from validated MD work. The presented methodology is also used to unravel other moisture-related mechanisms of wood, such as hysteretic behavior.

Keywords S2 layer composite material · Cellulose (crystalline and paracrystalline) · Hemicelluloses · Lignins · Hygromechanical behavior Hysteresis · Experimental data · Molecular dynamics (atomistic modeling) Sorption and swelling of wood

D. Derome (✉) · C. Zhang · M. Chen
Laboratory of Multiscale Studies of Building Physics, Empa, Dübendorf, Switzerland
e-mail: Dominique.Derome@empa.ch

K. Kulasinski
Department of Geochemistry, Lawrence Berkeley National Laboratory, Berkeley, CA, USA

C. Zhang · M. Chen · J. Carmeliet
Chair of Building Physics, Department of Mechanical and Process Engineering, ETH Zürich, Zürich, Switzerland

A. Geitmann and J. Gril (eds.), *Plant Biomechanics*,
https://doi.org/10.1007/978-3-319-79099-2_12

Introduction

Wood, an orthotropic cellular biomaterial, has the capacity of adsorbing water molecules from the surrounding environment into its hierarchical material structure. As water molecules attach themselves to the hydrophilic matrix in the cell walls, the induced fluid–solid interaction forces result in a swelling of the cell walls. Moisture-induced internal stresses due to restrained swelling highly influence the hygromechanical behavior of wood as observed at the macroscale. Adsorption of moisture in wood, in the hygroscopic range, i.e., until around 30% moisture content mass per mass, results in swelling up to 10% volumetrically and reduces the stiffness depending on the grain direction, with, e.g., a decrease of 20% in the longitudinal direction and by half in the radial and tangential directions between 5 and 25% MC for spruce (Neuhaus 1981, Forest Products Laboratory 2010).

Wood structural higher hierarchical levels can be considered as lumber, growth ring, cell, and layered cell wall material. The variation of the cellular structure across the growth ring has been found responsible, at least in part, for the anisotropy of swelling and stiffness (Boutelje 1962; Watanabe et al. 2000). From cellular and subcellular investigations, it has become clear that the microscopic origin of swelling lies at the scales of the cell wall composite material.

The cell wall of sclerenchyma cells is composed of four layers built as exoskeleton of the cell. The primary (P) layer is the first layer which is produced as the cell is growing. At mature stage, the cell builds its secondary (S) layer in three phases. The thin internal and external cell wall layers (i.e., S3 and S1) act as restraining corsets due to the winding of the cellulose microfibrils around the cell. In the central and, by far, thickest cell wall layer, namely S2, the cellulose microfibrils are almost parallel to the longitudinal axis of the cells, although the presence of an angle (called microfibril angle, MFA) results in a helicoidal organization of the fibrils. The hypothesis, that microfibrils were parallel to each other, is presently challenged as speculations of more complex configurations tend to be confirmed lately with tomographic transmission electron microscopy (Xu et al. 2007) or as developed by Salmén and Burgert (2009).

The cell wall composite material is made of the stiff cellulose microfibrils, mentioned above, embedded in a soft polymeric matrix, as shown in Fig. 1a. The microfibrils are made of crystalline cellulose (hydrophobic or hydrophilic depending on the surfaces of the crystal, where crystal dimensions are roughly 3 × 3 × 25 nm) bound by hydrophilic amorphous hemicellulose, thus making long thin filaments. Amorphized cellulose may occur at this crystal/hemicellulose interface. These microfibrils can form bundles, named aggregates. Microfibrils and aggregates lie within the hydrophilic amorphous matrix, composed of different hemicelluloses and lignins. Hemicelluloses of interest here are galactoglucomannan and arabinoglucoronoxylan (Salmén 2004). Measurements indicate that the components closest to the microfibrils are somewhat oriented along these, while an orientation is not found for components lying in the center of the inter-microfibril matrix (Salmén and Burgert 2009). Still, the object of debate (Eichhorn 2011; Cosgrove 2014), the chemical components of the

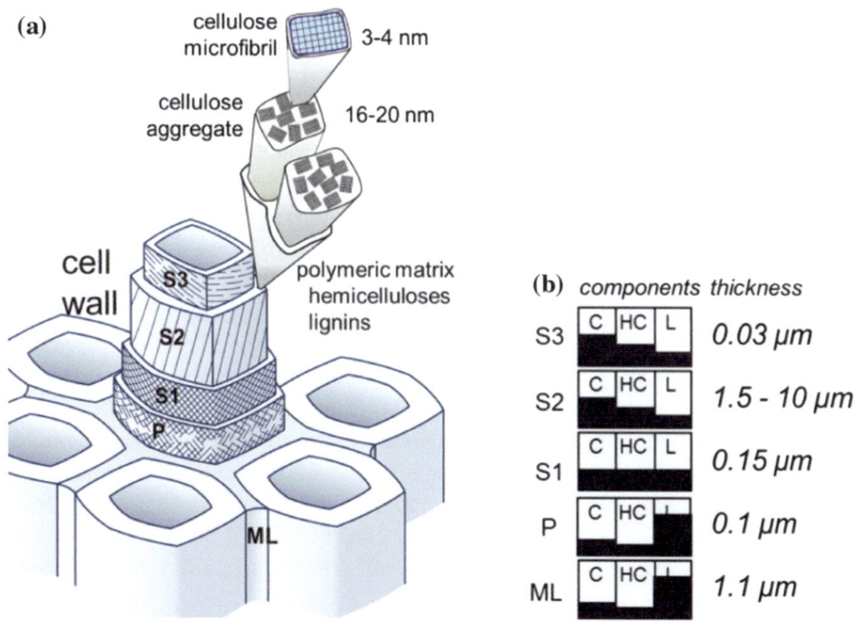

Fig. 1 **a** Schematic representation of the structure of the wood cell walls, with P primary layer, S secondary layer, ML middle lamella, **b** schematic representation of the proportion of cellulose (C), hemicelluloses (HC) and lignins (L) in the different cell wall layers and their thickness (Derome et al. 2012, with permission)

S2 layer and their relative configuration are progressively elucidated. Nevertheless, the distance between cellulose microfibrils is estimated to be in the order of 4–10 nm in dry cell walls and filled with the amorphous matrix (Dinwoodie 2000; Fahlén and Salmén 2005; Höfter et al. 2007). Proportion of the cellulose (C), hemicelluloses (HC), and lignins (L) are given schematically for each wall layer in Fig. 1b, showing that, overall, the cell wall material is composed in almost equal share of stiff cellulose microfibrils and soft polymeric matrix.

Experimental evidence of swelling has been long attained at wood scale. In particular, at the cellular tissue scale, a series of phase contrast synchrotron X-ray tomography (at the Tomcat beamline, Swiss Light Source, Paul Scherrer Institute) has been performed to identify the role of growth ring and cellular geometry in terms of effects on swelling, swelling anisotropy as function of density and hysteresis in swelling due to sorption hysteresis (Derome et al. 2011; Patera et al. 2013 and illustrated in Fig. 2a, b). Careful geometry analysis of the hierarchical structure of wood tissues is performed during moisture-induced deformations (swelling/shrinkage) at cellular and sub-cellular scales. Swelling in adsorption and shrinkage in desorption displays hysteresis, with higher deformation in desorption than in adsorption at the same relative humidity. Moisture content curves are also hysteretic with higher moisture content in desorption than in adsorption for the same relative humidity. However, softwood swelling is found to be non-hysteretic when considered in terms of moisture content highlighting the capacity of wood to deform similarly for the same moisture content and independently of moisture history. Further, these experimental results show that anisotropy in swelling in the cross-sectional directions is function of density, where the swelling of earlywood is restrained in the radial directions, whereas the latewood, given its bulky structure, deforms quite isotropically. Further, work using phase contrast synchrotron X-ray nanotomography captured the swelling of the isolated cell wall materials, sculpted as pillars with focused ion beam, using micropillars of the S2 layer (Fig. 2c and Rafsanjani et al. 2014) and of the middle lamella (ML) layer (Fig. 2d and Patera 2014) and subjecting them to different relative humidity. Observations reveal an anisotropic swelling in the S2 layer sample and an isotropic swelling in the ML sample, i.e. made of cell binding material which does not contain microfibrils. This indicates that swelling is driven from a sub-cellular scale. To summarize, Fig. 3 shows how wood structure constrains the swelling of its base material, with swelling coefficients for the S2 micropillars demonstrating a much higher volumetric swelling than those of the cellular tissues, either from earlywood or latewood. At cellular scale, the corset action of the S1 and S3 layers must enter in play. At timber scale, in presence of growth rings, this swelling is even further constrained.

Water molecules can easily form hydrogen bonds with the hydroxyl groups of the amorphous components. For example, amorphous cellulose is a strongly hydrophilic biopolymer (e.g., Beever and Valentine 1958) due to its open porous structure and the presence of exposed hydroxyl (O_2 and O_3) and hydroxymethyl (O_6) groups. Thus, as the adsorption process involves mainly atomistic interactions, molecular dynamics (MD) simulations are an appropriate tool for its investigation. Typical scales of the process range from angstroms to micrometers. Contrary to mesoscale methods, such as coarse grain MD, atomistic MD can capture hydrogen bonds formation and dipoles interaction and, unlike better resolution methods (e.g., DFT, generally limited to sizes of hundreds of atoms and time-scales of picoseconds), MD enables simulation of larger structures (up to millions of atoms) and at larger time scales (up to μs). For the problem of moisture-related behavior, atomistic MD offers several advantages. First, it allows full consideration of explicit water molecules. As such, MD can deliver information on hydrogen bonds (Kulasinski et al. 2015a; Nishiyama et al. 2002) and on dipoles of the water molecules (e.g., Teleman et al. 1987; Kulasinski et al. 2015c) and it can track the motion of molecules to document diffusion (Kulasinski et al. 2015d). Second, if needed, MD allows control of atoms as external forces can be applied to a group of atoms, hence enabling experimentally challenging operations, such as polymer chain pulling or free energy determination. Finally, as each component is homogeneous, the properties of each component can be determined by applying periodic boundary conditions on the representative volume element (RVE). This means that atoms in the system interact across the boundary, and they can exit at one end of the system and reenter from the other end. Furthermore, the application of periodic boundary conditions allows determining the long-range electrostatic forces by Ewald summation instead of truncation, which gives a better evaluation of the long-range electrostatic interactions (Oostenbrink et al. 2004; Lins and Hünenberger

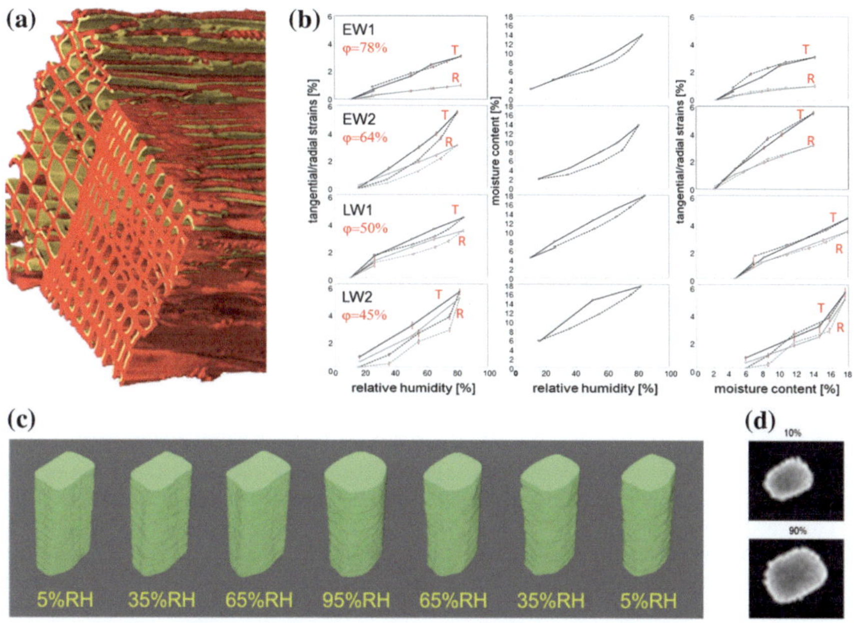

Fig. 2 Imaging using X-ray CT during swelling experiment of **a** cellular structure at growth ring border at 25% (red) and 50% RH (ochre), **b** swelling strains and moisture content versus relative humidity, and swelling strains versus moisture content for 4 homogeneous wood tissues of high to low porosity (φ) (EW earlywood, LW latewood, T tangential, R radial, dashed line in adsorption and solid line in desorption) (from Patera et al. 2013, with permission) **c** S2 micropillar during adsorption and desorption, **d** ML micropillar at 10 and 90% RH (from Patera 2014)

2005). For all these reasons, MD has been used toward studying the S2 cell wall material behavior.

The goal of this chapter is to convey the physical processes at play during water sorption in wood cell wall using the particular lens of recent MD investigations. The specific objectives are as follows:

1. To provide an overview of the method, capacities, and limits of molecular dynamics
2. To display the possible types of investigations and results achievable with this modeling approach when looking at the impact of water on the behavior of the S2 cell wall layer
3. To expose the underlying role of hydrogen bonding, as a unifying agent.

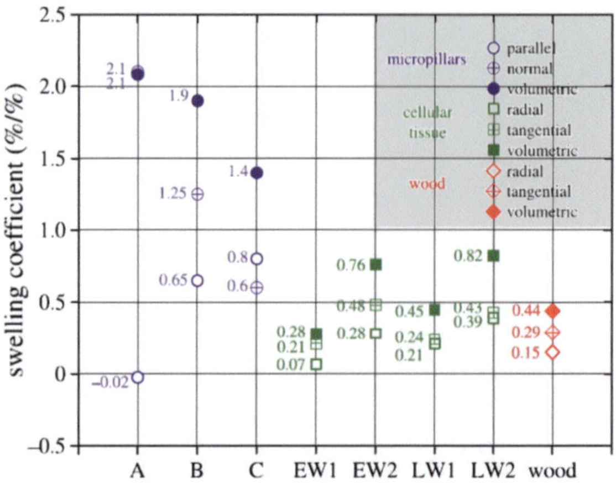

Fig. 3 Comparison of swelling coefficients at three scales: samples A, B, C are S2 micropillars (μm), the next four samples are pure early- or latewood samples (500 × 500 μm) (Patera et al. 2013) and for wood at plank scale cm (Rafsanjani 2013b), all of Norway spruce (from Rafsanjani et al. 2014, with permission)

Methods

Molecular Dynamics

Molecular dynamics simulations are carried out, for the most part, using Gromacs software (Hess et al. 2008) with leap-frog algorithm for integration of Newton's equations of motion and the Gromos 53a6 united-atom force field (Oostenbrink et al. 2004; Lins and Hünenberger 2005). In section "Hysteresis", molecular dynamics simulations are carried out using LAMMPS (Large-Scale Atomic/Molecular Massively Parallel Simulator) (Plimpton 1995) with velocity-Verlet algorithm for integration of Newton's equations of motion and the PCFF (Polymer Consistent Force Field) force field (Sun et al. 1994).

The size of the studied systems (RVE) is in the order of nanometers, counting 10^5 atoms at maximum and the sampling time could reach 20 ns. All MD measurements are carried out at a constant temperature, 300 K, and at constant pressure (using the isobaric–isothermal ensemble NPT, i.e., constant number of atoms, pressure, and temperature, while allowing variation in volume).

The initial configuration of the each polymer is constructed based on the position of atoms in the chains as known from experiments or other MD works, e.g., for xylan (Wagenführ and Scholz 2012; Jin et al. 2015) and for lignin (with predefined sequence used in Charlier and Mazeau 2012; Petridis and Smith 2009; Petridis et al. 2011a, b; Sangha et al. 2012; or dynamic bond-forming procedure used by Jin et al. 2015). This initial structure must then find an equilibrium configuration. The procedure of amorphization of polymeric chains has two stages (Kulasinski et al. 2014a).

First, the polymer is raised to a temperature above its melting point and equilibrated there in stress-free conditions until the chains lose their ordered structure, typically after 1 ns. Then, the amorphized chains are quenched to room temperature and equilibrated in stress-free conditions. For example, for single component, the number of polymer atoms for such simulation is around 3000–5000, resulting in a domain of $3 \times 3 \times 3$ nm^3 and the largest system built to represent the S2 materials, with four cellulose crystal separated by galactoglucomannan and surrounded by lignin, counts close to 38 000 atoms and has a volume of 516 nm^3. Periodic boundary conditions (PBC) are employed in order to mimic bulk material properties, despite the small size of the RVE. For those polymers with their size exceeding the MD box dimensions, the PBC is achieved by covalently bonding the molecules to themselves across the periodic boundary. For the S2 material, the RVE is configured as a unit which, when repeated via the periodic boundaries conditions, yields an acceptable S2 overall configuration.

In order to study the different responses of the polymers to water, e.g., the responses being here mechanical, topological, hygric or thermal, water molecules are introduced to the system (Smith and van Gunsteren 1994). The water model used is the simple point charge water model (SPC) (from Berendsen et al. 1984). The SPC water model has three interaction points corresponding to the three atoms of the water molecule, with point charge on each site. Starting from a dry system, water molecules are inserted one by one at random into the void space between the polymer chains, see Fig. 4. A successful insertion of a water molecule is followed by a relaxation run at stress-free conditions. Relaxation time and temperature usually need to be adapted to each system and moisture content. Once the desired number of water molecules has been inserted and the system allowed to relax, the system is interrogated during production runs (Kulasinski et al. 2014b, 2015a, b, c, d). A hydrogen bond can be recognized when the acceptor and the donor of a couple are within 0.35 nm from each other and the angle between the prolongation of the OH branch with the hydrogen bond is less than 30°, as shown in Fig. 5.

The production runs are used to study the moisture-related behavior of the different polymers and the results are compared with experimental data for validation, as explained below. In particular, the following properties can be determined and are described in more detailed below. Water adsorption isotherms, as shown in Figs. 6 and 7, are obtained using One-Step Perturbation method (Zwanzig 1954) for the determination of chemical potential. Swelling of the system at each water insertion leads to the determination of the dependence of swelling coefficient (Fig. 8) and porosity (Fig. 9) on moisture content. The mean square displacement of the water molecules is used to determine water diffusion coefficient versus moisture content (Fig. 10). Under mechanical loading, i.e., an anisotropic barostat is applied and the pressure is specified, the system finds its equilibrium and a corresponding strain can be measured. The bulk and shear moduli dependence on moisture content can be determined, as shown in Fig. 11.

An analysis of the nanoscopic behavior of polymer–water systems is possible given the unique level of information provided by MD simulations. For example, the shift in the diffusion coefficient dependence is seen to concur with the percolation

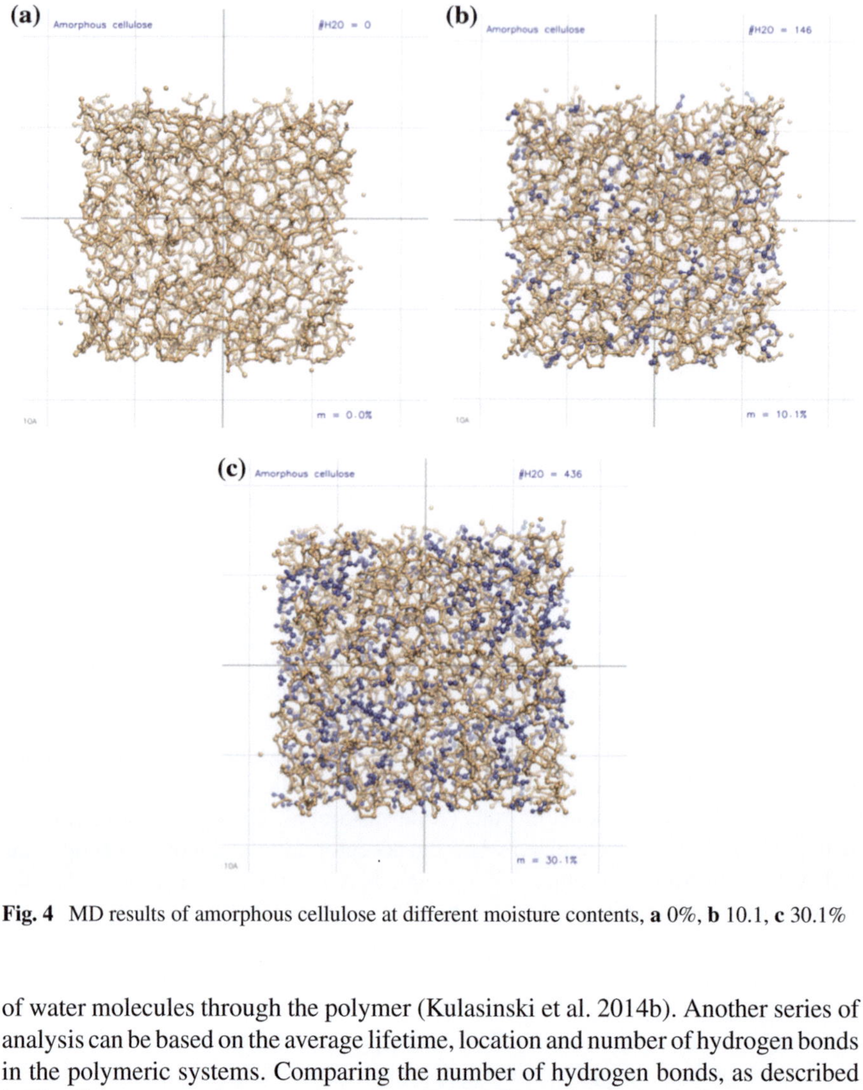

Fig. 4 MD results of amorphous cellulose at different moisture contents, **a** 0%, **b** 10.1, **c** 30.1%

of water molecules through the polymer (Kulasinski et al. 2014b). Another series of analysis can be based on the average lifetime, location and number of hydrogen bonds in the polymeric systems. Comparing the number of hydrogen bonds, as described in Fig. 5, between polymers versus the number of such bonds between polymer and water, can reveal insightful information as shown in Fig. 12.

As mentioned above, the methods described are used to determine the moisture adsorption isotherm, the moisture dependency of swelling, mechanical properties, and diffusion coefficient for all components and systems, in order to have a full characterization of the S2 layer, in addition to a better insight of S2 behavior in contact with moisture. The involved polymers are crystalline cellulose, amorphous cellulose, galactoglucomannan, arabinoglucuronoxylan, condensed lignin, and uncondensed lignin. Cellulose is a polysaccharide consisting of a linear chain of $\beta(1 \rightarrow 4)$ linked D-glucose units. It can form crystalline, paracrystalline, and amorphous states. There are two major species of hemicellulose in softwood, namely galactoglucoman-

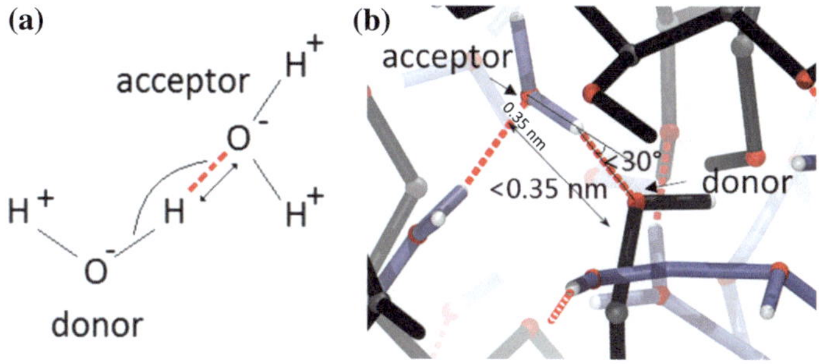

Fig. 5 Criteria of the existence of hydrogen bonds in terms of angle and distance, **a** schematic representation, **b** color visualization in a MD system with hydrogen bonds, shown in red, present between polymers (covalent bonds and carbon shown in black), between polymer and water (with blue bonds between oxygen in red and hydrogen in white) and between water molecules (from Kulasinski 2015 with permission)

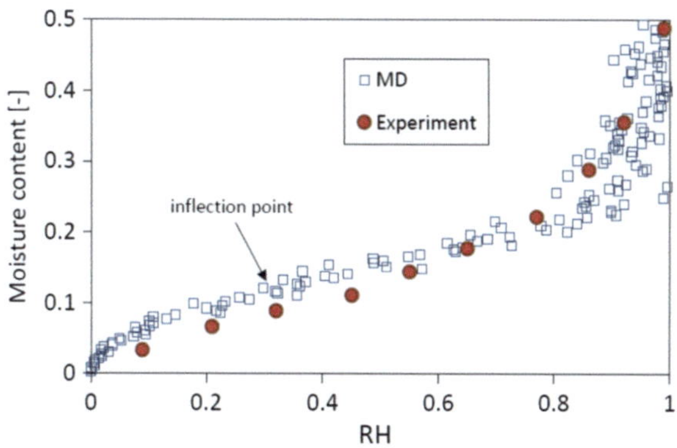

Fig. 6 Validation example of comparison of simulation and experiments results, taken from Beever and Valentine (1958), of the isotherm of amorphous cellulose of moisture content versus relative humidity (RH) (from Kulasinski et al. 2014b with permission). Given the periodic boundary conditions used in MD, the modeled system is considered bulk amorphous cellulose, thus comparable with experiments on bulk amorphous cellulose

nan, which consists of a backbone of randomly distributed ($1 \rightarrow 4$)-linked mannose and glucose units with ($1 \rightarrow 6$)-linked galactose units attached to mannose units, and arabinoglucuronoxylan which consists of a backbone of $\beta(1 \rightarrow 4)$ linked D-xylopyranose with MeGlcA and α-L-Araf residues attached at positions 2 and 3. Softwood lignin majorly consists of coniferyl units. Condensed lignin is a complex randomly branched polymer with several linkage types, e.g., β-O-4, α-O4, β-4, 5-5, and 5-O-4. Uncondensed lignin is more a linear polymer, with majorly β-O-4 linkages.

In addition to the individual components, we introduce three composite systems: a matrix mix of xylan and uncondensed lignin, microfibrils made of galactoglucomannan wrapping crystalline cellulose and S2 understood here to be microfibril

Table 1 Types of studied systems, full names, acronyms and symbols

Type	Class	Full name	Acronym	Symbol
Single component	Cellulose	Crystalline cellulose	CC	–[a]
Single component	Cellulose	Amorphous cellulose	AC	◇
Single component	Hemicellulose	Galactoglucomannan	GGM	● (grey)
Single component	Hemicellulose	Arabinoglucuronoxylan	Xyl	●
Single component	Lignin	Condensed lignin	cLGN	□
Single component	Lignin	Uncondensed lignin	uLGN	■
Composite	Matrix	Xyl + uLGN	Matrix	▲
Composite	Microfibrils	CC + GGM (red) as per inset (a) below	μF	✡
Composite	S2	CC + GGM (red) + LGN (black) as per inset (b) below	S2	✶

[a]no symbol as no result of single CC is presented in this chapter

aggregates, where four microfibrils are wrapped by lignin. The list of systems is found in Table 1.

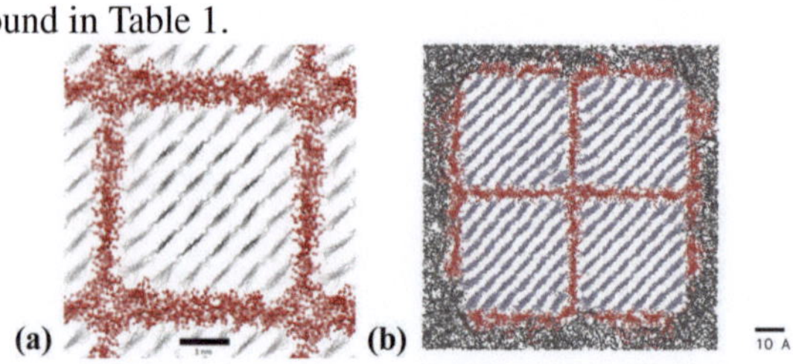

Grand Canonical Monte Carlo

A limitation of applying MD in sorption problems is that the random insertion of water molecules does not recognize the difference of potential fields within different pores. This makes it impossible to distinguish between adsorption and desorption as potential landscapes of the polymer system. To study sorption hysteresis, an alternative way is to include the Grand Canonical Monte Carlo method (GCMC). Instead of imposing a certain number of water molecules and then determining the chemical potential, GCMC operates in an opposite way, that is, the chemical potential is imposed as the boundary condition while the number of the adsorbed water molecules is determined based on statistics of the simulation results. Specifically, after imposing the chemical potential, the GCMC employs the Metropolis algorithm to perform a certain number of trial insertions and removals, which can be accepted or rejected. The acceptation/rejection probability is determined by the chemical potential and the potential landscape within the polymer system. In this way, the difference of potential landscapes between adsorption and desorption caused by different configurations is taken into consideration so that one can mimic the physical process of adsorption and desorption, respectively, thus with the possibility to describe the hysteretic behavior.

Validation

In the work presented below, the different systems are evaluated in terms of moisture-related behavior by studying the systems at different moisture content. As for any modeling investigation, validation of the work is required. This is done here by comparing the simulation results with experimental values. It was found that MD simulations provide results that are in agreement with experiments, even if the size of a MD model is rather limited, for adsorption isotherms (an example is given in Fig. 6 for amorphous cellulose), mechanical moduli, swelling coefficient, heat of sorption, and diffusion coefficients from experiments carried out using wood or components. This said we note that experimental data at comparable scales or for comparable chemical compositions, as some polymers may undergo modifications during wood extraction, are not abundant.

Results

Sorption

Isotherms provide the total moisture content ($m = (m_{wet} - m_{dry})/m_{dry}$) adsorbed at different relative humidity (RH, the conventional surrogate for chemical potential μ, and $k_B T \ln(RH) = \mu - \mu_{sat}$ with k_B Boltzmann constant, T temperature, μ_{sat} chemical potential of saturated vapor), determined as described above from the chemical potential resulting from the addition of water molecules in the polymeric systems.

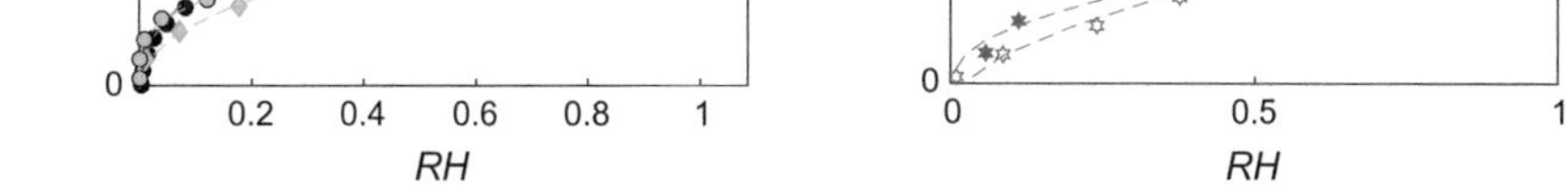

Fig. 7 Adsorption curves for **a** amorphous cellulose and two hemicelluloses, **b** microfibrils and S2 composite systems, in function of relative humidity (RH) (graph **b** from Kulasinski 2015), where m is moisture content

For the single components, we verify that water distribution within the polymer is quite uniform. The sorption isotherms agree with experimental results in terms of the shape of the curve and of the moisture content range attained (Fig. 7a). For the composite systems, we find that more water is adsorbed at the interface between two polymers showing higher water density profiles. Nevertheless, due to the presence of crystalline cellulose which does not absorb any water molecule, the total moisture contents are significantly lower in the microfibril systems than in the single systems (Fig. 7b).

Swelling

The introduction of water molecules results in a displacement of the polymeric systems, as visible in Fig. 4 with the reference geometry given by a red rectangle, leading to an overall material swelling. For the single components and the matrix system, made of amorphous components, swelling is found to be isotropic. Figure 8 shows that volumetric swelling strains (determined by $(V - V_0)/V_0$, where V, V_0 represent the current and dry volumes of the system, respectively) vary linearly with moisture content larger than 17%, the slope of this relationship yielding the swelling coefficient. Swelling coefficients are ranging from 1.07 for S2 to 2.0 for GGM. The different swelling coefficients stem from the combined effects of initial density and porosity, number of hydroxyl sites, and bulk modulus, resulting from the different monomers content and arrangement of each system. We note that the slope at low moisture content is always smaller and nonlinear, this aspect is further discussed in the section "Porosity". The volumetric swelling of S2 layer is found to be smaller than those of hemicellulose and lignin alone. The crystalline cellulose acts as a stiff skeleton preventing the system from deformation. This effect is particularly pro-

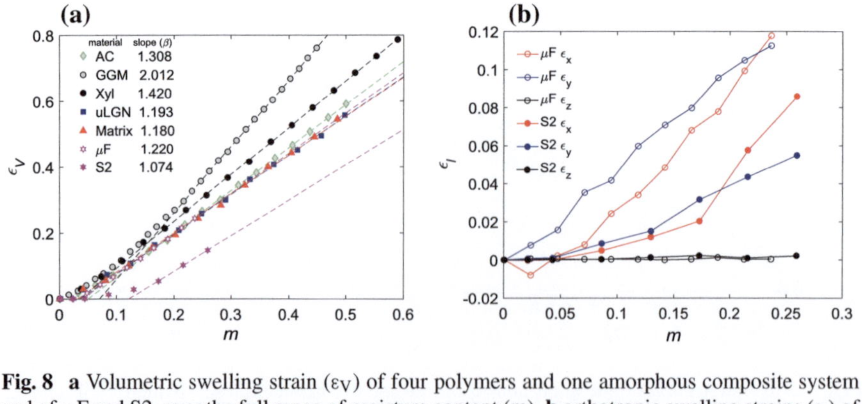

Fig. 8 **a** Volumetric swelling strain (ε_V) of four polymers and one amorphous composite system and of μF and S2, over the full range of moisture content (m). **b** orthotropic swelling strains (ε_l) of μF and S2, in function of moisture content (m)

nounced in the longitudinal direction, where due to cellulose covalent bonds, the swelling coefficient is two orders of magnitude smaller ($\epsilon_z = 2 \times 10^{-3}$ at maximum moisture content) than in the other directions. This indicates that, for the microfibril systems, anisotropy directly results from the presence of crystalline cellulose that restrains most of the deformation in the direction of the crystal.

Porosity

The porosity of a structure is estimated by sampling the volume with a probe, after removing the adsorbed water, without changing the position of the atoms (Kulasinski et al. 2015a). Pore volume is defined as the void space between polymer chains which is large enough to accommodate spheres of given radius or molecules, here we use water molecule. Pore size distribution is probed by spheres of varying radius, i.e., starting with a radius of 1 nm and gradually decreasing the radius of the probe by an interval of 0.1 nm until reaching a radius of 0.1 nm (Fig. 9a). The total porosity is

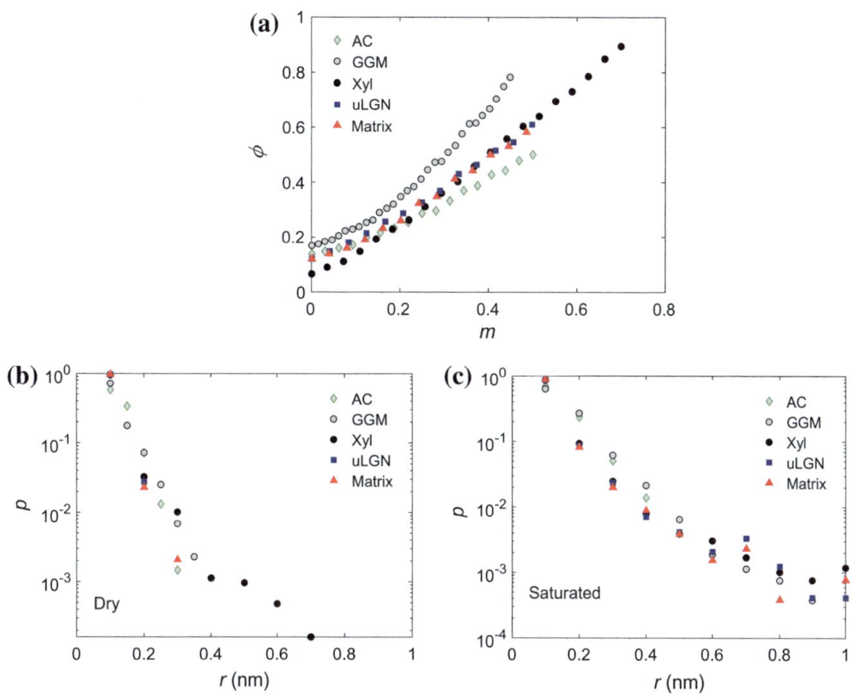

Fig. 9 **a** Porosity (ϕ) of homogeneous components and the amorphous composite as function of moisture content (m). Pores size distribution (p) of **b** dry and **c** saturated system, corresponding to moisture contents for AC of 50%, GGM 44.8%, Xyl 70%, uLGN 48.5% and matrix 49%, in function of equivalent pore radius (r)

found to increase linearly with moisture content above 10% moisture content (MC). The rate of increase is higher for GGM and lower for AC, a trend which follows the swelling results presented above. The slower rate of increase seen below 10%MC is due to the initial porosity. The pores present initially are filled with water that is barely displacing the polymers as seen in Fig. 8. Once the initial porosity is occupied, any further water molecule absorption results in a displacement of the polymeric chains, thus yielding porosity and swelling strains to increase linearly. In contrast, in the first part of the curves, the swelling strains and porosity are increasing nonlinearly.

The pore size distribution for each system is given in dry and wet states (Fig. 9b, c). The profiles are very similar for all polymers in dry or in wet states. However, a general shift of the profile toward larger pore size is clearly observed.

Diffusion

MD allows for the careful observation of the displacement of water molecules within the polymeric systems at different moisture content. It is seen that water molecules alternate between hydrogen-bonded states with little displacement, or waiting time, and "free" movement with the systems. Waiting time is determined by measuring the length of the sections in mean square displacement when the deviation of the displacement from its average value is smaller than 0.1 nm (OH bond length) and the rate of change of its squared displacement is smaller than 8×10^{-8} nm^2 ns^{-1}. Bonding energy is calculated using transition state theory with waiting time and separately determined characteristic oscillation frequency (Kulasinski 2015d). Sorption sites, i.e., hydroxyls, are strongly polarized and this electrostatic attractiveness is conducive to the formation of hydrogen bonds, however, the presence of neighbor water molecules can mitigate this attractiveness. When a system holds few water molecules, a water molecule bonded to a sorption site on a polymeric chain is not perturbed by

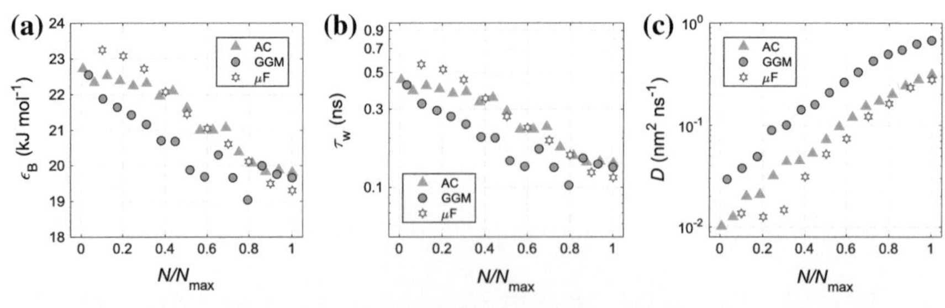

Fig. 10 **a** Bond energy of water with polymer, ε_B, **b** waiting time of bonded water molecule, τ_W, and **c** diffusion coefficient, D, versus normalized number of water molecules, where N_{max} is the maximum number of molecules that can be adsorbed in each system. N_{max} is determined from experimentally measured adsorption isotherms and is equal to 800, 2900, and 1000 for AC, HC, and μF, respectively (adapted from Kulasinski et al. 2015d)

neighbor water molecules, resulting in the hydrogen bond to be of higher bond energy and to undergo a longer waiting time, than seen of water–hydroxyl hydrogen bonds within water-abundant systems, as shown respectively in Fig. 10a, b. To calculate the diffusion coefficient, the mean square displacement (MSD) of the water molecules is followed over a sufficient time interval, i.e., larger than 1 ns, and its averaging, as per Einstein's equation, yields the diffusion coefficient. As the waiting time of bonded water molecules decreases, consequently, the global diffusion coefficient of water in the polymeric systems increases markedly with moisture content, Fig. 10c.

Mechanical Behavior

As moisture content increases, all investigated polymeric systems show a decrease in elastic moduli. In single systems, the decrease of the bulk moduli, determined by imposing tensile or compressive stress, is seen to be in the range of factor 3 or 4. Remarkably, the decrease in shear moduli is much more substantial, over one magnitude (Fig. 11a, b). These lower shear moduli at higher moisture content are in part due to the presence of water layers offering almost no resistance in shear. For anisotropic

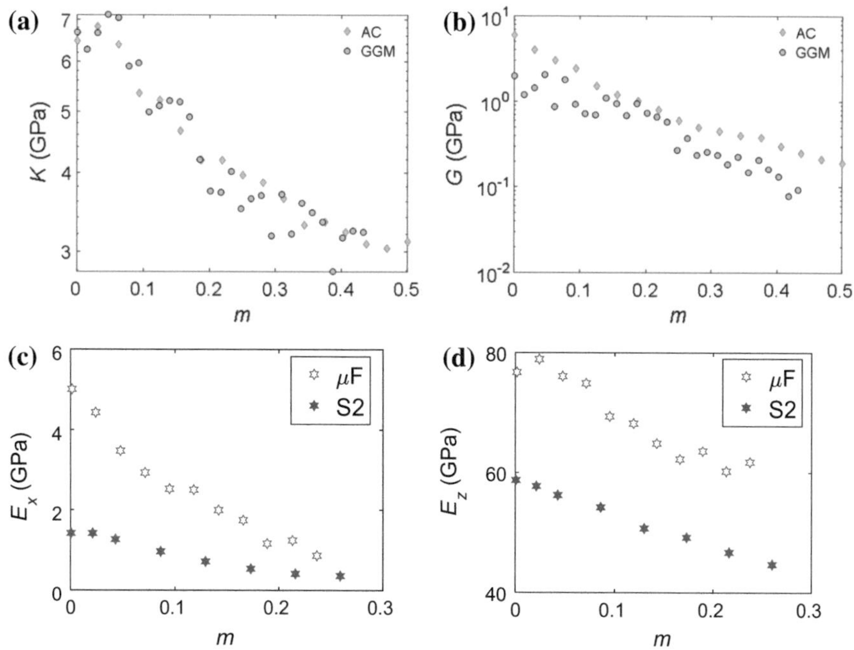

Fig. 11 Mechanical properties, **a** bulk (K) and **b** shear modulus (G) of AC and GGM, and Young's modulus of μF and S2 (E) **c** perpendicular and **d** parallel to crystalline cellulose, in function of moisture content (m) (adapted from Kulasinski 2015)

composite systems, Young's moduli are determined in the three principal directions. Results in the two directions x and y, perpendicular to the crystal orientation, are very similar. The weaker behavior with increasing moisture content is similar in all directions, but the magnitudes are very different for the behavior parallel to the crystal orientation (Fig. 11c, d). The difference between microfibril and S2 is in large part due to the lignin which makes about one-quarter of the S2 layer and which was here implemented with a low level of cross-polymerization, and as such is observed to weaken significantly.

Analysis Based on Hydrogen Bonds

Hydrogen bonds play a crucial role in the mechanical response of the polymeric systems studied here as they are exposed to increasing moisture content. As moisture content increases, hydrogen bonds linking polymer to polymer (HB^{PP}) are broken to create space for the water molecules and the adsorption sites that thus become available can undergo hydrogen-bonding polymer to water (HB^{PW}). As water content further increases, two water molecules in close neighborhood can also bond (HB^{WW}). With higher water content, swelling occurs and porosity increases. Figure 12a shows the power-law relationship between normalized number of hydrogen bonds and porosity. Further, Fig. 12b presents the clear relationship between the normalized bulk and shear moduli and the number of polymer–polymer hydrogen bonds, indicating that hydrogen-bond breaking, due to water molecules adsorption, is directly responsible for the weakening of the mechanical behavior of the polymers, either amorphous cellulose or hemicellulose (GGM).

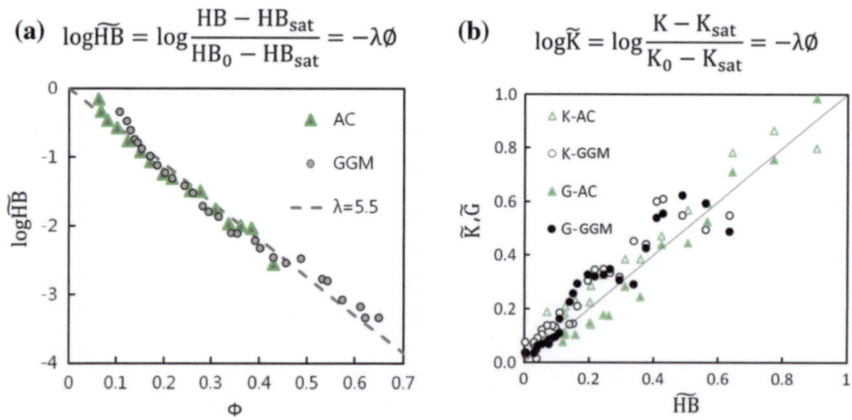

Fig. 12 Relationships between normalized, denoted with tilde, **a** number of polymer–polymer hydrogen bonds (HB) and porosity (ϕ), where λ is the slope, and **b** bulk (K) and shear (G) moduli and number of hydrogen bonds (adapted from Kulasinski et al. 2015a)

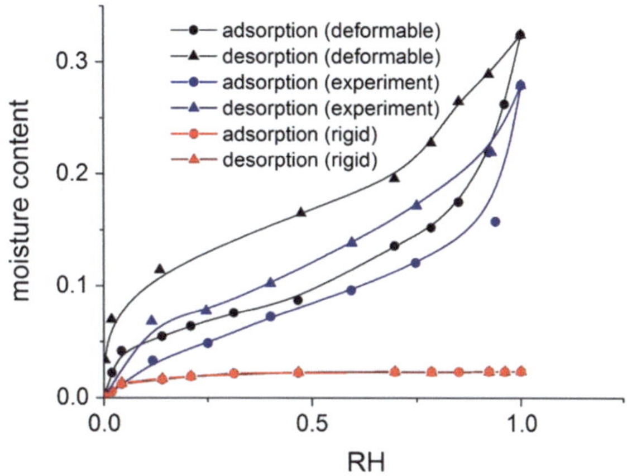

Fig. 13 Isotherms of moisture content versus relative humidity (RH) for a deformable (black) and a rigid (red) AC samples, with circle for adsorption and triangle for desorption compared with experimental data blue (Mihranyan et al. 2004)

Hysteresis

Simulations of adsorption and desorption can be conducted in sequence following the GCMC approach described in section "Grand Canonical Monte Carlo". For the adsorption branch, dry AC is considered as initial configuration and several simulations are conducted at different chemical potentials to obtain the adsorption isotherm, shown with circles in Fig. 13. In desorption, the saturated state is the initial configuration and another series of simulations is conducted to obtain the desorption isotherm, as shown with triangle. Agreement with experiment (blue curves) is found acceptable. When the polymeric system is rigidly maintained in place, once the preexistent pores are filled, no further moisture sorption occurrs as shown in red. This result displays that absorption as seen in such polymers is possible due to the occurrence of material swelling. Also, moisture content versus volumetric strain is found to follow a linear relationship except at low moisture content range, where the curve is convex.

As AC is a non-crosslinked polymer, hydrogen bonds play a crucial role in both sorptive and deforming process and this is studied for both adsorption and desorption curves. When a water molecule is absorbed, it can form hydrogen bonds with either the sorption sites on the cellulose chains or with other present water molecules. Hydrogen bonds, formed between polymer to polymer (HB^{PP}), cellulose to water (HB^{PW}), and water to water (HB^{WW}), are counted at each moisture content states. Hysteresis in number of HB is found in all cases with more HB^{PP} and HB^{WW} in adsorption than desorption but with more HB^{PW} in desorption than adsorption. This can be attributed to the fact that more hydroxyl sites are exposed and become new adsorption sites because of the breakage of HB^{PP} as the moisture content increases.

Upscaling MD Results

MD simulations yield results that are analogous to experimentally determined behavior. Actually, there is a significant lack of information of the behavior of the constituents of wood and such information would be needed in multiscale modeling of the moisture behavior of wood from nanoscale to macroscale taking into consideration its constituents. To circumvent this scarcity, some multiscale works rely on micromechanics. Hofstetter et al. (2005) developed a continuum micromechanics multiscale model for softwood, where the elastic behavior of the cell wall material, composed of crystalline and amorphous cellulose embedded in polymer matrix, is estimated by a Mori-Tanaka scheme. Bader et al. (2011) extended this micromechanics framework to study the poroelastic role of water in the cell walls of softwoods. Following poro-micromechanics of multiphase materials, they estimated the effective poroelastic properties of softwood from a hierarchical set of matrix-inclusion problems at different length scales. Qing and Michnaevsky (2009) developed a multiscale finite element model to study the effects of microstructure on the moisture-related properties of softwood. Rafsanjani et al. (2013a) developed a multiscale model starting from the cell wall components and upscaled over cell wall to cellular structure and geometry, where they used for the lower scales the tri-layered structure of the cell wall. The material properties of the components came from previous experimental characterization and micromechanical homogenization. However, in all these approaches, strong assumptions and simplified modeling are made, in particular for the lower scale material behavior.

In a natural prolongation of the work, the MD material properties determined above can be parametrized and used directly to determine the properties of materials at higher scale using a poroelastic framework described below, removing the need to revolve to the many assumptions used in micromechanical modeling, in particular, the geometry of the inclusions and the absence of interface.

In a nanopore, the fluid is not in its bulk state as its molecules interact with the atoms of the polymer, thus being in an adsorbed state. For such nanoporous materials, conventional poromechanics breaks down. Brochard et al. (2012) developed an adequate poromechanical approach, where the equations are validated using GCMC simulations and applied the model to the case of methane adsorption in coal. There is still very little work along these lines where the actual interaction of the fluid in a nanoporous material, such as wood cell wall material, is taken into account. Nevertheless, the MD results of this chapter have successfully been upscaled to a poromechanical framework (Kulasinski et al. 2015a, b, c, d, Rafsanjani et al. 2015 following the methodology of Carmeliet et al. 2013). This model leads to the explicit formulation of the dependence of the poroelastic properties on mechanical loading and number of water molecules, as follows:

$$d\varepsilon = C d\sigma + B dn \tag{1}$$

$$d\mu = -B d\sigma + M^{-1} dn, \tag{2}$$

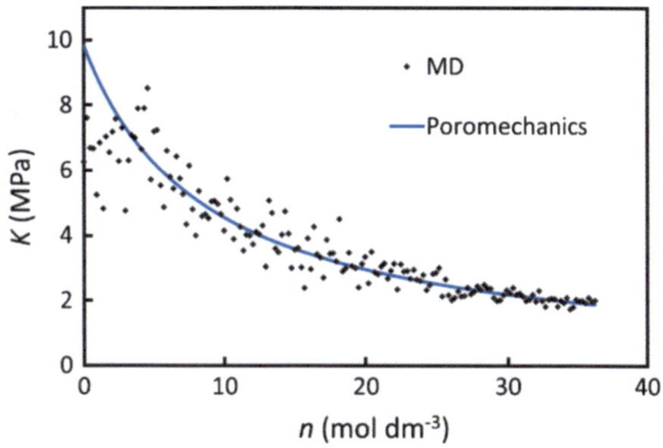

Fig. 14 MD results (dots) and poroelastic model results (blue line) of undrained bulk modulus (K) in function of moisture concentration (n) (from Kulasinski et al. 2015b with permission)

where ε is volumetric strain, C bulk compliance, σ volumetric stress, B a coupling coefficient, μ molar chemical potential, n molar concentration of water molecules, and M moisture capacity. To determine the necessary coefficients in the energy formulation, the MD results of adsorption, nonlinear swelling at zero stress and mechanical weakening of the bulk stiffness are used. The model can also describe mechano-sorptive effects, meaning sorption to be dependent on mechanical loading. Drained bulk modulus becomes stress dependent, even when the porous medium behaves linearly in undrained conditions. The poroelastic constitutive equations have been developed for a particular nanoporous material, amorphous cellulose, but this approach is generally applicable to the other components of wood or other materials (Fig. 14).

Outlook

The approach and work presented above allow detailed investigations of material behavior and physical phenomena. Only a small fraction of possibilities have been explored yet. Several paths are worthy and different scenarios could be tested. Possible outlook and perspectives could include:

1. Investigation of shape memory effect, another interesting hygromechanical behavior aside from swelling.
2. As computational capability grows, we can expect larger atomistic model, which reflects behaviors at larger scale, and reduces the disadvantages of present limited sized system.
3. Controlling chemical potential of water molecules, using the GCMC-MD hybrid method described above, for the determination of drained properties, which may also need longer simulations.
4. Keeping abreast of further chemical determination of the S2 polymeric constituents and configurations, including the shape of cellulose crystals (Fig. 15).

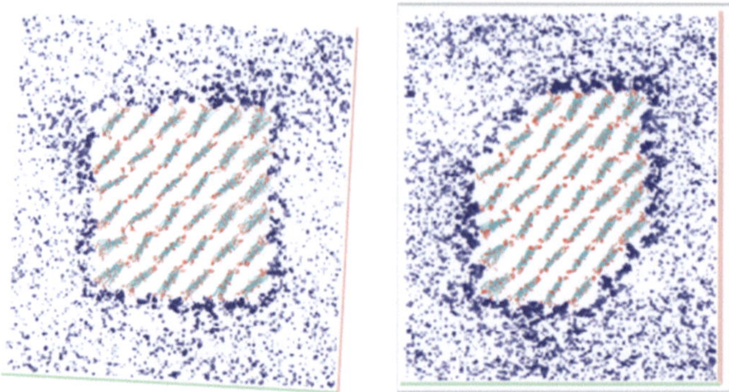

Fig. 15 Different water molecule profiles around two shapes of crystalline cellulose as example of further work. Crystalline cellulose is shown in cyan (carbon atom) and red (oxygen atom). Blue dots denote water molecule

5. Further development of a multiscale modeling framework from nano- to macroscale, with multiscale validation, combining experimental, MD and even mecoscopic modeling work.

Summary

Using atomistic modeling approach (molecular dynamics MD), we observed at nanoscale that, during adsorption of water molecules within hydrophilic polymers, water molecules first fill the existing pore space and afterwards displace the polymeric chains, leading to mostly linear swelling and weakening mechanical properties, in part due to the changes in hydrogen-bond network (Kulasinski et al. 2014b, 2015a, b, c, d). In addition to having demonstrated the capacity of MD to mimic macro phenomena at nanoscale, we show that 1. Moisture-related phenomena are scale-invariant, 2. HB dynamics play a key role in the control the moisture response, 3. Highly coupled phenomena are occurring. This work can lead to the construction of larger models, using MD as an input, to answer higher scale-related questions, such as shape memory effect known since long and seen experimentally at cellular scale in Derome et al. (2012).

Wood is almost ubiquitous as it is found in buildings structure, finishings, and furniture. Controlling swelling has been an everlasting challenge of the users of wood, therefore a better understanding of the swelling of wood can lead to better wood products design and longer durability. The understanding of the swelling of the S2 layer, a nanoporous composite material that interact strongly with moisture, can inspire the design of new materials where swelling would be desirable as with moisture-sensitive sensors, where anisotropy should be controlled either through careful design of the matrix, the fibrils or their interface, or where moisture-induced shape memory would be aimed, taking benefit of a network of the fibrils. As source of raw material, cellulose crystals (nanowhiskers) and cellulose fibrils transformation are being industrialized and industrial-scale production is now thinkable. Thus, the investigation of S2 layer system could provide ideas and insights for the development of new functional nanomaterials, coatings, and nanocomposites, for a huge variety of engineering and other applications, including biomedical applications, using renewables. An ability to deform material during sorption of fluids offers a lot of possible applications related to enhanced environmental problems such as energy, climatic change, water, and CO_2 sequestration. Therefore, a fundamental understanding and modeling of sorption and its coupling to swelling and weakening are mandatory for their reliable implementation in engineering applications and the work presented in this chapter offer one promising way to achieve such understanding.

References

Bader T, Hofstetter K, Hellmich C, Eberhardsteiner J (2011) The poroelastic role of water in cell walls of the hierarchical composite softwood. Acta Mechanica 217:75–100

Beever DK, Valentine L (1958) Studies on the sorption of moisture by polymers. Part IV. Interval and integral sorption of water vapor by polymers. J Polym Sci A 32:521–522

Berendsen HJC, POstma JPM, van Gunsteren WF, DiNola ARHJ (1984) Molecular dynamics with coupling to an external bath. J Chem Phys 81(8):3684–3690

Boutelje JB (1962) The relationship of structure to transverse anisotropy in wood with reference to shrinkage and elasticity. Holzforschung 16:33–46

Brochard L, Vandamme M, Pellenq RJM (2012) Poromechanics of microporous media. J Mech Phys Solids 60:606–622

Carmeliet J, Derome D, Dressler M, Guyer R (2013) Nonlinear poro-elastic model for unsaturated porous solids. J Appl Mech 80(2):020909

Charlier L, Mazeau K (2012) Molecular modeling of the structural and dynamical properties of secondary plant cell walls: influence of lignin chemistry. J Phys Chem B 116(14):4163–4174

Cosgrove DJ (2014) Re-constructing our models of cellulose and primary cell wall assembly. Curr Opin Plant Biol 22:122–131

Derome D, Griffa M, Koebel M, Carmeliet J (2011) Hysteretic swelling of wood at cellular scale probed by phase-contrast X-ray tomography. J Struct Biol 173:180–190

Derome D, Rafsanjani A, Patera A, Guyer R, Carmeliet J (2012) Hygromorphic behaviour of cellular material: hysteretic swelling and shrinkage of wood probed by phase contrast X-ray tomography. Phil Mag 92:3680–3698

Dinwoodie JM (2000) Timber, its nature and behaviour, 2nd. E & FN Spon, London, New York, England

Eichhorn S (2011) Cellulose nanowhiskers: promising materials for advanced applications. Soft Matter 7:303–315

Fahlén J, Salmén L (2005) Pore and matric distribution in the fiber wall revealed by atomic force microscopy and image analysis. Biomacromol 6(1):433–438

Forest Products Laboratory (2010) Wood handbook—wood as an engineering material. General Technical Report FPL-GTR-190. Madison, WI: U.S. Department of Agriculture, Forest Service, Forest Products Laboratory, 508 p

Hess B, Kutzner C, Van Der Spoel D, Lindahl E (2008) GROMACS 4: algorithms for highly efficient, load balanced, and scalable molecular simulation. J Chem Theory Comput 4:435–447

Hofstetter K, Hellmich C, Eberhardsteiner J (2005) Development and experimental validation of a continuum micromechanics model for the elasticity of wood. Eur J Mech A/Solids 24:1030–1053

Höfter H, Gonneau M, Vernhettes S (2007) 2.22 Biosynthesis of cellulose. In: Kalerming H (ed) Comprehensive glycoscience. Oxford Elsevier, pp 737–763

Jin K, Qin Z, Buhler M (2015) Molecular deformation mechanisms of the wood cell wall material. J Mech Behav Biomed Mater 42:198–206

Kulasinski K (2015) Physical and mechanical aspects of moisture adsorption in wood biopolymers investigated with atomistic simulations. Dissertation ETH No. 23046, ETH Zurich

Kulasinski K, Keten S, Churakov SV, Derome D, Carmeliet J (2014a) A comparative molecular dynamics study of crystalline, paracrystalline and amorphous states of cellulose. Cellulose 21:1103–1116

Kulasinski K, Keten S, Churakov S, Guyer R, Derome D, Carmeliet J (2014b) Molecular mechanism of moisture-induced transition in amorphous cellulose. ACS Macro Lett 3:1037–1040

Kulasinski K, Guyer R, Keten S, Derome D, Carmeliet J (2015a) Impact of moisture adsorption on structure and physical properties of amorphous biopolymers. Macromolecules 48:2793–2800

Kulasinski K, Guyer R, Derome D, Carmeliet J (2015b) Poroelastic model for adsorption-induced deformation of biopolymers obtained from molecular simulations. Phys Rev E 92:022605

Kulasinski K, Guyer R, Derome D, Carmeliet J (2015c) Water adsorption in wood microfibril: role of the crystalline-amorphous cellulose. Biomacromol 16:2972–2978

Kulasinski K, Guyer R, Derome D, Carmeliet J (2015d) Water diffusion in hydrophilic systems: a stop and go process. Langmuir 31:10843–10849

Lins RD, Hünenberger PH (2005) A new GROMOS force field for hexopyranose-based carbohydrates. J Comput Chem 26:1400–1412

Mihranyan A, Llogostera AP, Karmhag R, Stromme M, Ek R (2004) Moisture sorption by cellulose powders of varying crystallinity. Int J Pharm 269(2):433–442

Neuhaus FH (1981) Elastizitätszahlen von Fichtenholz in Abhängigkeit von der Holzfeuchtigkeit. PhD thesis, Ruhr-Universität Bochum, Bochum, Germany

Nishiyama Y, Langan P, Chanzy H (2002) Crystal structure and hydrogen-bonding system in cellulose Iβ from synchrotron X-ray and neutron fiber diffraction. J Am Chem Soc 124:9074–9082

Oostenbrink C, Villa A, Mark AE, Van Gunsteren WF (2004) A biomolecular force field based on the free enthalpy of hydration and solvation: the GROMOS force-field parameter sets 53A5 and 53A6. J Comput Chem 25:1656–1676

Patera A (2014) 3D experimental investigation of the hygro-mechanical behaviour of wood at cellular and sub-cellular scales. Diss. ETH no. 22230, ETHZ, Zurich

Patera A, Derome D, Griffa M, Carmeliet J (2013) Hysteresis in swelling and in sorption of wood tissue. J Struct Biol 182:226–234

Petridis L, Schulz R, Smith JC (2011a) Simulation analysis of the temperature dependence of lignin structure and dynamics. J Am Chem Soc 133:20277–20287

Petridis L, Pingali SV, Urban V, Heller WT, O'Neill HM, Foston M, Smith JC (2011b) Self-similar multiscale structure of lignin revealed by neutron scattering and molecular dynamics simulation. Phys Rev E 83(6):061911

Petridis L, Smith JC (2009) A molecular mechanics force field for lignin. Comp Chem 30:457–467

Plimpton S (1995) Fast parallel algorithms for ahort-range molecular dynamics. J Comput Phys 117:1–19

Qing H, Mishnaevsky L (2009) Moisture-related mechanical properties of softwood: 3D micromechanical modeling. Comput Mater Sci 46:310–320

Rafsanjani A, Derome D, Carmeliet J (2013a) Micromechanics investigation of hygro-elastic behavior of cellular materials with multi-layered cell walls. Compos Struct 95:607–611

Rafsanjani A, Lanvermann Derome D, Niemz P, Carmeliet J (2013b) Multiscale analysis of free swelling of Norway spruce. Compos Part A Appl Sci Manuf 54:70–78

Rafsanjani A, Stiefel M, Jefimovs K, Mokso R, Derome D, Carmeliet J (2014) Hygroscopic swelling of latewood cell wall micropillars reveals ultra-structural anisotropy. Interface 11:20140126

Rafsanjani A, Derome D, Carmeliet J (2015) Poromechanical modeling of moisture induced swelling anisotropy of cellular tissues of softwoods. RCS Adv 5:3560–3566
Salmén L (2004) Micromechanical understanding of the cell-wall structure. C.R. Biologies 327:873–880
Salmén L, Burgert I (2009) Cell wall features with regard to mechanical performance. A review. Cost action E35 2004-2008: wood machining—micromechanics and fracture. Holzforschung 63:121–129
Sangha AK, Petridis L, Smith JC, Ziebell A, Parks JM (2012) Molecular simulation as a tool for studying lignin. Environ Prog Sustain Energy 31(1):47–54
Smith PE, van Gunsteren WF (1994) Consistent dielectric properties of the simple point charge and extended simple point charge water models at 277 and 300 K. J Chem Phys 100:3169–3174
Sun H, Mumby SJ, Maple JR, Hagler AT (1994) An ab Initio CFF93 all-atom force field for polycarbonates. J Am Chem Soc 116:2978–2987
Teleman O, Jönsson B, Engström S (1987) A molecular dynamics simulation of a water model with intramolecular degrees of freedom. Mol Phys 60:193–203
Wagenführ A, Scholz F (2012) Taschenbuch des Holztechnik. Carl Hanser Verlag, Munich
Watanabe U, Norimoto M, Morooka T (2000) Cell wall thickness and tangential Young's modulus in coniferous early wood. J Wood Sci 46:109–114
Xu P, Donaldson LA, Gergely ZR, Staehelin LA (2007) Dual-axis electron tomography: a new approach for investigating the spatial organization of wood cellulose microfibrils. Wood Sci Technol 41:101–116
Zwanzig RW (1954) High-temperature equation of state by a perturbation method. I. nonpolar gases. J Chem Phys 22:1420–1426

Part III
Molecular Underpinnings of Cell Wall Mechanics

Calcium–Pectin Chemistry and Biomechanics: Biological Background and Mathematical Modelling

Mariya Ptashnyk and Henry R. Allen

Abstract In this chapter the impact of pectin and calcium-pectin chemistry on the mechanical properties, morphogenesis, and growth of plant cell walls and tissues is considered. We combine an overview of the underlying biological processes with the formulation of and numerical simulation results for mathematical models for interactions between the chemistry and mechanical properties and growth of plant cell walls. The role of the microscopic structure of plant cell walls and pectin-related cell wall signalling pathways in the interplay between the pectin chemistry and mechanics is also discussed.

Introduction

Mechanical properties, development and growth of plant tissues and organs are strongly determined by the mechanical properties of plant cell walls and the adhesion between the cells, facilitated by the middle lamella. Plant cell walls are highly heterogeneous and have a complex structure, which in growing cells have the remarkable property of combining extreme tensile strength with extensibility. The primary wall of a plant cell consists mainly of cellulose microfibrils, pectins, hemicellulose, structural proteins and water. The secondary cell wall, composed of cellulose, hemicellulose and lignin is deposited on the primary cell wall and consists of different layers with different orientations of microfibrils (Burgert and Fratzl 2009). Pectin is one of the main components of the primary (growing) cell walls and middle lamella

M. Ptashnyk (✉) · H. R. Allen
Division of Mathematics, University of Dundee, Fulton Building,
Dundee DD1 4HN, UK
e-mail: m.ptashnyk@dundee.ac.uk

M. Ptashnyk
Department of Mathematics, Heriot-Watt University, Colin Maclaurin Building,
Edinburgh EH14 4AS, UK
e-mail: m.ptashnyk@hw.ac.uk

A. Geitmann and J. Gril (eds.), *Plant Biomechanics*,
https://doi.org/10.1007/978-3-319-79099-2_13

in most plants. Pectin functions are not limited to cell structure but also play an important role in plant morphogenesis and growth, cell–cell adhesion, and cellular signalling processes (Mohnen 2008). The enzyme pectin methyl-esterase (PME) plays an important role in controlling the rheology of the pectin matrix. Homogalacturonan is the most abundant subclass of pectin. It is deposited into the cell wall in highly methyl-esterified form, where it is demethyl-esterified by PME. The demethyl-esterified pectin can create load-bearing cross-links with calcium molecules or decay due to the action of pectin degrading enzymes. The demethyl-esterification of pectin and calcium-pectin chemistry are associated with both the stiffening of the cell wall due to creation of calcium-pectin cross-links and cell wall loosening and expansion acceleration related to the deposition of new wall material and decay of demethyl-esterified pectin. Hence, growth control requires that the amount of pectin is precisely tuned, for example, through mechanisms that detect the presence of pectin in the cell wall and that control the deposition of pectin, PME activity and/or pectin turnover.

Biochemistry of Pectin

Pectin is structurally and functionally the most complex polysaccharide in plant cell walls and is the only major class of plant polysaccharide to be largely restricted to primary cell walls. Pectin constitutes $\approx$35% of the primary cell walls in dicots and non-graminaceous monocots and up to 5% of walls in woody tissue (Caffall and Mohnen 2009; Mohnen 2008). It is also present in the junction zone between cells with secondary walls including xylem and fibre cells in woody tissue. The middle lamella, connecting plant cells, is rich in pectin which plays an important role in the regulation of intercellular adhesion (Guillon et al. 2017; Jarvis et al. 2003; Knox 1992; Orfila et al. 2001; Willats et al. 2001b).

Pectin is a family of galacturonic acid-rich polysaccharides including homogalacturonan (HG), rhamnogalacturonan I (RG-I), and the substituted HGs rhamnogalacturonan II (RG-II) and xylogalacturonan (XGA), Fig. 1 (XGA is not specified in the schematic). The most abundant pectic polysaccharide is HG, a linear homopolymer of a-1,4-linked galacturonic acid (Ga1A) that comprises $\approx$65% of pectin. HG is estimated to range from 75 to 250 residues ($\approx$30–120 nm) in length (Cameron et al. 2011). Contrary to HG, RG-I is a highly branched polysaccharide and is variable both in its fine structure and in its occurrence within cell walls. RG-II, the most structurally complex pectin, contains 11 different sugar residues and forms dimers through borate esters. RG-II makes up $\approx$10% of total amount of pectin in plants and occurs widely in all primary cell walls, but may be absent from the middle lamella (Jarvis 1984). RG-I and RG-II are thought to be covalently attached to HG and possibly to each other. Pectin is synthesized in the Golgi apparatus by a large number of glycosyltransferases, methyltransferases and acetyltransferases and is inserted into the extracellular matrix by vesicle-mediated exocytosis (Mohnen 2008).

HG is deposited into cell walls in a highly methyl-esterified form, where it can be modified by the enzyme pectin methyl-esterase (PME), which removes methyl groups by breaking ester bonds, Fig. 2. The demethyl-esterified pectin is able to form load-bearing calcium-pectin cross-links and a low level of pectin methyl-esterification is often associated with reduced wall extensibility and the cessation of growth (Wolf and Greiner 2012; Zsivanovits et al. 2004). The strength of the interaction between Ca^{2+} and pectin increases with decreasing average degree of pectin methyl-esterification and increasing length of the unsubstituted galacturonan backbone. Pectin demethyl-esterification might also lead to wall loosening by enabling the action of HG degrading polygalacturonase. The removal of methyl H^+ groups during the interactions between pectin and PME leads to cell wall hydration and the decrease of pH value, which support the activity of cell wall remodelling enzymes. RG-II contributes to the cell wall structure and the integrity of the pectin network through borate ester dimerization of RG-II domains (O'Neill et al. 2004).

It is supposed that load-bearing during deformations of plant cell walls is shared by all three types of polysaccharides: cellulose, hemicellulose and pectin (Lin et al. 2016), and pectin acts as a distributor of mechanical loads in the cellulose–hemicellulose–pectin network. The interactions between cellulose and pectin are mediated by covalent pectin-xyloglucan links (Popper and Fry 2008) and the microfibrils, free of hemicellulose coating, can interact with pectin directly (Dick-Pérez et al. 2011; Wang et al. 2012, 2015). Thus, it is hypothesized that pectin plays a double role within primary cell walls: (i) in a xyloglucan-rich cell wall pectin is loosely bound to cellulose, whereas (ii) pectin interacts with cellulose directly in

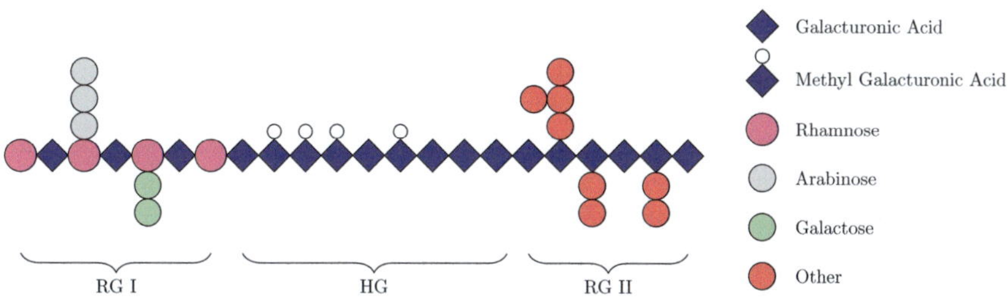

Fig. 1 Schematic representation of the molecular structure of pectins

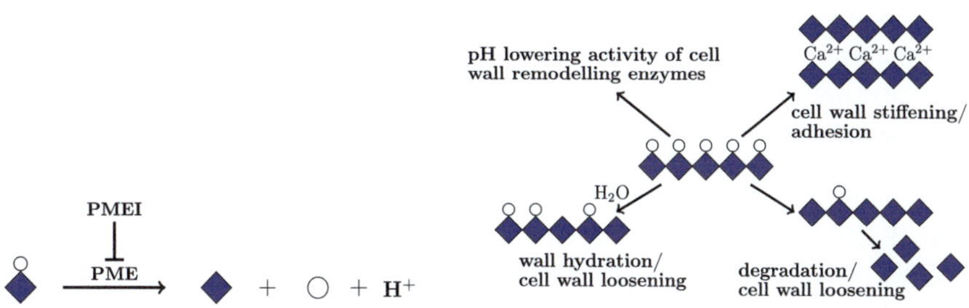

Fig. 2 Schematic representation of demethyl-esterification of HG pectin by PME. Adapted from Wolf and Greiner, Protoplasma (2012)

xyloglucan-deficient cell walls (Zykwinska et al. 2008). Pectin also controls the cell wall porosity through regulation of the interfibrillar spacing and relative orientations of the microfibrils, reduction of cellulose self-association and facilitation of cellulose slippage in expanding cell walls (Thimm et al. 2009). The extraction of pectin from a cell wall is observed to be associated with increased rigidity of the cell wall (Dick-Pérez et al. 2012), with microfibrils having a higher tendency to become denser and form aggregates (Thimm et al. 2009). The analysis of cobtorin, an inhibitor that perturbs the parallel relationship between cortical microtubules and cellulose microfibrils, identified that overexpression of PME and polygalacturonase suppressed the cobtorin-induced cell-swelling phenotype (Yoneda et al. 2010). This suggests that control over the properties of pectin is important for the deposition of cellulose microfibrils and/or the maintenance of their orientation parallel to cortical microtubules.

PMEs and Their Inhibitors PMEIs

It has been observed that both the degree of methyl-esterisfication and the demethyl-esterification patterns of pectin influence the mechanical properties of cell wall matrix and are spatially regulated in plant tissue (Peaucelle et al. 2011; Willats et al. 2001a, b, 2006). Spatial distribution patterns of methyl-esters within HG regions are related to the mechanism of demethyl-esterification: the enzymatic action by plant PMEs results mainly in an ordered block-wise demethyl-esterification, while the chemical demethyl-esterification and the demethyl-esterification associated with fungal PMEs produce random patterns of demethyl-esterification (Cameron et al. 2008). The ordered block-wise structure of the demethyl-esterified pectin has higher calcium ion binding capacity than random distribution of non-methoxylated carboxyl groups. Random demethyl-esterification makes HG susceptible to degradation by plant-derived pectic enzymes such as pectate lyases (PLs) and polygalacturonases (PGs) and contributes to cell wall loosening. Because the action of PGs in pectate gel, created by calcium-pectin cross-links, is limited, the ordered patterns of demethyl-esterification contribute to cell wall stiffening (Micheli 2001). Three types of mode of action of PME enzyme have been described: (i) a single-chain mechanism, where PME removes all contiguous methyl-esters from a single pectin chain before dissociating; (ii) a multiple-chain mechanism, where PME catalyses only one reaction before dissociating; and (iii) a multiple-attack mechanism, where PME removes a limited average number of methyl-esters before the enzyme–pectin complex dissociates (Cameron et al. 2008; Fries et al. 2007). By single-chain and multiple-attack mechanisms, PMEs remove methyl-esters in a blockwise fashion, creating long contiguous stretches of demethyl-esterified GalA residues on the pectin, whereas multiple-chain mechanism results in a ‘rather random’ distribution of demethyl-esterified GalA units. The mode of PMEs action depends on pH, and a single-chain mechanism occurs at pH 7.0 shifting to a multiple-chain mechanism at pH 4.5 (Denès et al. 2000). The activity of PMEs also depends on pH and the optimal pH for different

PMEs varies substantially (Denès et al. 2000). Plant PMEs may be poorly active at acidic pH values due to strong interactions between the positively charged enzyme and the free carboxyl groups of HG, whereas cations are important for enzymatic activity of PMEs. It is observed that PMEs activity in the cell wall is regulated by H^+ concentration in a cyclic manner during the cell growth. The enzyme activity is reduced by the local decrease of pH generated by protons H^+ released as a consequence of the reaction between PME and methyl-esterified pectin. In turn, the pH decrease activates glycosidases and glycosyltransferases, involved in cell wall extension. Then, the consequent dilution of negative charges and increase of local pH result in reactivation of PME, starting a new cycle (Willats et al. 2001a).

The expression of small proteinaceous PME inhibitors (PMEIs) may represent an additional control mechanism for PME activity (Camardella et al. 2000; Wolf et al. 2009). The inhibitory molecule was suggested to have a pectic configuration since it was deactivated by the pectic enzymes (PG and PL). The inhibition of PME occurs through the formation of a reversible PME-PMEI complex. The stability of the complex is strongly influenced by pH, indicating that PME activity can be modulated by pH either directly or by modulation of the affinity between the enzyme and its inhibitor. The affinity of the two proteins (PME and PMEI) is higher at a pH close to that of the apoplastic environment. Localized secretion of specific PMEIs could also represent an efficient mechanism for the precise temporal and/or spatial control of PME activity. PMEs are differentially regulated throughout development and can have different action patterns at different locations within cell walls. Hence, PMEs may have the potential to modify or fine-tune matrix properties at a local level in specific regions of the cell wall in response to functional requirements.

Biochemical properties of pectin and activity of PME also influence the cell separation in the root cap, having significant effect on the physical and biological properties of the rhizosphere. It has been shown that cell separation in the root cap can be promoted by the sequential action of PME and a polymer-cleaving enzyme such as PG and/or PL (Laskowski et al. 2006; Rhee et al. 2003; Stephenson and Hawes 1994).

Calcium–Pectin Chemistry and Growth

Many experimental results indicate that pectin is crucial for cell wall expansion and the amount of pectin and its chemical configuration are important for plant tissue growth and development (Pelletier et al. 2010). Due to the creation of load-bearing calcium-pectin cross-links, the increase in the density of demethyl-esterified pectin is associated with reduction of cell wall extensibility. However, it has been shown in experiments for *Chara corallina* cells that newly-supplied pectin and the interactions between calcium and pectin can also accelerate cell wall extension (Proseus and Boyer 2012). The new pectin, deposited into the cell wall, acts after it is demethyl-esterified as a chelator to remove calcium from calcium-pectin cross-links weakened by tension, and creates a molecular mechanism for loosening the wall structure and

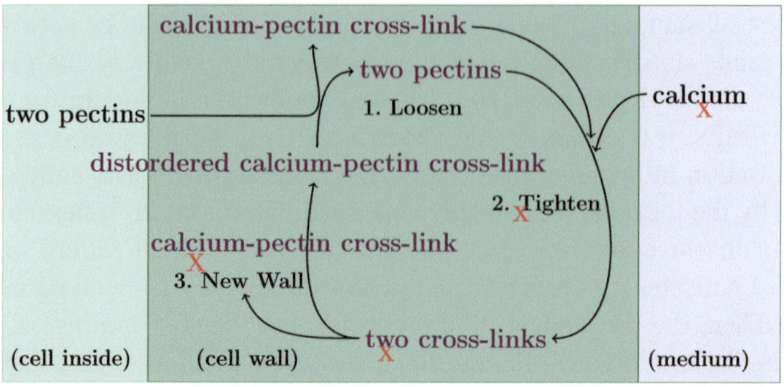

Fig. 3 Schematic representation of pectin-calcium cycle. The absence of new calcium coming from the medium, indicated by 'X' in the figure, prevents tightening of the cell wall and creation of new cell wall material, illustrated by 'X' in the figure. Adapted from Proseus and Boyer (2012)

inducing growth acceleration (Proseus and Boyer 2007). The cyclical interaction between pectin and calcium begins by removing calcium from distorted cross-links, which sets in motion accelerated wall extension, relaxation of the distorted bonds, and binding of the new undistorted pectin to the calcium in the cell wall, which then leads to new wall deposition, see Fig. 3. The increased extension of the cell wall distorts other calcium-pectin cross-links and starts the cycle again. Additional calcium is required to occupy the previously vacated calcium sites and to maintain the density of calcium-pectin cross-links holding the wall together. Thus, calcium is a critical constituent bearing the longitudinal load for growth and new calcium has to enter the wall from the external medium or cytosol to prevent a too long growth acceleration and disruption of the cell wall and tissue (Proseus and Boyer 2006, 2007). The exchange of Ca^{2+} in calcium-pectin cross-links for H^+ may explain, at least in part, the acid-promoted increase in cell wall extensibility (Virk and Cleland 1988). It has been shown that a pH shift in the physiological range from 6 to 5 simultaneously caused a strong increase in the extensibility, and a decrease in the yield threshold of the cell wall. Interestingly, high Ca^{2+} concentrations reverted this increase in extensibility but did not affect the yield threshold, and chelation of Ca^{2+} perfectly mimicked at pH = 6 the acid-induced increase in wall extensibility (Ezaki et al. 2005). The experimental results also indicate that pectin demethyl-esterification influences the timing of the acceleration of cell elongation in the sense that overexpression of PMEIs caused a delay in the growth acceleration (Pelletier et al. 2010). It is assumed that pectin demethyl-esterification controls the initiation of the growth acceleration but not the growth process itself.

Hence, depending on the context, demethyl-esterification can induce opposite changes in the visco-elastic properties of the plant cell wall. Demethyl-esterification can rigidify the cell wall by increasing the density of load-bearing calcium-pectin cross-links or make it more extensible in the absence of new calcium and due to degradation of demethyl-esterified pectin. The finding that pectin controls growth rate suggested a molecular link between growth and wall deposition. Also, any molecular mechanism for growth must account for the dependence of growth on the turgor pressure P since growth ceases below a critical threshold of P. Capacity of pectic network to be degraded by plant and microbial pectinases provides an environment for the activities of cell wall-modifying enzymes.

It has been shown that calcium-pectin chemistry not only controls growth rate and new wall deposition but also accounts for 'stored' growth and the ability of plants to compensate for brief exposure to low turgor pressure (Proseus and Boyer 2008), i.e. under the low turgor pressure the new wall materials must be accumulating unused and are then used to restore the normal growth rate when the turgor pressure is back to normal value. The level of turgor pressure required for growth is also required for rapid wall deposition in the cell walls. The turgor pressure provides the force for polymer deposition in the wall via P-induced deformation of the polymers and the threshold value for P appeared to increase as molecular weight of polymer increases, until normal P is insufficient to move the polymers into the wall. In this situation, the largest polymers remained pressed against the inner wall face where they would encase new cellulose microfibrils being produced at the plasma membrane.

Interactions Between Calcium–Pectin Chemistry, Mechanics and Growth of Plant Cell Walls: Mathematical Modelling

The growth of a plant tissue is controlled by the interplay between two opposing forces: the intracellular turgor pressure from one side and mechanical properties (stiffness) of cell walls and adhesion between the cells from another side.

The first model for plant cell wall growth was derived by Lockhart (1965), relating the growth rate to turgor pressure P:

$$\frac{d\varepsilon}{dt} = \Phi(P - P_c)^+, \tag{1}$$

where $d\varepsilon/dt = V^{-1}dV/dt$ defines the relative rate of change in cell wall volume V, Φ is the cell wall extensibility and P_c is a critical threshold value for the pressure. In order to consider possible variations in the turgor pressure during the growth processes an augmented Lockhart model was derived in Ortega (1990) by introducing an additional equation for the dynamics of turgor pressure in plant cells

$$\begin{aligned} \frac{d\varepsilon}{dt} &= \Phi(P - P_c)^+ + \frac{1}{\delta}\frac{dP}{dt}, \\ \frac{dP}{dt} &= \delta\left[L(\Delta\Pi - P) - \Phi(P - P_c)^+ - T\right], \end{aligned} \tag{2}$$

where $L(\Delta\Pi - P)$ corresponds to rate of water uptake, L is the relative membrane hydraulic conductance, $\Delta\Pi$ is the osmotic pressure difference across the membrane, T is the transpiration rate and δ is the volumetric elastic modulus. If neglecting the water uptake and transpiration, the model (2) simplifies to

$$\begin{aligned} \frac{d\varepsilon}{dt} &= \Phi(P - P_c)^+ + \frac{1}{\delta}\frac{dP}{dt}, \\ \frac{dP}{dt} &= -\delta\Phi(P - P_c)^+. \end{aligned} \tag{3}$$

Although the models (1)–(3) provide a good description of plant growth, for a better understanding of the mechanisms controlling plant growth and development as well as interactions between the growth and the molecular processes it is important to relate the equations for growth with chemical processes in plant cell walls.

Mathematical Modelling of Influence of Calcium and Pectin Dynamics on Growth of Pollen Tubes

It has been shown experimentally that pectin located at the tip of the growing pollen tube is highly methyl-esterified, while it is demethyl-esterified in areas away from the tip (Parre and Geitmann 2005). This indicates a clear role of pectin in the control of cell extension in the growing pollen tube and supports the hypothesis that PME activity is necessary to rigidify pollen tube cell walls.

Along with experimental studies, mathematical models have been derived to analyse and better understand the effect of changes in the chemical configurations of pectin (methyl-esterified versus demethyl-esterified) and calcium concentration on the behaviour and mechanical properties of a cell wall in a pollen tube (Kroeger et al. 2011; Rojas et al. 2011).

In Kroeger et al. (2011), the apex of the pollen tube is modelled as a thin shell of viscous material and the cell wall stress σ is related to the turgor pressure P by

$$\sigma = P\frac{r}{2h}, \tag{4}$$

where r is the radius of curvature of the tube and h is the cell wall thickness. Then, the Lockhart equation (1) for the pollen tube growth in terms of the cell wall stress reads

$$\frac{\partial\varepsilon}{\partial t} = F(\sigma - \sigma_c)^+, \tag{5}$$

where $F = 2h\Phi/r$ and $\sigma_c = P_c r/(2h)$. If v denotes the growth velocity of the pollen tube, then we obtain

$$\frac{\partial\varepsilon}{\partial t} = \frac{\partial v}{\partial s} \approx \frac{\overline{v}}{S},$$

where s is the curvilinear coordinate, S is the length of the arc and $\overline{v}$ is the average growth rate of the pollen tube, see Fig. 5.

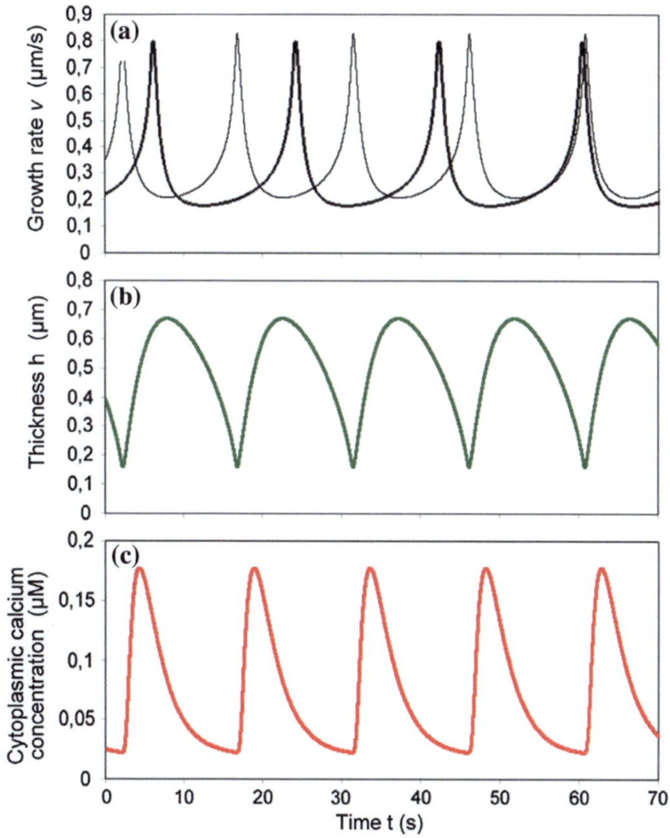

Fig. 4 Numerical simulations for mathematical model (5)–(9) predict oscillatory growth of a pollen tube. Frequency and amplitude depend on the extracellular calcium concentration C_0 (**a**) (thick line: $C_0 = 1$ μM, thin line: $C_0 = 1.5$ μM). Dynamics of cell wall thickness (**b**) and cytosolic calcium concentration (**c**). Reproduced with permission from Kroeger et al. (2011)

Fig. 5 Sketch of the parametrization of the pollen tube geometry

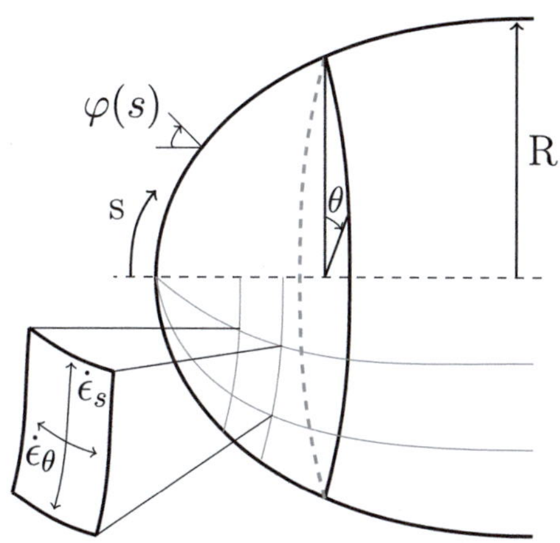

The elongation of the pollen tube leads to the stretching and thinning of the cell wall in the apical region of the tube and depends on the Poisson's ratio of the cell wall (ratio between transversal expansion and axial compression). To prevent the bursting of the growing pollen tube, the thinning due to stretching is balanced by the secretion and deposition of new cell wall material. Mathematically, the stretching of the cell wall and deposition of new wall material is modelled by the dynamics of the cell wall thickness: the rate of change of the cell wall thickness h is balanced by an increase through vesicle deposition $\mathcal{D}$ and a decrease due to the stretching resulting from growth

$$\frac{\partial h}{\partial t} = -\frac{3(r^2 - r_i^2)}{2r^2}\overline{v} + \mathcal{D}, \tag{6}$$

where r_i is the inner radius of the tube apex. The addition of cell wall material $\mathcal{D}$ through vesicle secretion is assumed to be proportional to the cytosolic calcium concentration C inside the pollen tube apex

$$\mathcal{D} = a_1 C,$$

where $a_1 = 4.1 \times 10^{-4}$ µm s^{-1} µM^{-1}. The dynamics of calcium concentration in cytosol are modelled by the reaction–diffusion equation

$$\frac{\partial C}{\partial t} = \mathrm{div}(D_c \nabla C) - a_2 C, \tag{7}$$

where D_c is the diffusion constant and a_2 is the natural decay rate. The change of calcium concentration in the pollen tube membrane is due to calcium influx through the membrane channels, denoted by J_m, and capture of calcium ions by available sites inside the apical and subapical cytoplasm modelled as $J_r = -a_2 C$:

$$\frac{\partial C}{\partial t} = \frac{A}{V}(J_m - J_r), \tag{8}$$

where A is the surface of the apex and V is the volume of the apex. It is assumed that stress-activated channel conductance depends on the membrane stress σ_{m}

$$J_m = D_m(\sigma_m)\frac{C_0 - C}{h},$$

where C_0 denotes the value of the extracellular calcium concentration,

$$D_m(\sigma_{\mathrm{m}}) = \frac{D_c}{1 + e^{-a_3(\sigma_m - \sigma_c)}} \quad \text{and} \quad \sigma_{\mathrm{m}} = \frac{2\eta}{h_m}\frac{\partial \varepsilon}{\partial t}.$$

Here, $h_m = 5$ nm denotes the membrane thickness and the dynamics of the wall viscosity η depends on the deposition of the new cell wall material

$$\frac{\partial \eta}{\partial t} = -\frac{\eta_s \mathcal{D}}{h} + k(\eta_{eq} - \eta), \tag{9}$$

where η_{eq} is the maximum value for viscosity that can be reached through demethyl-esterification of pectin, i.e. when all methyl groups have been removed from the pectin monomers, and η_s is the viscosity of the newly added material.

To better understand the oscillatory growth of the pollen tube, transient (short) perturbations in the turgor pressure were considered. To incorporate variations in the turgor pressure into the model, Eq. (5) is replaced by the augmented Lockhart model (3), written in terms of strain and stress as

$$\frac{\partial \varepsilon}{\partial t} = F\sigma + \frac{1}{E}\frac{\partial \sigma}{\partial t}, \tag{10}$$

where E is the Young-modulus of the cell wall of a pollen tube and the time dependence of turgor pressure is given explicitly by $P(t) = 0.5 + 0.8\exp(-0.2(t-21)^2)$, describing a short time (for <2 s) increase in turgor pressure.

The mathematical models (5)–(9) and (6)–(10) capture very well oscillatory growth of a pollen tube and oscillations in the apical cytosolic calcium concentration in a pollen tube, Fig. 4, and explain the lack of sensitivity of the average tube growth to global changes in the turgor pressure (Kroeger et al. 2011). It has been shown that the growth rate oscillates with a peak that precedes that of the cytosolic calcium concentration by 2 s, which is consistent with the experimental measurements for pollen tube growth. As a result of the mathematical model, it is shown that for small values of the extracellular calcium concentration ($C_0 = 0.3\,\mu$M), the growth rate increases as a function of turgor pressure until it reaches a plateau at 0.3 MPa. Above 0.3 MPa, the turgor pressure value has a minimal effect on the average growth rate. However, for large external calcium concentration ($C_0 = 12\,\mu$M), the average growth rate is proportional to the turgor pressure within the biologically relevant pressure range. This behaviour can be explained by the mathematical model by computing, using a slow–fast analysis, the relation between the averaged growth rate $\overline{v}$ and the turgor pressure

$$\overline{v} = v_c \left[1 + \frac{v_c \tau R}{r^2 F P}\right]^{-1}.$$

Also, for very short and transient turgor variations the instantaneous value of the growth is directly proportional to the turgor pressure. The dependence of the calcium channels on plasma membrane stress results in the increase in cytosolic calcium during the cell expansion and slow down of the growth. The deposition of new cell wall material reduces the cell wall strain and starts the new growth cycle. The results of the mathematical models suggest that the maximal value of the tube growth rate depends on the membrane strain rate at which the calcium channels open. Specifically, the cell wall expansion rate cannot increase much beyond the strain rate that will open the channels. The predictions of the mathematical model are consistent with experimental results on how sensitive the growth rate is to changes in the osmotic pressure of the surrounding medium.

It is also suggested that the oscillating growth of pollen tubes observed experimentally and captured by the mathematical models (5)–(9) and (6)–(10) can be also related to an oscillatory release of pectin by exocytosis (McKenna et al. 2009).

A similar approach to model oscillatory growth of a pollen tube was proposed in Rojas et al. (2011). In this work, along with Lockhart equation, the dynamics of calcium-pectin cross-links density is considered to model the chemically driven mechanical expansion of the cell wall. Following experimental results (Proseus and Boyer 2007), it is assumed that the deposition of new pectin induces dissociation of existing calcium-pectin cross-links and new cross-links are immediately reformed after they are dissociated. Changes in demethyl-esterified pectin density r_d are governed by (i) demethyl-esterification of methyl-esterified pectin r_e, (ii) formation of calcium-pectin cross-links, (iii) dilution due to deposition of new material and (iv) cell wall elongation. Changes in cell wall thickness h are given by deposition of new material balanced against mechanical thinning. Further, it is assumed that both the total density of methyl-esterified and demethyl-esterified pectin and pectin involved in calcium-pectin cross-links, and the areal density of calcium-pectin cross-links are constant, i.e. $\beta = r_e + r_d + 2\chi$ and $\chi_A = \chi h = \text{const}$, where χ is the density of calcium-pectin cross-links.

The pollen tube surface is parameterised by the arclength from the pole of the cell s, and the azimuthal angle θ, Fig. 5. The rotational symmetry allows us to describe the geometry of the cell by specifying the curvature of a single meridian κ_s where κ_s and the 'circumferential' curvature κ_θ are defined as

$$\kappa_s(t,s) = \frac{\partial\varphi(t,s)}{\partial s}, \qquad \kappa_\theta(t,s) = \frac{\sin(\varphi(t,s))}{r(t,s)},$$

with $r(t,s)$ being the radial distance to the axis of growth and $\varphi(t,s)$ is the angle between the vector normal to the meridian and the axis of the cell.

It is assumed that the strain rate of the wall of a pollen tube is proportional to the rate of calcium-pectin cross-link dissociation and the averaged strain $\langle\varepsilon\rangle$ of the polymer networks in the wall:

$$\frac{\partial\varepsilon}{\partial t} = k_d r_d \langle\varepsilon\rangle,$$

where k_d is a dissociation rate constant. The polymer network of pectin, which at the microscopic scale forms an amorphous cross-linked hydrogel, is modelled as a cubic lattice of linear elastic springs with lattice constant (mesh size) ξ and the spring constant $\tilde{k}$, such that the force applied to a spring is $F = \tilde{k}\,\varepsilon$. The averaged strain and the strain rate in each of two directions are related to the in-plane stresses by

$$\langle\varepsilon_i\rangle = \frac{1}{\tilde{k}\xi\chi}\left(\sigma_i - \nu\sigma_j\right), \qquad \frac{\partial\varepsilon_i}{\partial t} = k_d r_d\langle\varepsilon_i\rangle = \frac{k_d r_d}{\tilde{k}\xi\chi}\left(\sigma_i - \nu\sigma_j\right), \tag{11}$$

for $i, j = s, \theta$, where ν is Poisson's ratio of the cell wall. The stress is related to the turgor pressure P in the pollen tube by

$$\sigma_s = \frac{P}{2h\kappa_\theta}, \qquad \sigma_\theta = \frac{P}{2h\kappa_\theta}\left(2 - \frac{\kappa_s}{\kappa_\theta}\right).$$

It is considered that $\xi = 5\,\text{nm}$, $\tilde{k} = 2\,\text{nN}$, and $\nu = 0.78$. Then, the averaged areal strain $\langle \varepsilon_A \rangle = \langle \varepsilon_s \rangle + \langle \varepsilon_\theta \rangle$ is defined as the sum of the averaged strains in 'meridional' and 'circumferential' directions

$$\langle \varepsilon_A \rangle = \frac{P(1-\nu)}{2\tilde{k}\xi\chi_A}\frac{1}{\kappa_\theta}\left(3 - \frac{\kappa_s}{\kappa_\theta}\right),$$

and the areal strain rate is given as

$$\frac{\partial \varepsilon_A}{\partial t} = k_d r_d \langle \varepsilon_A \rangle = k_d r_d \frac{P(1-\nu)}{2\tilde{k}\xi\chi_A}\frac{1}{\kappa_\theta(s)}\left(3 - \frac{\kappa_s(s)}{\kappa_\theta(s)}\right). \tag{12}$$

The dynamics of the density of demethyl-esterified pectin r_d and the wall thickness h are modelled by transport equations with transport velocity v_s:

$$\begin{aligned} \frac{\partial r_d}{\partial t} + v_s(s)\frac{\partial r_d}{\partial s} &= k_e r_e - 2k_d r_d \langle \varepsilon_A \rangle - \frac{\mathcal{D}(s)}{h} r_d, \\ \frac{\partial h}{\partial t} + v_s(s)\frac{\partial h}{\partial s} &= \mathcal{D}(s) - h\frac{\partial \varepsilon_A}{\partial t}. \end{aligned} \tag{13}$$

Here, the rate of deposition of new material is defined as

$$\mathcal{D}(s) = f(s)(v_{\text{crit}} - v_{\text{pole}})^+,$$

where the critical elongation rate is equal to $v_{\text{crit}} = 0.41\,\mu\text{m/s}$, the elongation rate is related to the strain rate at the pole of the cell by

$$v_{\text{pole}} = \frac{\partial \varepsilon_s(0)}{\partial t}\frac{R}{1.73}, \qquad \text{with } R = \int_0^{\pi/2} \frac{\cos(\varphi)}{\kappa_s} d\varphi \quad \text{being the cell radius,}$$

and

$$f(s) = \begin{cases} \alpha \cos^2\left(\dfrac{\varphi(s)}{\varphi(a)}\right) & \text{if } \varphi(s) \le \dfrac{\pi}{2}, \\ 0 & \text{if } \varphi(s) > \dfrac{\pi}{2}. \end{cases}$$

Here, a denotes the length scale over which the deposition of new material (exocytosis) occurs and it is estimated that $a = 7.5\,\mu\text{m}$ and $\alpha = 0.19$. The growth velocity is defined as

$$v_s(s) = \frac{\partial_t \varepsilon_\theta}{\kappa_\theta \cot(\varphi(s))} = k_d r_d \frac{P}{\tilde{k}\xi \chi_A} \frac{r(s)^2}{\sin(2\varphi(s))} \left(2 - \frac{\kappa_s(s)}{\kappa_\theta(s)} - \nu\right).$$

The rates of strain are determined by the current stress state and mechanical properties so that large deformation of the cell surface can be simulated as a series of small growth increments. The stresses are computed from the current geometry, the values for turgor pressure and wall thickness, and substituted in the constitutive Eq. (11) to obtain the strain rates. The following kinematic relations between the strain rates and velocity vector are used to update the geometry of the pollen tube (Dumais et al. 2006):

$$\frac{\partial \varepsilon_s}{\partial t} = v_n \kappa_s + \frac{\partial v_t}{\partial s}, \qquad \frac{\partial \varepsilon_\theta}{\partial t} = v_n \kappa_\theta + \frac{v_t \cos(\varphi)}{r}. \tag{14}$$

Then, the velocity of every point on the meridian can be calculated directly from (14):

$$\begin{aligned} v_t(t,s) &= \sin(\varphi) \int_S^s \frac{1}{\sin(\varphi)} \left(\frac{\partial \varepsilon_s}{\partial t} - \frac{\kappa_s}{\kappa_\theta} \frac{\partial \varepsilon_\theta}{\partial t} \right) ds, \\ v_n(t,s) &= \frac{\partial \varepsilon_\theta}{\partial t} \frac{1}{\kappa_\theta} - \cos(\varphi) \int_S^s \frac{1}{\sin(\varphi)} \left(\frac{\partial \varepsilon_s}{\partial t} - \frac{\kappa_s}{\kappa_\theta} \frac{\partial \varepsilon_\theta}{\partial t} \right) ds. \end{aligned} \tag{15}$$

Here, S is the distance from the pole to the equator. Then, the new geometry is determined by finding the position of each point of the discretisation mesh after a small time interval.

The mathematical model (11)–(15) has both steady state and periodic solutions, dependent on the set of parameters, and is able to capture very well the in vivo dynamics of growing pollen tube cell walls, Fig. 6. The surface in parameter domain that separates those two regimes can be approximated by the plane $v_{avg}/v_{crit} \approx 0.7$, where v_{avg} is the average elongation rate. The spatial distribution in the deposition of new material and cell wall expansion, considered in the model, is a possible explanation for the oscillatory morphology of a pollen tube, observed experimentally (Rojas et al. 2011). The model also captures the negative strain rates in the meridional direction observed near the base of the cell apex, which is related to the circumferential stress anisotropy ($\sigma_\theta > \sigma_s$).

Mathematical Modelling of Nonlinear Deformations of Plant Cell Walls

The main force for cell elongation (the turgor pressure) acts isotropically, and so it is the microscopic structure of the cell wall, given by the cellulose microfibrils embedded in the cell wall matrix of pectin and hemicellulose that results in the anisotropic growth of plant cells and tissues (Cosgrove 2005), Fig. 7a. To account for the microscopic structure, determined by the orientation of cellulose microfibrils, a fibre-reinforced hyperelastic-viscoplastic model for growth of a plant cell wall was

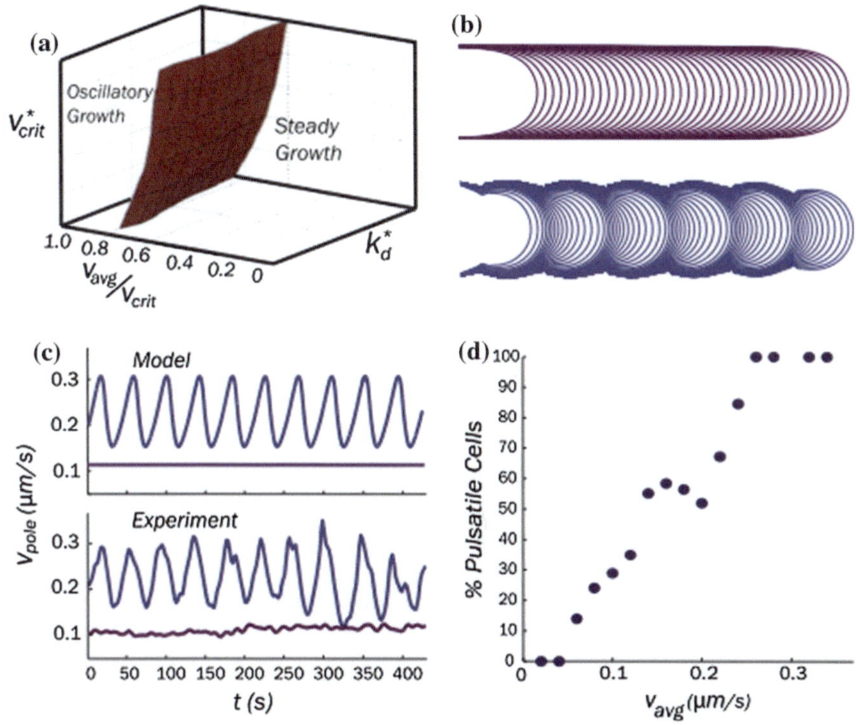

Fig. 6 Comparison between solutions of model (11)–(15) and experimental data on pollen tube growth. The parameter space for v_{crit}, k_d, $v_{\mathrm{avg}}/v_{\mathrm{crit}}$ is split into steady and oscillatory growth regimes (**a**). Numerical simulations of cell growth show two behaviours with different morphologies (**b**) and elongation rates (**c**). The results on the percentage of cells that are oscillating at each mean elongation rate (**d**) indicate that faster cells are more likely to oscillate. Reproduced from Rojas et al. (2011) with permission

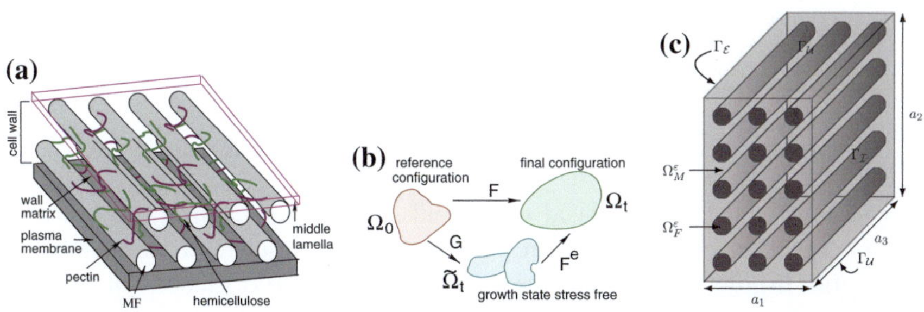

Fig. 7 **a** Schematic diagram of a plant cell wall. MF denotes cell wall microfibrils, CMT denotes cortical microtubules in a plant cell. **b** Schematic diagram of the multiplicative decomposition of the deformation gradient. **c** Schematic of the geometry of a part of a plant cell wall

considered in Huang et al. (2012b). In the mathematical model, the impact of the calcium-pectin cross-links on the growth rate of a cell wall, observed experimentally (Proseus and Boyer 2006, 2007), is reflected in the yield parameters and stiffening of the cell wall.

To derive a mathematical model for the growth of a part of a plant cell wall the concepts of finite deformations are considered and the deformation gradient is defined as

$$\mathbf{F}(X,t) = \frac{\partial \psi}{\partial X},$$

where $x = \psi(X,t)$ denotes the current position of a material point X. Using Rodriguez's approach, Rodriguez et al. (1994), the total deformation gradient $\mathbf{F}$ is decomposed as $\mathbf{F} = \mathbf{F}^e\mathbf{G}$, where $\mathbf{G}$ denotes the deformation due to the growth of the cell wall and $\mathbf{F}^e$ is the elastic deformation that ensures the compatibility of the total deformation, Fig. 7b. Then, the Cauchy-Green tensors in terms of elastic and total deformations, respectively, are given by

$$\mathbf{C} = (\mathbf{F}^e)^T\mathbf{F}^e \quad \text{and} \quad \mathbf{B} = \mathbf{F}\mathbf{F}^T.$$

To describe the fibrous structure of plant cell walls a family of fibres with the direction represented by a unit vector $a_0(X) \in S^2$, where $\tilde{a}(X,t) = \mathbf{F}(X,t)a_0(X)$, and $a(x,t) = \mathbf{G}a_0$, is considered. Then the deformation $\psi : \Omega_0 \to \Omega_t$ maps the pair of variables $(X, a_0) \in \Omega_0 \times S^2$ onto $(x, \tilde{a}) \in \Omega_t \times S^2$, where Ω_0 and Ω_t denote the reference and current configurations, respectively.

For the derivation of the momentum conservation equations, the following strain-energy function is considered

$$\begin{aligned}\mathbf{W} = \mathbf{W}_{\text{vol}}(J) + \mathbf{W}_{\text{iso}}(I_1, I_2, I_4, I_5) &= \frac{\kappa}{2}(J-1)^2 + \frac{\mu}{2}\left[J^{-\frac{2}{3}}\text{tr}(\mathbf{C}) - 3\right] \\ &\quad + \frac{1}{2}\frac{\kappa_1}{\kappa_2}\left[\exp(\kappa_2(I_4-1)^2) - 1\right],\end{aligned}$$

where

$$\begin{aligned}&J = \det(\mathbf{F}^e), && I_1 = J^{-\frac{2}{3}}\text{tr}(\mathbf{C}), \quad I_2 = \frac{1}{2}J^{-\frac{4}{3}}[(\text{tr}(\mathbf{C}))^2 - \text{tr}(\mathbf{C}^2)], \\ &I_4 = J^{-\frac{2}{3}}\mathbf{C} : \mathbf{A} = \lambda_{\mathbf{a}}, && I_5 = \mathbf{C}^2 : \mathbf{A}, \qquad \text{with } \mathbf{A} = a \otimes a,\end{aligned}$$

the constants κ and μ are related to the bulk and shear moduli of the cell wall matrix, respectively, and κ_1 and κ_2 are parameters of the elastic response of cellulose microfibrils. In order to account for the anisotropic growth, it is assumed that cellulose microfibrils have a much larger Young's modulus than the cell wall matrix.

Then, the second Piola–Kirchhoff stress tensor in the intermediate configuration $\widetilde{\Omega}_t$, reflecting changes in the cell wall due to growth, and the Cauchy stress tensor in the current configuration Ω_t are defined as

$$\mathbf{S} = 2\frac{\partial \mathbf{W}}{\partial \mathbf{C}} \quad \text{and} \quad \boldsymbol{\sigma} = J^{-1}\mathbf{F}^e\mathbf{S}(\mathbf{F}^e)^T$$

and the momentum conservation equations read

$$\operatorname{div} \boldsymbol{\sigma} = 0 \quad \text{in } \Omega_t, \quad \text{with} \quad \boldsymbol{\sigma} \cdot n = f \quad \text{on } \partial\Omega_t, \tag{16}$$

where f denotes the forces acting on the boundaries of the cell wall and n is the external normal vector.

The model is completed by the evolution equation for the growth tensor. The turgor-driven creep and new material deposition are widely accepted to be the fundamental aspects of the cell wall growth. In the proposed model, the new material deposition is considered as a process in which the new wall material is produced by a smooth volumetric source of mass at each point of Ω_t instead of a mass flux through the boundary of Ω_t. It is assumed that the growth processes can be decomposed into three parts: isochoric expansive growth, volumetric growth and irreversible microfibrils extension. If neglecting the plastic deformation of microfibrils, the equation for the growth tensor

$$\frac{\partial}{\partial t}\mathbf{G} = \left[\operatorname{sym}(\mathbf{R}^*)\frac{\phi_{\text{dev}}}{\eta} + \mathbf{R}^*_{\text{vol}}\frac{\phi_{\text{vol}}}{\eta/\beta_1} \right]\mathbf{G}.$$

accounts for isochoric expansive growth and volumetric growth, where η/β_1 is the wall matrix viscosity, η is a temperature-dependent viscosity defined as

$$\eta = \eta_0 \exp\left(-(Q/R)T^{-1}\right),$$

where T denotes the cell wall temperature, Q is the activation energy, and R is the gas constant, with $Q/R = 50\,°\text{C}$ and $\eta_0 = 10^{10} - 10^{13}$ Pa s, and

$$\operatorname{sym}(\mathbf{R}^*) = \frac{\partial \phi_{\text{dev}}}{\partial \mathbf{S}}, \qquad \mathbf{R}^*_{\text{vol}} = \frac{\partial \phi_{\text{vol}}}{\partial \mathbf{S}} = \operatorname{tr}(\mathbf{S}^A)(\mathbf{C} - (\mathbf{F}^e)^T\tilde{\mathbf{A}}\mathbf{F}^e) - \mathbf{C}\mathbf{S}^A\mathbf{C}.$$

To define tensors $\operatorname{sym}(\mathbf{R}^*)$ and $\mathbf{R}^*_{\text{vol}}$ and the functions ϕ_{vol} and ϕ_{dev} the tensors

$$\mathbf{S}^A = \mathbf{S} - \frac{1}{\lambda_a^2}[\mathbf{CSC}] : \mathbf{A}\,\mathbf{A}, \qquad \mathbf{S}_{\text{vol}} = \frac{1}{3}\operatorname{tr}(\mathbf{S}^A)\,\mathbf{C}^{-1},$$

and

$$\mathbf{S}^{\text{dev}} = \mathbf{S} - \frac{1}{2}\left[\operatorname{tr}(\mathbf{S}^A)\mathbf{C}^{-1} + \frac{3}{\lambda_a}\left(\mathbf{S} : \left[\frac{1}{\lambda_a}(\mathbf{F}^e)^T\tilde{\mathbf{A}}\mathbf{F}^e - \frac{1}{3}\mathbf{C}\right]\right)\mathbf{A}\right],$$

related to the strain tensor and orientation of microfibrils, are introduced. The impact of microfibrils orientation on cell wall growth is reflected in $\mathbf{A} = a \otimes a$ and $\tilde{\mathbf{A}} = \lambda_a^{-1}\tilde{a} \otimes \tilde{a}$, where the changes in the microfibrils orientation due to growth and total deformation are defined by $a = \mathbf{G}a_0$ and $\tilde{a} = \mathbf{F}^e a = \mathbf{F}a_0$, with a_0 denoting the orientation of microfibrils in the reference configuration, and $\lambda_a = \operatorname{tr}(\tilde{a} \otimes \tilde{a})$.

Then, the growth process is described by the overstress function defined in terms of the yield function

$$f_{\text{dev}} = \frac{J_1}{Y_1^2} + \frac{J_2}{Y_2^2} - (1+\alpha)$$

as

$$\phi_{\text{dev}} = \begin{cases} 0 & \text{if } f_{\text{dev}} \le 0, \\ \phi^m & \text{if } f_{\text{dev}} > 0, \end{cases}$$

where $\phi = -\theta\sqrt{J_1/(Y_1^2-\theta)^2 + J_2/(Y_2^2-\theta)^2}$, m being the rate-sensitivity exponent, and θ is one of the real roots of a quadratic equation

$$(1+\alpha)(\theta - Y_1^2)^2(\theta - Y_2^2)^2 - J_2 Y_2^2(\theta - Y_1^2)^2 - J_1 Y_1^2(\theta - Y_2^2)^2 = 0.$$

Here, Y_1 and Y_2 are yield parameters (Y_1 and Y_2 range from 3.5 to 7.0 MPa), J_1 and J_2 are fibre-dependence invariances defined as

$$J_1 = \frac{1}{\lambda_a}[\mathbf{C}(\mathbf{S}^{\text{dev}}\mathbf{C}\mathbf{S}^{\text{dev}})\mathbf{C}] : \mathbf{A} \text{ and } J_2 = \text{tr}(\mathbf{S}^{\text{dev}}\mathbf{C}\mathbf{S}^{\text{dev}}),$$

and α models hardening of the cell wall. The interactions between two different parts of the growth, i.e. isochoric expansive growth and volumetric growth, are reflected in the hardening law and the evolution of the cell wall thickness. The phenomenological equation for the hardening is given by

$$\frac{\partial}{\partial t}\alpha(t) = (h_1 + h_2\alpha)\frac{\phi_{\text{dev}}}{\eta}, \tag{17}$$

with two hardening parameters $h_1 \approx 20.6$ and $h_2 \approx 20.0$, estimated by fitting numerical solutions to experimental data from Proseus and Boyer (2006). Then, considering the evolution equation for the wall thickness

$$\frac{\partial}{\partial t}\left(\frac{\phi_{\text{vol}}}{\eta/\beta_1}\right) = -\frac{1}{t_c}\left(\frac{\phi_{\text{vol}}}{\eta/\beta_1} - \zeta(t)\frac{\phi_{\text{dev}}}{\eta}\right),$$

where

$$\zeta(t) = -\frac{(\mathbf{F}^e(t))^{-1}n \cdot \text{sym}(\mathbf{R}^*(t))(\mathbf{F}^e(t))^{-1}n}{(\mathbf{F}^e(t))^{-1}n \cdot \mathbf{R}^*_{\text{vol}}(t)(\mathbf{F}^e(t))^{-1}n}$$

and $t_c = 10\,\text{s}$ is a characteristic time, estimated using the experimental data from Proseus et al. (1999), we obtain the following relation between isochoric expansion and volumetric growth

$$\frac{\phi_{\text{vol}}(t)}{\eta/\beta_1} = \frac{1}{t_c}\int_0^t e^{(s-t)/t_c}\zeta(s)\frac{\phi_{\text{dev}}(s)}{\eta}ds.$$

The comparison of the numerical solutions of the mathematical model with experimental data from Proseus and Boyer (2006), Proseus et al. (1999) shows that the model is capable of describing the interplay between growth, turgor pressure and temperature (Huang et al. 2012b). The model solutions capture very well the main characteristics of the growth behaviour of a part of a plant cell wall due to changes in the turgor pressure P, observed experimentally (Proseus et al. 1999), i.e. instantaneous elastic response for step changes in values of P and changes of growth rate at different P levels.

The numerical results for the mathematical model indicate that there are three components in the elongation of a cell wall: (1) growth, (2) irreversible elongation and (3) elastic response. The numerical and experimental results for a mature cell, showing that the change of turgor pressure cannot alter the absence of growth in a mature cell, emphasise that growth is controlled by the biological mechanisms and

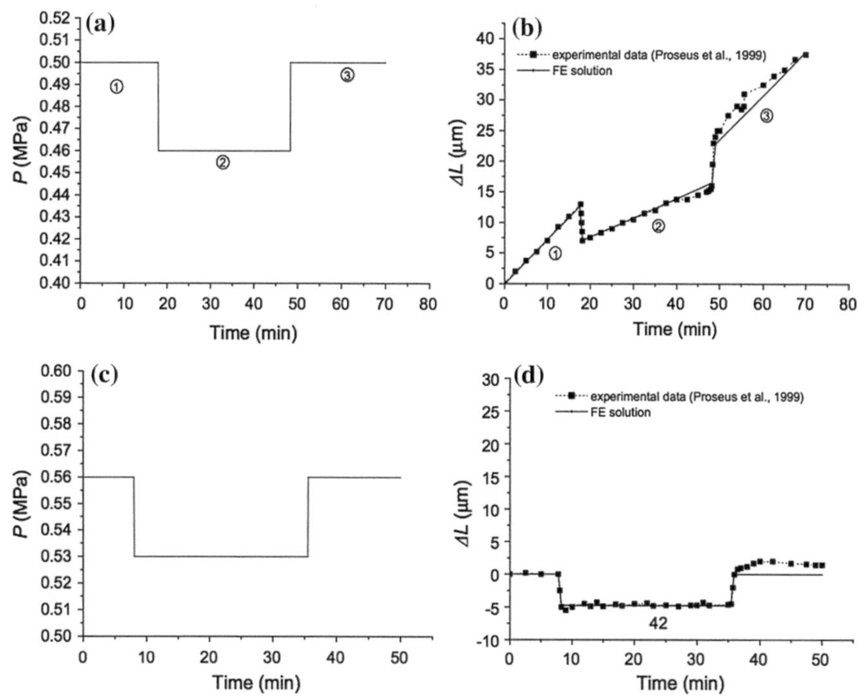

Fig. 8 The effect of changes in turgor pressure P on the deformation of young and mature cells. Changes in turgor pressure (**a**) and length (**b**) for young cells. Changes in turgor pressure (**c**) and length (**d**) for mature cells. For young cells Young's modulus for the cell wall matrix is $E = 1.65$ GPa, the parameters are $Y_1 = Y_2 = 6.125$ MPa. For mature cells Young's modulus is $E = 2.0$ GPa and the yield threshold is $Y_1 = Y_2 = 10$ MPa. In both cases, Poisson's ratio is $\nu = 0.3$, viscosity $\eta = 9.677745 \times 10^{12}$ Pa s, hardening parameters $h_1 = 20.6$, $h_2 = 20.0$, elastic parameters for microfibrils are $\kappa_1 = E/(2(1+\nu))$ and $\kappa_2 = 1.0$. Reproduced from Huang et al. (2012b) with permission

the turgor pressure is just a passive driving force (Huang et al. 2012b). There are two options in the present model: either raising the yield threshold or raising the viscosity. The two options, although involving different mechanisms from the point of view of solid mechanics, are capable of simulating the absence of cell wall loosening in a mature cell. Thus, a higher yield threshold is set as $Y_1 = Y_2 = 10$ MPa for a mature cell but the viscosity is kept the same as for a young cell. The numerical results show that the mature cell responds mainly in an elastic mode, Fig. 8.

The fact that at lower temperatures, e.g. 7.3 °C, growth vanishes but the elastic response is the same as at the higher temperature, e.g. 23.7 °C, is used to analyse the impact of stress-induced elastic deformations and growth separately, Fig. 9. The model can represent the experimental observation of the cell wall growth as a 'turgor pressure–temperature' interaction system. A good agreement between the numerical solutions of the mathematical model and the experimental results presented in Proseus and Boyer (2006), Proseus et al. (1999) indicates that the model can also accurately represent the effect of temperature on the cell growth, i.e. the decrease in the temperature is related to the decrease in the growth rate.

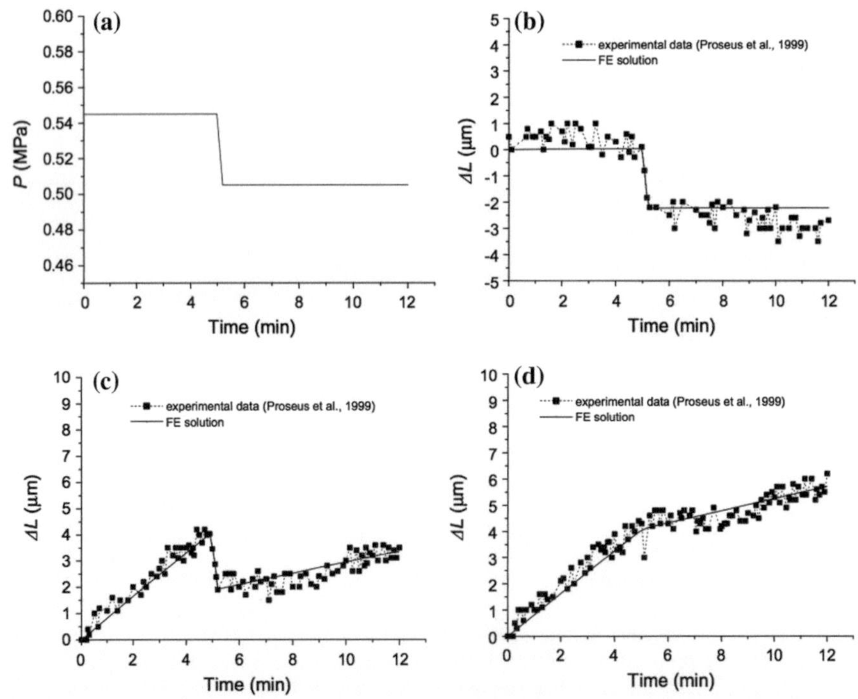

Fig. 9 The impact of turgor pressure and temperature on growth. **a** changes in turgor pressure P with time; **b** changes in cell length at 7.3 °C; **c** changes in cell length at 23.7 °C; **d** difference between length changes at 7.3 and 23.7 °C. Reproduced from Huang et al. (2012b) with permission

The model was also used to analyse the long-term growth of living cells and isolated cell wall. The model parameters for the living cell are given as follows the initial length of the cell is 12 mm, Young's modulus of the cell wall matrix is $E = 2.5$ GPa and Poisson's ratio $\nu = 0.3$, the parameters of the microfibrils elastic response are set as $\kappa_1 = E/2(1+\nu)$ and $\kappa_2 = 1$, the yield parameters are set as $Y_1 = Y_2 = 5.9$ MPa, the viscosity parameter $\eta = 2.0253 \times 10^{12}$ Pa s, turgor pressure is equal to $P = 0.5$ MPa, and hardening parameters are $h_1 = 20.6$ and $h_2 = 20$. For the isolated wall, all the parameters are the same as for the living cell except that $h_1 = 66$ and $h_2 = 0$. In living cells, initial fast growth changes into a stage of slower elongation or a steady growth rate is observed, whereas the isolated cell wall stops growing after a short period of fast growth. Numerical simulations of the mathematical model show that using the simple hardening law (17) it is possible to capture those differences in the growth behaviour observed experimentally.

Microscopic Modelling of Interactions Between Calcium–Pectin Chemistry and Elastic Deformations of Plant Cell Walls

For a better understanding of interactions between mechanical properties of the plant cell wall, calcium–pectin chemistry and microscopic structure, given by the cell wall microfibrils, a microscopic model for elastic deformations of plant cell walls on the scale of single microfibrils was derived in Ptashnyk and Seguin (2015).

In the mathematical model, the reference configuration of a flat section of a plant cell wall is denoted by $\Omega = (0, a_1) \times (0, a_2) \times (0, a_3)$, with some $a_1, a_2, a_3 > 0$, and the domains Ω_M and Ω_F denote the parts of Ω occupied by the cell wall matrix and microfibrils, respectively, Fig. 7c. Five species within cell wall matrix Ω_M are considered to describe the dynamics of calcium–pectin chemistry: methylestrified pectin galacturonic acid groups r_e, the enzyme PME p, demethyl-estrified pectin galacturonic acid groups r_d, calcium ions c, and calcium-pectin cross-links b. We assume that the densities n, with $n \in \{r_e, p, r_d, c, b\}$, are changing due to spatial movement (diffusion) and chemical reactions

$$\partial_t n - \operatorname{div}(\mathbf{D}_\mathrm{n} \nabla n) = \mathcal{R}_\mathrm{n} \quad \text{in } \Omega_M, \tag{18}$$

where $\mathcal{R}_n$ models the chemical reactions between the species and $\mathbf{D}_n$ is the diffusion coefficient. The momentum balance equation for the cell wall, if neglecting the effect of the external body forces and inertial forces, reads

$$\operatorname{div} \mathbf{S} = \mathbf{0} \quad \text{in } \Omega, \tag{19}$$

where $\mathbf{S}$ is the second Piola–Kirchhoff stress tensor. We consider the elastic behaviour of the cell wall or viscoelastic properties of the cell wall matrix and elastic behaviour of the microfibrils (Hayot et al. 2012a) and assume that the chemical processes in the wall matrix influence the mechanical properties of plant cell walls, see, e.g. Cosgrove (2005), Wolf and Greiner (2012). Then, the constitutive laws for the stress tensor are given by

$$\begin{aligned} \mathbf{S} &= \big(\mathbb{E}_M(b)\chi_{\Omega_M} + \mathbb{E}_F\chi_{\Omega_F}\big)\mathbf{e}(\mathbf{u}) \quad \text{or} \\ \mathbf{S} &= \big(\mathbb{E}_M(b)\mathbf{e}(\mathbf{u}) + \mathbb{V}_M(b)\mathbf{e}(\partial_t\mathbf{u})\big)\chi_{\Omega_M} + \mathbb{E}_F\mathbf{e}(\mathbf{u})\chi_{\Omega_F}, \end{aligned} \tag{20}$$

see, e.g. Ciarlet (1988), Fung (1993), where $\mathbb{E}_M(b)$ and $\mathbb{E}_F$ are elasticity tensors for cell wall matrix and microfibrils, respectively, $\mathbb{V}_M$ is a viscosity tensor for the cell wall matrix, and $\mathbf{e}(\mathbf{u}) = \frac{1}{2}(\nabla\mathbf{u} + \nabla\mathbf{u}^{\mathrm{T}})$ is the symmetric part of the displacement gradient, χ_{Ω_M} and χ_{Ω_F} are the characteristic function of the cell wall matrix and cell wall microfibrils, respectively.

We also assume that the stress or strain influences the chemical reactions and the dynamics of calcium-pectin cross-links, see e.g Proseus and Boyer (2012),

$$\mathcal{R}_{\mathrm{n}} = \mathcal{R}_{\mathrm{n}0}(r_e, p, r_d, c, b) + \mathbf{Z}_{\mathrm{n}}(r_e, p, r_d, c, b)\,\mathcal{N}(\mathbf{e}(\mathbf{u})) \quad \text{for } n \in \{r_e, p, r_d, c, b\},$$

where $\mathcal{R}_{\mathrm{n}0}$ and $\mathbf{Z}_{\mathrm{n}}$ define chemical reactions and $\mathcal{N}$ denotes a nonlinear function of the strain, specified below.

We assume that the cell wall matrix is isotropic, see, e.g. Zsivanovits et al. (2004), and, by using standard representation theorems for isotropic functions, see, e.g. Gurtin et al. (2009), obtain

$$\begin{aligned} \mathbb{E}_M(b)\mathbf{e}(\mathbf{u}) &= 2\mu(b)\,\mathbf{e}(\mathbf{u}) + \lambda(b)\,\mathrm{div}\,\mathbf{u}\,\mathbf{1}, \\ \mathbf{Z}_{\mathrm{n}}(r_e, p, r_d, c, b) &= z_{\mathrm{n}}(r_e, p, r_d, c, b)\mathbf{1}, \qquad \mathbf{D}_n = D_n\mathbf{1}. \end{aligned}$$

The calcium–pectin chemistry in the cell wall is defined by the following four interactions: (i) demethyl-esterification of pectin by PME, (ii) creation of calcium-pectin cross-links via binding of demethyl-estrified pectin and calcium ions, (iii) breakage of calcium-pectin cross-links under the presence of mechanical stress, and (iv) decay of demethyl-esterified pectin, see, e.g. Proseus and Boyer (2012), Wolf and Greiner (2012), Zsivanovits et al. (2004). Then, the reaction term $\mathcal{R}_{\mathrm{r_d}}$ for demethyl-esterified pectin r_d can be decomposed into three terms:

$$\mathcal{R}_{\mathrm{r_d}} = \mathcal{R}_{\mathrm{ep}} + \mathcal{R}_{\mathrm{dd}} + \mathcal{R}_{\mathrm{fb}}, \tag{21}$$

where $\mathcal{R}_{\mathrm{ep}}$ is the rate of change of r_d due to demethylesterification of r_e by PME, $\mathcal{R}_{\mathrm{dd}}$ is the rate of decay of r_d, and $\mathcal{R}_{\mathrm{fb}}$ is the rate of change of r_d associated with the formation and breakage of calcium-pectin cross-links. We have that

$$\mathcal{R}_{\mathrm{r_e}} = -\mathcal{R}_{\mathrm{ep}}. \tag{22}$$

We assume that both the binding of PME to and its dissociation from a pectin acid group are very fast and that the enzyme PME is not used-up during the demethylesterification process so that $\mathcal{R}_\mathrm{p} = 0$. From interaction (ii) we have

$$\mathcal{R}_\mathrm{b} = -\mathcal{R}_\mathrm{c} = -\frac{1}{2}\mathcal{R}_\mathrm{fb}. \tag{23}$$

The factor of a half in front of $\mathcal{R}_\mathrm{fb}$ reflects the fact that two demethyl-esterified galacturonic acids are needed to form a calcium-pectin cross-link. Interactions between demethyl-esterified pectin and calcium ions increase the number of cross-links, while stress can break the cross-links. Thus,

$$\mathcal{R}_\mathrm{b} = R_\mathrm{dc}(r_d, c) - k_b\, b\, \mathcal{N}(\mathbf{e}(\mathbf{u})), \tag{24}$$

where R_dc models the formation of cross-links through the interactions between demethyl-esterified pectin and calcium ions, $\mathcal{N}(\mathbf{e}(\mathbf{u})) = \big(\mathrm{tr}\,\mathbb{E}(b)\,\mathbf{e}(\mathbf{u})\big)^+$ or $\mathcal{N}(\mathbf{e}(\mathbf{u})) = \big(\mathrm{tr}\,\mathbf{e}(\mathbf{u})\big)^+$, and k_b is a positive constant. The reason for the choices for the function $\mathcal{N}$ is based on the idea that stretching, rather than compressing, of the cross-links will cause them to break. We assume that

$$\mathcal{R}_\mathrm{ep} = k_\mathrm{ep} r_e p, \quad \mathcal{R}_\mathrm{dd} = -d\, r_d, \quad R_\mathrm{dc}(r_d, c) = \frac{k_\mathrm{dc} c}{k_\mathrm{m} + c} r_d, \tag{25}$$

where k_ep, k_dc, k_m, and d are positive constants.

The boundary $\partial\Omega$ of Ω is decomposed into four disjoint surfaces: $\Gamma_\mathcal{I}$, $\Gamma_\mathcal{E}$, $\Gamma_\mathcal{U}$, and $\partial\Omega \setminus (\Gamma_\mathcal{I} \cup \Gamma_\mathcal{E} \cup \Gamma_\mathcal{U})$, where $\Gamma_\mathcal{I}$ is the part of $\partial\Omega$ in contact with the interior of the cell and $\Gamma_\mathcal{E}$ is the part of $\partial\Omega$ in contact with the middle lamella, $\Gamma_\mathcal{U}$ are upper and lower parts of the boundary, Fig. 7c. PME, produced in the Golgi apparatus of a plant cell, is deposited into the cell wall and diffuses through the cell wall into the middle lamella. PME can also diffuse back into the cell to degrade. Thus, we assume that the enzyme PME can enter or leave the cell wall through $\Gamma_\mathcal{I}$ but can only leave the wall through $\Gamma_\mathcal{E}$. To account for the mechanisms controlling the amount of PME in a cell wall Wolf et al. (2012a), we assume that the inflow of PME into the cell wall depends on methyl-esterified and demethyl-esterified pectin within the wall

$$D_\mathrm{p}\nabla p\cdot \boldsymbol{\nu} = \frac{\beta_\mathrm{p}\, r_e}{1+\alpha_\mathrm{p}\, r_d} - \zeta_\mathrm{p}\, p \quad \text{on } \Gamma_\mathcal{I}, \qquad D_\mathrm{p}\nabla p\cdot \boldsymbol{\nu} = -\gamma_\mathrm{p}\, p \quad \text{on } \Gamma_\mathcal{E}, \tag{26}$$

where α_p, β_p, ζ_p and γ_p are non-negative constants and $\boldsymbol{\nu}$ denotes the exterior unit-normal.

Methyl-esterified pectin is produced by the cell and then transported into the cell wall through $\Gamma_\mathcal{I}$, e.g. Wolf and Greiner (2012). To account for mechanisms controlling the amount of pectin in the cell wall, we assume that the inflow of new methyl-esterified pectin decreases with an increasing amount of pectin in the wall. Methyl-esterified pectin can leave the wall through $\Gamma_\mathcal{E}$ to enter the middle lamella:

$$D_{\mathrm{r_e}} \nabla r_e \cdot \boldsymbol{\nu} = \frac{\beta_{\mathrm{e}}\, r_e}{1 + \zeta_{\mathrm{e}}(r_e + r_d + b)} \qquad \text{on } \Gamma_{\mathcal{I}}, \quad D_{\mathrm{r_e}} \nabla r_e \cdot \boldsymbol{\nu} \;\; = -\gamma_{\mathrm{e}}\, r_e \qquad \text{on } \Gamma_{\mathcal{E}}, \tag{27}$$

where β_e, ζ_e, and γ_e are non-negative constants. We assume an outflow of demethyl-esterified pectin from the cell wall into the middle lamella and demethyl-esterified pectin cannot move into the cell:

$$D_{\mathrm{r_d}} \nabla r_d \cdot \boldsymbol{\nu} = 0 \qquad \text{on } \Gamma_{\mathcal{I}}, \qquad\qquad D_{\mathrm{r_d}} \nabla r_d \cdot \boldsymbol{\nu} = -\gamma_{\mathrm{d}}\, r_d \qquad \text{on } \Gamma_{\mathcal{E}}. \tag{28}$$

Calcium ions may enter or leave the cell wall through both $\Gamma_{\mathcal{I}}$ and $\Gamma_{\mathcal{E}}$ and the flow of calcium through $\Gamma_{\mathcal{I}}$ is controlled by stretch-activated calcium channels in the plasma membrane, see, e.g. Dutta and Robinson (2004), White (2001), i.e. the flow of calcium through $\Gamma_{\mathcal{I}}$ is assumed to depend on the local average of the stress:

$$\begin{aligned} &D_{\mathrm{c}} \nabla c \cdot \boldsymbol{\nu} = \beta_{\mathrm{c}} - \zeta_{\mathrm{c}}\, c \quad \text{on } \Gamma_{\mathcal{E}}, \\ &D_{\mathrm{c}} \nabla c \cdot \boldsymbol{\nu} = (\gamma_{\mathrm{c}} - \alpha_{\mathrm{c}}\, c) \fint_{B_\delta(x)} \big(\mathrm{tr}\, \mathbb{E}(b)\, \mathbf{e}(\mathbf{u})\big)^+ dx \quad \text{on } \Gamma_{\mathcal{I}}, \end{aligned} \tag{29}$$

with non-negative constants β_c, ζ_c, γ_c, and α_c and some $\delta > 0$. We consider zero-flux boundary conditions on the surface of the microfibrils and on the upper and lower boundaries $\Gamma_{\mathcal{U}}$

$$D_{\mathrm{n}} \nabla n \cdot \boldsymbol{\nu} = 0 \qquad \text{on } \Gamma, \qquad\qquad D_{\mathrm{n}} \nabla n \cdot \boldsymbol{\nu} = 0 \quad \text{on } \Gamma_{\mathcal{U}}, \qquad \text{for } n = r_e, p, r_d, c.$$

The traction boundary conditions

$$\mathbf{S}\boldsymbol{\nu} = -P\boldsymbol{\nu} \qquad \text{on } \Gamma_{\mathcal{I}}, \qquad\qquad \mathbf{S}\boldsymbol{\nu} = \mathbf{f} \qquad \text{on } \Gamma_{\mathcal{E}} \cup \Gamma_{\mathcal{U}}, \tag{30}$$

come from the turgor pressure P within the cell and the traction force $\mathbf{f}$, caused by surrounding cells. On $\partial\Omega \setminus (\Gamma_{\mathcal{I}} \cup \Gamma_{\mathcal{E}} \cup \Gamma_{\mathcal{U}})$ periodic boundary conditions for the densities and displacement are imposed. Using multiscale analysis techniques a macroscopic model, on the scale of a cell wall, for the coupling between elastic deformations and calcium-pectin chemistry can be derived (Ptashnyk and Seguin 2015).

Numerical simulations of the macroscopic model, derived from the microscopic model (18)–(30), see Ptashnyk and Seguin (2015), indicate the difference in the impact of cell wall stress and strain on the chemical processes in the cell wall matrix and demonstrate much stronger impact of cell wall stress on the deformation dynamics, Fig. 10. Those results, underpinned by further analysis and comparison with experimental data, would allow us to identify if the coupling between mechanics and chemistry is dominated by stress or by strain. Numerical solutions of the macroscopic model also reflect the stress-dependent distribution of the calcium–pectin cross-links and the stiffening of the cell wall material due to increase of cross-links density. Further analysis of the model and coupling with the growth processes will provide a good framework to analyse the interactions between calcium–pectin chemistry, mechanical properties, microscopic structure, and growth of plant cell walls and tissues.

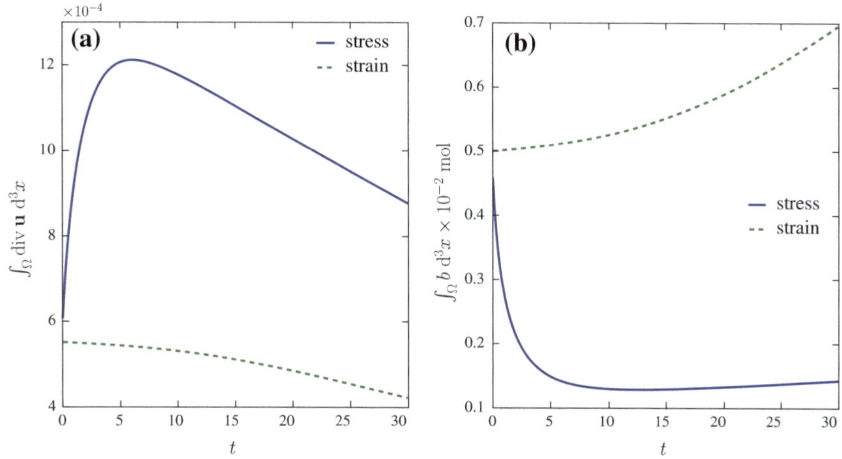

Fig. 10 Numerical solutions for the macroscopic model (18)–(30) for the cases when chemical reactions depend on the cell wall strain $\mathcal{N}(\mathbf{e}(\mathbf{u})) = (\mathrm{tr}\, \mathbf{e}(u))^+$ and stress $\mathcal{N}(\mathbf{e}(\mathbf{u})) = (\mathrm{tr}\, E(b)\mathbf{e}(u))^+$. **a** Dynamics of the average divergence of the deformation. **b** Dynamics of the averaged density of calcium-pectin cross-links. In numerical simulations, we consider the following parameter values: $D_{r_e} = D_{r_d} = 10^{-13}\,\mathrm{m/s^2}$, $D_p = D_c = 10^{-11}\,\mathrm{m/s^2}$, $k_{ep} = 9.0667 \times 10^{-6}\ \mathrm{m^3/U\ s^{-1}}$, $k_b = 0.14\,\mathrm{s^{-1}}$, $d = 0.01667\,\mathrm{s^{-1}}$, $k_{dc} = 0.18\,\mathrm{s^{-1}}$, $k_m = 0.08\ \mathrm{mol/m^3}$, $\beta_e = 2.575 \times 10^{-8}\,\mathrm{mol/m^2\,s^{-1}}$, $\zeta_e = 200\,\mathrm{m^3/mol}$, $\gamma_e = 2 \times 10^{-10}\,\mathrm{m/s}$, $\gamma_p = 2 \times 10^{-10}\,\mathrm{m/s}$, $\alpha_p = 50\,\mathrm{m^3/mol}$, $\beta_p = 0.04\ \mathrm{U\,m/mol\,s^{-1}}$, $\zeta_p = 3.33 \times 10^{-9}\,\mathrm{m/s}$, $\gamma_d = \beta_c = \zeta_c = \gamma_c = \alpha_c = 0$, $P = 0.209\,\mathrm{MPa}$, $\mathbf{f} = 0.6P$, and initial conditions $r_e = 3\,\mu\mathrm{M}$, $p = 0.3\,\mathrm{U/mm^3}$, $r_d = 5\,\mu\mathrm{M}$, $c = 10\,\mu\mathrm{M}$, $b = 5\,\mu\mathrm{M}$. For cell wall matrix, assumed to be isotropic, we consider $\nu_M = 0.3$, $E_M = 0.775b + 8.08\,\mathrm{MPa}$. For microfibrils, assumed to be transversally isotropic, we consider $E_F = 15 \times 10^3\,\mathrm{MPa}$, $\nu_{F1} = 0.3$, $\nu_{F2} = 0.06$, $n_F = 0.068$, $Z_F = 85 \times 10^3\,\mathrm{MPa}$

Pectin-Related Cell Wall Signalling Pathways

The morphogenesis and growth of plants are strongly controlled by the properties of their extracellular matrix (cell walls and middle lamella) and the integrity of the cell walls of growing plant cells requires a balance between processes that promote cell wall relaxation and stiffening. Hence, to control growth and to respond to external perturbations, the state of the cell wall is constantly monitored and information is conveyed to the cell interior in order to fine-tune the physico-chemical properties of the cell wall for optimal responses.

Over the last few years, much effort has been expended to try to better understand the relations between plant growth and cell wall-related signalling processes (Wolf et al. 2012a). A prominent example of how signalling processes and chemical modification of a component of the cell wall can influence its extensibility and growth rate is the demethyl-esterification of HG pectin by PME and calcium-pectin chemistry. The dynamics of calcium-pectin growth-controlling mechanism indicates that in the absence of appropriate compensatory mechanisms, plants would be very sensitive to environmental fluctuations in Ca^{2+} or other cations which would compete with Ca^{2+}

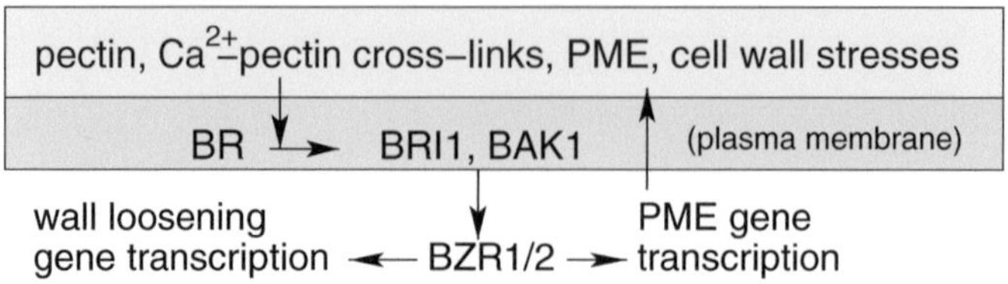

Fig. 11 Schematics of pectin-related cell wall signalling pathway. The cell wall signalling pathway involves the activation of gene expressions related to the production of PME and pectin by perception of brassinosteroid (BR) hormone or by mechanical activation of cell membrane receptors

for pectin binding and, hence, potentially interfere with the cell wall integrity. Therefore, a buffering capacity to compensate for changes in cation concentrations should be ensured by the large amounts of HG present in primary cell walls together with precise regulation of the degree of methyl-esterification, deposition and turnover of pectin. It has been shown that genetic or pharmacological interference with PME activity causes dramatic changes in growth behaviour, which are primarily the result of the activation of the brassinosteroid (BR) signalling pathway, one of the plant growth hormones. Interfering with PME activity (through EGCG treatment or PMEI expression) would lead to a reduction of demethyl-esterified pectin in the cell wall, which would cause the activation of the receptor BRASSINOSTEROID INSENSITIVE1 (BRI1) and the compensatory upregulation of cell wall remodelling agents including PME, Fig. 11. It is proposed that the activation of BR signalling is a part of a compensatory response, which protects the plants against the loss of cell wall integrity caused by the imbalance in pectin modification (Wolf and Greiner 2012; Wolf et al. 2012b).

In recent studies, a receptor-like protein (RLP44) is identified, which mediates the activation of BR signalling in response to PMEI-induced cell wall alterations through direct interaction with the BR co-receptor BAK1 (BRI1-ASSOCIATED RECEPTOR KINASE1), without the need for BR (Wolf et al. 2014). It is shown that loss of RLP44 function prevents activation of BR signalling pathway in response to overexpression of PMEI. Thus, RLP44 integrates cell wall surveillance with hormone signalling to control cell wall integrity and growth and is required for normal growth and response to stress in plant tissues. The experimental results show that RLP44 is not a constitutive component of the BR signalling pathway itself but sensitizes this pathway in response to cell wall-derived cues and indicate that RLP44 recruits the BR signalling module in response to PMEIox-induced cell wall perturbation at least in part through a direct interaction with BAK1. It is observed that RLP44 activation appeared to be sensitive to demethyl-esterified pectin concentration, although whether this is a direct causal relationship remains to be determined. Mathematical modelling of the BR signalling pathway coupled with the dynamics of RLP44 will help us to better understand the direct and indirect relationships between pectin, cell wall stress and activation of RLP44.

Pectin and Plant Morphogenesis

Plant cell expansion is mechanically limited by their wall, hence organ emergence requires changes in the mechanical properties of cell walls. Pectin is involved in morphogenesis though the effect of changes in the pectin methyl-esterification status on cell wall mechanics (Iwai et al. 2002; Palin and Geitmann 2012). It was shown that phyllotactic patterning at the shoot apical meristem is regulated by the demethyl-esterification of pectin in the cell walls (Peaucelle et al. 2008). Especially, it is observed that elevated PME levels resulted in an increased number of primordia and disturbed phyllotactic patterning, whereas overexpression of PMEI completely blocked lateral organ formation and led to a naked meristem phenotype. Hence, demethyl-esterification of HG not only accompanies but is also required for primordia formation in the inflorescence meristem and manipulation of pectins within the cell walls of the meristem affects organ emergence.

The important role of pectin in the morphogenesis is also indicated through its interaction with auxin. Plant organs are produced from meristems in a characteristic pattern, generated by local gradients of auxin. It is shown that the generation of auxin gradient is related to the demethyl-esterification of the HG pectin and the inhibition of HG demethyl-esterification disrupts organized polarity of the auxin transporter PIN1, while in the absence of auxin transport, demethyl-esterification of HG causes only local tissue growth but not full organ formation (Braybrook and Peaucelle 2013). This indicates a reverse interaction between pectin demethyl-esterification and auxin polarity and suggests that in the shoot apical meristem auxin-induced stiffness modulation is mediated through pectin demethyl-esterification, and altered cell wall mechanics affects the organization of PIN1 polarity.

New experimental results also suggest a two-step mechanism for achieving anisotropic plant cell growth stating that non-uniform softening through the modulation of pectin chemistry is a first step in the anisotropic expansion followed by stress-induced stiffening of the cell wall through cellulose microfibrils (Peaucelle et al. 2015).

Conclusion

The demethyl-esterification of pectin by PME and calcium–pectin chemistry have profound impact on plant growth, development, and morphogenesis. It has been observed experimentally that non-cross-linked pectins confer extensibility to the cell wall, whereas Ca^{2+} and borate cross-linking imparts rigidity of plant cell walls and tissues. The experimental results on the role of pectin in plant morphogenesis suggest that a very low degree of methyl-esterification does not necessarily lead to wall stiffening, and the turnover of the demethyl-esterified pectin and availability of calcium ions could possibly explain apparent discrepancies between expected and measured mechanical properties (Peaucelle et al. 2011). This multipurpose nature of pectin with properties that permit the cell to expand rapidly while still retaining enough strength to prevent cell bursting is fascinating from the experimental and modelling point of view, resulting in numerous experimental and a few modelling results.

While it is clear that PME activity must be tightly regulated to fine-tune properties of the pectin network in specific regions of the cell wall, the precise nature of the control mechanism in vivo requires further experimental studies supported by mathematical modelling. Experimental studies and mathematical modelling and analysis of cell wall signalling pathways will help to better understand the control mechanisms involved in plant morphogenesis, growth and homeostasis. Another important and not well-understood question is the exact relation between wall material deposition and growth. Future studies should also consider the interactions between pectin and the cellulose–hemicellulose network to provide a more detailed model of expansion of plant cell walls and tissues.

References

Braybrook SA, Peaucelle A (2013) Mechano-chemical aspects of organ formation in arabidopsis thaliana: the relationship between auxin and pectin. PloS ONE 8:e57813

Burgert I, Fratzl P (2009) Plants control the properties and actuation of their organs through the orientation of cellulose fibrils in their cell walls. Integr Comp Biol 49(1):69

Caffall KH, Mohnen D (2009) The structure, function, and biosynthesis of plant cell wall pectic polysaccharides. Carbohydr Res 344:1879–1900

Camardella L, Carratore V, Ciardiello MA, Servillo L, Balestrieri C, Giovane A (2000) Kiwi protein inhibitor of pectin methylesterase amino-acid sequence and structural importance of two disulfide bridges. Eur J Biochem 267:4561–4565

Cameron RG, Luzio GA, Goodner K, Williams MAK (2008) Demethylation of a model homogalacturonan with a salt-independent pectin methylesterase from citrus: I. Effect of ph on demethylated block size, block number and enzyme mode of action. Carbohydr Polym 71:287–299

Cameron RG, Luzio GA, Vasu P, Savary BJ, Williams MAK (2011) Enzymatic modification of a model homogalacturonan with the thermally tolerant pectin methylesterase from citrus: 1. Nanostructural characterization, enzyme mode of action, and effect of ph. J Agric Food Chem 59:2717–2724

Ciarlet PG (1988) Mathematical elasticity. Volume I: three-dimensional elasticity. North-Holland

Cosgrove DJ (2005) Growth of the plant cell wall. Nat Rev Mol Cell Biol 6:850–861

Denès J-M, Baron A, Renard CMGC, Péan C, Drilleau J-F (2000) Different action patterns for apple pectin methylesterase at ph 7.0 and 4.5. Carbohydr Res 327:385–393

Dick-Pérez M, Wang T, Salazar A, Zabotina OA, Hong M (2012) Multidimensional solid-state nmr studies of the structure and dynamics of pectic polysaccharides in uniformly 13^c-labeled arabidopsis primary cell walls. Magn Reson Chem 50:539–550

Dick-Pérez M, Zhang Y, Hayes J, Salazar A, Zabotina OA, Hong M (2011) Structure and interactions of plant cell-wall polysaccharides by two- and three-dimensional magic-angle-spinning solid-state nmr. Biochemistry 50:989–1000

Dumais J, Shaw SL, Charles RS, Long SR, Ray PM (2006) An anisotropic- viscoplastic model of plant cell morphogenesis by tip growth. Int J Dev Biol 50:209–222

Dutta R, Robinson KR (2004) Identification and characterization of stretch-activated ion channels in pollen proto- plasts. Plant Physiol 135:1398–1406

Ezaki N, Kido N, Takahashi K, Katou K (2005) The role of wall Ca^{2+} in the regulation of wall extensibility during the acid-induced extension of soybean hypocotyl cell walls. Plant Cell Physiol 46:1831–1838
Fries M, Ihrig J, Brocklehurst K, Shevchik VE, Pickersgill RW (2007) Molecular basis of the activity of the phytopathogen pectin methylesterase. EMOB J 26:3879–3887
Fung YC (1993) Biomechanics: mechanical properties of living tissues. Springer, New York
Guillon F, Mose A, Quemener B, Bouchet B, Devaux M-F, Alvarado C, Lahaye M (2017) Remodeling of pectin and hemicelluloses in tomato pericarp during fruit growth. Plant Sci 257:48–62
Gurtin ME, Fried E, Anand L (2009) The mechanics and thermodynamics of continua. Cambridge University Press
Hayot CM, Forouzesh E, Goel A, Avramova A, Turner J-A (2012a) Viscoelastic properties of cell walls of single living plant cells determined by dynamic nanoindentation. J Exp Biol 63:2525–2540
Huang R, Becker AA, Jones IA (2012b) Modelling cell wall growth using a fibre-reinforced hyperelastic-viscoplastic constitutive law. J Mech Phys Solids 60:750–783
Iwai H, Masaoka N, Ishii T, Satoh S (2002) A pectin glucuronyltransferase gene is essential for intercellular attachment in the plant meristem. PNAS 99:16319–16324
Jarvis MC (1984) Structure and properties of pectin gels in plant cell walls. Plant Cell Environ 7:153–164
Jarvis MC, Briggs SP, Knox JP (2003) Intercellular adhesion and cell separation in plants. Plant Cell Environ 26:977–989
Knox JP (1992) Cell adhesion, cell separation and plant morphogenesis. Plant J 2:137–141
Kroeger JH, Zerzour R, Geitmann A (2011) Regulator or driving force? The role of turgor pressure in oscillatory plant cell growth. PLoS ONE 6:e18549
Laskowski M, Biller S, Stanley K, Kajstura T, Prusty R (2006) Expression profiling of auxin-treated arabidopsis roots: toward a molecular analysis of lateral root emergence. Plant Cell Physiol 47:788–792
Lin D, Lopez-Sanchez P, Gidley MJ (2016) Interactions of pectins with cellulose during its synthesis in the absence of calcium. Food Hydrocoll 52:57–68
Lockhart JA (1965) An analysis of irreversible plant cell elongation. J Theor Biol 8:264–275
McKenna ST, Kunkel JG, Bosch M, Rounds CM, Vidali L, Winship LJ, Hepler PK (2009) Exocytosis precedes and predicts the increase in growth in oscillating pollen tubes. Plant Cell 21:3026–3040
Micheli F (2001) Pectin methylesterases: cell wall enzymes with important roles in plant physiology. Trends Plant Sci 6(5):414–419
Mohnen D (2008) Pectin structure and biosynthesis. Curr Opin Plant Biol 11:266–277
O'Neill MA, Ishii T, Albersheim P, Darvill AG (2004) Rhamnogalacturonan ii: structure and function of a borate cross-linked cell wall pectic polysaccharide. Annu Rev Plant Biol 55:109–139
Orfila C, Seymour GB, Willats WGT, Huxham IM, Jarvis MC, Dover CJ, Thompson AJ, Knox JP (2001) Altered middle lamella homogalacturonan and disrupted deposition of (1→5)-?-l-arabinan in the pericarp ofcnr, a ripening mutant of tomato. Plant Physiol 126(1):210–221
Ortega JKE (1990) Governing equations for plant cell growth. Physiol Plant 79:116–121
Palin R, Geitmann A (2012) The role of pectin in plant morphogenesis. BioSystems 109:397–402
Parre E, Geitmann A (2005) Pectin and the role of the physical properties of the cell wall in pollen tube growth of solanum chacoense. Planta 220(22):582–592
Peaucelle A, Braybrook SA, Le Guillou L, Bron E, Kuhlemeier C, Höfte H (2011) Pectin-induced changes in cell wall mechanics underlie organ initiation in arabidopsis. Curr Biol 21:1720–1726
Peaucelle A, Louvet R, Johansen JN, Höfte H, Laufs P, Pelloux J, Mouille G (2008) Arabidopsis phyllotaxis is controlled by the methyl-esterification status of cell-wall pectins. Curr Biol 18:1943–1948
Peaucelle A, Wightman R, Höfte H (2015) The control of growth symmetry breaking in the arabidopsis hypocotyl. Curr Biol 25:1746–1752
Pelletier S, Van Orden J, Wolf S, Vissenberg K, Delacourt J, Assoumou Ndong Y, Pelloux J, Bischoff V, Urbain A, Mouille G, Lemonnier G, Renou J-P, Höfte H (2010) A role for pectin

de-methylesterification in a developmentally regulated growth acceleration in dark-grown arabidopsis hypocotyls. New Phytol 188(3):726–739

Popper ZA, Fry SC (2008) Xyloglucan-pectin linkages are formed intra-protoplasmically, contribute to wall-assembly, and remain stable in the cell wall. Planta 227:781–794

Proseus TE, Boyer JS (2006) Calcium pectate chemistry controls growth rate of chara corallina. J Exp Bot 57(22):3989–4002

Proseus TE, Boyer JS (2007) Tension required for pectate chemistry to control growth in chara corallina. J Exper Bot 58(15/16):4283–4292

Proseus TE, Boyer JS (2008) Calcium pectate chemistry causes growth to be stored in chara corralling: a test of the pectate cycle. Plant Cell Environ 31:1147–1155

Proseus TE, Boyer JS (2012) Calcium deprivation disrupts enlargement of chara corralling cells: further evidence for the calcium pectate cycle. J Exp Bot 63(10):3953–3958

Proseus TE, Ortega JKE, Boyer JS (1999) Separating growth from elastic deformation during cell enlargement. Plant Physiol 110:775–784

Ptashnyk M, Seguin B (2015) Homogenization of a system of elastic and reaction-diffusion equations modelling plant cell wall biomechanics. ESAIM Math Modell Num Anal 50:593–631

Rhee SY, Osborne E, Poindexter PD, Somerville CR (2003) Microspore separation in the quartet 3 mutants of ara- bidopsis is impaired by a defect in a developmentally regulated polygalacturonase required for pollen mother cell-wall degradation. Plant Physiol 133:1170–1180

Rodriguez EK, Hoger A, McCulloch A (1994) Stress-dependent finite growth in soft elastic tissue. J Biomech 27:455–467

Rojas ER, Hotton S, Dumais J (2011) Chemically mediated mechanical expansion of the pollen tube cell wall. Biophys J 101:1844–1853

Stephenson MB, Hawes MC (1994) Correlation of pectin methylesterase activity in root caps of pea with root border cell separation. Plant Physiol 106:739–745

Thimm JC, Burritt DJ, Ducker WA, Melton LD (2009) Pectins influence microfibril aggregation in celery cell walls: an atomic force microscopy study. J Struct Biol 168:337–344

Virk SS, Cleland RE (1988) Calcium and the mechanical properties of soybean hypocotyl cell walls: possible role of calcium and protons in cell-wall loosening. Planta 176:60–67

Wang T, Bum Park Y, Cosgrove DJ, Hong M (2015) Evidence from solid-state nuclear magnetic resonance. Cellulose-pectin spatial contacts are inherent to never-dried arabidopsis primary cell walls. Plant Physiol 168:871–884

Wang T, Zabotina O, Hong M (2012) Pectin-cellulose interactions in the arabidopsis primary cell wall from two-dimensional magic-angle-spinning solid-state nuclear magnetic resonance. Biochemistry 51:9846–9856

White PJ (2001) The pathways of calcium movement to the xylem. J Exp Bot 52:891–899

Willats WGT, Knox P, Dalgaard Mikkelsen J (2006) Pectin: new insights into an old polymer are starting to gel. Trends Food Sci Technol 17:97–104

Willats WGT, McCartney L, Mackie W, Knox JP (2001a) Pectin: cell biology and prospects for functional analysis. Plant Mol Biol 47:9–27

Willats WGT, Orfila C, Limberg G, Buchholt HC, van Alebeek G-JWM, Voragen AGJ, Marcus SE, Christensen TMIE, Mikkelsen JD, Murray BS, Knox JP (2001b) Modulation of the degree and pattern of methyl-esterification of pectic homogalacturonan in plant cell walls. implications for pectin methyl esterase action, matrix properties, and cell adhesion. J Biol Chem 276(22):19404–19413

Wolf S, Greiner S (2012) Growth control by cell wall pectins. Protoplasma 249(2):169–175

Wolf S, Hématy K, Höfte H (2012a) Growth control and cell wall signaling in plants. Ann Rev Plant Biol 63:381–407

Wolf S, Mouille G, Pelloux J (2009) Homogalacturonan methyl-esterification and plant development. Mol Plant 2:851–860

Wolf S, Mravec J, Greiner S, Mouille G, Höfte H (2012b) Plant cell wall homeostasis is mediated by brassinosteroid feedback signaling. Curr Biol 22:1732–1737

Wolf S, van der Does D, Ladwig F, Sticht C, Kolbeck A, Schürholz A-K, Augustin S, Keinath N, Rausch T, Greiner S, Schumacher K, Harter K, Zipfel C, Höfte H (2014) A receptor-like protein mediates the response to pectin modification by activating brassinosteroid signaling. PNAS 111(42):15261–15266

Yoneda A, Ito T, Higaki T, Kutsuna N, Saito T, Ishimizu T, Osada H, Hasezawa S, Matsui M, Demura T (2010) Cobtorin target analysis reveals that pectin functions in the deposition of cellulose microfibrils in parallel with cortical microtubules. Plant J 64:657–667

Zsivanovits G, MacDougall AJ, Smith AC, Ring SG (2004) Material properties of concentrated pectin networks. Carbohydr Res 339:1217–1322

Zykwinska A, Thibault J-F, Ralet M-C (2008) Competitive binding of pectin and xyloglucan with primary cell wall cellulose. Carbohydr Polym 74:957–961

Cell Wall Expansion as Viewed by the Creep Method

Dmitry Suslov and Kris Vissenberg

Abstract After being formed, plant cells increase in volume during the cell expansion process in which the cell wall plays a prominent role. Cell wall extensibility, a crucial characteristic that defines plant cell growth rate at the biophysical level, is currently estimated by the use of several indirect techniques. One of them, the creep method, measures cell wall extension under a constant load over time in the principal direction of cell and organ growth and mimics the action of turgor on growing cell walls better than other techniques. This review provides the last technical advances and novel metrics of the creep method predicting cell wall extensibility. The potential of the creep method to uncover new cell wall-loosening/tightening proteins is discussed.

Keywords Expansive growth · Cell wall · Extensibility · Creep · Arabidopsis

Cell Wall Extensibility and Its Estimation

Plants have evolved a unique mechanism of cell enlargement, the so-called expansive growth, which is largely mediated by water uptake. Water is absorbed by growing plant cells as a result of higher osmotic concentrations of their contents compared with the extracellular solution. The resulting increase in volume is restrained by stiff yet extensible cell walls that encase plant cells. The wall resistance to the osmotically

D. Suslov (✉)
Faculty of Biology, Department of Plant Physiology and Biochemistry, Saint Petersburg State University, Universitetskaya emb. 7/9, 199034 Saint Petersburg, Russia
e-mail: suslov.dmitry75@gmail.com

K. Vissenberg
Biology Department, Integrated Molecular Plant Physiology Research, University of Antwerp, Groenenborgerlaan 171, 2020 Antwerp, Belgium

K. Vissenberg
Plant Biochemistry & Biotechnology Lab, Department of Agriculture, Technological Educational Institute of Crete, Stavromenos, 71410 Heraklion, Crete, Greece

A. Geitmann and J. Gril (eds.), *Plant Biomechanics*,
https://doi.org/10.1007/978-3-319-79099-2_14

mediated protoplast enlargement generates high turgor pressure, which in turn drives the irreversible cell wall expansion that accompanies growth. The involvement of cell walls in expansive growth was clarified by Lockhart (1965) and expressed in the following equation:

$$(1/V)\,(dV/dt) = \phi\,(P - Y)\,, \tag{1}$$

where $(1/V)(dV/dt)$ is the relative cellular expansive growth rate expressed as a change in cell volume (V), Φ is the wall yielding coefficient characterizing the sensitivity of growth rate to changes in turgor, P is the turgor pressure and Y is the yield threshold, i.e., the minimal turgor required for growth (Fig. 1). The Lockhart Eq. (1) postulates that the growth rate is a linear function of turgor in excess of the yield threshold. The terms Φ and Y, both characterizing growth-relevant cell wall properties, determine its extensibility, i.e., the ability of the cell wall to irreversibly increase in surface area during growth (Cosgrove 2016a). Numerous studies conducted in the 1970–1980s convincingly demonstrated that cell wall extensibility is a limiting biophysical factor which defines the rate of cellular expansive growth (reviewed in Cosgrove 1986). It was cell wall extensibility that changed rapidly under many different conditions/treatments that induced alterations in expansive growth rate. Thus, multiple signal transduction pathways involved in the control of cell expansion seem to converge at the cell wall level with cell wall extensibility being their most downstream point, the last factor that changes before a cell attains a new growth rate. Cell wall extensibility depends on the biochemical composition of wall polymers, as well as on numerous cell wall-loosening and tightening proteins that break and form covalent and non-covalent bonds between wall constituents (Chebli and Geitmann 2017).

Given the recognized pivotal role of cell wall extensibility in the biophysical control of cellular expansive growth, measuring or estimating this characteristic is of utmost importance. Cell wall extensibility was quantified in vivo by several methods developed in the 1960–1980s, such as "growth rate versus turgor", in vivo stress relaxation, and some others (reviewed in Cosgrove 1993). These techniques are not used at present anymore because they are laborious and incompatible with many experimental treatments required. Instead, cell wall extensibility is currently estimated indirectly by classical microtensile tests: the creep method (Cosgrove 1989; Thompson 2001; Suslov and Verbelen 2006), the stress/strain method (Ryden et al. 2003; Abasolo et al. 2009), the in vitro stress relaxation (Yamamoto 1996), as well as by more modern techniques based on nano- and microindentation of plant surfaces including atomic force microscopy (AFM) (Milani et al. 2013; Peaucelle et al. 2011, 2015) and cellular force microscopy (CFM) (Routier-Kierzkowska et al. 2012). All these techniques provide different metrics of cell wall biomechanics, and only some of them correlate with cell wall extensibility. Thus, the different biomechanical methods and their resulting metrics are not equivalent in their ability to explain growth rate in vivo.

This brief review is devoted to the creep (constant load) method and its use for the analysis of growth-relevant cell wall biomechanics in plants. We will compare it with

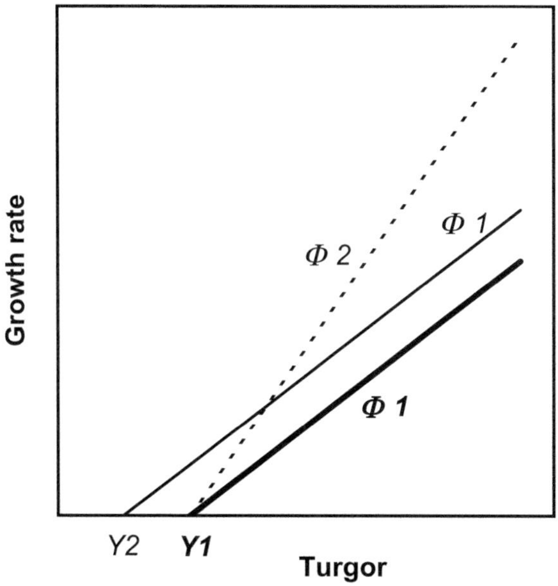

Fig. 1 Graphic representation of the Lockhart equation and two hypothetical ways of plant cell growth stimulation. The wall yielding coefficient (Φ) characterizes the sensitivity of growth rate to changes in turgor and corresponds to the slopes of the lines. The yield threshold (Y) shows the minimal turgor at which cell expansive growth starts and corresponds to the intercepts of the lines with the x axis. The original moderate growth rate (*thick line*) is limited by $\Phi 1$ and $Y1$. Growth stimulation via increasing Φ (*dashed line*) where $\Phi 2 > \Phi 1$ and Y does not change. Growth stimulation via decreasing Y (*thin line*) where $Y2 < Y1$ and Φ does not change. As a matter of fact, independent changes in Φ and Y values can occur during growth stimulation or suppression in vivo

alternative methods and consider practical aspects of its application and peculiarities of data interpretation, and discuss prospects for further development of this old, yet efficient technique.

We focus on the creep method because its outputs are easier to interpret and link with cell wall extensibility and in vivo growth rates when compared with the above-mentioned alternative techniques of plant biomechanics. The creep method is based on measuring the wall deformation under a constant load over time. The stress [i.e., the force on a body divided by the cross-sectional area across which the force acts (Baskin 2005)] generated in cell walls during creep measurements resembles the in vivo stress established by turgor in that they both are essentially constant over time, as turgor does not vary appreciably during cell expansion under normal conditions (Taiz 1984; Cosgrove 1986). In contrast to the creep method, execution of all the alternative techniques causes rapid and large changes in wall stress, which is clearly inconsistent with the in vivo conditions. Another advantage of the creep method is that the rate of wall deformation under a constant load is generally slow and comparable to the rate of cellular expansive growth in vivo (Richmond et al. 1980; Suslov et al. 2015). As a result, endogenous cell wall-loosening/tightening proteins that are thought to be implicated in wall extensibility control in vivo have sufficient time to exert their effects under the conditions of creep measurements. On the contrary, the alternative techniques use very rapid cell wall deformations being ill-suited for studying the growth-relevant effects of cell wall proteins. Although the spatial resolution of the creep method is much lower than that of AFM and CFM techniques, the former has two major advantages over the latter: (i) in creep measurements cell walls are extended along the principal direction of growth in vivo, while in AFM and CFM the walls are deformed orthogonal to the growth vector (Milani et al. 2013; Cosgrove 2016a); (ii) creep tests mostly reveal irreversible (plastic) cell wall deformations (Suslov and Verbelen 2006), which can be directly linked with growth, an irreversible process, while AFM and CFM measure predominantly reversible (elastic) cell wall deformations having no clear link with growth itself (Peaucelle et al. 2011).

All the abovementioned advantages of the creep method predetermined its main application in modern plant (cell wall) biology. The technique has been successfully used to confirm the involvement of a certain cell wall protein of interest in the control of cell wall extensibility. If the purified protein as such increases or decreases cell wall deformation in the creep test, this finding is considered as a direct confirmation of its role as a cell wall-loosening or cell wall-tightening protein, respectively. Thus, the considered protein is highly probable to work as a natural regulator of cell wall extensibility. This approach has shed light on the functions of expansins (McQueen-Mason et al. 1992), yieldins (Okamoto-Nakazato et al. 2000), lipid transfer proteins (Nieuwland et al. 2005), and xyloglucan endotransglucosylase/hydrolases (XTHs) (Van Sandt et al. 2007; Maris et al. 2009) in the control of cell wall extensibility and plant growth.

Preparation of Plant Samples for Creep Tests

Creep tests have been carried out using either living turgid plant organs or isolated cell walls. The most widely used approach for cell wall isolation involves freezing and thawing of plant organs or their segments. This treatment destroys cell membranes, eliminates turgor and yields only relaxed cell walls (Cosgrove 1989; Thompson 2001; Suslov and Verbelen 2006). The samples prepared in this way are considered to retain a significant part of cell wall protein activities implicated in the control of cell wall extensibility, and subsequent constant load application partially mimics the action of turgor on cell walls. This approach is convenient as it allows rapid fixation of large batches of plant material that is uniform by age, which is critical for data analysis and interpretation. The frozen samples can then be stored and analyzed within up to two weeks without any appreciable loss in cell wall properties (Miedes et al. 2013). However, this approach provides only a static picture of cell wall properties at the moment of freezing without any contribution from cellular metabolism (the secretion of new cell wall-loosening/tightening proteins, in vivo changes in apoplastic pH and reactive oxygen species production). Additionally, some cell wall proteins do not survive the freezing/thawing cycle, although this problem can be circumvented by an alternative protocol involving cell wall glycerination (Okamoto-Nakazato 2002). At first glance, the use of living plant organs in creep tests can address the above problems with frozen/thawed samples. In this case, the wall creep will also be controlled by metabolic inputs from living protoplasts (Kutschera and Schopfer 1986b; Nonami and Boyer 1990). However, the retained metabolic activity may create different problems. First, physiological stress responses will be induced as a result of mechanical disturbances of living experimental samples before and during creep measurements. Some of them, e.g., reactive oxygen species (ROS) production, could rapidly modify cell wall properties (Monshausen et al. 2009). Second, an extra load applied to the fully turgid living organs is physiologically irrelevant as changes in their growth rate are not normally caused by variations in the force (turgor) extending the wall in vivo (Cosgrove 1986). Finally, it is not easy to obtain many uniform living samples with respect to their developmental stage, as creep measurements are time-consuming, while age-related changes in cell expansion rates and growth zone positions can be very rapid (Bastien et al. 2016). Due to all these reasons, cell walls isolated after a freeze/thaw cycle are currently used more frequently.

Plant samples have to be highly permeable for exogenous buffers and proteins in many applications of the creep method. Traditionally this is achieved by peeling or abrading the sample surfaces (Rayle and Cleland 1992). The peeling procedure involves the complete removal of the epidermis with the use of a fine forceps, thereby exposing the hydrophyllic surfaces of underlying cell walls that are stretched under a constant load. Although peeling can be highly reproducible, it eliminates the tissue that plays a critically important role in growth control (Kutschera and Niklas 2007). So, at present it is more common to use epidermal peels in creep tests (Thompson 2001; Suslov et al. 2009), i.e., isolated epidermal cell layers, which are permeable even for exogenous proteins (Van Sandt et al. 2007; Maris et al. 2009). During the abrasion procedure, the hydrophobic cuticle covering the outer epidermal cell wall is removed by gentle rubbing with fine abrasive materials (Kutschera and Schopfer 1986a; Park and Cosgrove 2012a). This treatment does not damage the growth-controlling outer epidermal cell wall. However, abrasion may be less reproducible than peeling, when performed by different people. Finally, some organs like roots are highly permeable for exogenous solutes by their nature and can be used in creep tests without any treatments increasing their permeability (Büntemeyer et al. 1998).

Another preliminary treatment of frozen/thawed samples before creep tests involves their flattening under a high load to remove excessive water (Durachko and Cosgrove 2009), since water expelling from plant samples may affect their extension in the first minutes of creep. The flattening procedure is normally not required for experimental samples with a low volume per length ratio like epidermal peels or miniature Arabidopsis hypocotyls.

Estimating the effects of exogenous proteins on cell wall extensibility with the creep method usually requires the inactivation of endogenous proteins in experimental samples. Traditionally they are heat-inactivated by boiling in water for 15 s (McQueen-Mason et al. 1992; Van Sandt et al. 2007). A more gentle treatment is used with more fragile Arabidopsis hypocotyls that are heated under sub-boiling temperatures followed by rehydration at 4 °C (Boron et al. 2015; Suslov et al. 2015).

Comparisons of the mechanics of plant samples differing in their cross-sectional area are often made with the creep method (Miedes et al. 2013). In these specific cases, the same constant load will generate a different stress in the cell walls that need to be compared. As a consequence, their comparison will only be relevant if these stress values are taken into account, as the rate of sample deformation is proportional to the stress in it. To calculate stress in the walls during creep tests, the force arising from a constant load should be divided by the wall cross-sectional area. This force is directly defined with a load cell or calculated based on the fact that 1 N (Newton) force is generated by 102 g load under normal gravity. The cross-sectional area can be measured using electron micrographs of freeze-fractured samples (Derbyshire et al. 2007) in a direct, but very time-consuming approach, or alternatively, it can be calculated indirectly by measuring the wall dry weight per sample unit length, taking into account its geometry (Cleland 1967) and assuming that the density of wall polymers is 1.5 g cm^{-3} (Gibson 2012). This approach combined with exhaustive extraction of low molecular weight compounds has recently been adapted for cell walls of Arabidopsis hypocotyls (Miedes et al. 2013; Suslov et al. 2015).

Practical Aspects of Cell Wall Creep Measurements

Plant samples are extended in creep tests using extensometers. The setup we use is shown in Fig. 2, whereas other designs of extensometers were described in Kutschera and Schopfer (1986a), Cosgrove (1989), Okamoto and Okamoto (1994), and Takahashi et al. (2006). Experimental samples are secured in the setup with instant glue or mechanical clamps and placed in a watertight chamber, where they will be extended during creep tests while being submerged in buffers. Further technical details of tensile testing are elaborated in Chap. 10. Keeping the samples in a well-hydrated state during measurements is critically important, as this significantly affects their extension (Evered et al. 2007). Buffers are selected such that their maximal buffering capacity corresponds to physiological pH values found in the plant apoplast (pH from 4.5 to 7.0) (Bibikova et al. 1998; Monshausen et al. 2007). Additionally, the buffers should not have any significant calcium-chelating capacity, since this could influence cell wall mechanics (Cosgrove 2011). These conditions are met with buffers based on MES, acetate, Hepes, and histidine.

A constant load is applied to the sample secured in the extensometer using a load cell (Takahashi et al. 2006) or scale weights (Fig. 2; Durachko and Cosgrove 2009). The load chosen should ideally generate a wall stress that is comparable to those relevant in vivo. Excessive loads causing frequent cell wall failures have to be avoided. At the other extreme, too low loads resulting in cell wall deformation rates well below the in vivo growth rate are also irrelevant. Practically and as an example, in our work with Arabidopsis hypocotyls, we select loads such that the maximal load should induce no more than 30% of sample failures (Boron et al. 2015). The

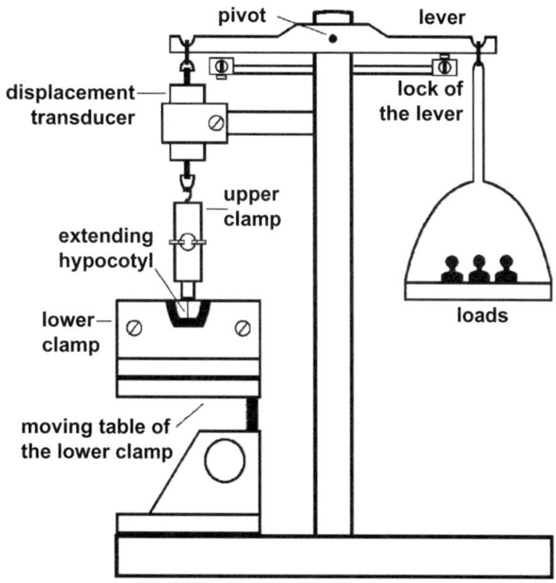

Fig. 2 Schematic drawing of the extensometer used in our onion epidermis and Arabidopsis hypocotyl work. The space where the hypocotyl segment resides is filled with buffers at different pH and can be supplemented with proteins of interest to reveal their effect on the measured extension of the segment

minimal load in our case is limited by the system resolution and normally produces creep rates that are four to eight times lower than the in vivo growth rate.

Standard creep tests give raw results in the form of biphasic extension curves including a high-amplitude initial deformation just after loading, followed by a slow time-dependent deformation referred to as "creep" (Fig. 3). Initial deformation is not a reliable predictor of cell wall extensibility, as it normally contains a very significant (Suslov and Verbelen 2006) or even predominant (Nolte and Schopfer 1997) elastic component. Still, initial deformation does relate to the rate of cell expansion under some conditions, e.g., in the case of auxin-induced growth (Kutschera and Schopfer 1986a) or age-related growth deceleration (Miedes et al. 2013). On the other hand, creep has long been considered as a process closely resembling in vivo extension of cell walls (Cosgrove 2005). So the majority of metrics obtained with the constant load method are currently based on the analysis of creep.

Traditional and Novel Metrics of the Creep Method

Creep rate (expressed in $\mu m\ h^{-1}$, $\%\ h^{-1}$ or h^{-1}) is the simplest and most commonly used metric of the constant load method. It is reliable when control and test cell wall samples have similar cross-sectional areas, i.e., any given constant load will generate the same stress in them. As was mentioned above, creep rates should ideally be compared under constant loads producing wall stress comparable to that existing in vivo. However, the actual in vivo wall stress is difficult to measure because it is not

Fig. 3 Typical rough deformation curve obtained using the creep method

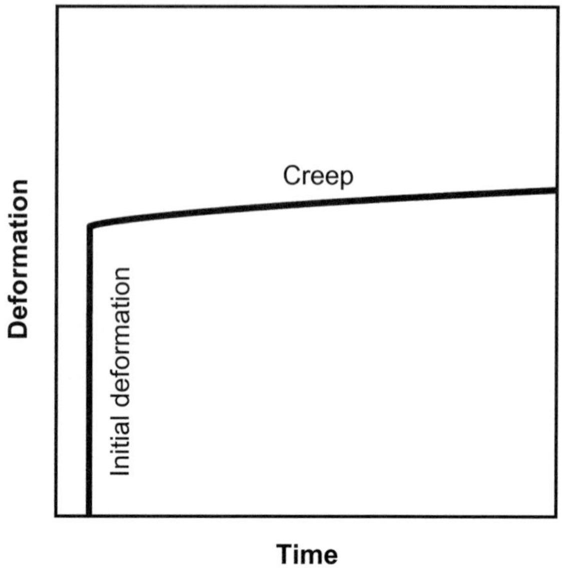

defined solely by turgor but also depends on tissue tensions (Hejnowicz and Sievers 1995), cell geometry (Baskin 2005) and stress anisotropy within multicellular plant organs (Baskin and Jensen 2013). To cope with this problem, creep rates are often compared under many different loads with the assumption that one of them does generate the wall stress close to that in vivo (Thompson 2001; Suslov and Verbelen 2006; Vandenbussche et al. 2011).

When cross-sectional areas of cell walls are different, creep rates are poor predictors for in vivo growth rates (Miedes et al. 2013). The simplest solution, in this case, is to compare the ratios creep rate $\times$ stress^{-1}. The fact that the ratios are defined by two independent variables (creep rate and stress) complicates their statistical comparisons. This problem has been recently solved by creating scripts that allow comparison of creep rate $\times$ stress^{-1} values using a novel algorithm (Suslov et al. 2015) based on bootstrapping.

Sometimes the creep rate depends on load (or wall stress) in a linear fashion (Okamoto and Okamoto 1994; Miedes et al. 2013; Suslov et al. 2015). In this particular case, two additional metrics of the creep method can be obtained: φ (in vitro cell wall extensibility) and y (in vitro cell wall yield threshold), i.e., in vitro analogs of the parameters Φ and Y from the Eq. (1). The value of φ characterizes the sensitivity of creep rate to changes in the load (or wall stress) and corresponds to the slope of the line in Fig. 4, while y shows the minimal load (or wall stress) at which creep starts and corresponds to the intercept of the line with the x axis (Fig. 4). The values of φ and y can be obtained based on serial loading (Okamoto and Okamoto 1994) or from independent creep measurements under different loads (Suslov et al. 2015). These alternative approaches give nonequivalent results (Takahashi et al. 2006; Suslov et al.

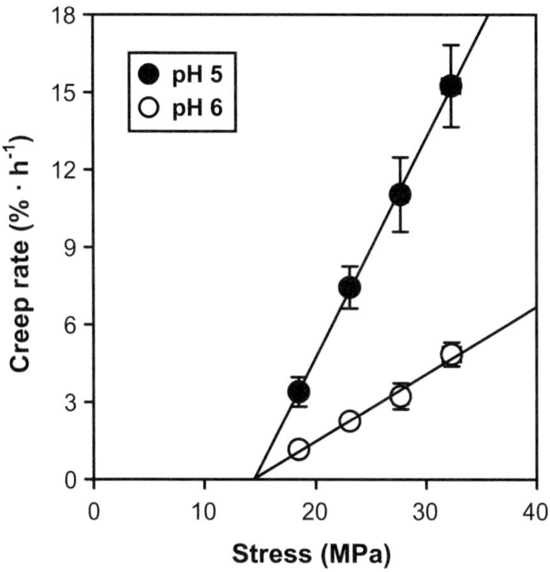

Fig. 4 Determination of φ (slopes of the lines) and y (intercepts with the x axis) values using independent creep measurements under different loads. Hypocotyls from frozen/thawed Columbia-0 Arabidopsis plants were extended at pH 5 or 6 under one out of four constant loads (400, 500, 600 and 700 mg). Average creep rate for each load was calculated. Then the loads were transformed into respective wall stress values based on cell wall cross-sectional area estimation. Average creep rates were plotted against average wall stress values generated under each constant load. The dependence of creep rate on wall stress was linear as determined by Deming regression taking into account variation in both creep rate and wall stress. The plot demonstrates that the acid-induced expansin-mediated cell wall extension in Arabidopsis hypocotyls results from an increase in φ

2015), and their critical comparison is given in Suslov et al. (2015). Although the latter approach (Suslov et al. 2015) is more laborious, its results are less affected by experimental conditions than in the former approach (Okamoto and Okamoto 1994; Takahashi et al. 2006). The advantage of φ and y over the other metrics of the creep method consists in their ability to give an insight into the mechanisms of changes in cell wall extension. For example, the expansin-mediated acid growth response is caused by increasing the φ values (Suslov et al. 2015), while XTH overexpression activates growth by lowering y (Miedes et al. 2013). Thus, φ and y determination may forecast that certain growth responses are induced by different cell wall proteins before they are revealed using transcriptomics or proteomics, purified, and their cell wall-loosening/tightening function is confirmed in the reconstitution creep assay.

Cell wall-loosening/tightening proteins may affect not only the rate but also the duration of creep in plant samples. For example, expansins seem to increase both of them (Cosgrove 1989). The temporal aspect of cell wall creep is often not taken into account, in part because its straightforward estimation requires hours-long creep recordings, which is very inconvenient. The importance of the temporal aspect was underscored by demonstrating that XTH overexpression renders creep curves more linear in time (Miedes et al. 2013). This finding has an implication that XTH enzymes may prolong cell expansive growth. Thus, there should be metrics describing creep curve curvature, i.e., changes in cell wall creep rates in time. For example, one can measure the first derivative of creep rate with respect to time for different points along a creep curve. This approach was applied for treatments stimulating, inhibiting or having no effect on growth of Arabidopsis hypocotyls (Suslov et al. 2012). In the case of growth stimulation, the derivative increased to values approaching nil demonstrating the practically linear dependence of creep rate on time. The growth inhibition rendered the derivative more negative, consistent with a faster decay of creep rate over time. In the absence of growth effect, no significant differences in the derivative were found. Further studies are needed to confirm that this metric of creep curve curvature can be used as a universal predictor of in vivo growth rate.

Cell wall extension involves plastic (irreversible) and elastic (reversible) components, which both can be estimated with the creep method. This is achieved by using loading/unloading cycles wherein one or more loads are first sequentially added to the wall sample and then sequentially removed from it, whilst the resulting cell wall extension/contraction is continuously recorded (Kutschera and Schopfer 1986a; Hejnowicz and Sievers 1996; Suslov and Verbelen 2006). These experiments provided rather conflicting results. Some authors argued that the plastic component represents in fact retarded elasticity (Hohl and Schopfer 1992; Nolte and Schopfer 1997), while others found true plastic deformations (Thompson 2001; Suslov and Verbelen 2006). It is not clear whether this disagreement reflects natural variation between the plant objects studied, or it is imposed by the different protocols. However, knowing the exact relationship between plastic and elastic components of cell wall extension is critically important for correct data interpretation. All modern nano- and microindentation-based techniques only measure cell wall elasticity (Braybrook 2017). The creep method can answer the question whether the plastic and elastic components change in a concerted manner during variations in growth rate. If these components do positively correlate, then the metrics of AFM and CFM techniques provide growth-relevant information. In the absence of such correlation, the outputs of nano- and microindentation tests should be interpreted with extreme care when attempting to link them with in vivo growth rates. Interestingly, the dynamic interconversion between plastic and elastic components of wall deformation was found during cell expansive growth (Edelmann and Köhler 1995). It was concluded that growth may be regulated by wall-loosening mechanisms which initially result in elastic wall extension. Later, this extension becomes irreversible via a fixation process (Edelmann 1995), for which the mechanism was also proposed (Hohl et al. 1995). To sum up, no generally accepted metrics are currently available to describe plastic versus elastic components of cell wall extension. The new metrics to be developed have to be universal for plant samples of different origin. Establishing the objective criteria for their direct comparisons is a challenging task because there are apparently no natural equivalents to the wall perturbations occurring during loading/unloading cycles. So it is not clear to what extent cell walls should be loaded and unloaded, and how long each loading/unloading step should last to provide accurate estimations of cell wall plastic versus elastic properties with the creep method. If no objective criteria exist, there should be a kind of convention on the measurement conditions between specialists in the field.

Novel Trends and Improvements of the Creep Method

Although the wall stress generated in a standard uniaxial creep test and the in vivo stress are similar in that they are essentially constant over time, their nature is different. Turgor induces biaxial stress in the plane of the wall, which limits its polymer reorientations due to geometrical constraints (Chanliaud et al. 2002). On the contrary, the uniaxial stress generated by a constant load results in major polymer reorientations in the direction of cell wall deformations. Normally this leads to much higher cell wall extension compared to that caused by equal biaxial stress. However, the action of in vivo biaxial stress and in vitro uniaxial stress along the main axis of elongation in organs that consist of highly anisotropically growing cells may not be so dramatically different. This is because the polymer reorientations induced by a constant load are very similar to those occurring in these organs in vivo (Refrégier et al. 2004; Suslov et al. 2009; Crowell et al. 2011). On the other hand, isotropic growth is not normally accompanied by polymer realignment. In organs composed of isotropically growing cells, the mechanical effect of uniaxial stretching could differ dramatically from the effect of turgor. This problem has long been recognized, and it was solved by designing the biaxial stretching systems. The first successful biaxial creep measurements were performed using giant internodal cells of *Nitella* (Métraux et al. 1980; Richmond et al. 1980). The walls of cylindrical segments of these cells containing intact apical nodes were inflated with mercury, and their resulting extension in length and in girth was monitored microscopically. This biaxial stretching system cannot be universally used, as it is based on the extraordinarily large size and specific morphology of *Nitella* internodal cells. Chanliaud et al. (2002) proposed an approach with more universal application possibilities. They designed a bulge-testing device in which a flat sample of *Acetobacter xylinus* composite (a biomimetic of plant cell walls) was placed over an orifice of a cylindrical chamber. This chamber was pressurized and the resulting sample bulging was measured. Essentially the same approach was recently applied to the walls of cherry epidermal peels (Brüggenwirth et al. 2014), and their creep resulting from biaxial stretching was studied under different conditions (Brüggenwirth and Knoche 2016). Biaxial cell wall stress generated in bulge-testing devices or their analogs mimics the action of turgor on growing cell walls much better than the uniaxial stress generated in traditional uniaxial creep tests. Until now, the bulging test was only used with mechanically isotropic cell walls. So this efficient technique should be further adapted for anisotropic samples and this will require new algorithms to analyze complex surfaces. One can imagine that the biaxial stretching of anisotropic samples will produce symmetric distortions of their bulging surfaces that should be quantified. At the next step of improvement, the bulging test should allow reconstitution assays with purified cell wall proteins. This would help confirm their role in the control of anisotropic cell expansion. Until now, there have been very scarce data on proteins that can differentially affect cell wall extensibility along two perpendicular axes in the plane of the cell wall (Wiedemeier et al. 2002; Van Sandt et al. 2007).

Traditional creep tests have limited spatial resolution. The minimal length of sample segments to be stretched is about 2 mm. This means that the sample should be at least 4–6 mm long, as extra length is needed to secure it in the setup. If one wants to study creep in a number of model plant objects, such as root growth zones, light-grown Arabidopsis hypocotyls and etiolated hypocotyls from different Arabidopsis mutants, these are shorter than the abovementioned limit and can therefore not be stretched. Thus, an increase in the spatial resolution of the creep method is highly demanded. Fortunately, this seems to be technically feasible as judged from recent advances in the stress/strain method. Zamil et al. (2013) achieved subcellular resolution with a novel tensile testing device based on a microelectromechanical system. They have obtained stress/strain curves from as small as 15×5 μm pieces of onion epidermal cell walls. Original measurements were performed on dried samples, as cell wall displacement was monitored by scanning electron microscopy. Later, the technique was improved and allowed the use of hydrated cell walls (Zamil et al. 2014, 2015; Kim et al. 2015). The ideas and technical solutions of this group can serve as a basis for the development of a novel high-resolution creep-testing device.

Despite the central role of *Arabidopsis thaliana* in modern plant biology, the creep method was first used with this popular object only quite recently and later than other biomechanical techniques (Vandenbussche et al. 2011). This is because its growing organs are small and fragile, which makes their usage rather difficult. The majority of technical issues have been successfully solved since then and the method is now routinely used in studies of Arabidopsis mutants (Park and Cosgrove 2012a, b) and overexpression lines (Miedes et al. 2013; Boron et al. 2015). The full potential of the creep method can be realized with cell wall mutants of Arabidopsis and their analysis should give insights into the functions of certain cell wall polymers, their interactions with each other, and this will finally help to create a realistic model of the cell wall architecture. The first steps on this way have already been done casting doubt on the traditional view of cell wall organization (Park and Cosgrove 2012a, b). Further progress in the field will require to increase the method resolution and to use new metrics for comparisons. The former will make new cell wall mutants available for analysis, while the latter will help reveal very subtle differences between them. The creep method is very promising for a screen of new cell wall-loosening/tightening proteins in Arabidopsis, keeping in mind that a huge amount of omics data is available for this plant species. The only pitfall for this application is the lack of an efficient reconstitution assay based on Arabidopsis organs. Although the abrasion technique has been used with Arabidopsis hypocotyls (Park and Cosgrove 2012b) to increase their permeability for exogenous proteins, this treatment may potentially damage these tender organs. Thus, an alternative and more gentle protocol should be developed. After it is available, a technology for homologous reconstitution with endogenous Arabidopsis cell wall proteins can be created. In this case, a knockout Arabidopsis plant for a protein of interest is produced, after which the purified protein of interest is applied to this knockout in a creep test to verify whether biomechanical complementation takes place. Finally, it would also be very productive to study Arabidopsis hypocotyls or roots in different growth states by using the creep method in combination with modern AFM or CFM techniques. This combination approach could demonstrate which AFM or CFM metrics are the best indicators of growth-relevant cell wall properties.

To sum up, despite the very simple principle and long history of use, the creep method remains evergreen in the field of plant biomechanics. The advent of novel instruments that will increase the method's spatial resolution along with better metrics for presenting its outputs will greatly improve our understanding of the mechanisms of cell expansive growth.

Acknowledgements The authors acknowledge Saint Petersburg State University for the research grant 1.40.492.2017, RFBR for the grant No 15-04-04075, the University of Antwerp, the Research Foundation-Flanders (FWO; grants G039815 N, G065613 N, G.0.602.11.N.10 and 1.5.091.11.N.00).

References

Abasolo W, Eder M, Yamauchi K et al (2009) Pectin may hinder the unfolding of xyloglucan chains during cell deformation: implications of the mechanical performance of Arabidopsis hypocotyls with pectin alterations. Mol Plant 2:990–999

Baskin TI (2005) Anisotropic expansion of the plant cell wall. Annu Rev Cell Dev Biol 21:203–222

Baskin TI, Jensen OE (2013) On the role of stress anisotropy in the growth of stems. J Exp Bot 64:4697–4707

Bastien R, Legland D, Martin M et al (2016) KymoRod: a method for automated kinematic analysis of rod-shaped plant organs. Plant J 88:468–475

Bibikova TN, Jacob T, Dahse I et al (1998) Localized changes in apoplastic and cytoplasmic pH are associated with root hair development in Arabidopsis thaliana. Development 125:2925–2934

Boron AK, Van Loock B, Suslov D et al (2015) Over-expression of AtEXLA2 alters etiolated arabidopsis hypocotyl growth. Ann Bot 115:67–80

Braybrook SA (2017) Analyzing cell wall elasticity after hormone treatment: an example using tobacco BY-2 cells and auxin. Methods Mol Biol 1497:125–133

Brüggenwirth M, Knoche M (2016) Factors affecting mechanical properties of the skin of sweet cherry fruit. J Am Soc Hortic Sci 141:45–53

Brüggenwirth M, Fricke H, Knoche M (2014) Biaxial tensile tests identify epidermis and hypodermis as the main structural elements of sweet cherry skin. AoB Plants 6:plu019

Büntemeyer K, Lüthen H, Böttger M (1998) Auxin-induced changes in cell wall extensibility of maize roots. Planta 204:515–519

Chanliaud E, Burrows KM, Jeronimidis G, Gidley MJ (2002) Mechanical properties of primary plant cell wall analogues. Planta 215:989–996

Chebli Y, Geitmann A (2017) Cellular growth in plants requires regulation of cell wall biochemistry. Curr Opin Cell Biol 44:28–35

Cleland R (1967) Extensibility of isolated cell walls: measurement and changes during cell elongation. Planta 74:197–209

Cosgrove D (1986) Biophysical control of plant cell growth. Ann Rev Plant Physiol 37:377–405

Cosgrove DJ (1989) Characterization of long-term extension of isolated cell walls from growing cucumber hypocotyls. Planta 177:121–130

Cosgrove DJ (1993) Wall extensibility: its nature, measurement and relationship to plant cell growth. New Phytol 124:1–23
Cosgrove DJ (2005) Growth of the plant cell wall. Nat Rev Mol Cell Bio 6:850–861
Cosgrove DJ (2011) Measuring in vitro extensibility of growing plant cell walls. Methods Mol Biol 715:291–303
Cosgrove DJ (2016) Plant cell wall extensibility: connecting plant cell growth with cell wall structure, mechanics, and the action of wall-modifying enzymes. J Exp Bot 67:463–476
Crowell EF, Timpano H, Desprez T, Franssen-Verheijen T, Emons AM, Höfte H, Vernhettes S (2011) Differential regulation of cellulose orientation at the inner and outer face of epidermal cells in the Arabidopsis hypocotyl. Plant Cell 23:2592–2605
Derbyshire P, Findlay K, McCann MC et al (2007) Cell elongation in *Arabidopsis* hypocotyls involves dynamic changes in cell wall thickness. J Exp Bot 58:2079–2089
Durachko DM, Cosgrove DJ (2009) Measuring plant cell wall extension creep induced by acidic pH and by alpha-expansin. JoVE 25:e1263
Edelmann HG (1995) Wall extensibility during hypocotyl growth: A hypothesis to explain elastic-induced wall loosening. Physiol Plant 95:296–303
Edelmann HG, Köhler K (1995) Auxin increases elastic wall-properties in rye coleoptiles: implications for the mechanism of wall loosening. Physiol Plant 93:85–92
Evered C, Majevadia B, Thompson DS (2007) Cell wall water content has a direct effect on extensibility in growing hypocotyls of sunflower (*Helianthus annuus* L.). J Exp Bot 58:3361–3371
Gibson LJ (2012) The hierarchical structure and mechanics of plant materials. J R Soc Interf 9:2749–2766
Hejnowicz Z, Sievers A (1995) Tissue stresses in organs of herbaceous plants II. Determination in three dimensions in the hypocotyl of sunflower. J Exp Bot 46:1045–1053
Hejnowicz Z, Sievers A (1996) Tissue stresses in organs of herbaceous plants—III. Elastic properties of the tissues of sunflower hypocotyl and origin of tissue stresses. J Exp Bot 47:519–528
Hohl M, Schopfer P (1992) Physical extensibility of maize coleoptile cell walls: apparent plastic extensibility is due to elastic hysteresis. Planta 187:498–504
Hohl M, Greiner H, Schopfer P (1995) The cryptic-growth response of maize coleoptiles and its relationship to H2O2-dependent cell wall stiffening. Physiol Plant 94:491–498
Kim K, Yi H, Zamil MS, Haque MA, Puri VM (2015) Multiscale stress-strain characterization of onion outer epidermal tissue in wet and dry states. Am J Bot 102:12–20
Kutschera U, Niklas KJ (2007) The epidermal-growth-control theory of stem elongation: an old and a new perspective. J Plant Physiol 164:1395–1409
Kutschera U, Schopfer P (1986a) Effect of auxin and abscisic acid on cell wall extensibility in maize coleoptiles. Planta 167:527–535
Kutschera U, Schopfer P (1986b) In vivo measurement of cell wall extensibility in maize coleoptiles: effects of auxin and abscisic acid. Planta 169:437–442
Lockhart JA (1965) An analysis of irreversible plant cell elongation. J Theor Biol 8:264–275
Maris A, Suslov D, Fry SC et al (2009) Enzymic characterization of two recombinant xyloglucan endotransglucosylase/hydrolase (XTH) proteins of Arabidopsis and their effect on root growth and cell wall extension. J Exp Bot 60:3959–3972
McQueen-Mason S, Durachko DM, Cosgrove DJ (1992) Two endogenous proteins that induce cell wall extension in plants. Plant Cell 4:1425–1433
Métraux JP, Richmond PA, Taiz L (1980) Control of cell elongation in *Nitella* by endogenous cell wall pH gradients: multiaxial extensibility and growth studies. Plant Physiol 65:204–210
Miedes E, Suslov D, Vandenbussche F et al (2013) Xyloglucan endotransglucosylase/hydrolase (XTH) overexpression affects growth and cell wall mechanics in etiolated Arabidopsis hypocotyls. J Exp Bot 64:2481–2497
Milani P, Braybrook SA, Boudaoud A (2013) Shrinking the hammer: micromechanical approaches to morphogenesis. J Exp Bot 64:4651–4662

Monshausen GB, Bibikova TN, Messerli MA et al (2007) Oscillations in extracellular pH and reactive oxygen species modulate tip growth of Arabidopsis root hairs. P Natl Acad Sci USA 104:20996–21001

Monshausen GB, Bibikova TN, Weisenseel MH et al (2009) Ca2 + regulates reactive oxygen species production and pH during mechanosensing in Arabidopsis roots. Plant Cell 21:2341–2356

Nieuwland J, Feron R, Huisman BAH et al (2005) Lipid transfer proteins enhance cell wall extension in tobacco. Plant Cell 17:2009–2019

Nolte T, Schopfer P (1997) Viscoelastic versus plastic cell wall extensibility in growing seedling organs: a contribution to avoid some misconceptions. J Exp Bot 48:2103–2107

Nonami H, Boyer JS (1990) Wall extensibility and cell hydraulic conductivity decrease in enlarging stem tissues at low water potentials. Plant Physiol 93:1610–1619

Okamoto H, Okamoto A (1994) The pH-dependent yield threshold of the cell wall in a glycerinated hollow cylinder (in vitro system) of cowpea hypocotyl. Plant Cell Environ 17:979–983

Okamoto-Nakazato A (2002) A brief note on the study of yieldin, a wall-bound protein that regulates the yield threshold of the cell wall. J Plant Res 115:309–313

Okamoto-Nakazato A, Nakamura T, Okamoto A (2000) The isolation of wall-bound proteins regulating yield threshold in glycerinated hollow cylinders of cowpea hypocotyls. Plant Cell Environ 23:145–154

Park YB, Cosgrove DJ (2012a) Changes in cell wall biomechanical properties in the xyloglucan-deficient *xxt1/xxt2* mutant of Arabidopsis. Plant Physiol 158:465–475

Park YB, Cosgrove DJ (2012b) A revised architecture of primary cell walls based on biomechanical changes induced by substrate-specific endoglucanases. Plant Physiol 158:1933–1943

Peaucelle A, Braybrook SA, Le Guillou L et al (2011) Pectin-induced changes in cell wall mechanics underlie organ initiation in Arabidopsis. Curr Biol 21:1720–1726

Peaucelle A, Wightman R, Höfte H (2015) The control of growth symmetry breaking in the Arabidopsis hypocotyl. Curr Biol 25:1746–1752

Refrégier G, Pelletier S, Jaillard D, Höfte H (2004) Interaction between wall deposition and cell elongation in dark-grown hypocotyl cells in Arabidopsis. Plant Physiol 135:959–968

Richmond PA, Métraux J-P, Taiz L (1980) Cell expansion patterns and directionality of wall mechanical properties in *Nitella*. Plant Physiol 65:211–217

Routier-Kierzkowska AL, Weber A, Kochova P et al (2012) Cellular force microscopy for in vivo measurements of plant tissue mechanics. Plant Physiol 158:1514–1522

Rayle D, Cleland R (1992) The acid growth theory of auxin-induced cell elongation is alive and well. Plant Physiol 99:1271–1274

Ryden P, Sugimoto-Shirasu K, Smith AC et al (2003) Tensile properties of Arabidopsis cell walls depend on both a xyloglucan cross-linked microfibrillar network and rhamnogalacturonan II-borate complexes. Plant Physiol 132:1033–1040

Suslov D, Verbelen J-P (2006) Cellulose orientation determines mechanical anisotropy in onion epidermis cell walls. J Exp Bot 57:2183–2192

Suslov D, Verbelen JP, Vissenberg K (2009) Onion epidermis as a new model to study the control of growth anisotropy in higher plants. J Exp Bot 60:4175–4187

Suslov D et al (2012) Cell wall mechanical properties of Arabidopsis hypocotyls in connection with growth and gravitropism: effects of xyloglucan endotransglucosylase/hydrolases, expansin-like proteins and brassinosteroids. In: 7th plant biomechanics international conference, Clermont-Ferrand, 20–24 Aug 2012

Suslov D, Ivakov A, Boron AK et al (2015) In vitro cell wall extensibility controls age-related changes in the growth rate of etiolated Arabidopsis hypocotyls. Funct Plant Biol 42:1068–1079

Taiz L (1984) Plant cell expansion: regulation of cell wall mechanical properties. Ann Rev Plant Physiol 35:585–657

Takahashi K, Hirata S, Kido N et al (2006) Wall-yielding properties of cell walls from elongating cucumber hypocotyls in relation to the action of expansin. Plant Cell Physiol 47:1520–1529

Thompson DS (2001) Extensiometric determination of the rheological properties of the epidermis of growing tomato fruit. J Exp Bot 52:1291–1301

Vandenbussche F, Suslov D, de Grauwe L et al (2011) The role of brassinosteroids in shoot gravitropism. Plant Physiol 156:1331–1336

Van Sandt V, Suslov D, Verbelen J-P et al (2007) Xyloglucan endotransglucosylase activity loosens a plant cell wall. Ann Bot 100:1467–1473

Wiedemeier AM, Judy-March JE, Hocart CH, Wasteneys GO, Williamson RE, Baskin TI (2002) Mutant alleles of *Arabidopsis RADIALLY SWOLLEN 4* and *7* reduce growth anisotropy without altering the transverse orientation of cortical microtubules or cellulose microfibrils. Development 129:4821–4830

Yamamoto R (1996) Stress relaxation property of the cell wall and auxin-induced cell elongation. J Plant Res 109:75–84

Zamil MS, Yi H, Haque MA, Puri VM (2013) Characterizing microscale biological samples under tensile loading: stress-strain behavior of cell wall fragment of onion outer epidermis. Am J Bot 100:1105–1115

Zamil MS, Yi H, Puri VM (2014) Mechanical characterization of outer epidermal middle lamella of onion under tensile loading. Am J Bot 101:778–787

Zamil MS, Yi H, Puri VM (2015) The mechanical properties of plant cell walls soft material at the subcellular scale: the implications of water and of the intercellular boundaries. J Mater Sci 50:6608–6623

Tensile Testing of Primary Plant Cells and Tissues

Amir J. Bidhendi and Anja Geitmann

Abstract The primary cell wall controls plant growth and morphogenesis but also determines the structural resilience of nonwoody plant organs. The predominant mechanical role of the primary cell wall lies in its ability to resist or conform to tensile forces. Assessing the tensile properties of the cell wall, therefore, is fundamental for both biomechanics and mechanobiology. Tensile testing is a classic approach used for the mechanical characterization of materials. Various loading strategies such as monotonic or cyclic loading or creep or relaxation allow for analysis of the material response in terms of elastic, viscoelastic, and failure properties. Here, we discuss tensile testing strategies for plant samples with primary cell walls with the aim to provide a practical guide that highlights challenges and offers solutions for the design, execution, and interpretation of such tests.

Keywords Mechanical characterization · Micromechanics · Cell wall mechanics · Tension test · Uniaxial tensile test · Biaxial testing · Young's modulus · Cellulose · Primary cell wall · Anisotropy · Inverse finite element analysis · Multiscale · *Arabidopsis*

Introduction

Mechanical testing of material is an essential element in the toolbox of biomechanical research. Understanding how wood breaks, how it resists compression and bending stresses is essential in its use as a construction material or to understand how trees respond to external stresses such as wind and gravity. The resistance to damage of

A. J. Bidhendi · A. Geitmann
Département de sciences biologiques, Institut de recherche en biologie végétale, Université de Montréal, Montreal, QC H1X 2B2, Canada

A. Geitmann (✉)
Department of Plant Science, McGill University, Macdonald Campus, 21111 Lakeshore, H9X 3V9 Ste-Anne-de-Bellevue, Québec, Canada
e-mail: geitmann.aes@mcgill.ca

A. Geitmann and J. Gril (eds.), *Plant Biomechanics*,
https://doi.org/10.1007/978-3-319-79099-2_15

fruit after harvest determines the storage and handling procedures. To be meaningful, mechanical tests should mimic relevant stress types that the biological material experiences during its lifetime. Typical tests for woody plant tissues are three-point bending or compression tests, whereas fruits are typically tested under compression or impact loading.

The mechanical properties of plant tissues are not only relevant for the plant to withstand and respond to externally applied forces but are also involved in internal processes such as plant development and differentiation. The primary cell wall, characteristic of growing cells, is typically between a few tens of nanometers and a few microns depending on the species, the organ of interest as well as cell type (Derbyshire et al. 2007; Zamil et al. 2013). During plant development, at a cellular scale, the forces driving cellular expansion are generated internally by the turgor pressure that acts on the inner face of primary cell walls. As a result of this, a net permanent expansion of the cell wall results under tension that is interpreted as growth. To be consistent with this stress type, the methods used to measure the mechanical properties of plant material in the context of developmental processes need to relate to the tensile behavior of the cell wall. Such pertinent mechanical tests do not only provide information on the mechanical behavior of the plant materials regarding stiffness or failure but also produce vital ingredients for mathematical models. Modeling approaches such as those based on finite element (FE) method attempt to explore phenomena such as cell growth, morphogenesis or organogenesis to elucidate the underlying biology or to guide future experimental strategies (Bidhendi and Geitmann 2018). The quality of the predictions made by models is greatly augmented if actual quantitative information can be obtained from mechanical tests. In this chapter, the focus is on tensile testing related to primary plant tissues. The concepts of tensile testing, an introductory guide to tensile experiment components, and the challenges involved in its implementation are discussed.

Woody plant tissues are formed by the secondary xylem, a tissue that depending on the species, comprises only a few different cell types most of which have lignified cell walls. From a macroscopic point of view, wood tissues are, therefore, relatively homogenous. Depending on the scale at which measurements are performed, a block of wood behaves relatively uniformly albeit with significant anisotropy because of the longitudinal arrangements of cells and the presence of growth rings. Interpretation of tensile test results of entire tissues can, therefore, be used to deduce the mechanical properties of single cells in relatively straightforward manner. Herbaceous plant organs, on the other hand, possess different types of tissues including turgid primary tissues such as parenchyma, collenchyma, and sclerenchymatous tissues such as primary xylem. Because of this heterogeneous composition, the analysis of tensile tests administered to whole primary organs needs to consider the spatial variation of mechanical behavior within the organ. Similar considerations are made in the chapter "Bending Stress in Plant Stems: Models and Assumptions" for bending tests.

The mechanics of woody tissue is dominated by the lignin and cellulose-rich secondary wall of the sclerenchymatous wood cells. Primary plant cell walls, on the other hand, are composed of several types of polysaccharides, proteins, ions, and a significant amount of water. Lignin is absent from the primary wall, and cellulose, while abundant, is less organized and dominant than in the secondary wall. Cellulose is still considered to be a major load-bearing component, however, since due to a low tensile compliance, its spatial arrangement can determine the magnitude and orientation of anisotropy in the cell wall (Baskin 2005; Cosgrove 2005). Because of the biochemical complexity of the material, the roles of polymers other than cellulose are rather prominent in defining the mechanics of primary plant cell wall (Bidhendi and Geitmann 2016). The chemical configuration of pectin, for instance, is associated with the regulation of cell and tissue growth (Bidhendi and Geitmann 2016; Palin and Geitmann 2012), organogenesis and the ripening of fruits (Braybrook and Peaucelle 2013; Prasanna et al. 2007). Therefore, the visualization of cell wall polysaccharides has been an important tool to understand the plant cell structural makeup. However, in many studies, the spatial information on the distribution of cell wall components has been used directly to make inferences on cell mechanics, which, unless accompanied by mechanical testing data, remain inconclusive. The reason is that the relationship between the chemical changes and mechanics is not straightforward. Highly methylesterified pectin in the tip, where the growth occurs in pollen tubes, is associated with a low stiffness as shown by microindentation techniques (Chebli et al. 2012; Zerzour et al. 2009). Yet, pectin de-esterification is found to be associated with local softening in the shoot apical meristem, where it seems to be a prerequisite for organogenesis or growth anisotropy (Braybrook and Peaucelle 2013; Peaucelle et al. 2015). This is not to mention the ultrastructural changes in the cell, such as changes in wall thickness or cell shape that are often not accounted for in studies focusing on the chemistry of wall polymer alone (Bidhendi and Geitmann 2018). Therefore, visualization of cell wall composition needs to be associated with mechanical tests and in silico experiments to determine the consequence of specific changes in cell wall chemistry (Bidhendi and Geitmann 2016).

The mechanics of the primary plant cell wall regulate both irreversible and reversible plant processes. Among the reversible processes are those regulating the opening and closing of stomatal pores or the pulvini-driven motion of plant leaves. While in silico modeling has helped to understand the mechanical underpinnings of these mechanisms (Bidhendi and Geitmann 2018; Cooke et al. 1976; Forterre et al. 2005), limited experimental work exists to determine the mechanical properties of these structures quantitatively. One of the primary challenges is the type of mechanical test and the scale at which relevant tests should be performed to yield meaningful information.

Mechanical testing is an essential tool for correlating cell wall chemistry and shape to growth and movement in plant cells and tissues. As a result, many mechanical testing techniques have been developed or adapted to study the mechanics of plant cell walls over the past two decades. Micro- and nanoindentation techniques (Bolduc et al. 2006; Milani et al. 2013; Peaucelle et al. 2015; Routier-Kierzkowska et al. 2012; Zerzour et al. 2009), tensile testing (Phyo et al. 2017; Saxe et al. 2016; Zamil et al. 2013), and various forms of acoustics-based microscopy (Gadalla et al. 2014) have been used to this end. These techniques offer a spectrum of force and spatial resolution and each is associated with its own strengths and limitations. For instance, if the deforming force has out-of-plane components with regard to the plane of the specimen, as in indentation experiments, correlating the measured stiffness with in-plane properties of the cell wall is not always straightforward (Eder et al. 2013; Milani et al. 2013). Tensile testing allows for an in-plane stretch of the specimen, and therefore, produces data that are more tangibly relevant for turgor-induced stretch of the cell wall during cell growth. This aspect makes tensile testing a major mechanical test for the characterization of thin anisotropic primary plant cell walls.

Tensile testing for the assessment of primary plant cell wall mechanics can be classified into two categories. In the first approach, the sample is stretched while either the force (creep test) or the length (relaxation test) is kept constant and the variations in the other parameter are monitored. This approach of tensile testing seeks to identify the time-dependent properties of the specimen. Creep test is the more prevalent type of testing in this category which is also referred to as extensiometry. The creep approach is often employed to investigate the role of enzymes, temperature, and other agents in the modulation of cell wall properties. Important conclusions have been drawn on the roles of agents such as expansins and pH (Cosgrove 1998; Durachko and Cosgrove 2009). The second approach in tensile testing is to stretch the specimen until failure with the aim to extract passive mechanical properties such as stiffness, yield or strength. The data from this approach are also relevant for cell growth since they can provide information on the orientation of cellulose and bonding of other wall polymers. For instance, tensile testing of adaxial onion epidermis has shown a higher stiffness parallel to the main axis of cell growth while in the transverse direction the tissue appeared to be more extensible. This tissue behavior is interpreted in terms of the mean longitudinal orientation of cellulose microfibrils in cells of this tissue (Vanstreels et al. 2005). Therefore, this approach of tensile testing can be used to study the cell wall anisotropy which, along with time-dependent enzymatic modification of the cell wall provides a full scenario under which the cell grows or the tissue achieves its functional mechanical properties. The first approach and experiments related to cell wall extensibility are covered in chapter "Cell wall expansion as viewed by the creep method".

In this chapter, we focus on the second approach, the classic tensile test. Tensile testing of plant specimens can be exploited to acquire essential information across multiple scales, from the organ-level to the subcellular scale. The plant root, for instance, requires cell wall pliability at the cellular level to grow, yet its overall material and structure should be able to withstand environmental stresses and be rigid enough to penetrate the soil without buckling. Therefore, the mechanical evaluation of root tissue can provide insight into its developmental stages and identify the parameters that a root should possess for the survival of the plant in a specific soil and environment. Chimungu et al. (2015) observed that the tensile strength of root specimens decreases with the diameter. It was shown that the tensile properties of the stele predominate the overall root tensile strength while the cortical properties such as cell count and thickness were shown to affect the bending and buckling properties of the root. Saxe et al. (2016) studied the effect of age on mechanics of the elongated zone in etiolated *Arabidopsis* hypocotyls. They correlated the mechanics of the elongated zone to material density, geometry, and cellulose content. Their results suggest that the tensile stiffness of the samples may increase over a period of growth with no apparent increase in cellulose content implying a role for other components such as pectin (for related reading refer to Bidhendi and Geitmann 2016; Phyo et al. 2017). Such an interesting approach combining tensile testing with biochemical analysis enables unraveling the contribution of different wall biopolymers and cell shape to macroscale organ properties.

Tensile Testing Parameters and Methods

Tensile testing requires gripping a specimen either at two ends (uniaxial testing) or its circumference or along two axes (biaxial testing) and applying a pulling force. Tensile testing can be performed under strain (or displacement) control or force (stress or load) control. In the former, the rate of opening of the device "jaw" or elongation of the specimen is regulated while in the latter the magnitude of the extending force is adjusted. These modes are not to be mistaken with creep or relaxation experiments where either the strain or stress is kept constant. Load application stretches the specimen, and upon release of the force it either returns to its original dimensions immediately (elastic deformation), remains at the stretch-induced dimension (plastic deformation), or slowly returns partially or fully to its original dimensions (viscoelastic or anelastic). During stretching, the force corresponding to the elongation is recorded. For a uniaxial tension test, the force–displacement data points are used to derive a stress–strain graph as

$$\sigma = \frac{F}{A_0} \text{ and } \varepsilon = \frac{L - L_0}{L_0}$$

σ and ε are engineering stress and strain. A_0 and L_0 are original cross-section area and length of the specimen prior to onset of the experiment. A typical engineering stress–strain graph is depicted in Fig. 1a. The curve may consist of three main zones. A transient "toe" region may indicate initial straightening of the specimen or rearrangement of its load-bearing components. The elastic regime in which deformations are recoverable is followed by the plastic zone, where elastic and permanent deformations occur in parallel. It should be noted, however, that many biological materials such as plant tissues, can exhibit considerably different stress–strain behaviors with the marked zones either hard to distinguish or entirely absent (see the engineering stress–strain curve from a tensile test on onion epidermis in Fig. 1b). From this graph, several mechanical parameters of the sample can be obtained. These parameters include the Young's modulus E, the yield strength σ_Y, the engineering ultimate strength σ_U and fracture strength σ_F. The Young's modulus indicates the material's resistance to deform linearly and is determined from the slope of the linear reversible portion of the stress–strain curve. Yield strength corresponds to the stress values at which the specimen begins to deform permanently. This measure is often difficult to pinpoint precisely since plastic deformations can occur even at low strains. Ultimate engineering tensile strength or simply engineering tensile strength corresponds to the highest stress in the engineering stress–strain graph. This corresponds to maximum force occurring prior to fracture divided by the original cross-section area. If the cross-section changes dramatically over the experiment, sometimes an average cross-section is used. Fracture strength corresponds to the stress at fracture point. For materials exhibiting brittle fracture, tensile and fracture strengths can be close to each other. However, for ductile failure, the fracture strength is lower than the ultimate strength on a tensile testing graph based on "engineering" stress and strains. Other than engineering stress–strain curves, true stress–strain calculation accounts for changes in cross-section area of the specimen (generally a reduction) to calculate the stresses. This becomes more prominent in the plastic zone, at least for ductile materials with a tendency for necking (a phenomenon characterized by narrowing and a decrease in the cross-section of the specimen prior to failure; this is not to be confused with reversible lateral contraction of the material referred to as the Poisson's effect that occurs even in the small deformations in the elastic regime. However, necking does not seem to be a characteristic of plant materials) (Fig. 1a). True stresses and strains can be calculated as:

$$\sigma_T = \sigma\ (1 + \varepsilon)\ \text{and}\ \varepsilon_T = ln\ (1 + \varepsilon)$$

In small strains, despite the lateral contraction of the specimen, the difference between the engineering and true graphs is negligible. However, beyond yielding and especially necking, the difference can be more dramatic. While the engineering stress in the specimen decreases after necking due to decrease in force required to stretch the sample, the true stress continues to increase since the effective cross-section is getting smaller in sample. Tensile tests allow for evaluation of the fracture behavior of the material and determination of fracture toughness. Fracture toughness is a measure of the resistance of the material against the propagation of existing cracks and tears. Such experiments can be carried out, for instance, by tension of a tissue with a preexisting notch. The area below the engineering stress–strain curve up to the elastic limit indicates the strain energy per volume stored and relates to "resilience", while if the curve is considered up to fracture, the area relates to tensile "toughness". Poisson's ratio, ν, measured at the elastic zone, provides information on the extent of the material shrinkage in directions perpendicular to that of the applied stress. Poisson's ratio can be determined based on:

$$\nu = -\frac{\varepsilon_{lateral}}{\varepsilon_{axial}}$$

$\varepsilon_{lateral}$ and ε_{axial} correspond to strains in perpendicular to and along the stress direction, respectively. Note that since most common materials contract laterally when stretched, $\varepsilon_{lateral}$ is negative, and therefore the Poisson's ratio holds a positive value. Most isotropic elastic materials hold a Poisson's ratio between 0 and 0.5. In case of a near 0 Poisson's ratio, the material does not considerably strain laterally when strained axially. This applies, for example, to cork and is exploited in its application to seal bottles (Silva et al. 2005). When squeezed, cork does not expand in other

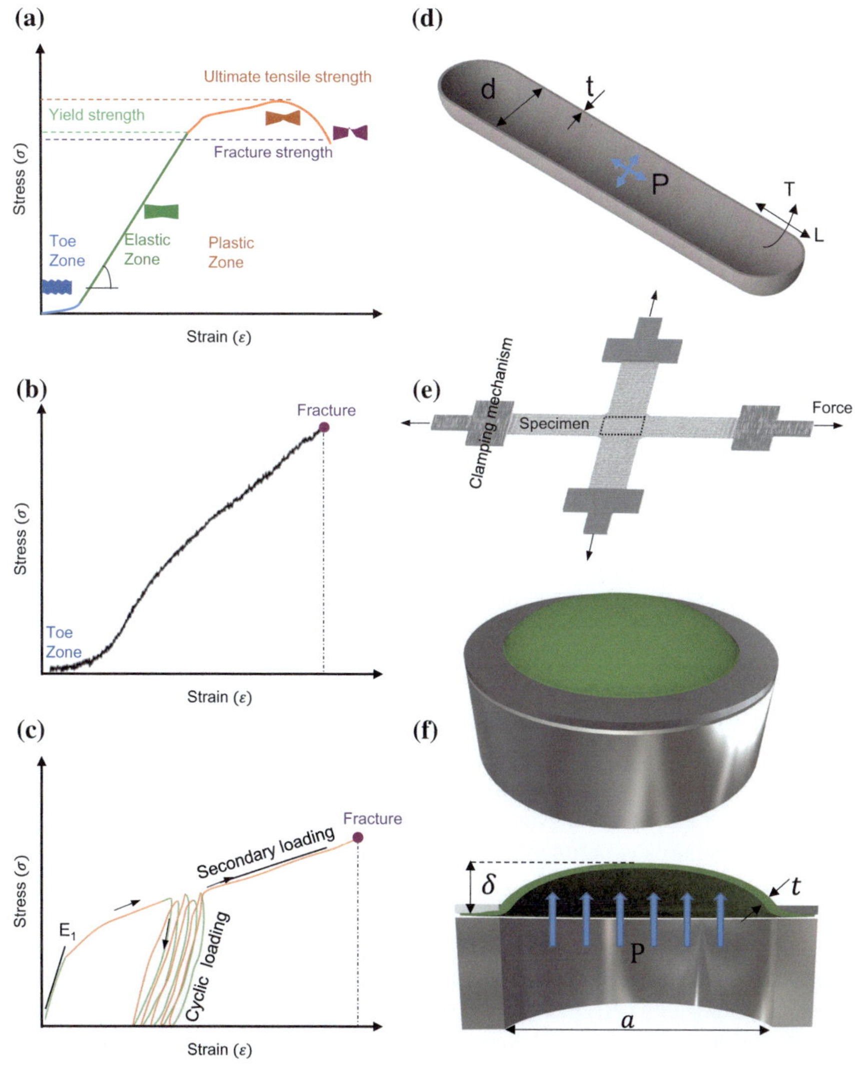

Fig. 1 **a** Schematic of a typical strain–stress graph obtained from the tensile test. Various zones of material behavior are indicated. **b** Graph of experimental tensile test carried out on an onion tissue up to failure. **c** A generic cyclic tensile testing pattern applied to onion epidermal specimen similar to the study conducted by Vanstreels et al. (2005). The initial loading is followed by a number of loading–unloading cycles. The specimen is eventually loaded up to failure. **d** Cutaway view of a thin-walled cylindrical vessel under pressure (P), with a thickness of (t) and a diameter (d). T and L represent transverse and longitudinal directions, respectively. **e** Schematic of a biaxial tensile test cruciform specimen. **f** Bulge test of a flat specimen. The pressure gradient working on the specimen results in a hemispherical bulge of the sample. The displacement of the specimen at the center (δ), thickness (t), diameter (a), and the hydraulic pressure (P) are used to derive the material behavior of the specimen

directions which facilitates its insertion into and removal from the bottle's neck. A Poisson's ratio of 0.5 refers to an incompressible material behavior, where the volume changes through longitudinal increase and lateral contraction cancel each other out. For anisotropic materials, the Poisson's ratio differs based on the direction and, unlike for isotropic materials, the value is not confined to a particular range (Norris 2006; Ting 2004; Ting and Chen 2005). It might be interesting to note that Poisson's ratio can even hold negative values. "Auxetic" materials with a negative Poisson's ratio, counterintuitively, expand in the transverse dimensions while being stretched uniaxially. This is often due to a particular internal structure, for example, that of the polymer network or shape of unit cells (Mir et al. 2014). Auxetic materials are, however, rare in nature and are mostly synthetic. Most common materials including plant materials contract to compensate for elongation in the other dimension to preserve the volume. The magnitude of this contraction is correlated with the microstructure but also with the state of hydration. The presence of water in the porous network renders the material less compressible if the short force exertion time or impermissible solid network prevents any water movement through the sample microstructure.

When plastic and elastic behaviors occur combined, the components need to be distinguished to correctly deduce the mechanical properties. Experimentally, this can be achieved by closing the "jaw" of the tensile tester until the force read from the sensor reaches zero. The difference between the length at which the force becomes zero upon unloading and the mechanical zero at which the experiment started constitutes the permanent deformation. If further recovery eventually occurs, but not instantly, the material exhibits viscoelastic behavior. Most biological materials do exhibit viscoelastic behavior if given enough time. Therefore, whether to account for viscous effects is a matter of the time frame of the experiment and whether such a time frame is of biological relevance (i.e., may occur in vivo).

Uniaxial testing is experimentally easiest to execute, and the obtained data are informative, certainly for plant organs that primarily grow longitudinally such as stems or roots. However, it is important to realize that even in uniaxially growing cells, the stress applied to the cell wall by the turgor pressure, and by the contact with other cells, is biaxial. Furthermore, a considerable mechanical anisotropy exists in the cell wall of elongating cells that promotes growth in preferential direction in the first place. Therefore, the mechanical characterization of plant materials and specifically plant cell wall by uniaxial testing may result in an incomplete representation of their overall mechanical behavior. Further, it has been shown for biological material such as arterial wall tissue—a collagen fiber-reinforced composite—that in uniaxial tests fibers can reorient toward the stress direction. This results in altered stiffness measurements in that direction when compared with biaxial tension test (Zemánek et al. 2009). Indeed, recent studies on onion epidermis have demonstrated that during the uniaxial stretch of the sample, cellulosic bundles passively reorient toward the main direction of the stretch (Kafle et al. 2017; Zhang et al. 2017). A similar result was shown for cellulose microfibrils in wood cells (Keckes et al. 2003). Another complicating factor in uniaxial tensile testing occurs when the stress is applied in a direction that does not align with one of the principal anisotropy axes. This can be the case when the anisotropy is not structurally evident, i.e., not correlated to the cell geometry. Off-axis loading of the anisotropic material results in the development of shear forces between the matrix components (e.g., between fibers and the matrix) in addition to tension. While this effect is also useful for the determination of shear strength of composite materials, it has been shown in a flax fiber composite that the failure mechanism of the sample may vary dramatically based on the angle between the loading and the direction of anisotropy (Shah et al. 2012). Therefore, the elastic modulus and failure parameters derived without knowledge of the tissue anisotropy require further attention.

Biaxial tension tests, in general, comprise two pairs of coaxially applied forces. Biaxial tension tests allow several important experiments such as the application of equal or different tension in two perpendicular directions or a stretch in one direction while the other direction is constrained. An in vivo way of applying biaxial tension to the walls of intact cells is to record the changes in cell dimension with and without turgor pressure. The changes in diameter and length of the cell can bear the elastic modulus (or moduli for anisotropic wall properties assumption) of the cell wall. For a thin shell pressure vessel ($\frac{t}{d} \ll 1$) (Fig. 1d), the resulting stresses along the two principal axes are

$$\sigma_L = \frac{Pd}{4t} \text{ and } \sigma_T = \frac{Pd}{2t},$$

where P is the internal (e.g., hydrostatic) pressure, t and d are the wall thickness and the diameter of the vessel, and σ_L and σ_T are longitudinal and transverse (hoop) stresses, respectively. It can be seen that the transverse stress is two times greater than the longitudinal stress simply due to the shape (for the same reason barbeque sausages split along their axis). From the inverse of the Hooke's law in three dimensions for a linear isotropic material and with plane stress assumption (variation in stress in thickness of the wall is negligible), it can be shown that

$$\varepsilon_L = \frac{1}{E}(\sigma_L - \nu\sigma_T) \text{ and } \varepsilon_T = \frac{1}{E}(\sigma_T - \nu\sigma_L),$$

where ε_L and ε_T are longitudinal and transverse strains. The strains can be determined from changes in dimension from microscopic images. The Poisson's ratio, however, needs to be measured separately or adopted from the literature. Approaching an incompressible behavior ($\varepsilon_L + \varepsilon_T + \varepsilon_{th} = 0$ with the subscript *th* denoting the direction of the wall thickness) corresponds to a Poisson's ratio of $\nu = 0.5$. However, as mentioned before, elongated cells rarely have an isotropic wall. In most cases, there exists a preferential orientation of cellulose microfibrils in the circumferential direction, although this may later change to a longitudinal direction due to passive reorientation during cell growth (Anderson et al. 2010; Green 1960). In such a case, the strains cannot be explained with a single Young's modulus. A special case of anisotropy, the transverse isotropy that corresponds to unidirectional composites can be relevant to the composition of plant cell walls with a relatively well-organized direction of fibers. Simplifying the inverse of Hooke's law written in form of strains for transverse isotropy and plane stress assumptions, the longitudinal and transverse strains can be written as

$$\varepsilon_L = \frac{1}{E_L}\left(\sigma_L - \nu_{TL}\sigma_T\right) \text{ and } \varepsilon_T = \frac{1}{E_T}\left(\sigma_T - \nu_{LT}\sigma_L\right),$$

where ν_{LT} and ν_{TL} are the Poisson's ratios. The subscript LT, for instance, denotes the Poisson's ratio effect in the transverse direction due to the strain in the longitudinal direction. In case of incompressible transverse isotropy, it can be shown that

$$\nu_{LT} = \frac{1}{2} \text{ and } \nu_{TL} = 1 - \frac{E_L}{2E_T}$$

Therefore, the strain relationships can be further simplified to (for a more in-depth reading refer to classic books on the theory of elasticity or, for instance, see Argatov and Mishuris 2015; Bernal et al. 2011)

$$E_L = \frac{\sigma_T}{2\varepsilon_L + \frac{4}{3}\varepsilon_T} \text{ and } E_T = \frac{3}{4}\frac{\sigma_T}{\varepsilon_T}$$

Therefore, having measured the axial and transverse strains, Young's moduli in two directions can be calculated based on the above relationships. While this method allows for measurement of the elastic moduli of the cell wall along the two axes, it requires a measurement or a justified assumption of the turgor pressure. In the absence of quantitative information on turgor, the ratio of anisotropy ($\frac{E_L}{E_T}$) can be evaluated using this approach.

A common way of performing biaxial tension test for planar materials is to prepare square or cruciform shapes of the sample to be stretched by a pair of perpendicularly positioned actuators (Hannon and Tiernan 2008; Lally et al. 2004) (Fig. 1e). The elastic and plastic deformations are then analyzed in the midsection of the specimen which should remain stationary by proper adjustment of displacement actuators. Bulge testing is yet another common technique for tensile characterization of both thin and thick films (Small et al. 1994; Srikar and Spearing 2003; Yu et al. 2016). It has the advantage of not requiring preparation of complex sample shapes or dealing with complex stress distributions in specimens. In the bulge test, the sample can be of circular or square shapes. The specimen is fixed at the open end of a tube and pressure is applied on one side (Fig. 1f). Usually, a positive pressure is applied, for instance, by using a viscous silicone oil that does not permeate the specimen (Chanliaud et al. 2002), resulting in the outward bulge of the specimen into a hemispherical profile. Depending on the shape of the membrane specimen, various analytical solutions have been developed to correlate the parameters such as pressure (P), the maximum deflection of the specimen (δ), its diameter (a), and thickness (t) (for instance, refer to Maier-Schneider et al. 1995). Alternatively, inverse FE modeling can be employed to identify the unknown parameters. The maximum deflection at the center of the specimen and the sample thickness can be measured using either tactile techniques or, without contact, through optical methods such as 3D digital image correlation

(DIC), and plotted against the pressure of the fluid in the chamber (Machado et al. 2012; Neggers et al. 2014; Nouira et al. 2014; Orthner et al. 2010). This technique also allows for the determination of the Poisson's ratio of the sample. If the pressure is kept constant, the increase in the bulge size versus time can be used as a creep test to acquire viscoelastic properties of the specimen. Adopting this strategy, Chanliaud et al. (2002) performed uni- and biaxial tension tests to study the mechanical behavior of the cell wall under turgor pressure using *Acetobacter xylinus* based cellulose as plant cell wall analog. Cellulose, cellulose/pectin, and cellulose/xyloglucan composites were tested under tension and compared with FE modeling used to interpret the pressure–displacement test data.

Tensile Testing Setup

The tensile setup generally consists of two distinct compartments: sensors and actuator units. In this section, we describe the essentials of a simple uniaxial tensing device. For biaxial testing, a significant number of considerably different configurations have been developed over the years (for instance, refer to a review by Hannon and Tiernan 2008). However, the concepts remain similar to a great extent. The displacement is generated by a displacement transducer. A wide range of actuators can be used to this end such as mechanical, piezoelectric or pneumatic mechanisms (Fig. 2). In motorized lead screw actuators, for instance, rotation of the lead screw by a stepper motor causes linear displacement, similar to the driving mechanism of many syringe pumps. Finely threaded screws in conjunction with precise stepper motors allow for the production of displacements of several centimeters in range and submicrometer in resolution well-suited to displacements required in many micromechanical tensile tests. Several methods exist for measurement of displacement or strain in the specimen. The simplest method is to rely on the displacement values given as inputs to the displacement transducer. However, different sources such as the existence of backlash in the gear system can cause a lag and result in the actual displacement to deviate from the input. As a result, the use of displacement input is not ideal, and instead, the device's "crosshead" or "jaw" motion which is equivalent to the actual displacement of the actuator and the gripping ends can be measured and used. Using gripping end displacement as a measure of sample strain is usually accomplished by readings from a position sensor such as a linear variable differential transformer (LVDT). LVDTs consist of a ferromagnetic core that can move freely without contact in a hollow cylinder equipped with an assembly of primary and secondary coils. The primary coil fed with an AC or DC input voltage induces a voltage in the secondary coils. The movement of the core, connected from one side to the displacement actuator or a gripping end, in or out of the hollow cylinder changes the amplitude and phase of the voltage induced in secondary coils which constitutes the output signal from the LVDT enabling measurement of that displacement. However, this reading does not take into account the compliance of the loading frame. The loading part of the device including the force sensor, extensions and fixtures can deform along with the specimen during the loading and result in an underestimation of the elastic modulus of the specimen (Sanders et al. 1997). The frame compliance can be corrected by using a rigid mock-sample of known elastic modulus (Turek 1993). Most times, the device compliance is considered to be a constant, meaning that a linear relationship between the force and the deformation of the loading apparatus is presumed (Sanders et al. 1997). However, this might not always be the case, and nonlinearity might exist resulting in deviation of the measurements based on the specimen being used (Kalidindi et al. 1997). Another method for strain measurements in tensile testing is the use of strain gauges. Strain gauges often but not always refer to thin film patches with an electrical resistance that changes when the patch is deformed (for an example of such a design, refer to Pang et al. 2012). Generally, strain gauges refer to the use of such contact patches although sometimes the term is also used for noncontact strain measurement approaches (e.g., see Yamaguchi 1981). There are numerous types of strain gauges based on their working mechanisms and the working environment they are intended for. However, the contact nature of the strain gauge requires it to be mounted and bonded to the surface of the specimen at various locations. The need for bonding to the specimen and the miniature size of most plant specimens render the use of strain gauges in plant tensile studies challenging. Further, the presence of strain gauges on the sample can affect the behavior of the specimen due to reinforcement or due to the weight of the patch and the connecting wires. Despite all these challenges, they only provide limited information on strains in the whole sample as the information is limited to the location of their installment. Therefore, strain fields and possible inhomogeneities and hotspots outside the installment zones remain undetected (Motra et al. 2014; Soons et al. 2012).

Extensometers can also make another class of strain measurement techniques. Basic contact extensometers rely on the attachment of a pair of "sensing arms" to two ends of the gauge length of the specimen averaging the strain over this region (for an example of its application refer to Boitier et al. 2000). For plant applications, this method suffers from similar shortcomings as the strain gauges in that it affects the delicate sample and that only average strain data are produced. Noncontact extensiometry approaches, on the other hand, seem very promising for specimens of various texture and sizes, especially in micro- and nanomechanical characterization. Two broad classes of these methods rely on either image matching algorithms or optical interferometry. Matching algorithms such as DIC or differential digital image tracking (DDIT) techniques (Sharpe et al. 2007) are based on analysis of consecutive images taken from the sample during the deformation. This is accomplished by means of one or more cameras (2D, 3D or volumetric) or by other techniques such as scanning electron microscopy (Kashfuddoja and Ramji 2013). A sufficient number of speckles or patterns, either artificial or digital, for DIC (Lava et al. 2009) or intensity peaks on the image for the DDIT are required for comparison and correlation of successive images to carry out full-field measurement of strains in the sample (e.g., Pan et al. 2009; Sutton and Reu 2017). DIC can provide a subpixel displacement/strain resolution (Hua et al. 2007; Zhou and Goodson 2001). Another advantage of an image processing-based approach is that rigid body motion (such as that occurring due to slippage of the specimen at clamps) can be accounted

Fig. 2 A miniaturized tensile testing device developed by Lynch and Lintilhac (1997) allowing for mechanical testing in air and liquid conditions. The testing chamber consists of (1) load shims, (2) LVDT (tube) (3) pneumatic bellows that act as displacement actuator opening or closing the frame driven by air flow (4) adjusted by a control unit based on displacement or force feedback. The deformation of the strain gauges (5) adhered to the load shims is read as the change in the output voltage and translated to forces acting to bend the shims. The displacement of the LVDT core (6) inside the tube reads the displacements

and corrected for. Various interferometry techniques have been developed for strain measurement. Electronic speckle pattern interferometry (ESPI), for instance, is one of the common interferometry methods for strain or displacement measurements. In this technique, the sample surface is illuminated with a coherent laser beam. The reflected laser forms a pattern on the image recording setup. Topographical information of the specimen can be acquired based on changes in the phase difference between the reflected and original laser beams that is caused by sample deformation or displacement (Soons et al. 2012; Yang and Ettemeyer 2003).

Other than displacement actuator and sensors, force sensors comprise another compartment of the tensile device. Forces are measured by one or two load cells that are installed on the gripping ends. Load cells are commonly based on one or more strain gauges. The deflection of the load cell beam results in changes in the electrical resistance of the strain gauge incorporated into it and, in turn, to a change in the output voltage that is read through an amplifier. The compliance of the load cells can be measured for different loads during the calibration stage and prior to the experiment and be accounted for in the displacement measurements. It is good practice to calibrate the device for force and displacement after mounting all the extensions on which the sample will be fixed, or in case of any change in environmental temperature or humidity. Excluding the weight of the extensions and loading frame from the force and displacement measurements is even more important in vertical setups.

Common Challenges Associated with Tensile Testing

While some mechanical tests such as nanoindentation are relatively easy to perform, correlating the force–displacement data from these tests to in-plane properties of the cell wall, and accounting for complex probe–sample interactions is not straightforward and sometimes demands cumbersome calculations. Tensile testing, on the other hand, is considered a test requiring relatively simple calculations for extraction of the parameters from the experimental data. However, carrying out the experiment itself can be accompanied by many challenges that begin with sample preparation and handling and extend to sample alignment and gripping. Some of these challenges are briefly discussed.

Measurement of the Original Values

Obtaining the mechanical parameters of the sample in a tensile test relies on the measurement of several independent variables such as displacement, force, and dimensions of the specimen that serve as inputs for the calculations. The eventual outputs of the tensile test, i.e., stresses and strains, can be only as accurate as the measurements of the independent values allow. Above, we commented on the practical aspects of the force and strain measurements. An important set of input variables used to calculate the tensile testing parameters includes the original length and cross-section area (through diameter or width and thickness) of the specimen. With minute plant specimens such as roots or shoots of *Arabidopsis*, measurements of the original dimensions are typically carried out using optical techniques such as microscopic visualization. The issue here is that other than the considerable biological variability among samples, the errors in measurement of these dimensions can dramatically alter the conclusions of the study. This becomes significant in situations where small changes in the mechanics of the cells or tissues are being investigated through tensile testing. The changes in mechanical properties of the material over close time points might not be dramatic and could be masked by errors or uncertainties in the measurement of original dimensions. An example of how important the measurement of the initial dimensions can be, even if they add complexity, is provided in a study by Saxe et al. (2016), which tested *Arabidopsis* hypocotyls for their tensile properties. In this study, a subset of hypocotyl samples exhibited decreasing cross-sections with age while in other datasets, no significant age-dependency was observed. The environmental and growth parameters such as humidity and temperature, while able to affect the length of the growing hypocotyls, were shown to not significantly affect the cross-section. The authors calculated the stress based on dividing the force by the cross-section of hypocotyls, as it should be done. However, the biologically atypical reduction in diameter, whether caused by biological variation or other, may significantly influence the determined stress values, thus making a clear-cut interpretation more difficult. An increase in the number of biological replicates will increase the reliability of the conclusions drawn from experimental data. It will also be helpful to define methods and protocols that ascertain precise, consistent and reproducible quantification of dimensions and other parameters with potential influence on the measured mechanical values.

Gripping

One of the prominent technical challenges associated with tensile testing, particularly for soft and small tissue specimens, is fixing the ends of the sample. Stress concentration and damage near the grip can result in premature failure of the specimen near the gripping ends. With cylindrical samples such as root, shoot or hook structures such as prickles and spines, the ends of the specimen can be tied using a resistant nondeforming thread such as Kevlar (for instance, refer to Gallenmüller et al. 2015). However, this requires a certain minimum size of the sample and the thread compliance needs to be accounted for in calculations. Clamping or grips are one of the common methods to hold the ends of the tensile specimen. Various grades of sandpaper can be used between the specimen and the clamp to prevent slippage. Sometimes, a layer of a softer material is placed between the sample and camping mechanism to reduce the risk of specimen damage (Hervy et al. 2017). Clamp gripping methods tend to induce damage and are particularly prone to slippage, due to reasons such as the non-flat geometry of the samples and moisture. Further, in case of the model plant *Arabidopsis*, the minute size of its organs make solutions such as tying unviable. Piercing structures such as fishing hooks have also been used for tensile testing of soft tissue such as porcine arterial specimens (Lally et al. 2004). More invasive gripping techniques such as fishing hooks inevitably result in failure of the tissue around the fixture and are better suited for determination of tensile modulus and not strength. Due to these reasons, application of fast-curing glues such as cyanoacrylate can represent an attractive alternative. A high viscosity glue is preferable to prevent diffusion into the specimen as it can dramatically alter the mechanical properties of the tissue. Still, using a glue is not without its challenges. The glued area tends to shrink upon curing which causes a prestress in the sample. This is easily observable as the force sensors begin to register increasing tensile forces as the glue is curing. Therefore, it is necessary to release this tension prior to the onset of the tensile test. This can be done by closing the "jaw" of the device up to a point that the force reads zero which can be set as the mechanical zero for displacement. Further, moisture on the hydrated tissue can hinder curing of the glue. Since the tissue needs to be kept constantly hydrated, if not thoroughly submerged in liquid, acquiring a perfect fusion in a reasonable time may not be trivial. As the dimensions get smaller in micromechanical tests, the importance of applying glue droplets small enough to prevent bleeding into the gauge length of the specimen becomes significant. In such scenarios, techniques such as platinum deposition have been used to fix the microscale specimen on the loading frame, for instance, to study a subcellular patch of the onion cell wall under the scanning electron microscope (Zamil et al. 2013). Handling and mounting thin film specimens such as an epidermis is challenging since they are prone to wrinkles and folding that can affect the stress distribution in the sample. Water surface provides a substrate to hold the specimen without wrinkles until mounting. In a recent study on tensile testing of ultrathin films, it was shown that stiction forces such as van der Waals could constitute a gripping mechanism (Kim et al. 2013), a promising approach for ultrathin plant layers.

Biphasic Behavior, Effects of the Loading Rate and Preconditioning of the Cell Wall

An interesting note related to the tensile behavior of the plant cell wall is a reported "biphasic" or "bilinear" behavior in the stress–strain curve. Such behavior is described as the existence of two different slopes in the stress–strain curve that in some cases are associated with a primary and a secondary Young's moduli (Pieczywek and Zdunek 2014; Spatz et al. 1999; Vanstreels et al. 2005) (Fig. 1c). While the existence of a strain-dependent and nonlinear elasticity in the plant cell wall, as also observed in mammalian cells and tissues (e.g., refer to Fabry et al. 2001; Guilak et al. 2014; Kollmannsberger and Fabry 2011; Mofrad 2009), is expected, the approach adopted to demonstrate this "bilinear" behavior seems unclear. As seen in Fig. 1c depicting a generic form of such graphs, in these studies the second Young's modulus suggesting a secondary elastic behavior is defined after a turning in the graph (likely due to yielding) and on the loading (as opposed to unloading) part of the stress–strain curve. First, the slope of engineering stress–strain curves obtained through tensile testing of many materials does change after yielding. The existence of a slope change in the graph per se does not indicate a secondary elastic modulus. Further, Young's modulus is defined based on relatively small strains prior to yielding or upon separating the plastic deformations from the total strain (thus using the unloading data) and is strictly indicative of deformations that are elastic, ε_e (where total strain $\varepsilon_t = \varepsilon_e + \varepsilon_p$). However, in these studies, the loading path is marked for the secondary Young's modulus during a monotonic loading and the deformation seems to be associated with considerable plastic strains. For regions that are not purely elastic, fitting of appropriate elastoplastic models or presenting the data in terms of moduli other than Young's modulus such as tangent modulus, therefore, is more appropriate. Second, while in some studies the transition between the primary and secondary elastic behavior is demonstrated under relatively small strains (less than 5% strain in Spatz et al. 1999), others have demonstrated these two values in a range of over 20% strains (Vanstreels et al. 2005). Besides the monotonic tensile loading, cyclic loading is performed in some studies and seems to suggest that when unloading paths are considered, the slope of the stress–strain curve remains relatively unaffected and comparable to initial cycles (Vanstreels et al. 2005). Therefore, the existence of such bilinearity in the stress–strain behavior of the wall material needs further assessment through analysis of reversible deformations. Moreover, studies are required to look into underlying wall composition and structures that result in such a putatively biphasic behavior.

Viscoelastic materials exhibit a rate dependency in their deformation behavior. This means, the faster the deformation is applied, the stiffer the tissue appears to be. Such a strain-rate dependency can affect the measurement of stiffness, but also the strength and maximum strain at fracture. Higher loading rates can result in increased apparent stiffness, increased stress, but less strain at fracture (the specimen appears less extensible). It is therefore interesting to investigate to what extent the results obtained by stretching the plant cell wall at different loading rates are comparable. Further, in many cases involving tensile testing of soft tissues, the specimen is preconditioned prior to actual data collection. In many animal tissues, it can be observed that the loading and unloading paths do not coincide (hysteresis). The area between them corresponds to the dissipated energy due to viscous or plastic effects. Further, for some materials, the peak load for the same amount of strain appears to differ in each cycle. Therefore, to obtain reproducible results, the specimen is sometimes "preconditioned" prior to measurements, for calculation of the elastic modulus and tensile strength. This is not to be confused with preconditioning term used in other contexts such as cold acclimation of the plant or removal of turgor pressure before tensile testing. The preconditioning is carried out by repeating the loading and unloading cycles for a certain number of times to reduce the peak force variations and also to some degree the hysteresis prior to actual measurements (Lee et al. 1984). For tendons and ligaments, it has been shown that preconditioning results in an increase in both stiffness and strength of the samples. This has been suggested to be due to an increase in recruitment or reorientation of collagen fibers resisting the tensile loads toward the force field (Miller et al. 2012; Schatzmann et al. 1998). To verify whether preconditioning has caused damage or a permanent deformation in the sample, the toe region of the force–displacement curve can be analyzed. An increase in the length of the toe region (i.e., the existence of some strain without a corresponding force) may indicate permanent deformations in the specimen. The reorientation of fibers that might also depend on their initial distribution, to begin with, may render the preconditioning a questionable practice if original properties of the tissue are of interest. There is a considerable variation between the strains used to precondition the samples among various studies. While some advocate a fraction of the real testing strains for preconditioning strains, others suggest a full experimental strain to be used before actual data collection and, obviously, this treatment must be reported along with other experimental conditions (Cheng et al. 2009). In case that preconditioning strains were observed to induce damage, either due to fatigue (failure in the material at stresses significantly lower than the ultimate and fracture strengths of the material due to repetitive loading) or simply exceeding the fracture strength, there is a need to settle for lower strains in the actual experiments.

Tensile testing of onion epidermis was also demonstrated to exhibit hysteresis in loading–unloading paths (Vanstreels et al. 2005). The existence of such a load history dependency of the plant tissue is suggested to stem from the presence of wrinkles or adhesion of other structures on the onion sample that unfold and resolve over a number of cycles (Wei et al. 2001). However, in addition to these potential artifacts, the hysteresis and rate dependency seem to be an inherent part of most biological tissues, including plant materials. Tensile testing of onion epidermis has also shown a considerable transient response in the first cycle; a significant difference in stress–strain values between the initial and later loading cycles (Kerstens et al. 2001; Vanstreels et al. 2005; Wei et al. 2001). This phenomenon that might be due to plastic deformations of the matrix occurring in the first cycle, becomes even more interesting when considering the contrasting results reported for the onion epidermis. From the work of Wei et al. a more significant transient response is observable in onion epidermal strips stretched perpendicular to the cell axis compared to the longitudinal one. Other studies (e.g., Kerstens et al. 2001) report the opposite. Since in these studies, this behavior is correlated with the composite material behavior of the cell wall, e.g., the orientation of cellulose microfibrils, further studies are required to address these different observations.

Effect of Cellularity

A major factor that distinguishes the use of tensile testing for the mechanical characterization of plant specimens from common industrial materials is that plant tissues are made up of cells. Whether the aim is to determine tissue-level mechanical properties or to deduce the cell-level mechanical properties, the cellularity of the material incorporates a number of factors that need to be accounted for when interpreting the experimental data. An immediate consequence of the cellularity is the difference between the effective and geometrical cross-sections of samples. The effective cross-section is the one that bears the loads. For plant tissues, this area is not the same as the tissue cross-section. If cell wall-level properties are to be deduced from the results, measuring only the portion containing the cell walls by subtracting the void space of cell lumina and intercellular spaces from geometrical (tissue-level) cross-section can be beneficial. However, instead, this is often avoided and the whole cross-section is used in the calculations. In cases where factors such as cell type, cell size, and cell wall thickness are expected to be identical between samples, this is probably an acceptable simplification. In cases where any of these factors are different, a more granular approach should be considered. Saxe et al. (2016) estimated the cell wall density based on the hypocotyl's dry weight and its volume in the hydrated state.

Aside from the measurement of the effective cross-section outlined above, inferring the mean mechanical properties of individual cell walls from tissue-level tensile testing is not straightforward. First, due to the shape of the specimen, cellularity, and presence of borders in the material, the stresses may not be distributed uniformly across the cells. This means that different cells in the tissue can experience different levels of stresses and strains. As a result, the properties of some cells may affect the calculations disproportionately, and the outcome of the tensile test may be a weighted rather than the arithmetic average of the properties of all cells. Additionally, even within a single cell, not all the walls experience the same share of stresses and can be stressed to a varying degree depending on their orientation with regard to the stress field and the cell geometry (Fig. 3). To what extent the mechanical properties of individual cells of different shapes and locations affect the overall elasticity of the tissue warrants further studies. Nevertheless, a study on onion epidermal cells has suggested that the growth anisotropy of the cells is correlated with tissue stiffness, with tissue samples with more elongated, higher aspect ratio cells appearing stiffer (Vanstreels et al. 2005). While such differences are generally attributed to the anisotropic composition of the cell wall, it might be also due to cells shape. It should be noted that the apparent stiffness of a structure is determined in part by the properties of its building material, e.g., Young's modulus, porosity, and degree of anisotropy, but also by the geometry of the structure. To better illustrate this, consider a beam with a rectangular cross-section under bending. It can be easily illustrated with a plastic ruler that the beam behaves stiffer under bending when placed with the longer edges of its cross-section vertical or parallel to the force (due to an increase of the second moment of area). Similarly, it is expected that the individual cells with different shapes resist the tissue-level deformation differently due to their geometry and orientation, affecting their contribution to overall mechanical properties of the tissue. This can confound comparison of the mechanical properties of different tissues with considerably different cell geometries if cell-level mechanical properties are to be inferred and correlated to tissue-level behavior. This issue concerns both classes of tensile testing outlined previously. Therefore, studies are required to untangle the contribution of cell geometry and material properties of the cell wall in the mechanical behavior that the tissue exhibits.

Tensile testing can be performed on both turgid and plasmolyzed cells. Turgid tissues are stiffer and reportedly fracture in a brittle manner. On one hand, tensile testing of turgid tissues is interesting since it more closely mimics the state of living plants. On the other hand, discriminating cell wall properties per se from the properties measured when under turgor induced tension requires a comparison between the two hydration states of the sample (Pieczywek and Zdunek 2014).

Unraveling the mechanical properties of cell walls from the apparent tensile modulus obtained through tissue-scale mechanical characterization requires to also consider the intercellular interface, known as the middle lamella (Pieczywek and Zdunek 2014; Zamil and Geitmann 2017; Zamil et al. 2017). In terms of tensile strength, the role of the middle lamella is intuitive. If tissue failure occurs by the failure at

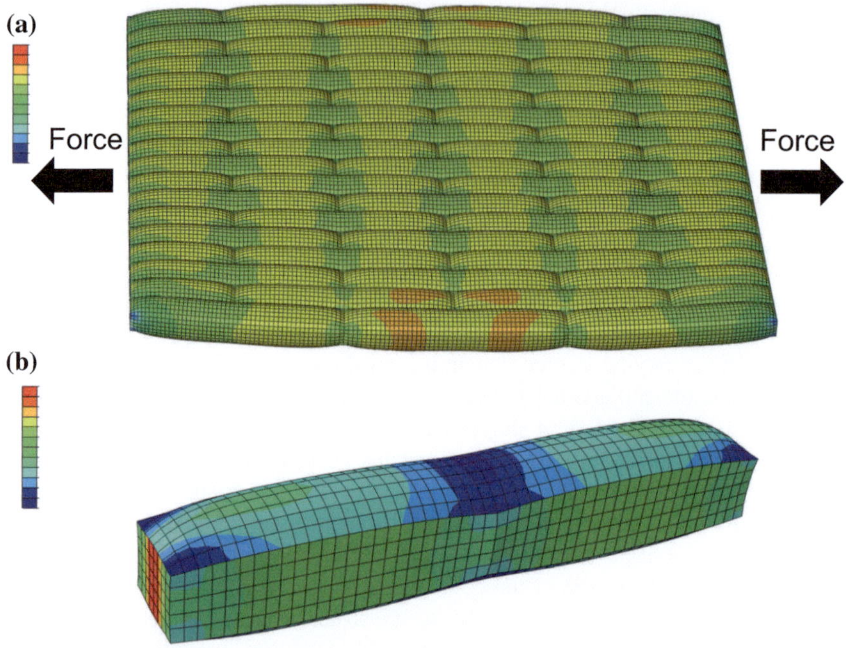

Fig. 3 Finite element model of a tissue consisting of pressurized elongated cells. The distribution of stresses and strains is not uniform within the tissue (**a**), and can even vary in different walls of a single cell at subcellular scale (**b**). Color map represents the magnitude of principal stresses

the middle lamella, i.e., cell separation or delamination, the strength of this structure determines the strength of the whole tissue. As explained by the weakest link concept, the fracture of the specimen is determined by the strength of its weakest spot (Freundthal 1968). If and how the stiffness of the middle lamella also contributes to the tensile stiffness of the whole tissue remains elusive considering its negligible thickness.

To eliminate the compounding features such as cell shape and middle lamella, tensile testing on subcellular patches of the cell wall have been attempted (for instance, refer to Wei et al. 2006; Zamil et al. 2013). Besides the usual technical hurdles, manipulation and observation of the sample while keeping it hydrated are the major challenges at the micron scale. Given the challenges at each length scale, a combination of approaches incorporating macroscale response and micromechanics at the cell level is likely the most promising avenue in many cases. Multiscale experimental and computational strategies that combine tests both at the tissue and subcellular scales, giving consideration to the osmotic status, and integrating the experimental approach with numerical modeling allows for evaluation of the contribution of single cells to tissue-level mechanical properties and, conversely, determination of the cell-level properties from the tissue-level tension experiments. These considerations are equally important in the context of extensiometry and time-dependent studies, where the goal is to infer the effect of cell wall modifying treatments from the creep or relaxation of the whole tissue.

Nonuniform Strain Fields and Application of Optical Strain Measurement Methods

Tensile testing of common industrial materials is often well-documented and protocols regarding dimensions of the testing specimen, mounting and loading conditions are well-defined. These dimensions and proportions are available in various standards provided by organizations such as ISO (International Organization of Standardization) or ASTM (American Society for Testing and Materials). Standards often ask for certain shapes of the specimen such as a so-called "dogbone" geometry or rectangular specimens of certain aspect ratios. The dogbone geometry incorporates a gauge length of a reduced cross-section in the middle part of the specimen and ensures that strains are mainly concentrated and uniform in that region rather than near the clamps. The specimen shape, boundary conditions, and loadings prescribed in these standards ensure accurate, and importantly, reproducible results. In plants, however, the delicate nature of the samples, and the minute dimensions in many cases preclude production of such sample geometries. While some standards exist for tensile testing of thin papers and cardboards such as BS EN ISO 1924, relatively large dimensions, in the order of a few centimeters, are required that do not often correspond well with dimensions of plant tissues (Hervy et al. 2017). Further, if strips need to be prepared, cutting the samples needs to be accurate without tears and flaws at edges that would cause stress concentration, premature failure and influence the mechanical parameters obtained from the test. All samples to be compared should be of same dimensions or at least the gauge length undergoing tension needs to be of same dimensions (Carew et al. 2003; Tsuchiya et al. 1998).

Due to lack of standardized protocols, a great variety of sample shapes (e.g., aspect ratios), often arbitrarily chosen, and even more, variable testing conditions have been used for plant materials. Several studies have demonstrated the sensitivity of tensile testing results to sample dimensions, aspect ratio and geometry (Anssari-Benam et al. 2012; Carew et al. 2003; Hervy et al. 2017). Further, due to errors introduced when measuring the strain based on the motion of the displacement transducer or the crosshead of the tensile device, different sample shapes can result in considerable differences in the calculated tensile modulus even within otherwise standardized samples. Hervy et al. demonstrated that measurement of stiffness of cellulose nanopapers shows a considerable dependence on sample geometry (Hervy et al. 2017). The authors demonstrated that using the optical measurement of strain instead of the opening of the tensile device crosshead reduces the geometry dependence in the estimated mechanical properties. This suggests that size dependence in tensile testing is, at least in part, related to strain measurement (e.g., slippage). The tensile strength defined at failure theoretically does not depend on the specimen size. However, in practice, it has been observed that an increase in sample width can result in a decrease in tensile strength. This can be explained by the weakest link theory. As the width of the specimen increases so does the likelihood of inclusion of weak spots or defects in the sample (Hervy et al. 2017). To add icing on the cake, the cellular nature of the plant tissues, as mentioned previously, and material anisotropy in individual cells, make the strain fields of plant specimens susceptible to inhomogeneity. As a result, the values reported by the force–displacement sensors may not be able to explain directly the phenomena occurring during a tensile test and comparing the mechanical properties of the biological specimens harvested at different stages of development or from tissues featuring intrinsically different cellular arrangement may be problematic. To fully address this concern, noncontact full strain field measurements can be used to acquire local information on the deformation of the sample. Optical extensiometry methods provide an attractive tool for the investigation of deformation within anisotropic and discontinuous materials such as intragranular and boundary movements in granular materials (Hall et al. 2010). The displacement field in subregions of the specimen can be acquired by optical imaging and combined to construct the local strain information. The strain field data can be used to determine the mechanical parameters of the material using approaches such as virtual fields (Grédiac et al. 2002; Promma et al. 2009) or inverse FE method. Through comparing full-field displacement data with the predictions made by the FE model, material parameters can be reverse engineered. An FE model of the tensile test can be developed with inputs of the specimen dimensions (length, width, and thickness) and force recorded from the sensors. The FE-produced displacement field is then compared with the displacement field of the landmarks or fiducial markers. The objective can be defined as the minimization of an error function based on the difference between the in silico and experimentally obtained values which results in the identification of the optimized parameters.

Optical extensiometry techniques are broad in their possible configurations as mentioned for DIC or interferometry approaches (Sutton and Reu 2017). In the simplest form, they can be carried out by monitoring a few fiducial points on the specimen (such as two pairs of points in axial and transverse directions by Hervy et al. 2017; or four symmetrically placed points in the middle of the square biaxial test specimen by Lally et al. 2004). To increase the number of data points, fiducial markers such as fluorescent beads can be sprayed or fixed on the surface, ideally with a higher density close to the middle and ends of the specimen. The tensile testing device can be placed under a stereo- or confocal microscope to record the displacement of fiducial markers versus the force exerted at two ends of the specimen. Kim et al. (2015) used fluorescent nanobeads as fiducial markers to calculate the microscale deformation of onion epidermis under tension. Application of optical methods in displacement/strain measurements is not limited to tensile testing. For instance, Armour et al. (2015) used fluorescent fiducial markers on the leaf epidermis of *Arabidopsis* to perform a time-lapse study of growth in epidermal pavement cells or Kuchen et al. (2012) used cell corners as landmarks to assess the local growth behavior of leaves.

Acknowledgements Research in the Geitmann lab is funded by Discovery and Accelerator Grants from the National Science and Engineering Research Council of Canada and by the Canada Research Chair Program.

References

Anderson CT, Carroll A, Akhmetova L, Somerville C (2010) Real-time imaging of cellulose reorientation during cell wall expansion in *Arabidopsis* roots. Plant Physiol 152:787–796

Anssari-Benam A, Legerlotz K, Bader DL, Screen HR (2012) On the specimen length dependency of tensile mechanical properties in soft tissues: gripping effects and the characteristic decay length. J Biomech 45:2481–2482

Argatov I, Mishuris G (2015) Contact mechanics of articular cartilage layers. In: Asymptotic models. Springer

Armour WJ, Barton DA, Law AM, Overall RL (2015) Differential growth in periclinal and anticlinal walls during lobe formation in *Arabidopsis* cotyledon pavement cells. Plant Cell 27:2484–2500

Baskin TI (2005) Anisotropic expansion of the plant cell wall. Annu Rev Cell Dev Biol 21:203–222

Bernal M, Urban MW, Rosario D, Aquino W, Greenleaf JF (2011) Measurement of biaxial mechanical properties of soft tubes and arteries using piezoelectric elements and sonometry. Phys Med Biol 56:3371

Bidhendi AJ, Geitmann A (2016) Relating the mechanics of the primary plant cell wall to morphogenesis. J Exp Bot 67:449–461

Bidhendi AJ, Geitmann A (2018) Finite element modeling of shape changes in plant cells. Plant Physiol 176:41–56

Boitier G, Chermant J, Vicens J (2000) Understanding the creep behavior of a 2.5 DC f–SiC composite: II. Experimental specifications and macroscopic mechanical creep responses. Mater Sci Eng, A 289:265–275

Bolduc J-F, Lewis LJ, Aubin C-É, Geitmann A (2006) Finite-element analysis of geometrical factors in micro-indentation of pollen tubes. Biomech Model Mechanobiol 5:227–236

Braybrook SA, Peaucelle A (2013) Mechano-chemical aspects of organ formation in Arabidopsis thaliana: the relationship between auxin and pectin. PLoS ONE 8:e57813

Carew E, Patel J, Garg A, Houghtaling P, Blackstone E, Vesely I (2003) Effect of specimen size and aspect ratio on the tensile properties of porcine aortic valve tissues. Ann Biomed Eng 31:526–535

Chanliaud E, Burrows KM, Jeronimidis G, Gidley MJ (2002) Mechanical properties of primary plant cell wall analogues. Planta 215:989–996

Chebli Y, Kaneda M, Zerzour R, Geitmann A (2012) The cell wall of the *Arabidopsis* pollen tube—spatial distribution, recycling, and network formation of polysaccharides. Plant Physiol 160:1940–1955

Cheng S, Clarke EC, Bilston LE (2009) The effects of preconditioning strain on measured tissue properties. J Biomech 42:1360–1362

Chimungu JG, Loades KW, Lynch JP (2015) Root anatomical phenes predict root penetration ability and biomechanical properties in maize (Zea Mays). J Exp Bot 66:3151–3162

Cooke JR, De Baerdemaeker JG, Rand RH, Mang HA (1976) A finite element shell analysis of guard cell deformations. Trans ASAE 19:1107–1121

Cosgrove DJ (1998) Cell wall loosening by expansins. Plant Physiol 118:333–339

Cosgrove DJ (2005) Growth of the plant cell wall. Nat Rev Mol Cell Biol 6:850–861

Derbyshire P, Findlay K, McCann MC, Roberts K (2007) Cell elongation in Arabidopsis hypocotyls involves dynamic changes in cell wall thickness. J Exp Bot 58:2079–2089

Durachko DM, Cosgrove DJ (2009) Measuring plant cell wall extension (creep) induced by acidic pH and by alpha-expansin. J Vis Exp JoVE

Eder M, Arnould O, Dunlop JW, Hornatowska J, Salmén L (2013) Experimental micromechanical characterisation of wood cell walls. Wood Sci Technol 47:163–182
Fabry B, Maksym GN, Butler JP, Glogauer M, Navajas D, Fredberg JJ (2001) Scaling the microrheology of living cells. Phys Rev Lett 87:148102
Forterre Y, Skotheim JM, Dumais J, Mahadevan L (2005) How the Venus flytrap snaps. Nature 433:421–425
Freundthal AM (1968) Statistical approach to brittle fracture. In: Lievowitz H (ed) Fracture: an advanced treatise, vol 2. Academic Press, New York, pp 591–619
Gadalla A, Dehoux T, Audoin B (2014) Transverse mechanical properties of cell walls of single living plant cells probed by laser-generated acoustic waves. Planta 239:1129–1137
Gallenmüller F, Feus A, Fiedler K, Speck T (2015) Rose prickles and Asparagus spines–different hook structures as attachment devices in climbing plants. PLoS ONE 10:e0143850
Grédiac M, Toussaint E, Pierron F (2002) Special virtual fields for the direct determination of material parameters with the virtual fields method. 1—Principle and definition. Int J Solids Struct 39:2691–2705
Green PB (1960) Multinet growth in the cell wall of Nitella. J Cell Biol 7:289–296
Guilak F, Butler DL, Goldstein SA, Baaijens FP (2014) Biomechanics and mechanobiology in functional tissue engineering. J Biomech 47:1933–1940
Hall SA, Muir Wood D, Ibraim E, Viggiani G (2010) Localised deformation patterning in 2D granular materials revealed by digital image correlation. Granul Matter 12:1–14
Hannon A, Tiernan P (2008) A review of planar biaxial tensile test systems for sheet metal. J Mater Process Technol 198:1–13
Hervy M, Santmarti A, Lahtinen P, Tammelin T, Lee K-Y (2017) Sample geometry dependency on the measured tensile properties of cellulose nanopapers. Mater Des 121:421–429
Hua T, Xie H, Pan B, Qing X, Dai F, Feng X (2007) A new micro-tensile system for measuring the mechanical properties of low-dimensional materials—Fibers and films. Polym Test 26:513–518
Kafle K, Park YB, Lee CM, Stapleton JJ, Kiemle SN, Cosgrove DJ, Kim SH (2017) Effects of mechanical stretching on average orientation of cellulose and pectin in onion epidermis cell wall: a polarized FT-IR study. Cellulose 1–10
Kalidindi S, Abusafieh A, El-Danaf E (1997) Accurate characterization of machine compliance for simple compression testing. Exp Mech 37:210–215
Kashfuddoja M, Ramji M (2013) Whole-field strain analysis and damage assessment of adhesively bonded patch repair of CFRP laminates using 3D-DIC and FEA. Compos B Eng 53:46–61
Keckes J, Burgert I, Frühmann K, Müller M, Kölln K, Hamilton M, Burghammer M, Roth SV, Stanzl-Tschegg S, Fratzl P (2003) Cell-wall recovery after irreversible deformation of wood. Nat Mater 2:810–813
Kerstens S, Decraemer WF, Verbelen J-P (2001) Cell walls at the plant surface behave mechanically like fiber-reinforced composite materials. Plant Physiol 127:381–385
Kim J-H, Nizami A, Hwangbo Y, Jang B, Lee H-J, Woo C-S, Hyun S, Kim T-S (2013) Tensile testing of ultra-thin films on water surface. Nat Commun 4
Kim K, Yi H, Zamil MS, Haque MA, Puri VM (2015) Multiscale stress–strain characterization of onion outer epidermal tissue in wet and dry states. Am J Bot 102:12–20
Kollmannsberger P, Fabry B (2011) Linear and nonlinear rheology of living cells. Annu Rev Mater Res 41:75–97
Kuchen EE, Fox S, de Reuille PB, Kennaway R, Bensmihen S, Avondo J, Calder GM, Southam P, Robinson S, Bangham A (2012) Generation of leaf shape through early patterns of growth and tissue polarity. Science 335:1092–1096
Lally C, Reid A, Prendergast P (2004) Elastic behavior of porcine coronary artery tissue under uniaxial and equibiaxial tension. Ann Biomed Eng 32:1355–1364
Lava P, Cooreman S, Coppieters S, De Strycker M, Debruyne D (2009) Assessment of measuring errors in DIC using deformation fields generated by plastic FEA. Opt Lasers Eng 47:747–753

Lee JM, Courtman DW, Boughner DR (1984) The glutaraldehyde-stabilized porcine aortic valve xenograft. I. Tensile viscoelastic properties of the fresh leaflet material. J Biomed Mater Res Part A 18:61–77

Lynch TM, Lintilhac PM (1997) Mechanical signals in plant development: a new method for single cell studies. Dev Biol 181:246–256

Machado G, Favier D, Chagnon G (2012) Membrane curvatures and stress-strain full fields of axisymmetric bulge tests from 3D-DIC measurements. Theory and validation on virtual and experimental results. Exp Mech 52:865–880

Maier-Schneider D, Maibach J, Obermeier E (1995) A new analytical solution for the load-deflection of square membranes. J Microelectromech Syst 4:238–241

Milani P, Braybrook SA, Boudaoud A (2013) Shrinking the hammer: micromechanical approaches to morphogenesis. J Exp Bot 64:4651–4662

Miller KS, Edelstein L, Connizzo BK, Soslowsky LJ (2012) Effect of preconditioning and stress relaxation on local collagen fiber re-alignment: inhomogeneous properties of rat supraspinatus tendon. J Biomech Eng 134:031007

Mir M, Ali MN, Sami J, Ansari U (2014) Review of mechanics and applications of auxetic structures. Adv Mate Sci Eng

Mofrad MR (2009) Rheology of the cytoskeleton. Annu Rev Fluid Mech 41:433–453

Motra H, Hildebrand J, Dimmig-Osburg A (2014) Assessment of strain measurement techniques to characterise mechanical properties of structural steel. Eng Sci Technol Int J 17:260–269

Neggers J, Hoefnagels J, Hild F, Roux S, Geers M (2014) Direct stress-strain measurements from bulged membranes using topography image correlation. Exp Mech 54:717–727

Norris A (2006) Extreme values of Poisson's ratio and other engineering moduli in anisotropic materials. J Mech Mater Struct 1:793–812

Nouira H, Salgado J, El-Hayek N, Ducourtieux S, Delvallée A, Anwer N (2014) Setup of a high-precision profilometer and comparison of tactile and optical measurements of standards. Meas Sci Technol 25:044016

Orthner M, Rieth L, Solzbacher F (2010) High speed wafer scale bulge testing for the determination of thin film mechanical properties. Rev Sci Instrum 81:055111

Palin R, Geitmann A (2012) The role of pectin in plant morphogenesis. Biosystems 109:397–402

Pan B, Qian K, Xie H, Asundi A (2009) Two-dimensional digital image correlation for in-plane displacement and strain measurement: a review. Meas Sci Technol 20:062001

Pang C, Lee G-Y, T-i Kim, Kim SM, Kim HN, Ahn S-H, Suh K-Y (2012) A flexible and highly sensitive strain-gauge sensor using reversible interlocking of nanofibres. Nat Mater 11:795–801

Peaucelle A, Wightman R, Höfte H (2015) The control of growth symmetry breaking in the Arabidopsis hypocotyl. Curr Biol 25:1746–1752

Phyo P, Wang T, Kiemle SN, O'Neill H, Pingali SV, Hong M, Cosgrove DJ (2017) Gradients in wall mechanics and polysaccharides along growing inflorescence stems. Plant Physiol 175:1593–1607

Pieczywek PM, Zdunek A (2014) Finite element modelling of the mechanical behaviour of onion epidermis with incorporation of nonlinear properties of cell walls and real tissue geometry. J Food Eng 123:50–59

Prasanna V, Prabha T, Tharanathan R (2007) Fruit ripening phenomena–an overview. Crit Rev Food Sci Nutr 47:1–19

Promma N, Raka B, Grediac M, Toussaint E, Le Cam J-B, Balandraud X, Hild F (2009) Application of the virtual fields method to mechanical characterization of elastomeric materials. Int J Solids Struct 46:698–715

Routier-Kierzkowska A-L, Weber A, Kochova P, Felekis D, Nelson BJ, Kuhlemeier C, Smith RS (2012) Cellular force microscopy for in vivo measurements of plant tissue mechanics. Plant Physiol 158:1514–1522

Sanders PG, Eastman J, Weertman J (1997) Elastic and tensile behavior of nanocrystalline copper and palladium. Acta Mater 45:4019–4025

Saxe F, Weichold S, Reinecke A, Lisec J, Döring A, Neumetzler L, Burgert I, Eder M (2016) Age effects on hypocotyl mechanics. PLoS ONE 11:e0167808

Schatzmann L, Brunner P, Stäubli H (1998) Effect of cyclic preconditioning on the tensile properties of human quadriceps tendons and patellar ligaments. Knee Surg Sports Traumatol Arthrosc 6:S56–S61

Shah DU, Schubel PJ, Clifford MJ, Licence P (2012) The tensile behavior of off-axis loaded plant fiber composites: An insight on the nonlinear stress–strain response. Polym Compos 33:1494–1504

Sharpe W, Pulskamp J, Gianola D, Eberl C, Polcawich R, Thompson R (2007) Strain measurements of silicon dioxide microspecimens by digital imaging processing. Exp Mech 47:649–658

Silva S, Sabino M, Fernandes E, Correlo V, Boesel L, Reis R (2005) Cork: properties, capabilities and applications. Int Mater Rev 50:345–365

Small MK, Daniels BJ, Clemens BM, Nix WD (1994) The elastic biaxial modulus of Ag–Pd multilayered thin films measured using the bulge test. J Mater Res 9:25–30

Soons J, Lava P, Debruyne D, Dirckx J (2012) Full-field optical deformation measurement in biomechanics: digital speckle pattern interferometry and 3D digital image correlation applied to bird beaks. J Mech Behav Biomed Mater 14:186–191

Spatz H, Kohler L, Niklas K (1999) Mechanical behaviour of plant tissues: composite materials or structures? J Exp Biol 202:3269–3272

Srikar V, Spearing S (2003) A critical review of microscale mechanical testing methods used in the design of microelectromechanical systems. Exp Mech 43:238–247

Sutton M, Reu PL (eds) (2017) international digital imaging correlation society. In: Proceedings of the first annual conference 2016. Springer International Publishing

Ting T (2004) Very large Poisson's ratio with a bounded transverse strain in anisotropic elastic materials. J Elast 77:163–176

Ting T, Chen T (2005) Poisson's ratio for anisotropic elastic materials can have no bounds. Q J Mech Appl Math 58:73–82

Tsuchiya T, Tabata O, Sakata J, Taga Y (1998) Specimen size effect on tensile strength of surface-micromachined polycrystalline silicon thin films. J Microelectromech Syst 7:106–113

Turek DE (1993) On the tensile testing of high modulus polymers and the compliance correction. Polym Eng Sci 33:328–333

Vanstreels E, Alamar M, Verlinden B, Enninghorst A, Loodts J, Tijskens E, Ramon H, Nicolaï B (2005) Micromechanical behaviour of onion epidermal tissue. Postharvest Biol Technol 37:163–173

Wei C, Lintilhac LS, Lintilhac PM (2006) Loss of stability, pH, and the anisotropic extensibility of Chara cell walls. Planta 223:1058–1067

Wei C, Lintilhac PM, Tanguay JJ (2001) An insight into cell elasticity and load-bearing ability. Measurement and theory. Plant Physiol 126:1129–1138

Yamaguchi I (1981) A laser-speckle strain gauge. J Phys E Sci Instrum 14:1270

Yang L, Ettemeyer A (2003) Strain measurement by three-dimensional electronic speckle pattern interferometry: potentials, limitations, and applications. Opt Eng 42:1257–1266

Yu Z, Xu H, Chen H, Pei Y, Fang D (2016) characterization method of thick films using the bulge test technique. Exp Mech 56

Zamil M, Geitmann A (2017) The middle lamella—more than a glue. Phys Biol 14:015004

Zamil MS, Yi H, Haque M, Puri VM (2013) Characterizing microscale biological samples under tensile loading: Stress–strain behavior of cell wall fragment of onion outer epidermis. Am J Bot 100:1105–1115

Zamil MS, Yi H, Puri VM (2017) A multiscale FEA framework for bridging cell-wall to tissue-scale mechanical properties: the contributions of middle lamella interface and cell shape. J Mater Sci 13:7947–7968

Zemánek M, Burša J, Děták M (2009) Biaxial tension tests with soft tissues of arterial wall. Eng Mech 16:3–11

Zerzour R, Kroeger J, Geitmann A (2009) Polar growth in pollen tubes is associated with spatially confined dynamic changes in cell mechanical properties. Dev Biol 334:437–446

Zhang T, Vavylonis D, Durachko DM, Cosgrove DJ (2017) Nanoscale movements of cellulose microfibrils in primary cell walls. Nat Plants 3:17056

Zhou P, Goodson KE (2001) Subpixel displacement and deformation gradient measurement using digital image/speckle correlation (DISC). Opt Eng 40:1613–1620

Part IV
Water Transport, Mechanosensing and Biomimetics

Water Motion and Sugar Translocation in Leaves

Tomas Bohr, Hanna Rademaker and Alexander Schulz

Abstract We give an overview of the current understanding of the coupled water— and sugar flows in plants with special emphasis on the leaves. We introduce the Münch mechanism and discuss the particularities of osmotically driven flow in the phloem and the consequences for the allometry of the vasculature. This is first done in the context of the entire tree, where we discuss the optimum radius for the phloem tubes, and later for a single needle, where we give a more detailed solution of the osmotic flow profile, allowing us to understand the constraints on needle sizes. We then discuss recent results from microscopy of cross sections along the midvein of a birch leaf, allowing us to measure how the number and radius of the sieve elements depend on the distance from the petiole and compare this to the available area and the minor vein endings in the entire leaf. We finally discuss the pre-phloem water flow in the leaf, i.e. the coupled water/sugar transport from the mesophyll via the bundle sheath into the sieve tubes. We review the distinct sugar loading mechanisms with special emphasis on active symplasmic loading ('polymer trapping'), where one needs to compute water and sugar flow through extremely narrow channels.

Introduction

The ability to provide water for all vital parts is crucial for the survival of a plant. The predominant solvent in all cells is water, which allows intescellular transport by diffusion or through membrane proteins. In addition, plants depend upon water for transpiration, for photosynthesis and for transport of photosynthates—mainly sugars. Since water is predominantly taken up in the roots and sugars are produced in the leaves, the plant has a great need of long-distance translocation of sap—water with solutes—and thus for a vascular system. The vascular system basically consists

T. Bohr (✉) · H. Rademaker
Department of Physics, Technical University of Denmark, 2800 Kgs, Lyngby, Denmark
e-mail: tomas.bohr@fysik.dtu.dk

A. Schulz
Department of Plant and Environmental Sciences, University of Copenhagen, Copenhagen, Denmark

A. Geitmann and J. Gril (eds.), *Plant Biomechanics*,
https://doi.org/10.1007/978-3-319-79099-2_16

of two parts, the *xylem* and the *phloem*, as sketched in Fig. 1, and although the driving mechanisms for the flow in these two strongly interconnected circuits are completely different, they both are activated by the leaves. See, e.g. Taiz and Zeiger (2010)) for a thorough introduction to the vascular system of plants and (Jensen et al. 2016) for a recent review of the physics of sap-translocation.

The fact that plants, through transpiration from the leaves, reemit most of their precious water uptake is quite surprising, but the acquisition of CO_2, which occurs by diffusion from the surrounding air through the stomata in the leaf surface, inevitably causes great water loss due to the small concentration of CO_2 in the air and the large water concentration (essentially 100% humidity) in the leaf. This creates a need for large flows from the roots up to—and through—the leaves, which takes place in the tracheids and vessels of the xylem. This flow is believed to be mainly driven by suction in the leaves—a surprisingly strong suction. The direct measurement of pressures inside leaf cells is extremely difficult, due to the seemingly unavoidable issues of leakage at such strong suction rates; but results from pressure bombs (Taiz and Zeiger 2010) indicate that leaves are able to create extremely low pressures (or indeed large suctions, since these pressures are well below zero) down to, say, -80 bar (-8 MPa). Both the generation of these large suction rates and the dependability

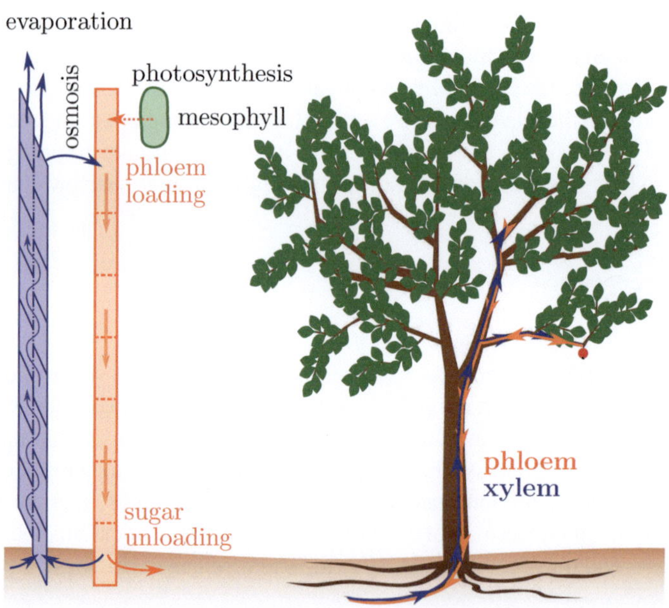

Fig. 1 The vascular system of a plant consists of two parts—the *xylem* transports water from the roots up to the leaves, where most of the water is evaporated. A fraction of the delivered water is used in photosynthesis inside the green mesophyll cells. Another portion is used to drive the flow in the other part of the vascular system—the *phloem*, which distributes the sugars produced in photosynthesis to other parts of the plant, e.g. roots, growing shoots and fruits

of such obviously unstable flows are not well understood. The mechanism for the generation of negative pressures is usually explained in terms of surface tension, i.e. capillary rise in extremely narrow (nanometric) pores between the cell walls of the mesophyll cells (Taiz and Zeiger 2010). It is probably more likely that it is the gelatinous polymers, e.g. the pectins, which provide the water suction, and thus that the cell wall structure is of paramount importance, as it is for growth of the leaves (see, e.g. Jensen et al. 2016 for a discussion). For the stability of the flow in the xylem, the organization of the tubes (vessels and tracheids) through which it travels is very important. These tubes are laterally connected and the flow moves between them through the so-called pits—valves that close if an embolism (an air bubble) occurs, thereby preventing the air from spreading.

The other part of the vascular system, the phloem, is quite different on almost all counts. The flow volume in the sieve tubes of the phloem is an order of magnitude smaller, it contains photosynthates in concentrations up to about 1 M, and the sap flows under positive pressure, generated by osmosis, stealing the water out of the xylem. The so-called Münch hypothesis explains the flow as caused by the loading of sugars into the sieve tubes in the minor veins in mature leaves, which generates a large osmotic pressure. Similarly, sugars are released from the sieve tubes in the roots, whereby the osmotic pressure becomes lower. The result is a pressure difference of up to around 10 bars (1 MPa) driving the flow, corresponding to a concentration difference of around 0.5 M. The phloem tubes in the stem of a tree of height, say, $H = 50\,\mathrm{m}$ have a typical radius $a \approx 20\,\mu\mathrm{m}$ and the sap viscosity η is about twice that of water, i.e. $\eta = 2 \times 10^{-3}\,\mathrm{Pa\,s}$. With these values, we can estimate a typical flow velocity in a cylindrical phloem (sieve) tube with stationary Poiseuille flow:

$$u \approx \frac{a^2}{8\eta H}\Delta p \approx 5 \times 10^{-4}\ \mathrm{m/s} \approx 180\ \mathrm{cm/h}, \tag{1}$$

which corresponds reasonably well to measured flows, being on the high side. In fact, the sieve plates that separate the individual sieve 'elements' (or sieve cells) contribute a friction which effectively doubles the viscosity (Jensen et al. 2012c). Also, there are significant differences between angiosperms (broad-leaved trees) and gymnosperms (conifers), whose flows are slower (Liesche et al. 2015). So although it seems counter-intuitive that osmotic pressure differences between leaves and roots in a 50 m high tree can generate the measured flows, actual measurements of flow rates, concentrations and the diameters of the sieve tubes indicate that this is indeed feasible (Jensen et al. 2016). In particular, there is an indirect, but quite strong, argument coming from the allometry of the phloem vascular system (Jensen et al. 2011). Indeed, the osmotic flows generated in the leaves would predict a specific scaling of the typical sieve tubes radius a with the height H of the stem and the length l of the leaves. One has to remember that the flow is generated in the leaves, by osmotic water uptake across the surface of the sieve tubes, whereas the resistance to the flow mainly comes from the stem as in (1). Modelling the phloem as long tubes going from leaf to root without change of radius, one can ask what the optimal radius a^* would be, that is, the radius which makes the flow velocity largest (Jensen et al.

2016). If the radius is very small, the osmotic 'pump' in the leaf is very efficient, since the surface to volume ratio of the tube is large; but the resistance through the stem also becomes large since the tube is narrow. On the other hand, if the radius is large, the resistance through the stem is small, but now the osmotic pump is less efficient. In conclusion, we might expect an optimum somewhere in between.

To investigate this quantitatively, we can, following (Jensen et al. 2012b, 2016), represent the different elements of the tube as resistors, yielding a certain velocity, when a certain pressure difference is applied. For the stem, the relation would be given by (1), with Δp representing the pressure drop Δp_s from top to bottom of the stem, but for the leaves, the relation would be very different: The flux would be proportional to the *surface area* $2\pi al$. Dividing by the tube cross section πa^2, the velocity would be

$$u \approx \frac{2l}{a} L_p \Delta\pi_l, \tag{2}$$

where L_p is the membrane conductance or permeability (pr. unit area), and $\Delta\pi_l$ the osmotic pressure difference between the inside and the outside of the tube, namely $RT\Delta c$, where Δc is the (sugar) concentration difference across the tube membrane. Since the velocities in the leaf and the root should be the same, occurring in the same tube, the pressures are related by

$$\frac{\Delta\pi_l}{\Delta p_s} \approx \frac{a^3}{16\eta L_p l H} \tag{3}$$

and, as one can easily verify, the largest velocity is achieved when this ratio is 2 which determines an optimal tube radius

$$a^* = (2H\, l\, l_{\mathrm{m}})^{1/3} \tag{4}$$

i.e. the cube root of the product of three lengths: the height of the tree, the length of the leaves and a 'membrane length' $l_{\mathrm{m}} = 16\eta L_p$. In Fig. 2a, we show this prediction compared with data from field measurements on a broad range of plants—with heights from 10 cm to 50 m, and as one can see, they compare reasonably well. Here, a geometrical factor G is included (Jensen et al. 2012b) taking into account the non-cylindrical cross section of some sieve tubes (especially in conifers), and a is thus also an effective radius, $a = \sqrt{A/\pi}$.

One might speculate why the roots were not included in the analysis, when, in fact, they play a role very similar to the leaves, just unloading sugar and water instead of letting them in. Thus, the length of the roots should also appear in (4). As one can easily verify (Jensen et al. 2016), the inclusion of a root section with length l_r would simply change the l appearing in (4) to $(l^{-1} + l_r^{-1})^{-1}$ and the result we gave above was actually valid for the case $l \ll l_r$, which is the case for most trees. One can also note the very small value of the 'membrane length'l_{m}. For typical sieve tube membranes, the permeabilities are very small and $\eta L_p \approx 10^{-16}$ m. One should, however, not think of this as a length. If we think of the pores in the membranes as narrow channels of radius a_p, length (thickness of the membrane) d and covering fraction ϕ, we would find

$$\eta L_p \sim \frac{\phi a_p^2}{d} \tag{5}$$

and with $d \approx 5\,\text{nm}$, $\phi \approx 10^{-4}$ we find such values for ηL_p, when a_p is of the order of an Ångström, which corresponds to the size of the aquaporins that transport the water. As a final comment of the scaling relation, one might note that it is the *velocity* of the phloem, not the *flux*, which is maximized. One can easily verify that the fluxes, obtained by multiplying the velocities in (1) and (2) by the cross section πa^2 would not have a maximum for any finite a, but would increase without bounds with a. If, as is more realistic, we fix the total cross-sectional area available for transport, we do get maximal flux for the a^* predicted above.

Another interesting size scaling relates the radius of the sieve tubes to the height of the tree. As shown in Fig. 2b, this radius, or, more precisely the cross-sectional area A which is preferred since the tubes are not all cylindrical, increases with height for small plants, but at a height of around 10 m it saturates at a value corresponding to a tube radius of around 20 μm. The reason for this is currently unknown, but since each section of the sieve tube (the sieve tube element) is a single cell, there might be good architectural reasons for not increasing its size.

An interesting corollary of the size scalings of Fig. 2a, b is brought out in Jensen and Zwieniecki (2013), where it is shown that large trees have small leaves. Indeed, if the sieve tubes of large trees are fixed at $a \approx 20\,\mu\text{m}$, and these radii are also optimal as in (4), the leaf length l and the tree height must be linked as $l \sim 1/H$. There could of course be other reasons why tall trees have small leaves, e.g. the increased wind

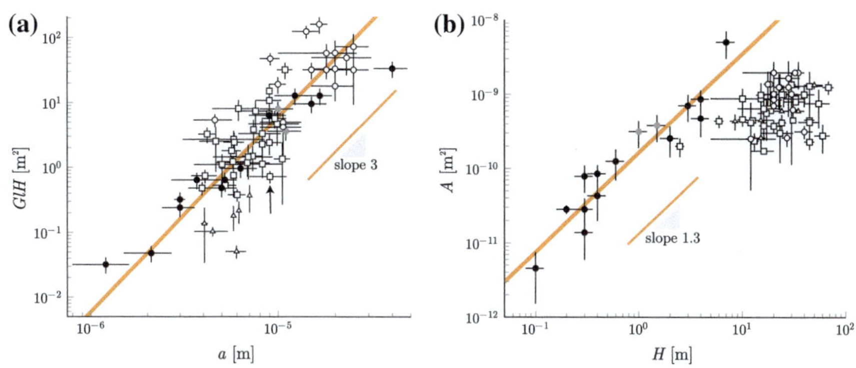

Fig. 2 **a** The optimal sieve tube radius (a^*) versus the values of the sieve tube radii from field measurements in mature stems of angiosperm herbs (solid circles), angiosperm trees (open circles), angiosperms shrubs (grey circles), gymnosperm trees (open squares) and gymnosperm trees with scales (open triangles). G is geometrical factor taking into account the non-cylindrical cross section of some sieve tubes (Jensen et al. 2012b). **b** Sieve tube cross-sectional area $A \approx \pi a^2$ as function of the height H of the tree. From Jensen et al. (2012b)

shear at higher altitudes. The more conventional arguments describing the mechanical constraints on leaf sizes are given in the classical text by Niklas (1994).

The Münch hypothesis has recently been tested directly on living plants (Knoblauch et al. 2016). Measuring the pressure directly in sieve tubes of the vine 'Morning Glory', it was verified that the pressure differential between the lowest leaves and the roots was sufficient to drive the flow, and that it increased for older plants, where these leaves were at a larger height. Interestingly enough this pressure differential does not grow linearly with height, but the plant compensates by the increased conductivity of the growing sieve tubes.

The scaling described above (represented by (4)) is based on the simplistic model of a tree as a collection of uniform tubes, which of course is very unrealistic. The radii for the sieve tubes of trees appearing in Fig. 2 were measured on mature plants, typically at a height of 1.3 m. In leaves, the sieve tube radii are actually considerably smaller, and in the veins of broad leaves, with a hierarchical venation pattern, the number of sieve tubes increases from order to order while their size decreases (Carvalho et al. 2017a, b). How, or whether, this can lead to a resistance proportional to the leaf length l as used in the simple model, is currently not understood.

The efficiency of the osmotically driven flows in the phloem depends on the leaves' ability to load sugars into the sieve elements of the minor veins and transport it away rapidly. For this, the vein architecture as well as the 'loading strategy' is of major importance. In the following, we shall review the current understanding of the flow of sugar in the leaf, from the mesophyll cells to the petiole. We shall start with the flow in the veins, and discuss two cases: first the flow in a leaf with a linear architecture—a conifer needle—and second, the architecture and flow of a birch leaf. Then, we shall discuss the pre-phloem pathway—from the mesophyll, through the bundle sheath and into the sieve tubes—with particular emphasis on the role of the water.

The Simplest Osmotic Flows with Conifer Needles as Example

We shall start with a somewhat more detailed picture of an osmotic flow. Instead of trying to model the entire tree, we shall focus only on the leaf-part, and only do that in the simplest possible configuration, which is a linear leaf like a conifer needle. Typically, there are one or two vascular bundles at the centre of the needle cross section, running from tip to petiole (Fig. 3). There is no hierarchical branching structure as in many broad leaves (see section "The Architecture of Broad Leaves"). Ronellenfitsch et al. (2015) measured the numbers and radii of sieve elements in needles of four conifer species. For their study, they chose needles of very different lengths, ranging from 1 to 35 cm. In all four species, they found the radius of the sieve elements to be almost constant from tip to base. They further found the number of sieve elements to increase from tip to base in a square-root shape, where most of the sieve elements start already close to the tip of the needle. Based on these findings,

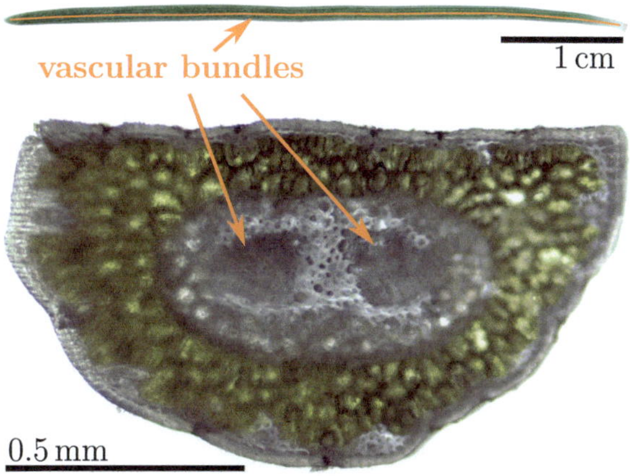

Fig. 3 Linear, unbranched venation in needles. Outside view and cross section of a conifer needle with two vascular bundles. The vascular bundles, and the conducting sieve tubes within, are running all the way from the tip to the base of the needle. The green cells surrounding the vascular bundles are sugar-producing mesophyll cells. From Rademaker et al. (2017)

we model a conducting tube as a single, unbranched conduit of the same length as the needle. We shall show that this way of modelling will allow us to understand why needles are quite restricted in length compared to broad leaves.

As discussed above, osmotic flows are driven by the concentration differences of solutes (mostly sugars) across the semipermeable membranes of the tubes. We shall assume that the plant is able to uphold a constant concentration difference between the inside and the outside of the tube—using one of the sugar loading mechanisms described in section "Water Flow Inside the Leaf". As a consequence, water is drawn in from the surroundings and creates a net axial flow as indicated in Fig. 4. In this case, the direction of the flow is from left to right, since we are assuming an impenetrable wall at $x = 0$, corresponding to the tip of the needle. As explained in detail in Jensen et al. (2016), such flows are well described by the so-called Münch-Horwitz equations, in terms of only the mean axial velocity u, the mean concentration c and pressure p. For *stationary flows*, to which we shall restrict our attention in the following, the Münch-Horwitz equations are: first, the osmotic water uptake, for a tube of fixed radius a and membrane permeability L_p, giving the incremental change in fluid flux $Q(x)$ along the tube in terms of the water potential difference $\Delta\Psi$ across the membrane surface as

$$dQ = L_p dA \Delta\Psi = L_p dA(RTc(x) - p(x)), \tag{6}$$

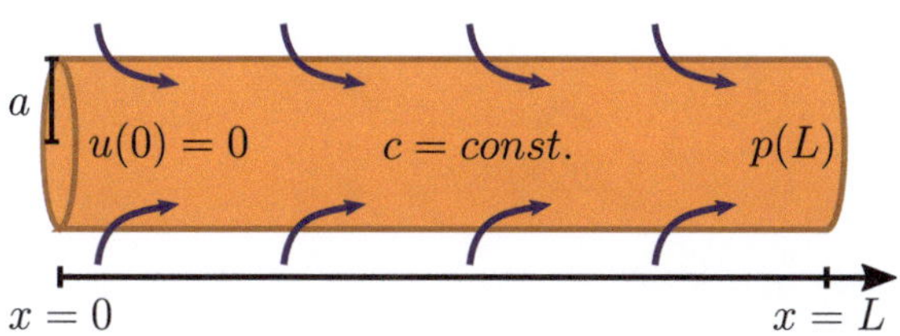

Fig. 4 The simplest osmotic flow. In the context of a conifer needle, $x = 0$, where the flow begins, is the tip and the outlet $x = L$, where the pressure prescribed would be the base

where $c(x)$ and $p(x)$ are *differences* between the concentrations and pressures inside and outside the tube and $dA = 2\pi a dx$ is the differential surface area. This means that the axial velocity satisfies

$$\frac{du}{dx} = \frac{1}{\pi a^2}\frac{dQ}{dx} = \frac{2L_p}{a}(RTc(x) - p(x)), \tag{7}$$

The second equation is Darcy's law (for a Hagen–Poiseuille flow), relating the pressure gradient to the velocity through the viscosity η of the fluid

$$\frac{dp}{dx} = -\frac{8\eta}{a^2}u(x). \tag{8}$$

These equations should be complemented by an equation for the sugar loading, the reaction–diffusion equation

$$\frac{d(uc)}{dx} = D\frac{d^2c}{dx^2} + \Upsilon(x), \tag{9}$$

defining the loading function $\Upsilon(x)$. In the following, we shall, however, following (Rademaker et al. 2016), assume that the loading is able to keep the concentration $c(x)$ constant $= c_0$ throughout the tube. This does not seem far from the situation in many plants and it is close to the situation obtained from 'target concentration' models (Jensen et al. 2012a). It can further be shown (Rademaker et al. 2016) that this choice of $c(x) = c_0$ is *optimal*, in the sense that no concentration profile, limited everywhere by the value c_0, can generate larger flows. If we are interested in limits to sap flow efficiency this simplification is thus relevant. Differentiating (7) and inserting (8) we find

$$\frac{d^2u}{dx^2} = -\frac{2L_p}{a}\frac{dp}{dx} = \frac{16\eta L_p}{a^3}u(x) \tag{10}$$

and introducing the characteristic length

$$L_{\text{eff}} = \frac{a^{3/2}}{(16L_p\eta)^{1/2}} = \frac{a^{3/2}}{l_{\text{m}}^{1/2}}, \tag{11}$$

we can write (10) as

$$\frac{d^2u}{dx^2} = \frac{1}{L_{\text{eff}}^2}u(x) \tag{12}$$

which indicates an exponential behaviour as $e^{\pm x/L_{\text{eff}}}$. In fact, when $u(x=0) = 0$, the solution is

$$u(x) = A\sinh\left(x/L_{\text{eff}}\right). \tag{13}$$

The fact that there exists a characteristic scale L_{eff} different from the length L of the tube is very important. Their ratio can be expressed in terms of the dimensionless Münch number (Jensen et al. 2016)—the ratio of axial resistance in the tube and radial resistance through the surface.

$$M = \frac{16\eta L_p L^2}{a^3} = \frac{L^2}{L_{\text{eff}}^2}. \tag{14}$$

That the length scale L_{eff} plays a special role was first pointed out by Landsberg and Fowkes (1978) in the context of water uptake in the roots. In this case, the concentration is practically zero, and thus, the roots are dragging in water, mostly through suction generated by the negative pressures in the leaves. Since the concentration c_0 does not appear in (12), the same equation holds.

The flow profile along a tube, which is formed by chains of sieve cells in gymnosperm needles, strongly depends on $L_{\text{eff}}/L = 1/\sqrt{M}$. In Fig. 5, we show the velocity profile along the tube for three values of L_{eff}/L. For tube lengths not much larger than L_{eff}, the velocity grows linearly along the tube, but for long tubes (where L_{eff}/L is small), it becomes strongly non-linear. In fact, it almost vanishes near the tip and only reaches appreciable values in a region of the order of L_{eff} near the base. For a pine needle, a large value of M would thus not be very good. As shown in the lower curve in Fig. 5, a large region of the tube near the tip ($x = 0$) would be stagnant, and the sugar produced there would never get to the outlet.

The flow at the outlet $x = L$ (the base of the needle) is similarly found as Rademaker et al. (2016)

$$u(L) = \frac{1}{2}\sqrt{\frac{rL_p}{\eta}}\,(RTc_0 - p(L))\tanh\frac{L}{L_{\text{eff}}}, \tag{15}$$

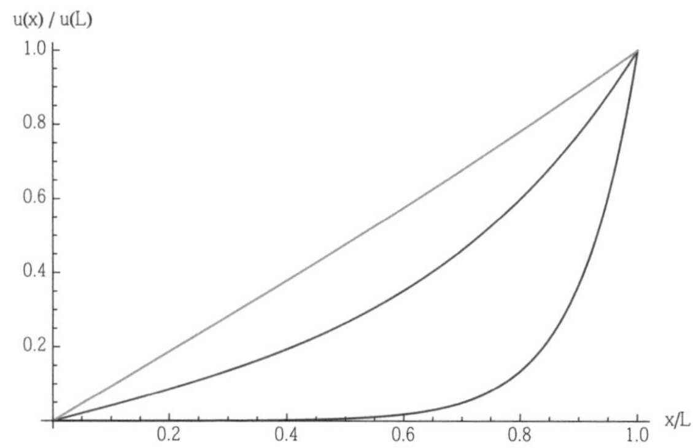

Fig. 5 The velocity profile along the tube formed by continuous sieve cells in a needle for three values of L_{eff}/L: 1.4 (top, yellow), 0.4 (middle, red) and 0.1 (bottom, blue). The tip of the needle would be towards the left at $x = 0$ and the base to the right at $x = L$. For large Münch numbers M (bottom curve), a zone of stagnant fluid forms at the tip of the needle

where $p(L)$ is the specified pressure at the outlet. For long tubes (i.e. large M), this approaches

$$u(L) = \frac{1}{2}\sqrt{\frac{rL_p}{\eta}}\,(RTc_0 - p(L)) \tag{16}$$

which is *independent* of L.

This result does of course depend on the assumption that the sugar concentration is constant in the tube. If the conifer could increase the concentration near the needle tip, larger pressures could build up and the sap would start to flow. But there must be a limit to the concentration a given species is able to build up. If we call that concentration c_0, then one can show (Rademaker et al. 2017) that the largest exit velocity is generated for the situation, where $c(x) = c_0$ along the whole tube. So, e.g. for the lowest curve (with $L_{\text{eff}}/L = 0.1$) the exit velocity is limited by (16) even if parts of the tube have a lower concentration. The exit velocity will thus not increase with L. This would mean that a needle would not gain much in sugar transport by making its needles longer than L_{eff}, so if no other concerns (e.g. competition for sun or elastic properties) are equally important this length should limit needle sizes. In fact, as shown in Fig. 6, the distribution of needle lengths seems roughly exponential with 75% of needles no longer than 6 cm in accordance with the value of L_{eff}. There could of course be other reasons, why long needles would not be optimal, e.g. their ability for self-support (Tadrist and Darbois-Texier 2016). Also, it would be very interesting to know how the 25% of species with longer needles have dealt with the flow limitations described above.

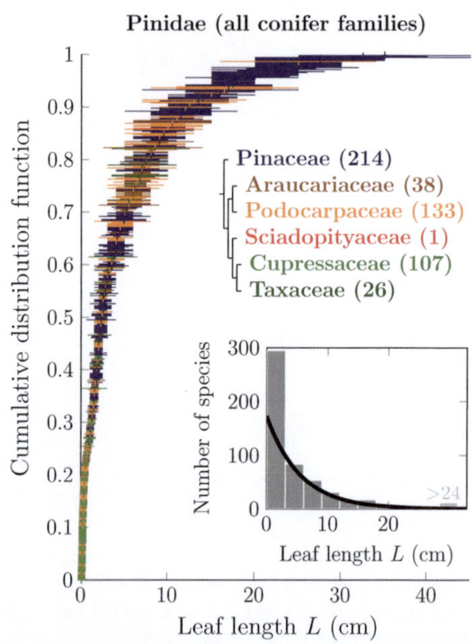

Fig. 6 The distribution of needle lengths of 519 conifer species. 75% of needles are no longer than 6 cm. Longer needles might be disadvantageous because of the formation of a stagnant zone in needles longer than L_{eff}, as predicted by Eq. 11 From Rademaker et al. (2017)

The Architecture of Broad Leaves

Leaves have two tissue types that are relevant for understanding the movement of water and sugars: mesophyll tissue and vascular tissue. Mesophyll cells contain chloroplasts and are therefore able to do photosynthesis to produce sugars. These sugars are then transported towards the vascular tissue, the veins, and taken up into the phloem tissue inside the veins. The main functions of the vascular tissue are mechanical stability and transport. Here, we will concentrate on the transport function.

A typical spatial arrangement of mesophyll and vascular tissue can be seen in Figs. 7 and 8. A dense layer of mesophyll cells at the upper side of the leaf supports the optimal absorption of sunlight for photosynthetic activity. The spongy arrangement of mesophyll cells towards the lower side of the leaf enables optimal gas exchange with the surrounding air spaces, securing sufficient supply of carbon dioxide. Embedded in these green cells are the veins, which are enclosed by a single layer of bundle sheath cells, shielding the vein from the surrounding air spaces.

Inside the veins are several cell types, two of which are directly relevant to transport: first, the large water-conducting vessels of the xylem and second, the thick-walled sieve elements of the phloem, responsible for the distribution of photosynthates and signalling molecules. The mature vessels are dead cells, which reduces the resistance to flow; and even in sieve tubes, which are living cells, only few organelles are present. This makes them dependent on neighbouring companion cells, to which they have numerous symplasmic connections, the so-called plas-

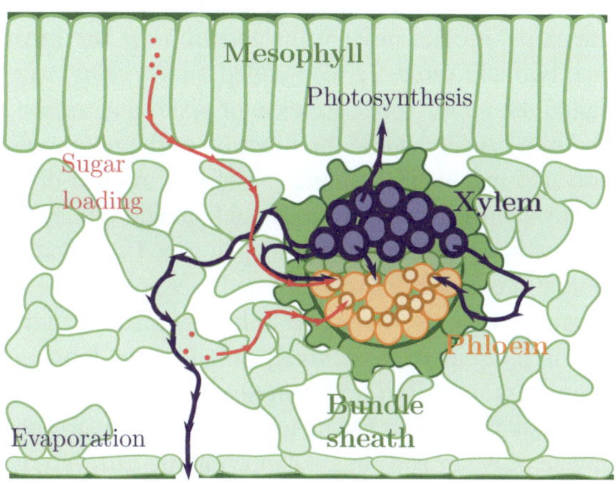

Fig. 7 Leaf cross section showing phloem and xylem conduits inside the vein, enclosed by the bundle sheath, which protects the vascular bundle from air. The surrounding mesophyll cells produce sugars (red) in photosynthesis, which are then transported towards the vein and loaded into the phloem. Water (blue) from the xylem is mostly evaporated through stomatal pores on the lower side of the leaf. A small portion of the water is used in photosynthesis and to drive the flow in the phloem

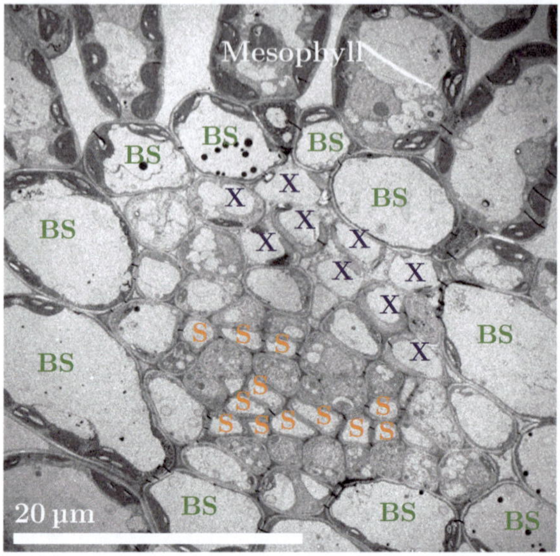

Fig. 8 Transmission electron microscopy image of a vein cross section, taken from an apple (*Malus sylvestris*) leaf. The image shows the xylem vessels (X) and the sieve elements (S) of the phloem inside the vein. A layer of bundle sheath (BS) cells encloses the vein and shields it from the surrounding air spaces. See also Fig. 7

modesmata. It is through these plasmodesmata that the sieve cells are supplied with substances they cannot produce themselves. Companion cells also play an important role in sugar loading, the process of taking up sugars into the phloem (see section "Water Flow Inside the Leaf").

Veins are organized in a hierarchical manner, where the largest (or *first order*) vein is the continuation of the vascular tissue in the petiole. Veins branching off from the first-order vein are called second-order veins and so forth, until the smallest veins. Veins of order one, two and three are also called *major veins*, whereas the higher order veins are called the *minor veins*. Release of water and uptake of sugars takes place mainly in the minor veins, while the function of the major veins is thought to be mostly transport and mechanical stability. The hierarchical organization allows for efficient access to the minor veins from the whole leaf blade (Fig. 9). Leaves of different species show different vein branching patterns. Understanding what makes these patterns optimal has been strongly debated (see, e.g. Jensen et al. 2016 and the references therein). However, there are only few studies addressing the inner structure of veins—in particular the relatively small part belonging to the phloem. The two basic parameters related to transport are the *number* of conduits and their *radius*. Conduits with smaller radius have a larger surface area to volume ratio. Since the release and uptake of water and sugars happen at the cell wall, we would expect the conduits inside minor veins to have smaller radii than inside major veins and a much greater number. Recently, these important parameters were carefully measured for leaves of poplar (Carvalho et al. 2017b) and of gingko (Carvalho et al. 2017a). In poplar, which has a classical hierarchical venation pattern bifurcating from the midvein towards the edge of the leaf it was found that the radii of the individual sieve elements decrease from the petiole to the 7th order veins by roughly a factor of 3,

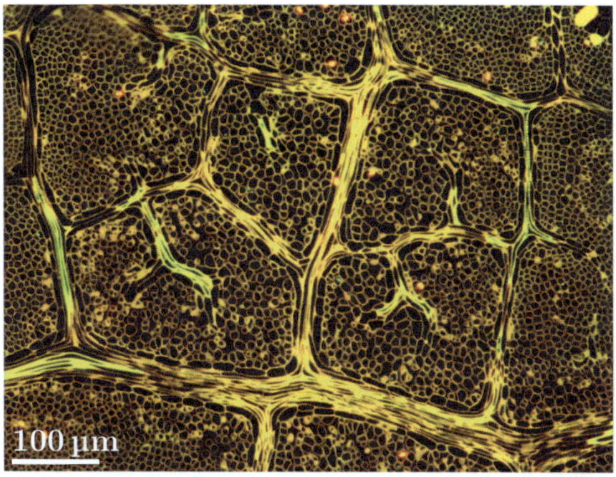

Fig. 9 Paradermal section of a young, fixed birch leaf stained with the fluorescent dye Coriphosphine O. The image shows veins of different orders from the smallest minor veins to a third-order vein. The veins are surrounded by mesophyll cells

whereas the total cross-sectional area (not surface area) of sieve elements *increases* roughly exponentially with order.

Our own study on birch leaves (Rademaker 2016) shows similar behaviour. Analysing ten cross sections along the midvein of a birch leaf, this study found that the radius of sieve tubes in the petiole is about twice as large as at the tip of the leaf ($(1.9 \pm 0.6)\,\mu\text{m}$ vs. $(1.0 \pm 0.2)\,\mu\text{m}$), where the midvein is almost reduced to the size of a minor vein ($(0.7 \pm 0.2)\,\mu\text{m}$). Looking at cross sections of minor veins, we found on average three sieve elements per smallest minor vein. The total number of minor vein endings per birch leaf was estimated from pieces of cleared leaves (Martens 2017). These two informations combined lead to an estimate of about 150,000 sieve elements in total in all minor vein endings of one leaf. Comparing this with the approximately 570 sieve elements found inside the petiole shows a massive (roughly 260-fold) reduction in the number of sieve tubes from minor to major veins. Moreover, the total cross-sectional area of sieve elements measured at points along the midvein, was found to increase from tip to base, correlating closely with the blade area of the part of the leaf, which would be expected to export sugar through this point (i.e. the leaf area measured from the point on the midvein, where the section was taken, to the tip of the leaf, choosing the boundary parallel to the second-order veins).

Water Flow Inside the Leaf

As discussed in the previous section, the leaves are the endpoint of the long-distance transport of water in the xylem and the starting point of the long-distance transport of sugars from the site of their production in the mesophyll to the site of consumption and storage in the sink organs of the plant, such as roots, immature leaves and reproductive organs (flowers and cones). Significantly, the leaves also provide the main driving force for both, xylem and phloem transport. We will first consider the pathway of water from the xylem through the cell wall and air space in the leaf to the stomata, which regulate the rate of evaporation, and then the opposite pathway of sugars from the mesophyll cells via the bundle sheath to the phloem as sketched in Fig. 7. When discussing transport in the leaf one distinguishes between two basic pathways: symplasmic and apoplasmic. The symplasmic pathway goes through the interior of the cells via plasmodesmata, while the apoplasmic pathway is on the outside of the cells, through the cell walls. While the water, after leaving the xylem, is spreading out to different destinations—to be evaporated from the apoplast, to enter photosynthesis in the mesophyll and to enter the phloem osmotically—the sugar transport from the mesophyll cells is merely funnelled into the phloem. One can think of the water going into the phloem as a turning point of water flow in the vascular water circulation. While sugars follow a one-way path from source to sink, water circulates in the phloem back to the root where it is released from the phloem to travel up in the xylem together with freshly acquired water up to the leaves again.

Transpiration and Stomata

After exiting the xylem vessels, water might stay in the apoplast or enter the living cells and diffuse symplasmically. Apoplasmic water can follow the cell wall contacts via the bundle sheath into the mesophyll from which it can evaporate into the large intercellular air spaces (Fig. 10). The cell wall material consists of a number of different carbohydrates, the most important of which are cellulose, hemicelluloses and pectins. Those parts are hydrophilic and able to bind water molecules strongly. Water release from leaves takes a phase transition from liquid to vapour at the interface of cell walls and air spaces inside the leaf, and is controlled by a large number of stomata, openings that connect the internal air spaces with the environment. Each stomatal opening is controlled by a pair of guard cells which are able to respond to internal and environmental cues such as plant hormones, light conditions and CO_2 concentration. Their increase in volume opens stomata, while shrinkage closes them. The leaf surface is covered by a lipophilic (hydrophobic) layer and does not contribute to the evaporation (Lu et al. 2012).

In terms of biomechanics, the collection of stomatal openings act as a resistor for evaporation—and thus for water transport from roots to leaves—and limits at the same time the uptake of CO_2 through these openings which again limits photosynthesis. According to textbook knowledge, the driving force for transpiration is the difference in water vapour concentration between the leaf air spaces and the external air. This again is regulated by the diffusional resistance of the pathway through the stomatal openings. These values, though different for plants adjusted to different environments, can easily be quantified. By contrast, it is difficult to estimate the share of water that takes the apoplasmic pathway and evaporates compared to the two others: the share that is taken up by mesophyll cells for photosynthesis and the share that is osmotically taken up by the phloem directly and enters circulation. For

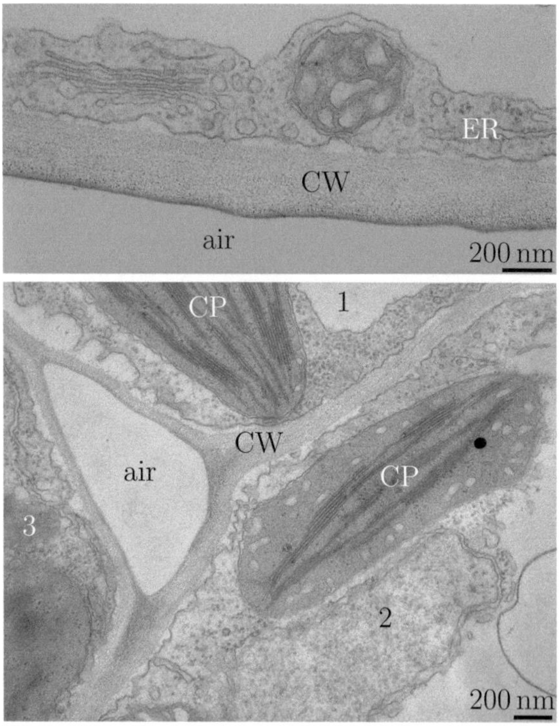

Fig. 10 Transmission electron microscopy images showing (top) cell walls (CW) and adjacent endoplasmic reticulum (ER) within the cell at the water/air interface and (bottom) intercellular air space and cell walls at the point of contact of three mesophyll cells (1, 2, 3) with chloroplasts (CP) in an *Arabidopsis thaliana* leaf

entering the symplasm of phloem and mesophyll cells water channels (aquaporins) have to be present, since the cell membrane consists of phospholipid bilayers that offer high resistance to water molecules.

Balance of water partitioning from xylem to evaporation, mesophyll uptake and phloem circulation is crucial for the adaption of plants to different growth conditions and climates. Indeed, the molecular biological removal of plasma membrane aquaporins in poplar by RNAi led to changes in the leaf shape, higher evaporation rates, larger stomatal opening and a restructuring of the cell wall as indicated by changes in gene expression for the respective membrane transporters and phytohormones (Bi et al. 2015). Binding of water molecules to the hydrophilic cellulose, hemicellulose and pectin compounds of the cell wall plays a large role at the cell wall–intercellular space interface. Evaporation will lead to an increasingly negative pressure between the structural cell wall compounds at the interface. This negative pressure is assumed to pull the water from the xylem to the evaporating cell wall surface in the mesophyll (Taiz and Zeiger 2010), i.e. a quite long distance in the range of 50–300 μm, including passage of the bundle sheath. It is unclear, though, how these events on the surface of the cell wall translate to the inner layers of the 200–500 nm thick mesophyll cell wall (Fig. 10) which abuts the plasma membrane. Under drought conditions, their aquaporins are closed ('gated') in order to limit the intracellular water loss, which would lead to loss of turgor of the mesophyll cells and, thus, leaf wilting (Törnroth-Horsefield et al. 2006). The question to be raised here is how the evaporation-generated water tension in the cell wall of a given mesophyll cell translates into the negative pressure in xylem vessels. It is suggestive to postulate an unbroken apoplasmic water continuum between mesophyll and xylem, necessary to pull water up into the leaves. The evaporation-generated water loss in the mesophyll will however also have a direct influence on the water content in the mesophyll cells as long as aquaporins are open. It is also conceivable that the intracellular water content of mesophyll cells is supplied for via the symplasmic pathway, i.e. plasmodesmata between the bundle sheath and the mesophyll (Fig. 7). A pivotal role of the bundle sheath for water export from the vascular bundles and sugar import into vascular bundles is given in plant taxa that have an endodermis-like bundle sheath, as, e.g. the gymnosperms (Liesche et al. 2011). Here, water from the xylem has to cross the plasma membrane of the bundle sheath and might well be transported on symplasmically, i.e. through plasmodesmata (Schulz 2015).

Pre-phloem Pathways of Sugars

Evolution of the long-distance transport of assimilates has led to three different strategies of phloem loading, two active ones and a passive one. The active ones involve an accumulation of sugars in the sieve element–companion cell complex, and consumption of energy. By contrast, the passive one does not show accumulation of sugars in the phloem, but has the highest sugar concentration in the mesophyll. While the active apoplasmic strategy is characterized by an isolated phloem configuration with few if any plasmodesmata between the sieve element–companion cell complex and the bundle sheath, both symplasmic strategies depend upon phloem cells which are well coupled to the bundle sheath via plasmodesmata (Fig. 11). The pathway from the mesophyll to the bundle sheath, and where existent, further to the phloem parenchyma (pre-phloem pathway), is symplasmic for any loading strategy, i.e. bound to plasmodesmata (Schulz 2015).

Mechanistically, the apoplasmic loading mode is the easiest to understand. Since the sieve element–companion cell complexes are symplasmically isolated, they can accumulate sugar without leakage—once in, it can only follow the sieve tube systems towards the sinks. The interesting interface for this strategy which is most established in herbs is the wall isolating the sieve element–companion cell complex from neighbouring cells such as the bundle sheath. For the release of sucrose, the main transport sugar in those plant families, there are membrane transport proteins of the SWEET type allowing facilitated diffusion. Release through theses permeases is downhill and does not require energy (Chen et al. 2012). Sucrose in the apoplasmic interface between bundle sheath and complex has then to be actively taken up by sucrose-proton cotransporters called SUTs that are indirectly fuelled by ATP hydrolysis driving the plant proton pump. Both membrane steps involved, facilitated diffusion from the bundle sheath into the apoplast and active uptake into the phloem, are saturable and mainly depend on their respective maximal reaction rate V_{max} and

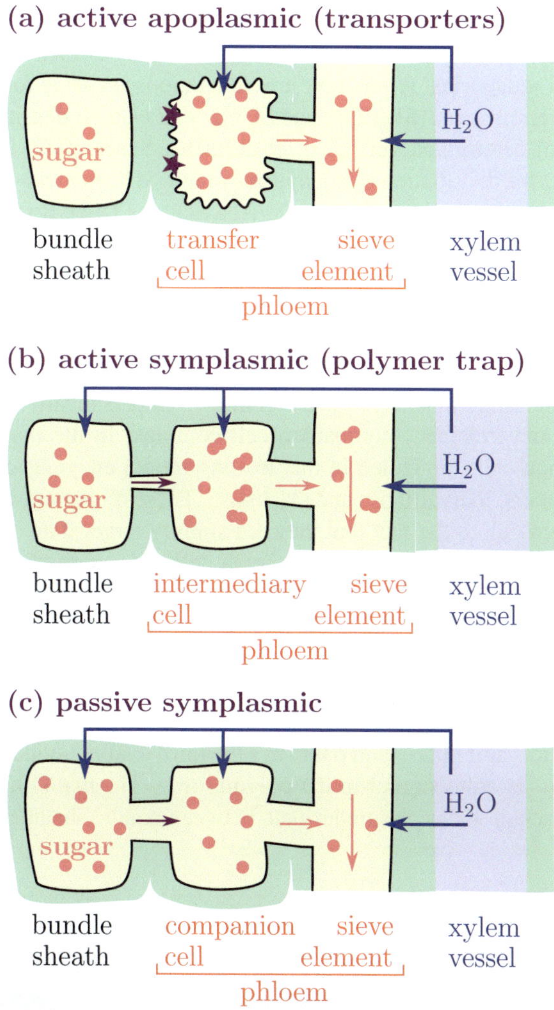

Fig. 11 The three known modes of sugar loading into the phloem and possible water pathways. **a** In active apoplasmic loading, sugars are taken up from the apoplast (cell wall space) into transfer cells with the help transporter proteins. Water can enter the transporting sieve elements either directly from the xylem or via the transfer cells. **b** In active symplasmic loading, the sugar uptake from bundle sheath into intermediary cell happens via the symplast. The loading mode is considered active due to the enzymatic transformation of sucrose to larger sugar molecules, like raffinose and stachyose, inside the intermediary cell. The larger sugars are thereby trapped inside the phloem, which facilitates an increased sugar concentration in the phloem. The trapping is only possible, because the symplasmic connections (plasmodesmata) between bundle sheath and intermediary cell are very narrow in this loading mode. Compared to apoplasmic loading, water can additionally enter the phloem symplasmically through these narrow plasmodesmata. **c** In passive symplasmic loading, sugar follows a downhill gradient from bundle sheath into the phloem. All cells are well connected with plasmodesmata

number per membrane area. It can be postulated that the high sucrose concentration in the sieve element–companion cell complex osmotically attracts water which can move in through aquaporins. While there are calculations available for the $V_{\max}$ of the sucrose transporter proteins (SWEETs and SUC2/SUT1), their density in the plasma membrane is not determined yet (Chen et al. 2012; Sauer 2007; Kühn and Grof 2010). Neither is the distribution of aquaporins in the sieve element–companion cell complex or their gating behaviour assessed (see Schulz 2015). It is far from trivial to keep track of the water being co-transported by the active sugar transporter. Many membrane pumps carry water through the membrane in addition to their target 'substrate' molecule. In this way, water can move against chemical potential gradients of substantial size as seen, e.g. in the intestinal sodium/glucose cotransporter SGLT1 (Zeuthen et al. 2016) as shown pictorially in Fig. 12.

In active symplasmic loading, sugars can move symplasmically from the bundle sheath into the sieve element–companion cell complex. In the intermediary cells (as the companion cells are called in this loading mode) enzymes convert sucrose into sugar oligomers, like raffinose and stachyose. These oligomers are then trapped inside the phloem due to the fact that they are slightly larger than sucrose and that the plasmodesmal pores between bundle sheath cell and intermediary cell are very narrow. This is why the mechanism is also called the *polymer trap*. In contrast to the apoplasmic loading, water can in this mechanism also enter the phloem from the bundle sheath through the plasmodesmata. This bulk flow of water even contributes to the loading of sugars into the phloem, so that the uptake of sugars is not purely diffusive, but also partly advective.

With respect to water uptake and pathways the third loading mode—passive symplasmic loading—is quite similar to the polymer trap. The main differences to the polymer trap case are wider plasmodesmata between bundle sheath and companion

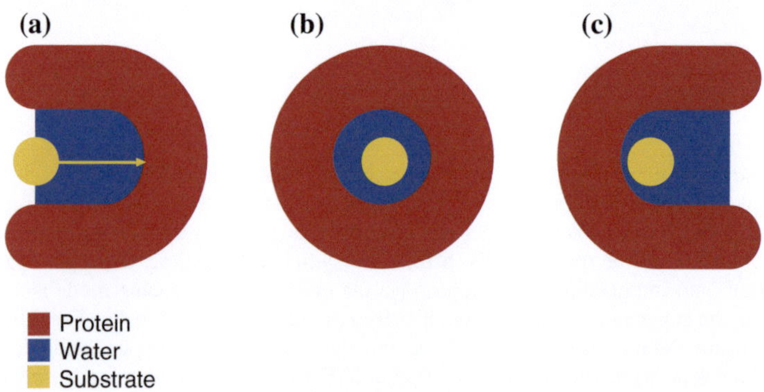

Fig. 12 Cartoon of a membrane protein that co-transports water in the process of pumping the substrate molecule shown in yellow. There will be a net water transport if some water is expelled after step (c), which can occur, e.g. by conformational changes of the protein. This mechanism allows for 'uphill' water transport—against a pressure gradient or a water concentration (osmotic) gradient. From Zeuthen and MacAulay (2012)

cells, and no enzymatic sugar conversion in the passive symplasmic loading mode. The concentration of sugars is highest in the mesophyll cells and sugars follow a downhill gradient into the phloem, in contrast to both active loading modes.

Mathematically the flux of water and of sugar through plasmodesmata can be described by the Kedem–Katchalsky equations (Kedem and Katchalsky 1958):

$$J_V = L_p(\Delta p - \sigma RT \Delta c) \tag{17}$$

$$j_s = \omega RT \Delta c + (1 - \sigma)\bar{c} J_V, \tag{18}$$

where $\bar{c}$ is the mean solute concentration, ω is the mobility of the solute and σ is the reflection coefficient. (17) contains two causes for the movement of water: a hydrostatic pressure potential Δp or a difference in osmotic potential $RT\Delta c$. According to (18), sugar is advected with the bulk volume flow J_V, and diffusing due to the difference in sugar concentration. Here J_V is a flux pr. area with SI unit $m^3\ s^{-1}\ m^{-2} = m\ s^{-1}$ (as a velocity) composed of

$$J_V = j_w \bar{v}_w + j_s \bar{v}_s, \tag{19}$$

where j_w and j_s are molar fluxes of water and solutes, respectively, with SI unit mol $s^{-1}\ m^{-2}$, and $\bar{v}_w$ and $\bar{v}_s$ are the corresponding molar volumes. In situations where the solute concentrations are low, it is usually a good approximation to replace J_V with the water part $j_w \bar{v}_w$, an approximation which is often made.

The meaning of reflection coefficient σ and mobility ω can be best understood by thinking of the porous interface as a membrane. If this membrane is ideally semipermeable, then $\sigma = 1$ and $\omega = 0$, meaning that the solute size is larger than the pore diameter and the solute cannot pass through the pore. In the other extreme, the pores are much larger than the solute, implying that $\sigma = 0$ and ω is directly proportional to the free diffusion coefficient ($\omega = \frac{A}{dRT} D_{\mathrm{free}}$, with A and d the area and thickness of the membrane, and assuming a constant concentration gradient).

The challenge in understanding the polymer trap loading mechanism is that the plasmodesmatal pores should be extremely selective. Indeed, they should let sucrose through without too much trouble, but they should stop raffinose and stachyose, although the hydrodynamic radius of a raffinose molecule is only 25% larger than sucrose molecules. It is well known that the plasmodesmata connecting the companion cell to the bundle sheath are unusually narrow and branched towards the companion cell end. In addition, the endoplasmic reticulum (ER), which permeates the plasmodesmata, makes the estimate of the available space for transport very uncertain (Volk et al. 1996; Fisher and Gifford 1986; Waigmann et al. 1997). Aside from partially blocking the plasmodesmata, the role of the ER is not well understood. Recently, Nixon-Abell et al. (2016) have obtained spectacular views of the packed tubular arrays of the ER inside animal cells, and it is obvious that many important functions could be carried out within these structures—something which remains to be explored.

Some progress has been made in using simple models of the unblocked part of the plasmodesmata through which the fluid can flow to study the sugar/water transport (Dölger et al. 2014; Comtet et al. 2017). In Waigmann et al. (1997), it was suggested to model them as circular slits with half-width around 1 nm, as shown in Fig. 13, and this suggestion is followed in Dölger et al. (2014).

With this geometry, one can estimate the parameters in the Kedem–Katchalsky equations as

$$\omega = n_{\mathrm{PD}} \frac{4\pi r_{\mathrm{PD}} h}{dRT} H(r_{\mathrm{solute}}/d) D_{\mathrm{cyt}}, \tag{20}$$

where, from the Einstein relation, the molecular diffusivity outside of the pores is

$$D_{\mathrm{cyt}} \approx \frac{kT}{6\pi \eta r_{\mathrm{solute}}} \tag{21}$$

and $H(\lambda = r_{\mathrm{solute}}/d)$ is a 'hindrance factor' for passage through a narrow tube (Dechadilok and Deen 2006).

Similarly, the membrane permeability is

$$L_p \approx n_{\mathrm{PD}} \frac{4\pi r_{\mathrm{PD}} h^3}{3\eta d} \tag{22}$$

and with similar estimates for the 'convective' hindrance factor $W(\lambda) = 1 - \sigma$ (Dechadilok and Deen 2006), one can compute the water and solute fluxes for given concentration gradients. Inside the pores one can decompose the solute flux into a diffusive and a convective part:

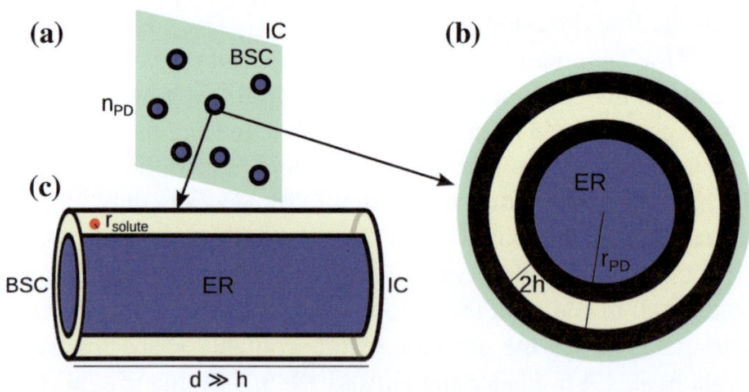

Fig. 13 A possible structure for a plasmodesmal pore: slit pores as suggested by Waigmann et al. (1997). **a** Part of the cell wall between the bundle sheath (BSC) and the intermediary cell (IC) with plasmodesmata (PD) density n_{PD}. **b** The assumed substructure of a PD shown in cross section and **c** in three-dimensional view. The cytoplasmic sleeve (light yellow) available for water and sugar transport is restricted by the desmotubule of the endoplasmic reticulum (ER, blue) and electron-dense particles (black) attached to the membrane, and is assumed to take the form of a circular slit with radius r_{PD}, half-width h and length d. From Dölger et al. (2014)

$$j_s = -D\frac{\partial c}{\partial x} + uc, \tag{23}$$

where, comparing to (18), $u = (1-\sigma)J_V$ and $D \approx \omega RTd$ for a pore with length d. Finally, sugar conservation gives

$$\frac{\partial c}{\partial t} = -\nabla \cdot j_s = -\frac{\partial j_s}{\partial x} \tag{24}$$

and for the stationary case, where $\frac{\partial c}{\partial t} = \frac{\partial j_s}{\partial x} = 0$, j_s is constant. The linear relation between solute flux j_s and concentration differences Δc in (18) assumes that the Peclet number $\mathrm{Pe} = ud/D = d(1-\sigma)J_V/D$ is small. In general, one can simply solve (23) for c with constant j_s, i.e.

$$\frac{\partial c}{\partial s} = \mathrm{Pe}\ c - n, \tag{25}$$

where $n = j_s d/D$ and s is the scaled variable $s = x/d$. The solution with $c(x = 0) = c_0$ and $c(x = d) = c(s = 1) = c_1$ is

$$\frac{c(s)}{c_0} = \left(1 - \frac{n}{\mathrm{Pe}}\right)e^{s\mathrm{Pe}} + \frac{n}{\mathrm{Pe}} \tag{26}$$

giving

$$\frac{c_1}{c_0} = \left(1 - \frac{n}{\mathrm{Pe}}\right)e^{\mathrm{Pe}} + \frac{n}{\mathrm{Pe}} \tag{27}$$

so that we can express $j_s = nD/d$ in terms of the concentrations on the two sides of the membrane as

$$j_s = (1-\sigma)J_V\left(c_0 + \frac{\Delta c}{e^{\mathrm{Pe}} - 1}\right), \tag{28}$$

where $\Delta c = c_0 - c_1$. For small Pe, this reduces to

$$j_s \approx (1-\sigma)J_V c_0 + \frac{D}{d}\Delta c \tag{29}$$

in agreement with (18) with c_0—the 'upwind' concentration—replacing $\bar{c}$. This approach was used in Comtet et al. (2017), who also included a description of the enzymatic reaction oligomerizing the sugars. As one can see, the effect of including large values of Pe (or small values of D) in (28) is very small since $\mathrm{Pe} \gg 1$ leads to

$$j_s \approx (1-\sigma)J_V c_0 + (1-\sigma)J_V e^{-\mathrm{Pe}}\Delta c, \tag{30}$$

where the diffusive contribution is exponentially small. Indeed, the conclusions of Dölger et al. (2014) and of Comtet et al. (2017) are quite compatible. To avoid the oligomers (in particular raffinose) from flowing back into the bundle sheath, the slit half-width has to be less than around 6.0 Å, but to allow sucrose to flow through it has to be larger than around 4.2 Å. For molecules (sucrose) passing such very narrow tubes, the theory summarized in Dechadilok and Deen (2006) for the hindrance factors $H(\lambda)$ and $W(\lambda)$ is at the limit of its validity—in addition to the fact that the molecules are treated as spherical. If these approximations are used anyway, it is found that it is possible to get enough sugar through, and in addition (Dölger et al. 2014) that enough water will likely follow along with this sugar flow to drive the phloem, and thus, that no additional water intake is needed through aquaporins in the membranes. It is also remarked that efficient blocking of raffinose will only take place when the pores are very close to the 5.2 Å. One might have speculated that the water current was sufficiently strong to keep out the raffinose of even larger pores, but this does not seem feasible, at least with the parameters coming from *Cucumis melo*. In Comtet et al. (2017), it is speculated that *some* backflow of raffinose might actually occur, since it might be degraded once it reaches the bundle sheath. This might indeed be so, but even with the very narrow pores assumed in Dölger et al. (2014), the sucrose flow rate seems substantially larger than observed in *Cucumis melo* (Schmitz et al. 1987). This might be due to the very complicated and mostly unknown structure and function of the ER as alluded to above.

Conclusion

In this short review, we have tried to present an overview of essential points in the current knowledge and mode of description of the coupling of water motion and sugar translocation in leaves—both in the sieve tubes of the veins, where the sugar is transported to the rest of the plant, and through the pre-phloem pathway that leads from the sugar producing mesophyll cells via the bundle sheath to the sieve cells. The challenges in understanding the full circuit of water in the leaf are considerable. From the xylem tubes inside the vascular bundle where water exits the vascular system, most of it will evaporate out through the stomata in the leaf surface, but a small amount has to be left behind, and take part in the photosynthetic sugar production and the subsequent sugar transport. Presumably, the large negative pressures, dragging the water all the way up to the leaves, originate in the cell walls of the mesophyll, and thus, the mesophyll cells must be able to balance the water potential very delicately, so that enough water is retained to carry out these tasks. Being part of the pre-phloem, the mesophyll cells are believed to typically have positive pressures (turgor) inside and this means that they have to be able to create very large water potential gradients from high concentrations of solutes. How this is done, and how it plays together with the different loading mechanisms, is not known in detail. When plasmodesmata are important for the sugar translocation, they obviously have to be able to conduct sugar efficiently enough while presumably keeping out many other similarly sized (or even smaller) molecules which the cell wants to keep. How this is done is not known, but very likely the endoplasmic reticulum plays an important role to be explored in the future.

References

Bi Z, Merl-Pham J, Uehlein N, Zimmer I, Mühlhans S, Aichler M, Walch AK, Kaldenhoff R, Palme K, Schnitzler J-P, Block K (2015) RNAi-mediated downregulation of poplar plasma membrane intrinsic proteins (PIPs) changes plasma membrane proteome composition and affects leaf physiology. J Proteomics 128:321–332

Carvalho MR, Turgeon R, Owens T, Niklas KJ (2017a) The hydraulic architecture of ginkgo leaves. Am J Bot 104(9):1285–1298

Carvalho MR, Turgeon R, Owens T, Niklas KJ (2017b) The scaling of the hydraulic architecture in poplar leaves. New Phytol 214:145–157

Chen L-Q, Qu X-Q, Hou B-H, Sosso D, Osorio S, Fernie AR, Frommer WB (2012) Sucrose efflux mediated by sweet proteins as a key step for phloem transport. Science 335(6065):207–211

Comtet J, Turgeon R, Stroock AD (2017) Phloem loading through plasmodesmata: a biophysical analysis. Plant Physiol 175(2):904–915

Dechadilok P, Deen WM (2006) Hindrance factors for diffusion and convection in pores. Ind Eng Chem Res 45(21):6953–6959

Dölger J, Rademaker H, Liesche J, Schulz A, Bohr Tomas (2014) Diffusion and bulk flow in phloem loading: a theoretical analysis of the polymer trap mechanism for sugar transport in plants. Phys Rev E 90(4):042704

Fisher DB, Gifford RM (1986) Accumulation and conversion of sugars by developing wheat grains vi. gradients along the transport pathway from the peduncle to the endosperm cavity during grain filling. Plant Physiol 82(4):1024–1030

Jensen KH, Berg-Sørensen K, Bruus H, Holbrook NM, Liesche J, Schulz A, Zwieniecki MA, Bohr T (2016) Sap flow and sugar transport in plants. Rev Mod Phys 88:035007 (1–63)

Jensen KH, Lee J, Bohr T, Bruus H, Holbrook NM, Zwieniecki MA (2011) Optimality of the Münch mechanism for translocation of sugars in plants. J R Soc Interface 8(61):1155–1165

Jensen KH, Berg-Sørensen K, Friis SMM, Bohr T (2012a) Analytic solutions and universal properties of sugar loading models in münch phloem flow. J Theor Biol 304:286–296

Jensen KH, Liesche J, Bohr T, Schulz A (2012b) Universality of phloem transport in seed plants. Plant Cell Environ 35:1065–1076

Jensen KH, Mullendore DL, Holbrook NM, Bohr T, Knoblauch M, Bruus H (2012c) Modeling the hydrodynamics of phloem sieve plates. Front Plant Sci 3

Jensen KH, Zwieniecki MA (2013) Physical limits to leaf size in tall trees. Phys Rev Lett 110(1)

Kedem O, Katchalsky A (1958) Thermodynamic analysis of the permeability of biological membranes to non-electrolytes. BBA—Biochim Et Biophys Acta 27:229–246

Knoblauch M, Knoblauch J, Mullendore DL, Savage JA, Babst BA, Beecher SD, Dodgen AC, Jensen KH, Holbrook NM (2016) Testing the Munch hypothesis of long distance phloem transport in plants. Elife 5

Kühn C, Grof CPL (2010) Sucrose transporters of higher plants. Curr Opin Plant Biol 13(3):287–297

Landsberg JJ, Fowkes ND (1978) Water movements through plant roots. Ann Bot 42:493–508

Liesche J, Martens HJ, Schulz A (2011) Symplasmic transport and phloem loading in gymnosperm leaves. Protoplasma 248(1):181–190

Liesche J, Windt C, Bohr T, Schulz A, Jensen KH (2015) Slower phloem transport in gymnosperm trees can be attributed to higher sieve element resistance. Tree Physiol (in press)

Lu SY, Zhao HY, Des Marais DL, Parsons EP, Wen XX, Xu XJ, Bangarusamy DK, Wang GC, Rowland O, Juenger T, Bressan RA, Jenks MA (2012) Arabidopsis eceriferum9 involvement in cuticle formation and maintenance of plant water status. Plant Physiol 159:930–944

Martens HJ (2017) Private correspondence

Niklas KJ (1994) Plant allometry: the scaling of plant form and process. University of Chicago Press

Nixon-Abell J, Obara CJ, Weigel AV, Li D, Legant WR, Xu CS, Pasolli HA, Harvey K, Hess HF, Betzig E, Blackstone C, Lippincott-Schwartz J (2016) Increased spatiotemporal resolution reveals highly dynamic dense tubular matrices in the peripheral ER. Science 354, aaf3928–1–12

Rademaker H (2016) Microfluidics of sugar transport in plant leaves and in biomimetic devices. PhD thesis, Technical University of Denmark

Rademaker H, Jensen KH, Bohr T (2016) Osmotically driven flows and maximal transport rates in systems of long, linear porous pipes. arXiv:1610.09175

Rademaker H, Zwieniecki MA, Bohr T, Jensen KH (2017) Sugar export limits size of conifer needles. Phys Rev E 95:042402

Ronellenfitsch H, Liesche J, Jensen Kaare H, Holbrook NM, Schulz A, Katifori E (2015) Scaling of phloem structure and optimality of photoassimilate transport in conifer needles. In: Proceedings of the royal society of london B: biological sciences, vol 282(1801)

Sauer N (2007) Molecular physiology of higher plant sucrose transporters. FEBS Lett 581(12):2309–2317

Schmitz K, Cuypers B, Moll M (1987) Pathway of assimilate transfer between mesophyll-cells and minor veins in leaves. Cucumis melo L. Planta 171(1):19–29

Schulz A (2015) Diffusion or bulk flow: how plasmodesmata facilitate pre-phloem transport of assimilates. J Plant Res 128(1):49–61

Tadrist L, Darbois-Texier B (2016) Are leaves optimally designed for self-support? an investigation on giant monocots. J Theor Biol 396:125–131

Taiz L, Zeiger E (2010) Plant physiology, 5th edn. Sinauer Associates Inc, Sunderland, MA

Törnroth-Horsefield S, Wang Y, Hedfalk K, Johanson U, Karlsson M, Tajkhorshid E, Neutze R, Kjellbom P (2006) Structural mechanism of plant aquaporin gating. Nature 439:688–694

Volk GM, Turgeon R, Beebe DU (1996) Secondary plasmodesmata formation in the minor-vein phloem. Cucumis Melo L and Cucurbita pepo L Planta 199(3):425–432

Waigmann E, Turner A, Peart J, Roberts K, Zambryski P (1997) Ultrastructural analysis of leaf trichome plasmodesmata reveals major differences from mesophyll plasmodesmata. Planta (Heidelberg) 203(1):75–84

Zeuthen T, Gorraitz E, Her K, Wright EM, Loo DDF (2016) Structural and functional significance of water permeation through cotransporters. Proc Nat Acad Sci (USA) 113(44):E6887–E6894

Zeuthen T, MacAulay N (2012) Transport of water against its concentration gradient: fact or fiction? WIREs membr transp signal 2012. https://doi.org/10.1002/wmts.54

Molecular Mechanisms of Mechanosensing and Mechanotransduction

Masatsugu Toyota, Takuya Furuichi and Hidetoshi Iida

Abstract Mechanical stimuli, such as touch, bending, gravity, and wounding, influence plant growth and development through the activation of intracellular signaling pathways and gene expression. Therefore, mechanosensing and mechanotransduction are of vital importance and have been attracting the attention of many plant scientists for nearly 150 years. Based on recent molecular and cellular approaches, candidates for mechanosensors have been discovered. These include mechanosensitive (MS) channels, such as MscS-like (MSL) proteins, *mid1*-complementing activities (MCAs), and reduced hyperosmolality-induced $[Ca^{2+}]_i$ increase 1 (OSCA1), which generate intracellular ionic signals and receptor-like kinases that trigger the activation of regulatory proteins or enzymes, including Ca^{2+}-binding proteins, protein kinases, protein phosphatases, and transcription factors. Other possible groups of mechanosensors are intracellular filamentous structures in the cytoskeleton, such as microtubules and actin filaments, which may directly act as sensors for the deformation of intracellular structures. In this chapter, we discuss the mechanisms by which plants sense and respond to mechanical stimuli by focusing on mechanosen-

M. Toyota and T. Furuichi contributed equally to this chapter.

M. Toyota
Department of Biochemistry and Molecular Biology, Saitama University, 255 Shimo-Okubo, Sakura-ku, Saitama City, Saitama 338-8570, Japan

M. Toyota
Department of Botany, University of Wisconsin, Madison, WI 53706, USA

M. Toyota
Japan Science and Technology Agency (JST), Precursory Research for Embryonic Science and Technology (PRESTO), 4-1-8 Honcho, Kawaguchi, Saitama 332-0012, Japan

T. Furuichi
Department of Human Life Science, Nagoya University of Economics, 61-1 Uchikubo, Inuyama City, Aichi 484-8504, Japan

H. Iida (✉)
Department of Biology, Tokyo Gakugei University, 4-1-1 Nukuikita-machi. Koganei City, Tokyo 184-8501, Japan
e-mail: iida@u-gakugei.ac.jp

A. Geitmann and J. Gril (eds.), *Plant Biomechanics*,
https://doi.org/10.1007/978-3-319-79099-2_17

sors along with their downstream signaling molecules, such as auxin and reactive oxygen species (ROS).

Keywords Mechanosensitive channels · Organelles · Cytoskeleton
Intracellular signals · Long-distance signals

List of Abbreviations

CBLs	Calcineurin B-like proteins
CICR	Ca^{2+}-induced Ca^{2+} release
CIPKs	CBL-interacting protein kinases
FtsZ	Filamentous temperature-sensitive Z
GLR	Glutamate receptor-like channels
$InsP_3$	Inositol 1,4,5-trisphosphate
MCA1	mid1-complementing activity 1
MCAs	mid1-complementing activities
MS	Mechanosensitive
MscL	Mechanosensitive channel of large conductance
MscS	Mechanosensitive channel of small conductance
MSL	MscS-like
OSCA1	Reduced hyperosmolality-induced $[Ca^{2+}]_i$ increase 1
RAL	Rapid alkalinization factor
ROS	Reactive oxygen species
SA	Stretch-activated
TM	Transmembrane
TPC1	Two-pore channel 1

Introduction

Plants are continuously exposed and respond to external mechanical stimuli, such as wind, gravity, and touch, which sometimes signal threats. For example, by perceiving a strong one-directional wind daily, trees generally grow leeward because they are grounded and unable to voluntarily move (Fig. 1). If they were to grow straight upward without recognizing the force and direction of wind, they may be fatally blown down. Therefore, mechanosensing and subsequent responses are particularly important for plants to survive under mechanically stressful conditions because of their sessile nature. In addition to wind stress, external mechanical stimuli may also include feeding and damage caused by herbivores as well as the hyphal invasion of fungi, all of which are threats to plants.

However, mechanical stimuli are not only restricted to external forces, but also they include internal forces. The activities of individual cells, including cell division,

Fig. 1 Trees responding to wind, a mechanical stress. Photographed at Seven Sisters at Exceat, East Sussex, UK, by Hidetoshi Iida

differentiation, and expansion, generate mechanical forces (Robinson et al. 2013; Hamant and Moulia 2016). Cells in tissues receive force from surrounding cells, and interactions between these cells are important for development and morphogenesis. In other words, plants must possess the ability to alter their developmental and environment-responsive programs in response to the strength, direction, and frequency of mechanical stimuli.

Mechanosensing and mechanoresponses have long been investigated. For example, 130 years ago, Darwin found that when he rubbed the petiole of a young leaf of *Clematis glandulosa* a few times at any angle, it eventually bent to that side in a few hours (Darwin 1888). Prior to this, Ciesielski discovered that by removing the root tip, the tip alone was sensitive to gravity. He also found that some "influences" (presumably mobile signals/hormones) were transmitted back to the elongation zone of the root from the tip, resulting in root bending. Darwin repeated and confirmed Ciesielski's observation 9 years later in 1880 (Ciesielski 1871; Darwin 1880). Their findings are remarkable in that they not only identified the position of gravity sensing in roots but also they predicted the presence of a long-distance signal transduction mechanism. Current studies on the mechanical regulation of plant growth have advanced to a considerable extent and have recently been reviewed (Moulia et al. 2015; Groover 2016).

Studies on mechanosensing and mechanotransduction at the molecular and cellular levels have been progressing rapidly since a quarter of a century ago. One of the pioneering discoveries was an immediate and transient increase in the cytoplasmic Ca^{2+} concentration, $[Ca^{2+}]_{cyt}$ in response to touch. This was demonstrated with transgenic tobacco seedlings expressing the Ca^{2+}-binding photoprotein aequorin of the jellyfish *Aequorea victoria* (Knight et al. 1991). A touch stimulus appears to be sensed and responded to by tobacco seedlings in seconds, even if they do not exhibit any obvious or visible growth/developmental responses. This also suggests that one of the mechanosensors is a Ca^{2+}-permeable mechanosensitive (MS) channel. The involvement of Ca^{2+} in the early phase of mechanoresponses is supported by the discovery that the expression of the Ca^{2+}-binding regulatory proteins, calmodulin and calmodulin-related proteins, is induced by rain, wind, and touch within 0.5 h of the mechanical stimulation (Braam and Davis 1990).

According to previous physiological/genetic studies, there are three principally different groups of mechanosensors in plants: MS channels, receptor-like kinases, and the cytoskeleton. We discuss recent advances in our understanding of the molecular and cellular aspects of mechanosensing and subsequent mechanotransduction by referring to regulatory molecules and cellular components, including mechanosensors, membrane lipids, ROS, and organelles such as chloroplasts.

Mechanosensitive Channels as Sensors

The function of MS channels or stretch-activated (SA) channels is to sense mechanical stresses given to the cell and to generate biochemical and electrical signals. In bacteria, these channels function as safety valves, releasing intracellular ions and small organic molecules such as amino acids upon hypoosmotic shock (Levina et al. 1999). MS channels may be opened by two mechanisms. The first mechanism is the force-from-lipid mechanism, which has been unambiguously demonstrated in bacterial MS channels of small conductance (MscS) and large conductance (MscL) as well as vertebrate MS K^+ channels (TREK1 and TRAAK) (Anishkin et al. 2014; Brohawn 2015; Teng et al. 2015). In this mechanism, the stretching of the lipid bilayers of cellular membranes alters force vectors on MS channels to induce conformational changes, thereby opening the channel (Fig. 2a). The other mechanism is tether-dependent; tethering proteins receive a mechanical force and then pull an MS channel protein directly or the lipid bilayers surrounding the MS channels (Teng et al. 2015) (Fig. 2b). In the latter case, the tethering proteins cause membrane distortion to open the MS channels. Animal tethers include microtubules, actin filaments, and ß spectrin, which link the plasma membrane to the actin cytoskeleton (Krieg et al. 2015).

In plants, MS channels were initially discovered using electrophysiological tools in tobacco protoplasts (Falke et al. 1988), and genes encoding MscS-like (MSL) proteins were then found in the *Arabidopsis thaliana* genome through comparative protein sequence analyses (Kloda and Martinac 2002). Similar analyses identified plant homologs of the animal MS channels, Piezo1 and Piezo2 (Coste et al. 2010). Genetic screening also revealed additional plant MS channels. The Arabidopsis MS channel, *mid1*-complementing activity 1 (MCA1), was established by screening cDNA that complements the conditionally lethal phenotype of the *mating pheromone-induced death 1* (*mid1*) mutant of the yeast *Saccharomyces cerevisiae*, defective in Ca^{2+} influx (Nakagawa et al. 2007). Another genetic screening involving Arabidopsis mutants

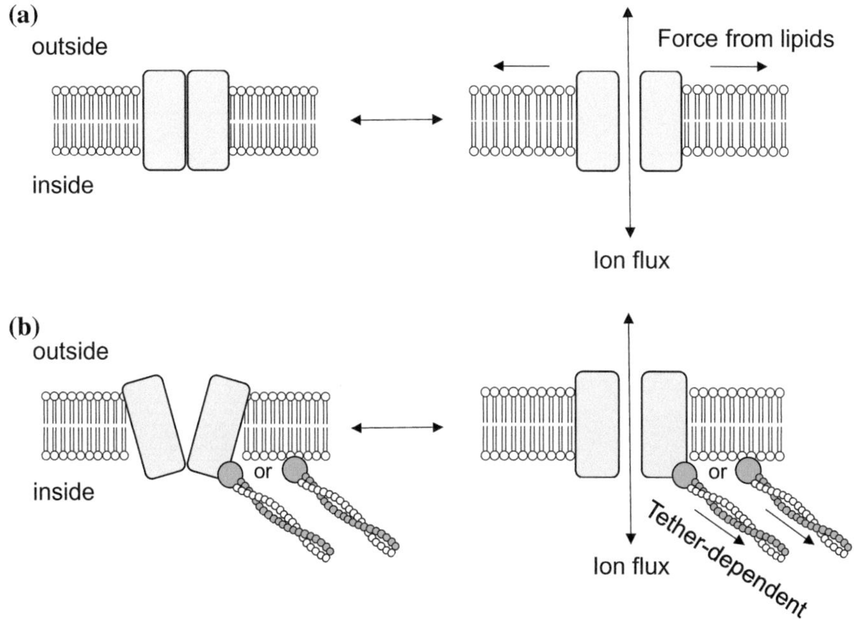

Fig. 2 Possible models for how MS channels are gated. **a** The force-from-lipid mechanism. **b** The tether-dependent mechanism. The closed state (left) and gated state (right) are shown

with low hyperosmolality-induced $[Ca^{2+}]_{cyt}$ increases resulted in the discovery of hyperosmolality-gated Ca^{2+}-permeable channel, OSCA1 (Yuan et al. 2014).

Studies on the four MS channel families will contribute to elucidating the molecular mechanisms underlying mechanosensing and mechanotransduction that influence a number of cellular and systemic responses to mechanical stimuli in plants. In this section, we will discuss current knowledge as well as the fundamentals. Some additional significant features of the families will also be addressed in the next section.

MSL Family

There are ten members of the MSL protein family in *A. thaliana* (*MSL1-MSL10*), seven in *Populus trichocarpa*, six in *Oryza sativa*, and two in *Zea mays* (Haswell et al. 2008). While being structurally homologous to each other, MSL proteins are not homologous to the bacterial MscS for the overall amino acid sequence, sharing only a limited conserved region, which includes the C-terminal transmembrane (TM) segment corresponding to the pore-forming TM3 segment of MscS.

Among plant MSL proteins, Arabidopsis MSL proteins have been studied intensively by Haswell and co-workers (Wilson et al. 2013; Hamilton et al. 2015). Based on their primary structures, they are sorted into three phylogenetic groups that cor-

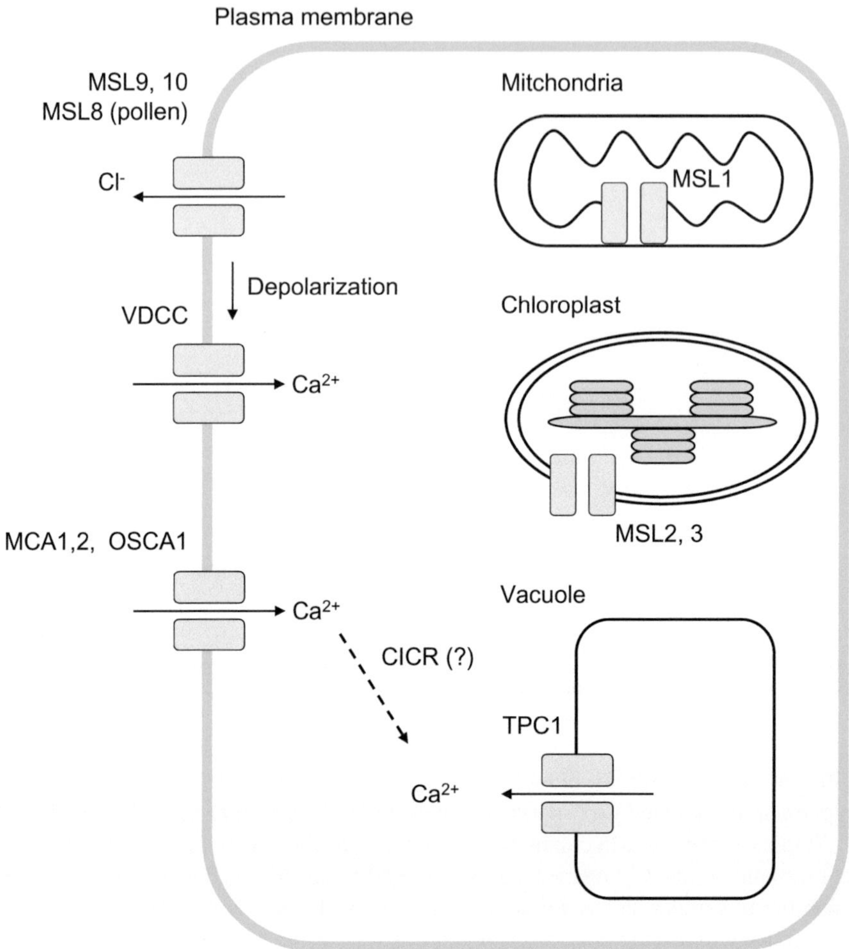

Fig. 3 Subcellular localization of MS channels in Arabidopsis cells. Voltage-gated Ca^{2+} channels (VDCCs) in the plasma membrane and two-pore channel 1 (TPC1) in the vacuolar membrane, which are expected to be secondarily activated by mechanical stresses, are also shown

respond well to three subcellular localizations (Fig. 3). Groups I and II have five putative TM segments and an N-terminal transit sequence. Group I proteins are MSL1, which are localized in the inner membrane of the mitochondrion. Group II proteins are MSL2 and MSL3, which are localized in the chloroplast. Group III proteins include all other MSL proteins: MSL4-10, all of which have six putative TM segments. MSL9 and MSL10 are localized in the plasma membrane.

Although it has been established that the bacterial MscS protein assembles into a homo-heptamer to form the channel protein (Booth and Blount 2012), the higher order structure of Arabidopsis MSL proteins is still unknown. Thus, difficulties are associated with clarifying the regulatory mechanisms by which MSL channels sense mechanical stress and how they transduce primary signals to cells. Due to diversity in subcellular localization and the number of TM segments, Arabidopsis MSL proteins appear to be regulated by a number of components present intracellularly and extracellularly. Further structural studies are anticipated.

In contrast to structural studies, physiological and electrophysiological studies have already made considerable advances. The currently characterized MSL proteins mostly form nonselective ion channels with a slight preference for anions over cations, including Ca^{2+} and Na^{+}. However, the efflux of anions such as Cl^{-} causes the depolarization of the plasma membrane, leading to the activation of voltage-dependent cation channels (VDCCs) that mediate Ca^{2+} influx (Fig. 3).

MSL1, with a single channel conductance of 1.2 nS, which is similar to that of *E. coli* MscS, is involved in the dissipation of the mitochondrial membrane potential and maintenance of redox homeostasis under abiotic stress (Lee et al. 2016).

MSL2 and MSL3 have common functions. They are components of the chloroplast division machinery involved in the size control of chloroplasts and nongreen plastids under hypoosmotic stress during normal plant growth (Wilson et al. 2011; Veley et al. 2012). MSL2 and MSL3 have conductances of ~45 and 140 pS, respectively, in root protoplasts (Haswell et al. 2008).

The genes encoding MSL4, MSL5, MSL6, MSL9, and MSL10 were disrupted to produce the *msl4 msl5 msl6 msl9 msl10* quintuple mutant. This mutant shows no MS channel activity in root protoplasts, but responds normally to a number of mechanical and osmotic stresses (Haswell et al. 2008). MSL7 has not yet been examined.

MSL8 is localized in the plasma membrane and endomembranes of pollen grains and germination tubes (Hamilton et al. 2015). It has a unitary conductance of 57 pS under negative membrane potentials and 39 pS under positive membrane potentials and is an anion-selective channel with a 6.3-fold preference for Cl^{-} over Na^{+}. MSL8 is required for pollen germination suppression, cellular integrity during germination, and the survival of mature pollen grains under hypoosmotic shock during rehydration.

MSL9, localized in the plasma membrane and endomembranes, has MS channel activity of ~45 pS in root protoplasts (Haswell et al. 2008). The physiological roles of this channel currently remain unknown.

MSL10 is localized in the plasma membrane and endomembranes of root protoplasts (Haswell et al. 2008) and has a conductance of ~100 pS MS with a 5.9-fold preference for Cl^{-} over Na^{+} when expressed in *Xenopus* oocytes (Maksaev and Haswell 2012). MSL10 has two distinct functions: as an MS channel and a cell death inducer. As described previously, even though the *msl4 msl5 msl6 msl9 msl10* quintuple mutant shows no detectable abnormalities in mechanical and osmotic stress responses, the cytoplasmic N-terminal domain of MSL10 with no MS channel activity is sufficient to induce cell death (Maksaev and Haswell 2012).

MCA Family

The MCA family is found exclusively in the plant kingdom (Nakagawa et al. 2007; Kurusu et al. 2013). There are two paralogous proteins of the MCA family in Arabidopsis, MCA1 and MCA2 (Nakagawa et al. 2007; Yamanaka et al. 2010). Both proteins are located in the plasma membrane and have 73% amino acid sequence identity as well as several structural motifs, including the EF hand-like, coiled-coil, and plac8 motifs. In contrast to the other plant MS channel proteins reported to date, MCA proteins bear unique structural features: they are small proteins composed of ~420 amino acid residues with only one TM segment near the N terminus (Kamano et al. 2015) and form a homo-tetramer (Nakano et al. 2011; Shigematsu et al. 2014). Only the N-terminal region (1-173 amino acids) of MCA1 and MCA2 lacking the coiled-coil and plac8 motifs is sufficient for Ca^{2+} influx when expressed in yeast cells (Nakano et al. 2011).

MCA1 has a conductance of 34 pS when expressed in *Xenopus* oocytes (Furuichi et al. 2012) and is required for Ca^{2+} influx, touch sensing in Arabidopsis roots, and the hypoosmotic response in Arabidopsis seedlings (Nakagawa et al. 2007). On the other hand, MCA2 is not required for touch sensing in roots under the conditions demonstrated by MCA1 (Yamanaka et al. 2010). This difference may be due to their expression regions: *MCA1* promoter::*GUS* is expressed in the root tip, while *MCA2* promoter::*GUS* is not. MCA1 is required for the initiation of lignin production and the regulation of osmosensitive changes of carbohydrate metabolism in response to cellulose biosynthesis inhibition (Denness et al. 2011; Wormit et al. 2012).

Nicotiana tabacum NtMCA1 and NtMCA2, which are found in the plasma membrane and Hechtian strands, mediate Ca^{2+} influx in cultured tobacco cells and $[Ca^{2+}]_{cyt}$ increases upon hypoosmotic shock when expressed in yeast cells (Kurusu et al. 2012c). *O. sativa* OsMCA1, also localized in the plasma membrane, mediates Ca^{2+} influx upon hypoosmotic shock and the addition of the membrane distortion reagent, trinitrophenol, to cultured rice cells, and is responsible for the hypoosmotic shock-induced generation of ROS (Kurusu et al. 2012a, b). *OsMCA1*-knockdown plants exhibit retarded growth and have shortened rachises (Kurusu et al. 2012b). A recessive point mutation in *OsMCA1* results in severe dwarf phenotypes with shorter and stunted leaves as well as fewer secondary branches during the vegetative and reproductive growth stages, primarily due to the inhibition of cell expansion caused by the upregulation of genes related to gibberellin deactivation (Liu et al. 2015). The maize (*Zea mays*) homolog of MCA proteins, cell number regulator13 (CNR13)/narrow odd dwarf (NOD), regulates cell division, expansion, and differentiation; the *nod-1* mutant exhibits an overall reduction in cell size and organ patterning defects and appears to have a constitutive upregulation of pathogen response pathways (Rosa et al. 2017).

OSCA Family

The OSCA family is found throughout eukaryotes. Arabidopsis and *O. sativa* have 15 and 11 paralogous proteins, respectively (Yuan et al. 2014; Li et al. 2015). Arabidopsis OSCA1, localized in the plasma membrane, is composed of 772 amino acid residues, has nine or ten putative TM segments, and forms a unique, hyperosmolarity-gated nonselective cation channel permeable to Ca^{2+}, which is presumably activated by the distortion of lipid bilayers. This channel has a conductance of ~49 pS, and the following permeability sequence: $K^+ > Ba^{2+} \approx Ca^{2+} > Na^+ = Mg^{2+} = Cs^+$. OSCA1 is required for abscisic acid-induced stomatal closure, water transpiration, and root growth under hyperosmotic conditions.

Piezo Family

Highly conserved during evolution, Piezos are very large proteins with ~2,500 amino acid residues that constitute functionally diverse MS channels (Coste et al. 2010). Mouse Piezo1 forms a trimeric propeller-like structure of ~900 kDa (Ge et al. 2015). Mammalian Piezo channels are involved in many cellular processes, including vascular development, gentle touch sensation, and dehydrated hereditary stomatocytosis (Volkers et al. 2015). Arabidopsis and *O. sativa* have both been reported to have a single Piezo; however, there is currently no detailed description of it.

Subcellularly Distributed Mechanosensors in Plants

Mechanosensors in the Plasma Membrane

Regarding external rapid mechanical stimuli such as osmotic shock or touch/bending, the MS channel at the plasma membrane plays a crucial role in the early events of mechanosensing and mechanotransduction because this type of mechanical stress directly causes the deformation of plant tissues/cells and changes the tension of the membrane (Toyota and Gilroy 2013). Osmotic shock and physical touch/bending are known to induce $[Ca^{2+}]_{cyt}$ increases in Arabidopsis (Nakagawa et al. 2007; Monshausen et al. 2009; Yuan et al. 2014). *MCA1* overexpression lines exhibit enhanced $[Ca^{2+}]_{cyt}$ increases in response to hypoosmotic shock (Nakagawa et al. 2007; Yuan et al. 2014), while *osca1* mutants show reduced $[Ca^{2+}]_{cyt}$ responses to hyperosmotic stresses. In addition, *osca1* mutants are susceptible to hyperosmolality conditions (Yuan et al. 2014), suggesting that the plasma membrane MS channels, MCA1 and OSCA1, are immediately activated by membrane stretching upon osmotic shock, and the resultant Ca^{2+} signal is significant for adapting to changes in osmotic conditions.

Pollen grains undergo dehydration and rehydration during their maturation and germination, respectively, throughout which the plasma membrane is subjected to various mechanical stresses due to osmotic shock. MS channels have been suggested to regulate the marked changes that occur in membrane dynamics during pollen development. Dutta and Robinson (2004) analyzed MS channel activities in the pollen grains and tubes of *Lilium longiflorum* in combination with the venom of the spider, *Grammostola spatulata*, which is known to inhibit MS channel activity in animals via the action of a peptide called GsMTx-4 (Chen et al. 1996; Suchyna et al. 2000; Bowman et al. 2007). Two distinct venom-sensitive Ca^{2+}- and K^{+}-permeable MS channels were electrophysiologically characterized in pollen protoplasts. Furthermore, the inhibitory effect of the venom on Ca^{2+} influx to the pollen tube was detected using a self-referencing microelectrode (Messerli et al. 1999). The venom also exerted physiological effects such as the inhibition of pollen tube elongation, disruption of the growth-related accumulation of vesicles at the pollen tube tip, the loss of cytoplasmic streaming, and eventual plasmolysis. These findings strongly suggest GsMTx-4-sensitive MS channel activity in plants and also that this venom toxin has potential as a powerful tool for identifying the associated activity, considering that the molecular nature of MS channel activities has not been identified for a long time. As previously discussed, the MS channel MSL8 was recently found to be localized at the plasma membrane and endomembranes in Arabidopsis pollen (Hamilton et al. 2015). Mature pollen grains in *msl8* mutants showed lower viability under hypoosmotic shock during rehydration and a higher germination rate than the wild type, suggesting that MSL8 plays a critical role in responding to changes in membrane tension during pollen hydration and germination.

The bending-induced $[Ca^{2+}]_{cyt}$ increase in Arabidopsis roots is a biphasic response consisting of two Ca^{2+} peaks and blocked by La^{3+}, the putative plasma membrane MS Ca^{2+} channel inhibitor (Monshausen et al. 2009). Although the MS channels required for the first Ca^{2+} peak have not yet been identified, the second Ca^{2+} peak involves the receptor-like kinase *FERONIA* in the plasma membrane (Shih et al. 2014). RALF (rapid alkalinization factor) is a secreted peptide hormone that binds to FERONIA and triggers a $[Ca^{2+}]_{cyt}$ increase and kinase activity (Haruta et al. 2014). However, the C-terminal intracellular kinase domain is not needed for the bending-induced $[Ca^{2+}]_{cyt}$ increase (Shih et al. 2014). Plants may have different types of mechanosensing and mechanotransduction systems for bending such as a channel type and kinase-independent hormone-receptor type.

Previous electrophysiological studies reported large variations in MS channel activities in the plasma membrane (Cosgrove and Hedrich 1991; Ding and Pickard 1993). Pharmacological analyses also suggested the involvement of plasma membrane MS channels in mechanosensing and mechanotransduction such as gravitational and rotational responses (Toyota et al. 2008b, 2013). Other MS channels including MCA1, MSL9, and MSL10 are expressed at the plasma membrane and have roles in the regulation of physiological processes. Since the plasma membrane, which is connected to the cell wall, is the boundary of plant cells, it is reasonable to assume that MS channels at the plasma membrane work as a mechanosensor converting external mechanical stimuli rapidly into intracellular signals.

Mechanosensors in Organelles

Wind-induced $[Ca^{2+}]_{cyt}$ increases are blocked by the endomembrane Ca^{2+} permeable channel inhibitor, ruthenium red (Knight et al. 1992), and since MSL8 is localized not only to the plasma membrane but also to endomembranes, it implies the contribution of intracellular mechanosensors to mechanosensing and mechanoresponses in plants. MSL2 and MSL3 are expressed in the plastid envelope (Haswell and Meyerowitz 2006) and spatially overlap with the plastid division protein, AtMinE (Maple et al. 2002). *msl2 msl3* double mutants possess enlarged spherical chloroplasts with multiple abnormal filamentous temperature-sensitive Z (FtsZ) rings (Wilson et al. 2011), while most wild-type plants have small ovoid chloroplasts with a single FtsZ ring. Since AtMinE and FtsZ rings are crucial for the proper division of chloroplasts, MSL2 and MSL3 are involved in the mechanisms underlying chloroplast division/formation, possibly via MS channel activity. The shapes and sizes of *msl2 msl3* double mutant chloroplasts were restored by increasing cytosolic osmolarity (Veley et al. 2012), and MSL3 rescues hypoosmotic shock-induced cell death in *E. coli* (Haswell and Meyerowitz 2006). This finding supports the model in which MSL2 and MSL3 regulate the size and shape of chloroplasts by controlling ion flux in response to membrane stretching, thereby protecting them from hypoosmotic shock.

Consistent with these ideas, a homolog of MscS, termed *MSC1*, was also cloned in *Chlamydomonas* and functionally reconstituted in *E. coli* (Nakayama et al. 2007). Although the full-length MSC1 was not as active as a channel in *E. coli*, when its N-terminal signal sequence was deleted, MSC1 exhibited SA channel activity. MSC1 appears to be an anion-selective SA channel, most probably passing Cl^-. Previous findings on its RNAi suppression are consistent with a role in plastid function (Nakayama et al. 2007).

MSL1, unlike MSL2 and MSL3, is localized in the inner mitochondrial membrane (Lee et al. 2016). MSL1 partially rescues hypoosmotic shock-induced cell death in *E. coli*, but does not affect the shape or size of isolated mitochondria under hypoosmotic conditions. *msl1* mutants showed a higher membrane potential in isolated mitochondria only under the condition in which F_1F_o-ATP synthase was inactivated due to ADP depletion or a treatment with an ATP synthase inhibitor. *msl1* mutants also showed a greater oxidative state for the mitochondrial glutathione pool under high-temperature and heavy-metal stress conditions. Taken together, these findings indicate that MSL1 dissipates the membrane potential and maintains the redox status in mitochondria perturbed by specific abiotic stressors such as high-temperature, heavy-metals, or reduced ATP synthase conditions.

Mechanosensors in the Cytosol

Do plants have additional mechanosensors in the cytosol? The cytoskeleton (microtubules and actin filaments) represents a good candidate for cytosolic mechanosensors. Microtubules are critical for the regulation of cell expansion through their interplay with cellulose synthesis. Therefore, it must play an integral role in the regulation of asymmetrical growth responses during mechanoresponses such as gravitropism. Primary cell walls are composed of cellulose microfibrils embedded in a polysaccharide matrix, providing strength and flexibility to plants (Carpita and Gibeaut 1993). Cell expansion, by loosening the cell wall, plays an essential role in plant morphogenesis; its direction strongly depends on the orientation of cellulose microfibrils (Taiz 1984; Baskin 2005).

Cortical microtubules beneath the plasma membrane most likely regulate the orientation of cellulose microfibril deposition by aligning it with the microtubule array, possibly by constraining the movement of the cellulose synthase complexes in the plasma membrane (Baskin 2001; Paredez et al. 2006; Gutierrez et al. 2009). Therefore, the orientation of cortical microtubules is important for assessing the direction of cell expansion in plants (Baskin et al. 1994). When plants are horizontally placed (gravistimulated), the principal orientation of cortical microtubules at the epidermis of the upper and lower flanks changes to longitudinal and transverse directions, respectively, in maize and sunflower (Nick et al. 1990; Blancaflor and Hasenstein 1993; Himmelspach et al. 1999). The longitudinal and transverse orientations of cortical microtubules are assumed to cause the slower and faster elongation of the organs, respectively, representing asymmetric growth leading to gravitropic bending. Thus, microtubules are the responsive element in the machinery that controls asymmetrical growth to the mechanical stress caused by gravity.

The direction of mechanical force was recently shown to be closely related to the orientation of microtubules (Moulia et al. 2015). When maize coleoptiles and azuki bean epicotyls were forced to bend downward, the orientation of cortical microtubules in extended (upper) and compressed (lower) flanks was transverse and longitudinal, respectively (Zandomeni and Schopfer 1994; Fischer and Schopfer 1997; Ikushima and Shimmen 2005). No change in the orientation of cortical microtubules was observed in coleoptiles or epicotyls that were kept straight, even during a gravistimulation. These findings suggest that the orientation of microtubules is directly/indirectly regulated by mechanical force in plant cells, and this is supported by another study using the shoot apical meristem of Arabidopsis (Hamant et al. 2008; Burian et al. 2013). The microtubule-depolymerizing drug, oryzalin, was found to inhibit the generation of anisotropic geometry in the meristem such as cell shape and boundary formation between the meristem and young primordia. The same treatment did not affect differentiation, phyllotaxis, or cell identity in the boundary domain, suggesting that microtubules are required for certain auxin-independent morphogenetic processes, particularly for the anisotropic structure of meristems. An image analysis showed that microtubules were highly dynamic (random) at the meristem summit and were circumferentially oriented at the periphery of the meristem, which is consistent with the direction of mechanical stress on a surface according to a pressure vessel model in which the meristem is regarded as an elastic shell inflated by an inner pressure. These findings suggest that the microtubule orientation reflecting maximal stress directions is critical for regulating the morphogenesis of meristems, possibly via the modulation of newly deposited cellulose microfibrils.

Actin filaments also play an important role in mechanoresponses including gravitropism. The actin filament-depolarizing drug, latrunculin B, attenuated $[Ca^{2+}]_{cyt}$ responses to a gravistimulation (Toyota et al. 2008b) and affected gravitropism (Hou et al. 2003, 2004; Palmieri and Kiss 2005; Nakamura et al. 2011). Although the directional relationship between actin filaments and mechanical force and whether actin filaments function as a tether that activates MS channels (Fig. 2b) remain unknown, actin filaments may regulate mechanosensing and mechanotransduction in plants.

Signal Transduction for Mechanosensing

Outline of Cell Signaling

Plants respond to various environmental stresses including mechanical stresses such as touch, wounding, and gravity. When cells perceive extracellular information, this information is immediately converted into intracellular signals that regulate and promote specific cellular responses. This conversion is mediated by receptors mainly localized at the plasma membrane: receptor kinases, G-protein coupled receptors (GPCR), and ion channels, including MS channels. Intracellular signaling molecules are composed of two groups: proteins and chemical compounds. Key members of the former are enzymes and regulatory proteins that change the state of activity through the phosphorylation/dephosphorylation and/or binding of ligands. Key members of the latter, termed second messengers, are cyclic nucleotides, phospholipids, and ions, such as cyclic adenosine monophosphate (cAMP), phosphatidylinositol 4,5-bisphosphate (PIP_2), and Ca^{2+}. These substances function as ligands that change their cytoplasmic concentrations through the activation of enzymes and channels. Thus, signals are not only transmitted but also enhanced by second messengers playing central roles in the promotion of specific cellular and, as a consequence, whole body responses to perceived information.

As described earlier, mechanical stresses, such as touch and hypotonicity, activate MS channels, resulting in transient increases in $[Ca^{2+}]_{cyt}$, and the depletion of Ca^{2+} attenuates subsequent plant responses. When $[Ca^{2+}]_{cyt}$ is transiently increased, Ca^{2+}-binding proteins such as calmodulin (CaM), the most abundant regulator proteins with multiple functions in the cell, are reversibly activated. Notably, previous studies revealed that Ca^{2+} signals are involved in almost all cell responses, growth, and metabolism. Thus, recent studies have focused on the mechanisms by which specific patterns of Ca^{2+} signals are generated and those that promote individual responses, termed the encoding and decoding of Ca^{2+} signals. A key feature of encoding is the oscillation of $[Ca^{2+}]_{cyt}$, a central regulator of cell signaling (Berridge 1997). The manipulation of $[Ca^{2+}]_{cyt}$ oscillations within defined ranges of frequency, transient number, duration, and amplitude mimicking abscisic acid (ABA)-induced oscillations promotes steady-state stomatal closure in Arabidopsis (Allen et al. 2001). The activation of a few Ca^{2+}-permeable channels, or even only one, may augment the mean Ca^{2+} signal because of the extremely low cytoplasmic Ca^{2+} concentration (approximately 10 nM). A local Ca^{2+} signal centered by a single MS channel has been observed in human endothelial cells (Hayakawa et al. 2008). Plant Ca^{2+}-permeable channels including MS channels are distributed in the plasma and endomembranes, are activated by individual conditions, and then promote specific cellular responses. Thus, the spatial distribution (locality) of Ca^{2+} signals should be crucial in the encoding of Ca^{2+} signals.

In the decoding of a Ca^{2+} signal, components of downstream signaling and response events caused by mediators and/or effectors, respectively, are expected to be localized near Ca^{2+}-permeable channels and/or form a complex with these channels. Calcineurin B-like proteins (CBLs) and the protein kinases of their partners (CIPKs, CBL-interacting protein kinases) have recently been considered as new members of mediators for Ca^{2+} signals. CBL–CIPK complexes are only formed when CBLs bind Ca^{2+}, then CIPKs exhibit phosphorylation activity (Weinl and Kudla 2009). Following the discovery of 10 CBLs and 25 CIPKs in Arabidopsis, their physiological roles, functions, and subcellular localizations are under investigation. Although the involvement of CBLs and CIPKs in mechanosensing remains undefined, the specific roles of some members have been identified. Under salt-stress conditions, the CBL1/4–CIPK24 complex localizes in the plasma membrane, and then activates the SOS1 Na^+/H^+ antiporter in the plasma membrane (Guo et al. 2001; Qiu et al. 2002). Interestingly, CBL10 also forms a complex with CIPK24, localizes in the tonoplast, and excludes Na^+ from the cytosol presumably by activating a Na^+/H^+ antiporter in the vacuolar membrane (Kim et al. 2007). Thus, CBLs guide CIPKs to promote the cellular responses precisely against the stresses.

Damage Sensing

Physical damage in nature, particularly by herbivorous insects, is one of the most severe stresses reducing crop yields. Consequently, plants show systemic, whole body responses to survive under these conditions by expressing a series of proteins, such as a proteinase inhibitor that tastes bitter and promotes diarrhea in herbivores. Key features of plant responses against physical damage are the production and emission of volatile compounds, such as jasmonates (JAs), salicylates (SAs), and green leaf volatiles, with multiple functions: the promotion of systemic responses, antibacterial activity, and herbivore avoidance. Volatiles induced by herbivore damage promote defense responses not only in the attacked plant but also in the neighboring plants through the volatile-induced Ca^{2+}-signal, indicating that plants use smell to communicate with nearby plants (Asai et al. 2009). Herbivore damage promotes the emission of terpenoids, while mechanical wounding promotes C_6-alcohol emission in lima beans; thus, the composition of volatiles varies depending on physiological damage (Arimura et al. 2000). While herbivore feeding markedly increases Ca^{2+} signals limited in a few cell layers near the wounded area, neither single nor repeated mechanical wounding induces such a significant response. Thus, it is suggested that oral factors, or herbivore-associated elicitors, of the herbivorous insects promote a herbivore-specific Ca^{2+} signal. (Zebelo and Maffei 2015) (and the references therein). In Arabidopsis, AtSR1 (SIGNAL RESPONSIVE1), a CaM-binding transcription factor that regulates glucosinolate metabolism, is an important component for plant resistance to insect herbivory (Laluk et al. 2012). In tea leaves, the expression of the CIPK gene, *CsEv9*, was significantly increased due to herbivory (Yang et al. 2011), suggesting that the CBL-CIPK pathway is also involved in decoding the herbivory-induced Ca^{2+} signal.

In addition to the transport of signal molecules, electrical responses (transmission of the membrane potential) and changes in hydraulic pressure are critical for systemic responses to physical damage, such as cutting or burning (Wildon et al. 1992; Stanković and Davies 1997, 1998). Fast electrical responses are expected to be initiated by ion leakage from disrupted cells and the subsequent change in the plasma membrane potential of neighboring cells, while relatively slow responses are transferred by the interplay between voltage-gated Ca^{2+}-permeable channels and NADPH oxidase, both of which are located in the plasma membrane. The Ca^{2+} signal activates NADPH oxidase, and vice versa, while extracellular ROS generated by NADPH oxidase activates Ca^{2+}-permeable channels in neighboring cells, which is similar to paracrine signaling in the synapses of the nervous system (Jeandroz et al. 2013; Steinhorst and Kudla 2013; Gilroy et al. 2014).

As a downstream event of a long-distance signal, recent studies revealed that wounding and herbivory both promote systemic Ca^{2+} signaling when the midrib is wounded (Kiep et al. 2015). Thus, the activation of MS channels in cells surrounding the vein in response to changes in hydraulic pressure may be involved; however, the underlying molecular mechanisms have not yet been elucidated in detail (Farmer et al. 2014). In addition to MS channels, the glutamate receptor-like channels, GLR3.3 and GLR3.6, and the vacuolar voltage-gated channel, two-pore channel 1 (TPC1), play an important role in long-distance signaling following wounding (Mousavi et al. 2013; Kiep et al. 2015). GLR may trigger cell-to-cell Ca^{2+} transmission (Farmer et al. 2014), and TPC1 may then maintain it via Ca^{2+}-induced Ca^{2+} release (CICR), which may amplify the Ca^{2+} signal (Evans et al. 2016) (Fig. 3).

Gravity Sensing

Gravity is a ubiquitous force on Earth that affects plant growth and development. As demonstrated by Charles Darwin and pioneers, the most famous plant response against gravity is the regulation of growth direction, termed gravitropism (or geotropism), in which the aerial parts of plants grow against the direction of gravity, whereas the roots grow toward it. The upward bending response of stems was observed within 30 min of plants being laid down horizontally. Even though this finding was reported more than one hundred years ago, the underlying molecular mechanism is still one of the most interesting topics in basic science, and numerous studies have been performed to elucidate how plants sense and respond to gravity.

The asymmetric transport and distribution of auxin, a plant hormone regulating cell elongation, are crucial in gravitropism. However, the key molecules for graviperception, or gravity sensors, have yet to be identified. A number of studies that focused on root gravitropism revealed that Ca^{2+} is involved; however, phenotypic observations under low Ca^{2+} or Ca^{2+}-depleted conditions need to be interpreted carefully. Since Ca^{2+} plays key roles in the building and remodeling of the cell wall and intercellular fibers, its depletion and shortage strongly inhibit cell division and elongation. By columella cells, root tips sense gravity, and gravitropic bending occurs in the elongation zone located shootward from the root tip. Gravity sensing, gravitropism, and their relationship need to be investigated and discussed separately, and the information obtained may then be combined.

Changes in plant positions with respect to the gravity vector promote the sedimentation of organelles with high density, i.e., amyloplasts (Nakamura et al. 2011; Toyota et al. 2014) and the distortion of plasma and endomembranes, leading to the activation of MS channels, including Ca^{2+}-permeable channels. By using fluorescent dyes and transgenic plants expressing aequorin, a Ca^{2+}-reporting photoprotein, a transient Ca^{2+} signal was immediately observed after an upsidedown reorientation with respect to the gravity vector. The upsidedown reorientation promoted a biphasic Ca^{2+} signal: the first spike signal within seconds originates from mechanical stress by turning, while the latter, lasting for 10–15 min, is caused by the upsidedown reorientation being dependent on the angular degree of this reorientation (Toyota et al. 2013). Ruthenium Red (RR), a potential inhibitor of endomembrane Ca^{2+}-permeable channels, inhibits the latter peak, but not the first spike, suggesting that these two Ca^{2+}-signals are operated in different manners, and the latter, a gravity-induced Ca^{2+}-signal, is enhanced by CICR (Fig. 3).

The intensity of the latter Ca^{2+} signal varied with the angular degree of reorientation, being maximal at 135°, suggesting that plants accurately sense their orientation with respect to the gravity vector (Plieth and Trewavas 2002). A biphasic Ca^{2+} signal with similar kinetics was also promoted by hypergravity stimuli using centrifugation (Toyota et al. 2007). Gravity-induced Ca^{2+} signals are attenuated by La^{3+} and Gd^{3+}, blockers of MS channels (Toyota et al. 2007, 2008b), indicating that changes in both the direction and intensity of gravity are perceived by common mechanisms, and that Ca^{2+}-permeable MS channels may play key roles in the gravity sensing of plants.

Inhibitors of auxin transport and vesicle trafficking have been shown to attenuate initial and secondary Ca^{2+} signals, while the onset time and rate of the rise were not significantly altered (Toyota et al. 2008a). In transgenic Arabidopsis plants expressing inositol polyphosphate 5-phosphatase, which hydrolyzes inositol 1,4,5-trisphosphate ($InsP_3$), basal $InsP_3$ levels were reduced, and consequently, a delay and reduction in gravitropism were observed (Perera et al. 2006). The secondary Ca^{2+} signal described above was markedly attenuated by an inhibitor of Ca^{2+} release from intracellular stores and an extracellular Ca^{2+} chelator, suggesting that the Ca^{2+} signal induced by the upsidedown reorientation is mediated by several Ca^{2+} entry paths, including CICR, the release of Ca^{2+} from internal stores activated by an influx of extracellular Ca^{2+}. As another initial event in response to the reorientation of gravity, intracellular ROS production has been observed in the convex endodermis of maize primary roots within 5 min (Joo et al. 2001). Gravitropism was attenuated by an antioxidant, and the asymmetric application of H_2O_2 promoted the curvature of horizontal roots, demonstrating that ROS is crucially involved in gravitropism.

Recent studies strongly suggest that plants show another response against gravity, called gravi-resistance. The spindly (elongated and slender) growth of the inflorescent stems of Arabidopsis has frequently been observed under microgravity in the International Space Station. In contrast, shorter and thicker inflorescent stems were noted in Arabidopsis grown under hypergravity by centrifugation (Hoson and Soga 2003). A rapid change in gravity intensity by centrifugation also promoted a Ca^{2+} signal with similar kinetic patterns to those when the plant was turned upsidedown. Although a series of MS channels, including Ca^{2+}-permeable MCA1 and MCA2, have been identified in the Arabidopsis genome along with their isolated mutants, the MS channels responsible for the perception of gravity have not yet been identified. A long-distance signal transmission pathway from columella cells, present in the root tip as sensory cells, to the elongation zone as the effector organ is also still obscure. The inactivation of CRK5, a CDPK-related protein kinase that phosphorylates the PIN2 auxin transporter, has been shown to inhibit elongation and gravitropism in the primary roots of Arabidopsis (Rigo et al. 2013). Furthermore, the extracellular application of auxin promotes Ca^{2+} signaling (Joo et al. 2001). This evidence suggests that Ca^{2+} signals play multiple roles in the perception of and responses to gravity.

Applications for Agriculture

Mild and severe mechanical stresses may affect plant growth and morphology through mechanosensing and its subsequent signaling events. Farmers empirically discovered the efficiency of applying mechanical stresses to crops and cultivars more than 300 years ago. One process, called "mugifumi" in Japanese, was to tread on the seedlings of autumn-sown wheat and barley to prevent spindly growth, while increasing tillers and ear length, leading to a better yield (Iida 2014). Another example in modern agriculture is the use of ceiling fans in greenhouses with a high planting density. The wind generated by fans not only exchanges gases (removes CO_2 and supplies O_2) but also exposes crops and vegetables to a mechanical stress, which attenuates spindly growth, resulting in good yields and quality.

A manned mission to Mars will take 300 days, during which time astronauts will be expected to eat at least 1.5 kg of food per day. Thus, our demands on "space agriculture" are increasing. In space, ceiling fans are essential to exchange gases because there is no heat convection under microgravity conditions. Due to microgravity conditions, plants grown on the International Space Station exhibited typical spindly growth and a longer height, but lower weight and fertility, leading to reduced productivity. In order to support sustainable food production, fine-tuning of the ceiling fans, namely, their frequency, duration, and amplitude, is needed in order to promote plant responses that inhibit spindly growth and boost yield, quality, and fertility. Techniques and devices to farm crops and vegetables in spacecrafts and/or a base on the Moon, which represent perfectly closed environments, will be very useful for developing plant factories of the next generation, particularly in desert and polluted areas.

Takemura et al. (2017) recently revealed that hypergravity conditions increased the biomass and photosynthetic activity of the moss, *Physcomitrella patens*. Taken together with plant phenotypes under microgravity conditions, the optimization of mechanosensing by the hyper- and/or desensitization of mechanosensors using genetic approaches and drugs may be useful for improving crop yields in fields as well as conserving energy in space agriculture.

References

Allen GJ, Chu SP, Harrington CL, Schumacher K, Hoffmann T, Tang YY et al (2001) A defined range of guard cell calcium oscillation parameters encodes stomatal movements. Nature 411:1053–1057

Anishkin A, Loukin SH, Teng J, Kung C (2014) Feeling the hidden mechanical forces in lipid bilayer is an original sense. Proc Natl Acad Sci USA 111:7898–7905

Arimura G, Ozawa R, Shimoda T, Nishioka T, Boland W, Takabayashi J (2000) Herbivory-induced volatiles elicit defence genes in lima bean leaves. Nature 406:512–515

Asai N, Nishioka T, Takabayashi J, Furuichi T (2009) Plant volatiles regulate the activities of Ca^{2+}-permeable channels and promote cytoplasmic calcium transients in Arabidopsis leaf cells. Plant Signal Behav 4:294–300

Baskin TI (2001) On the alignment of cellulose microfibrils by cortical microtubules: a review and a model. Protoplasma 215:150–171

Baskin TI (2005) Anisotropic expansion of the plant cell wall. Annu Rev Cell Dev Biol 21:203–222

Baskin TI, Wilson JE, Cork A, Williamson RE (1994) Morphology and microtubule organization in Arabidopsis roots exposed to oryzalin or taxol. Plant Cell Physiol 35:935–942

Berridge MJ (1997) The AM and FM of calcium signalling. Nature 386:759–760

Blancaflor EB, Hasenstein KH (1993) Organization of cortical microtubules in graviresponding maize roots. Planta 191:231–237

Booth IR, Blount P (2012) The MscS and MscL families of mechanosensitive channels act as microbial emergency release valves. J Bacteriol 194:4802–4809

Bowman CL, Gottlieb PA, Suchyna TM, Murphy YK, Sachs F (2007) Mechanosensitive ion channels and the peptide inhibitor GsMTx-4: history, properties, mechanisms and pharmacology. Toxicon 49:249–270

Braam J, Davis RW (1990) Rain-, wind-, and touch-induced expression of calmodulin and calmodulin-related genes in Arabidopsis. Cell 60:357–364

Brohawn SG (2015) How ion channels sense mechanical force: insights from mechanosensitive K2P channels TRAAK, TREK1, and TREK2. Ann NY Acad Sci 1352:20–32

Burian A, Ludynia M, Uyttewaal M, Traas J, Boudaoud A, Hamant O et al (2013) A correlative microscopy approach relates microtubule behaviour, local organ geometry, and cell growth at the Arabidopsis shoot apical meristem. J Exp Bot 64:5753–5767

Carpita NC, Gibeaut DM (1993) Structural models of primary cell walls in flowering plants: consistency of molecular structure with the physical properties of the walls during growth. Plant J 3:1–30

Chen Y, Simasko SM, Niggel J, Sigurdson WJ, Sachs F (1996) Ca^{2+} uptake in GH3 cells during hypotonic swelling: the sensory role of stretch-activated ion channels. Am J Physiol 270:C1790–C1798

Ciesielski T (1871) In Abwärtskrümmung der Wurzel. Inaug Dissert, Breslau

Cosgrove DJ, Hedrich R (1991) Stretch-activated chloride, potassium, and calcium channels coexisting in plasma membranes of guard cells of Vicia faba L. Planta 186:143–153

Coste B, Mathur J, Schmidt M, Earley TJ, Ranade S, Petrus MJ et al (2010) Piezo1 and Piezo2 are essential components of distinct mechanically activated cation channels. Science 330:55–60
Darwin C (1880) The power of movement in plants. John Murray, London
Darwin C (1888) The movements and habits of climbing plants. John Murray, London
Denness L, McKenna JF, Segonzac C, Wormit A, Madhou P, Bennett M et al (2011) Cell wall damage-induced lignin biosynthesis is regulated by a reactive oxygen species- and jasmonic acid-dependent process in Arabidopsis. Plant Physiol 156:1364–1374
Ding JP, Pickard BG (1993) Mechanosensory calcium-selective cation channels in epidermal cells. Plant J 3:83–110
Dutta R, Robinson KR (2004) Identification and characterization of stretch-activated ion channels in pollen protoplasts. Plant Physiol 135:1398–1406
Evans MJ, Choi WG, Gilroy S, Morris RJ (2016) A ROS-assisted calcium wave dependent on the AtRBOHD NADPH oxidase and TPC1 cation channel propagates the systemic response to salt stress. Plant Physiol 171:1771–1784
Falke LC, Edwards KL, Pickard BG, Misler S (1988) A stretch-activated anion channel in tobacco protoplasts. FEBS Lett 237:141–144
Farmer EE, Gasperini D, Acosta IF (2014) The squeeze cell hypothesis for the activation of jasmonate synthesis in response to wounding. New Phytol 204:282–288
Fischer K, Schopfer P (1997) Interaction of auxin, light, and mechanical stress in orienting microtubules in relation to tropic curvature in the epidermis of maize coleoptiles. Protoplasma 196:108–116
Furuichi T, Lida H, Sokabe M, Tatsumi H (2012) Expression of arabidopsis MCA1 enhanced mechanosensitive channel activity in the Xenopus laevis oocyte plasma membrane. Plant Signal Behav 7:1022–1026
Ge J, Li W, Zhao Q, Li N, Chen M, Zhi P et al (2015) Architecture of the mammalian mechanosensitive Piezo1 channel. Nature 527:64–69
Gilroy S, Suzuki N, Miller G, Choi WG, Toyota M, Devireddy AR et al (2014) A tidal wave of signals: calcium and ROS at the forefront of rapid systemic signaling. Trends Plant Sci 19:623–630
Groover A (2016) Gravitropisms and reaction woods of forest trees—evolution, functions and mechanisms. New Phytol 211:790–802
Guo Y, Halfter U, Ishitani M, Zhu JK (2001) Molecular characterization of functional domains in the protein kinase SOS2 that is required for plant salt tolerance. Plant Cell 13:1383–1400
Gutierrez R, Lindeboom JJ, Paredez AR, Emons AM, Ehrhardt DW (2009) Arabidopsis cortical microtubules position cellulose synthase delivery to the plasma membrane and interact with cellulose synthase trafficking compartments. Nat Cell Biol 11:797–806
Hamant O, Heisler MG, Jonsson H, Krupinski P, Uyttewaal M, Bokov P et al (2008) Developmental patterning by mechanical signals in Arabidopsis. Science 322:1650–1655
Hamant O, Moulia B (2016) How do plants read their own shapes? New Phytol 212:333–337
Hamilton ES, Jensen GS, Maksaev G, Katims A, Sherp AM, Haswell ES (2015) Mechanosensitive channel MSL8 regulates osmotic forces during pollen hydration and germination. Science 350:438–441
Haruta M, Sabat G, Stecker K, Minkoff BB, Sussman MR (2014) A peptide hormone and its receptor protein kinase regulate plant cell expansion. Science 343:408–411
Haswell ES, Meyerowitz EM (2006) MscS-like proteins control plastid size and shape in Arabidopsis thaliana. Curr Biol 16:1–11
Haswell ES, Peyronnet R, Barbier-Brygoo H, Meyerowitz EM, Frachisse JM (2008) Two MscS homologs provide mechanosensitive channel activities in the Arabidopsis root. Curr Biol 18:730–734
Hayakawa K, Tatsumi H, Sokabe M (2008) Actin stress fibers transmit and focus force to activate mechanosensitive channels. J Cell Sci 121:496–503
Himmelspach R, Wymer CL, Lloyd CW, Nick P (1999) Gravity-induced reorientation of cortical microtubules observed in vivo. Plant J 18:449–453
Hoson T, Soga K (2003) New aspects of gravity responses in plant cells. Int Rev Cytol 229:209–244

Hou G, Kramer VL, Wang YS, Chen R, Perbal G, Gilroy S et al (2004) The promotion of gravitropism in Arabidopsis roots upon actin disruption is coupled with the extended alkalinization of the columella cytoplasm and a persistent lateral auxin gradient. Plant J 39:113–125

Hou G, Mohamalawari DR, Blancaflor EB (2003) Enhanced gravitropism of roots with a disrupted cap actin cytoskeleton. Plant Physiol 131:1360–1373

Iida H (2014) Mugifumi, a beneficial farm work of adding mechanical stress by treading to wheat and barley seedlings. Front Plant Sci 5:453

Ikushima T, Shimmen T (2005) Mechano-sensitive orientation of cortical microtubules during gravitropism in azuki bean epicotyls. J Plant Res 118:19–26

Jeandroz S, Lamotte O, Astier J, Rasul S, Trapet P, Besson-Bard A et al (2013) There's more to the picture than meets the eye: nitric oxide cross talk with Ca^{2+} signaling. Plant Physiol 163:459–470

Joo JH, Bae YS, Lee JS (2001) Role of auxin-induced reactive oxygen species in root gravitropism. Plant Physiol 126:1055–1060

Kamano S, Kume S, Iida K, Lei KJ, Nakano M, Nakayama Y et al (2015) Transmembrane topologies of Ca^{2+}-permeable mechanosensitive channels MCA1 and MCA2 in Arabidopsis thaliana. J Biol Chem 290:30901–30909

Kiep V, Vadassery J, Lattke J, Maass JP, Boland W, Peiter E et al (2015) Systemic cytosolic Ca^{2+} elevation is activated upon wounding and herbivory in Arabidopsis. New Phytol 207:996–1004

Kim BG, Waadt R, Cheong YH, Pandey GK, Dominguez-Solis JR, Schultke S et al (2007) The calcium sensor CBL10 mediates salt tolerance by regulating ion homeostasis in Arabidopsis. Plant J 52:473–484

Kloda A, Martinac B (2002) Common evolutionary origins of mechanosensitive ion channels in Archaea, Bacteria and cell-walled Eukarya. Archaea 1:35–44

Knight MR, Campbell AK, Smith SM, Trewavas AJ (1991) Transgenic plant aequorin reports the effects of touch and cold-shock and elicitors on cytoplasmic calcium. Nature 352:524–526

Knight MR, Smith SM, Trewavas AJ (1992) Wind-induced plant motion immediately increases cytosolic calcium. Proc Natl Acad Sci USA 89:4967–4971

Krieg M, Dunn AR, Goodman MB (2015) Mechanical systems biology of C. elegans touch sensation. BioEssays 37:335–344

Kurusu T, Kuchitsu K, Nakano M, Nakayama Y, Iida H (2013) Plant mechanosensing and Ca^{2+} transport. Trends Plant Sci 18:227–233

Kurusu T, Iida H, Kuchitsu K (2012a) Roles of a putative mechanosensitive plasma membrane Ca^{2+}-permeable channel OsMCA1 in generation of reactive oxygen species and hypo-osmotic signaling in rice. Plant Signal Behav 7:796–798

Kurusu T, Nishikawa D, Yamazaki Y, Gotoh M, Nakano M, Hamada H et al (2012b) Plasma membrane protein OsMCA1 is involved in regulation of hypo-osmotic shock-induced Ca^{2+} influx and modulates generation of reactive oxygen species in cultured rice cells. BMC Plant Biol 12:11

Kurusu T, Yamanaka T, Nakano M, Takiguchi A, Ogasawara Y, Hayashi T et al (2012c) Involvement of the putative Ca^{2+}-permeable mechanosensitive channels, NtMCA1 and NtMCA2, in Ca^{2+} uptake, Ca^{2+}-dependent cell proliferation and mechanical stress-induced gene expression in tobacco (Nicotiana tabacum) BY-2 cells. J Plant Res 125:555–568

Laluk K, Prasad KV, Savchenko T, Celesnik H, Dehesh K, Levy M et al (2012) The calmodulin-binding transcription factor SIGNAL RESPONSIVE1 is a novel regulator of glucosinolate metabolism and herbivory tolerance in Arabidopsis. Plant Cell Physiol 53:2008–2015

Lee CP, Maksaev G, Jensen GS, Murcha MW, Wilson ME, Fricker M et al (2016) MSL1 is a mechanosensitive ion channel that dissipates mitochondrial membrane potential and maintains redox homeostasis in mitochondria during abiotic stress. Plant J 88:809–825

Levina N, Totemeyer S, Stokes NR, Louis P, Jones MA, Booth IR (1999) Protection of Escherichia coli cells against extreme turgor by activation of MscS and MscL mechanosensitive channels: identification of genes required for MscS activity. EMBO J 18:1730–1737

Li Y, Yuan F, Wen Z, Li Y, Wang F, Zhu T et al (2015) Genome-wide survey and expression analysis of the OSCA gene family in rice. BMC Plant Biol 15:261

Liu Z, Cheng Q, Sun Y, Dai H, Song G, Guo Z et al (2015) A SNP in OsMCA1 responding for a plant architecture defect by deactivation of bioactive GA in rice. Plant Mol Biol 87:17–30

Maksaev G, Haswell ES (2012) MscS-Like10 is a stretch-activated ion channel from Arabidopsis thaliana with a preference for anions. Proc Natl Acad Sci USA 109:19015–19020

Maple J, Chua NH, Moller SG (2002) The topological specificity factor AtMinE1 is essential for correct plastid division site placement in Arabidopsis. Plant J 31:269–277

Messerli MA, Danuser G, Robinson KR (1999) Pulsatile influxes of H^+, K^+ and Ca^{2+} lag growth pulses of Lilium longiflorum pollen tubes. J Cell Sci 112:1497–1509

Monshausen GB, Bibikova TN, Weisenseel MH, Gilroy S (2009) Ca^{2+} regulates reactive oxygen species production and pH during mechanosensing in Arabidopsis roots. Plant Cell 21:2341–2356

Moulia B, Coutand C, Julien JL (2015) Mechanosensitive control of plant growth: bearing the load, sensing, transducing, and responding. Front Plant Sci 6:52

Mousavi SA, Chauvin A, Pascaud F, Kellenberger S, Farmer EE (2013) GLUTAMATE RECEPTOR-LIKE genes mediate leaf-to-leaf wound signalling. Nature 500:422–426

Nakagawa Y, Katagiri T, Shinozaki K, Qi Z, Tatsumi H, Furuichi T et al (2007) Arabidopsis plasma membrane protein crucial for Ca^{2+} influx and touch sensing in roots. Proc Natl Acad Sci USA 104:3639–3644

Nakamura M, Toyota M, Tasaka M, Morita MT (2011) An Arabidopsis E3 Ligase, SHOOT GRAVITROPISM9, Modulates the Interaction between Statoliths and F-Actin in Gravity Sensing. Plant Cell 23:1830–1848

Nakano M, Iida K, Nyunoya H, Iida H (2011) Determination of structural regions important for Ca(2+) uptake activity in Arabidopsis MCA1 and MCA2 expressed in yeast. Plant Cell Physiol 52:1915–1930

Nakayama Y, Fujiu K, Sokabe M, Yoshimura K (2007) Molecular and electrophysiological characterization of a mechanosensitive channel expressed in the chloroplasts of Chlamydomonas. Proc Natl Acad Sci USA 104:5883–5888

Nick P, Bergfeld R, Schafer E, Schopfer P (1990) Unilateral reorientation of microtubules at the outer epidermal wall during photo- and gravitropic curvature of maize coleoptiles and sunflower hypocotyls. Planta 181:162–168

Palmieri M, Kiss JZ (2005) Disruption of the F-actin cytoskeleton limits statolith movement in Arabidopsis hypocotyls. J Exp Bot 56:2539–2550

Paredez AR, Somerville CR, Ehrhardt DW (2006) Visualization of cellulose synthase demonstrates functional association with microtubules. Science 312:1491–1495

Perera IY, Hung CY, Brady S, Muday GK, Boss WF (2006) A universal role for inositol 1,4,5-trisphosphate-mediated signaling in plant gravitropism. Plant Physiol 140:746–760

Plieth C, Trewavas AJ (2002) Reorientation of seedlings in the earth's gravitational field induces cytosolic calcium transients. Plant Physiol 129:786–796

Qiu QS, Guo Y, Dietrich MA, Schumaker KS, Zhu JK (2002) Regulation of SOS1, a plasma membrane Na^+/H^+ exchanger in Arabidopsis thaliana, by SOS2 and SOS3. Proc Natl Acad Sci USA 99:8436–8441

Rigo G, Ayaydin F, Tietz O, Zsigmond L, Kovacs H, Pay A et al (2013) Inactivation of plasma membrane-localized CDPK-RELATED KINASE5 decelerates PIN2 exocytosis and root gravitropic response in Arabidopsis. Plant Cell 25:1592–1608

Robinson S, Burian A, Couturier E, Landrein B, Louveaux M, Neumann ED et al (2013) Mechanical control of morphogenesis at the shoot apex. J Exp Bot 64:4729–4744

Rosa M, Abraham-Juarez MJ, Lewis MW, Fonseca JP, Tian W, Ramirez V et al (2017) The Maize MID-COMPLEMENTING ACTIVITY homolog CELL NUMBER REGULATOR13/NARROW ODD DWARF coordinates organ growth and tissue patterning. Plant Cell 29:474–490

Shigematsu H, Iida K, Nakano M, Chaudhuri P, Iida H, Nagayama K (2014) Structural characterization of the mechanosensitive channel candidate MCA2 from Arabidopsis thaliana. PLoS ONE 9:e87724

Shih HW, Miller ND, Dai C, Spalding EP, Monshausen GB (2014) The receptor-like kinase FERONIA is required for mechanical signal transduction in Arabidopsis seedlings. Curr Biol 24:1887–1892

Stanković B, Davies E (1997) Intercellular communication in plants: electrical stimulation of proteinase inhibitor gene expression in tomato. Planta 202:402–406

Stanković B, Davies E (1998) The wound response in tomato involves rapid growth and electrical responses, systemically up-regulated transcription of proteinase inhibitor and calmodulin and down-regulated translation. Plant Cell Physiol 39:268–274

Steinhorst L, Kudla J (2013) Calcium and reactive oxygen species rule the waves of signaling. Plant Physiol 163:471–485

Suchyna TM, Johnson JH, Hamer K, Leykam JF, Gage DA, Clemo HF et al (2000) Identification of a peptide toxin from Grammostola spatulata spider venom that blocks cation-selective stretch-activated channels. J Gen Physiol 115:583–598

Taiz L (1984) Plant cell expansion: regulation of cell wall mechanical properties. Annu Rev Plant Physiol 35:585–657

Takemura K, Kamachi H, Kume A, Fujita T, Karahara I, Hanba YT (2017) A hypergravity environment increases chloroplast size, photosynthesis, and plant growth in the moss Physcomitrella patens. J Plant Res 130:181–192

Teng J, Loukin S, Anishkin A, Kung C (2015) The force-from-lipid (FFL) principle of mechanosensitivity, at large and in elements. Pflugers Arch 467:27–37

Toyota M, Furuichi T, Tatsumi H, Sokabe M (2007) Hypergravity stimulation induces changes in intracellular calcium concentration in Arabidopsis seedlings. Adv Space Res 39:1190–1197

Toyota M, Furuichi T, Tatsumi H, Sokabe M (2008a) Critical consideration on the relationship between auxin transport and calcium transients in gravity perception of Arabidopsis seedlings. Plant Signal Behav 3:521–524

Toyota M, Furuichi T, Tatsumi H, Sokabe M (2008b) Cytoplasmic calcium increases in response to changes in the gravity vector in hypocotyls and petioles of Arabidopsis seedlings. Plant Physiol 146:505–514

Toyota M, Furuichi T, Sokabe M, Tatsumi H (2013) Analyses of a gravistimulation-specific Ca^{2+} signature in Arabidopsis using parabolic flights. Plant Physiol 163:543–554

Toyota M, Gilroy S (2013) Gravitropism and mechanical signaling in plants. Am J Bot 100:111–125

Toyota M, Ikeda N, Tasaka M, Morita MT (2014) Centrifuge microscopy to analyze the sedimentary movements of amyloplasts. Bio-protocol 4:e1229

Veley KM, Marshburn S, Clure CE, Haswell ES (2012) Mechanosensitive channels protect plastids from hypoosmotic stress during normal plant growth. Curr Biol 22:408–413

Volkers L, Mechioukhi Y, Coste B (2015) Piezo channels: from structure to function. Pflugers Arch 467:95–99

Weinl S, Kudla J (2009) The CBL-CIPK Ca^{2+}-decoding signaling network: function and perspectives. New Phytol 184:517–528

Wildon D, Thain J, Minchin P, Gubb I, Reilly A, Skipper Y et al (1992) Electrical signalling and systemic proteinase inhibitor induction in the wounded plant. Nature 360:62–65

Wilson ME, Jensen GS, Haswell ES (2011) Two mechanosensitive channel homologs influence division ring placement in Arabidopsis chloroplasts. Plant Cell 23:2939–2949

Wilson ME, Maksaev G, Haswell ES (2013) MscS-like mechanosensitive channels in plants and microbes. Biochemistry 52:5708–5722

Wormit A, Butt SM, Chairam I, McKenna JF, Nunes-Nesi A, Kjaer L et al (2012) Osmosensitive changes of carbohydrate metabolism in response to cellulose biosynthesis inhibition. Plant Physiol 159:105–117

Yamanaka T, Nakagawa Y, Mori K, Nakano M, Imamura T, Kataoka H et al (2010) MCA1 and MCA2 that mediate Ca^{2+} uptake have distinct and overlapping roles in Arabidopsis. Plant Physiol 152:1284–1296

Yang H, Xie S, Wang L, Jing S, Zhu X, Li X et al (2011) Identification of up-regulated genes in tea leaves under mild infestation of green leafhopper. Sci Hortic 130:476–481

Yuan F, Yang H, Xue Y, Kong D, Ye R, Li C et al (2014) OSCA1 mediates osmotic-stress-evoked Ca^{2+} increases vital for osmosensing in Arabidopsis. Nature 514:367–371

Zandomeni K, Schopfer P (1994) Mechanosensory microtubule reorientation in the epidermis of maize coleoptiles subjected to bending stress. Protoplasma 182:96–101

Zebelo SA, Maffei ME (2015) Role of early signalling events in plant-insect interactions. J Exp Bot 66:435–448

Biomechanics and Functional Morphology of Plants—Inspiration for Biomimetic Materials and Structures

Thomas Speck, Georg Bold, Tom Masselter, Simon Poppinga, Stefanie Schmier, Marc Thielen and Olga Speck

Abstract During the last decades, biomimetics has attracted increasing attention as well from basic and applied research as from various fields of industry. Biomimetics has a high innovation potential and offers the possibility for the development of sustainable technical products and production chains. Novel sophisticated methods for quantitatively analyzing and simulating the form–structure–function relationship on various hierarchical levels allow new fascinating insights into multi-scale mechanics and other functional parameter spaces of biological materials systems. On the other hand, new production methods enable for the first time the transfer of many outstanding properties of the biological role models into innovative biomimetic products at reasonable costs. Presented examples of biomimetic developments and products inspired by plants include branched and unbranched fiber-reinforced lightweight composite materials, structural materials with a high energy dissipation capacity as fiber-reinforced graded foams and compound materials, solutions for elastic architecture as the biomimetic façade-shading systems Flectofin® and Flectofold inspired by the bird of paradise flower and the waterwheel plant, respectively. Finally, a short overview of bioinspired self-repairing materials is given and a short discussion of the potential of biomimetic products to contribute to sustainable material development is presented.

Keywords Biomimetics · Fiber-reinforced composite · Self-repairing materials systems · Damping material and fracture toughness · Elastic architecture · Façade shading

---

T. Speck (✉) · G. Bold · T. Masselter · S. Poppinga · S. Schmier · M. Thielen · O. Speck
Plant Biomechanics Group Freiburg, Faculty of Biology, Botanic Garden, University of Freiburg, Freiburg im Breisgau, Germany
e-mail: thomas.speck@biologie.uni-freiburg.de

T. Speck · G. Bold · T. Masselter · S. Schmier · M. Thielen · O. Speck
Freiburg Center for Interactive Materials and Bioinspired Technologies (FIT), Freiburg im Breisgau, Germany

T. Speck · T. Masselter · S. Poppinga · M. Thielen
Freiburg Materials Research Center (FMF), Freiburg im Breisgau, Germany

T. Speck · T. Masselter · O. Speck
Competence Network Biomimetics, Freiburg im Breisgau, Germany

A. Geitmann and J. Gril (eds.), *Plant Biomechanics*,
https://doi.org/10.1007/978-3-319-79099-2_18

Introduction

Plants have evolved an incredible plethora of functions and underlying structures in adaptation to various aut- and synecological constraints. After 3.8 billion years of biological evolution, plants have not only developed a huge number of different life forms and colonized nearly all ecological niches but by having done so, plants can also provide answers to many technical problems which are of increasing interest for engineering and architecture.

The aim of biomimetic research is to analyze and tap nature's huge reservoir of such solutions. Biomimetics has a high innovation potential and offers the possibility of developing novel technical products and production chains. Mechanical features of plants represent one central field of interest in biomimetic research as plants and their organs have evolved many fascinating features to deal with different mechanical loads they are exposed to. The scale-overarching mechanical features are typically based on the pronounced hierarchical structure of plants which may include more than seven hierarchical levels and cover up 11 orders of magnitude (e.g., Fratzl et al. 2016; Knippers et al. 2016; Knippers and Speck 2012).

During the last decade, especially plant surfaces and materials systems were the focus of interest in biomimetic research. Bioinspired developments and products based on inspirations from structural and mechanical properties of plant surfaces are among the most impressive and successful examples of biomimetic research and include self-cleaning surfaces, attachment and anti-attachment structures, as well as special coatings for air-retention and antifouling (Barthlott et al. 2016, 2017; Gorb 2009; Prüm et al. 2012, 2013). In this chapter, however, we will concentrate on plant materials and structures as concept generators for novel bioinspired materials systems and structures which also represent a quickly emerging field in biomimetics research not least because of the fascinating mechanical properties found in the biological role models (Fratzl et al. 2016; Knippers and Speck 2012). As biological solutions are considered as cost-efficient, multifunctional, and environmentally friendly, they have attracted increasing attention as well from basic and applied research as from various fields of industry, architecture, and building construction. However, the aspect of sustainability has to be tested for each biomimetic product separately (Antony et al. 2014; Speck et al. 2017).

Novel sophisticated methods for quantitatively analyzing and simulating the form–structure–function relationship on various hierarchical levels allow new fascinating insights into multi-scale mechanics and other functions of biological materials and surfaces. On the other hand, new production methods enable for the first time the transfer of many outstanding properties of the biological role models into innovative biomimetic products at reasonable costs.

In this chapter, we present examples of plant-inspired biomimetic material systems and products developed during the last decade in the Plant Biomechanics Group Freiburg in close collaboration with colleagues from applied research and industry. The main focus is laid on bioinspired lightweight and damping materials and structures as well as on solutions for elastic architecture and self-repairing materials systems. Examples for lightweight materials with excellent mechanical properties are branched and unbranched fiber-reinforced composite materials and jackets for concrete pillars inspired by bamboo, horsetails, and branched stems of dragon trees, dwarf umbrella trees, and columnar cacti. For the development of structural materials with a high energy dissipation capacity and fracture toughness, role models include fiber-reinforced graded foams and compound materials which are inspired by fruit peels (pomelo, coconut) and tree bark (giant sequoia). As examples for elastic architecture, the biomimetic façade-shading systems Flectofin® and Flectofold are presented which are inspired by the bird of paradise flower and the waterwheel plant, respectively. In the last two paragraphs, a concise overview of bioinspired self-repairing materials systems is presented and a short discussion of the potential of biomimetic products to contribute to sustainable material development.

Branched and Unbranched Biomimetic Fiber-Reinforced Composites: Methods for Analysis and Technical Implementation

For several decades, plant stems have been analyzed as to their potential as role models for the development of bioinspired linear fiber-reinforced composites. For example, culms of the giant reed and of horsetails served as biological concept generators for the so-called technical plant stem (Milwich et al. 2006) which was custom built for the use in automotive industry and aerospace.

In these application areas as well as in building construction, branched technical structures typically are assembled by connecting linear elements by welding, soldering, gluing, riveting, etc. These connection nodes are often complicated to produce and notoriously prone to failure (Mattheck 2011). For this reason in the past 10 years, special focus was laid on investigating branched plant stems as inspiration for the development of branched biomimetic fiber-reinforced composites (Masselter et al. 2011, 2013, 2016a, b; Schwager et al. 2013; Haushahn et al. 2014; Born et al. 2016; Hesse et al. 2016), and on optimizing complex interconnected structures in architecture and building construction. These studies complement other novel approaches as described in, e.g., Falk and von Buelow (2008), Richardson et al. (2014) or in Ren and Galjaard (2015). In the following, a selection of various methodic approaches for analyzing the biological concept generators is introduced. Additionally, some examples for biomimetic implementation are provided for unbranched and branched fiber-reinforced composites.

Analyses of Morphology and Anatomy of the Biological Concept Generators

A detailed knowledge of the hierarchical structuring of studied plant specimens is a prerequisite for a correct interpretation of biomechanical results. After a screening process, the biological role models with the highest biomimetic potential were analyzed using various methods such as light, confocal laser and scanning electron microscopy, 3D light scans and μ-computer tomography scans (μ-CT) as well as magnetic resonance imaging (MRI). Key role models for unbranched fiber-reinforced composites were different bamboo and horsetail species (*Equisetum* spp.) as well as the giant reed (*Arundo donax*) (Spatz et al. 1997; Speck et al. 1998). Among the investigated branched plants were various monocotyledons with a shrub- or tree-like habit as, e.g., *Dracaena*, *Freycinetia*, and *Pandanus* (Haushahn et al. 2014). Other interesting structures were found in columnar cacti (Schwager et al. 2013), as well as in some genera of Araliaceae like *Schefflera*, *Polyscias*, *Fatsia*, and *Hedera* (Born et al. 2016).

The "classic technique" of light microscopy is still indispensable for the analysis of the inner structure and outer form of plants as, e.g., the course and arrangement of the fibrous bundles in *Dracaena reflexa* via the alignment of sectional images (Fig. 1a). Scanning laser microscopy has also proven as an important tool for scanning the outer surface and inner structures of plants in high resolution. The high energy of the laser beam permits for a detailed scan of fine structures as, e.g., the inner arrangement of the fibrous bundles in *Dracaena reflexa* (Fig. 1b). 3D light scans represent a fast and easy method to acquire information on morphology and outer form of branched plant specimens. Based on light scans, the outer form of a branched node in *Semiarundinaria* sp. can be digitized and used for further analyses (Fig. 1c).

As mentioned above, the inner arrangement of different plant structures can be observed well via light microscopy. However, it is very time-consuming to acquire a three-dimensional model of the inner plant structure by aligning a high number of thin sections. Therefore, the use of μ-CT scans was a major step toward mapping the full structure and morphology of plants as shown here for the branching region of *Schefflera arboricola* (Fig. 1d). A drawback of this method is its high invasiveness, medium resolution, and the need for structures with a rather high density difference. Magnetic resonance imaging (MRI) represents the last method for analyzing morphology and anatomy of plants, as shown for the branching region of *Dracaena marginata* (Fig. 1e, f). MRI also allows for medium resolution but is not invasive, i.e., can be used for living plants (Hesse et al. 2016), and allows for additional analyses as, e.g., water displacement.

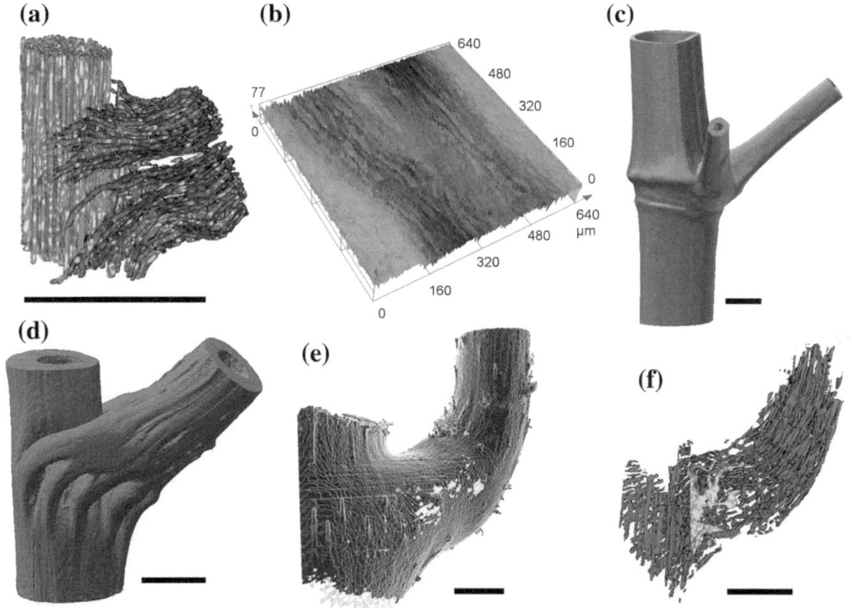

Fig. 1 Morphology and anatomy of various biological concept generators. **a** 3D reconstruction of the arrangement and course of the fibrous bundles in *Dracaena reflexa*. **b** Confocal laser scan of the arrangement of two fibrous bundles in *Dracaena reflexa*. **c** 3D scan of the outer form of the branching region of *Semiarundinaria* sp. **d** 3D reconstruction of the branching region of *Schefflera arboricola* by using μ-CT imaging. **e**, **f** Reconstruction of outer shape and the arrangement of fibrous bundles in a branching of *Dracaena marginata* by using magnetic resonance imaging. Scale bars: 5 mm

Biomechanical Analyses of Plant Structures

Universal or custom-built tension–compression machines were mostly used for investigating the mechanical parameters of plant organs and tissues such as stiffness in bending, compression or tension, and the corresponding moduli. Most often, these tests were performed within the elastic range and in several planes of the plant organs for assessing the stresses and strains of organs and/or tissues as shown, e.g., in Masselter et al. (2016a) (Fig. 2a). Other experiments were carried out beyond the elastic range for assessing the strength of the materials as, e.g., the tensile energy absorption or the fracture toughness of branched specimens (Masselter et al. 2011) (Fig. 2b).

Tensile Energy Absorption

The tensile energy absorption (TEA) up to the maximal force (F_{max}) or up to the force at fracture (F_{fract}) can be calculated by the integral under the force–displacement

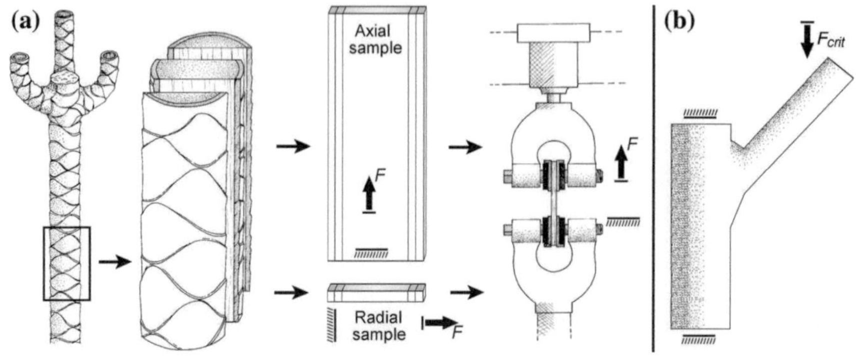

Fig. 2 Biomechanical testing. **a** Plant specimens are cut longitudinally and transversally to acquire sections in different planes that can then be tested via a tension–compression device. **b** Schematic setup for the testing of a stem–branch attachment until failure, i.e., until a critical force (F_{crit}) is reached

curve, with a function f(x) from 0 to x_{max} or 0 to x_{fract} in the force–displacement diagrams (Eq. 1).

$$\text{TEA} = \int_0^{x_{max}} f(x)\, dx \quad \text{or} \quad \int_0^{x_{fract}} f(x)\, dx \quad [\text{J}] \tag{1}$$

Fracture Toughness

In biological materials and structures not failing in a brittle manner, fracture toughness (FT) can be calculated in good approximation by dividing the tensile energy absorption by the cross-sectional area *A* of the fracture surface (Eq. 2).

$$\text{FT} = \frac{TEA}{A} \quad \left[\frac{J}{m^2}\right] \tag{2}$$

Magnetic Resonance Imaging

This method has proven to be highly useful as it allows not only for mapping the inner and outer structures of a plant but also for carrying out biomechanical experiments in vivo. With MRI for the first time, the strains of inner tissues such as the fibrous bundles in a branching of *Dracaena marginata* could be observed when the plant is subjected to mechanical loading (Hesse et al. 2016) (Fig. 1e, f).

Biomimetic Implementation

Based on quantitative analyses of the form–structure–function relationship in branched and unbranched plant stems, the relevant parameters as to outer form and inner structure have to be abstracted. In this abstraction process, the most important parameters for the intended transfer are translated into technical "language" often with help of analytical formulas or numerical simulations (Speck and Speck 2008). If the abstraction process was successful, the development and fabrication of the bioinspired products can start.

Fabrication of Linear Fiber-Reinforced Composites

The manufacturing of linear fiber-reinforced composites is a well-established process. There exist numerous braiding and pultrusion techniques for the production of linear fiber-reinforced components (e.g., Fig. 3). These techniques allow for a selection of various angles in which the fibers are laid on the braiding core. Thereby, the arrangement and distribution of the fibers that are present in the biological concept generators can be transferred in order to optimize the fiber angles of the biomimetic technical fiber-reinforced composite. The "technical plant stem" (Fig. 3d) is a prime example of such a structure (Milwich et al. 2006).

Fabrication of Branched Fiber-Reinforced Composites

In contrast to linear braiding, the production of branched braidings as "skeleton" for branched fiber-reinforced composites is a complex challenge as fiber angles and density of the fiber overlay is changing while the branched braiding core is transported through the braiding machine. In the course of the Priority Program SPP1420 and the Collaborative Research Centre TRR141, several novel approaches have been developed and adopted (Müller et al. 2013; Born et al. 2016; Masselter et al. 2016b). In one process, the braiding core is handled by a robot arm which is transporting the braiding core through the braiding machine. In this process, first, the stem of the branching is braided (Fig. 3b, e), and then the legs one after the other. Another process uses two braiding machines that braid the two legs in parallel. The manufacturing of threefold branchings (Fig. 3c, f) is especially complicated as the braiding eye through which the core is transported has to be wide enough to allow for the three legs to travel through but should be narrow enough for a close and dense laying of the fibers. This problem was solved by the development of a variable patented braiding eye which allows for a change in fiber angle and density in the legs of the branching according to the findings in the biological role models (Hufenbach et al. 2013) (Fig. 3c). All the techniques have advantages and drawbacks for the arrangement and distribution of fibers but help for the transfer of bioinspired design principles.

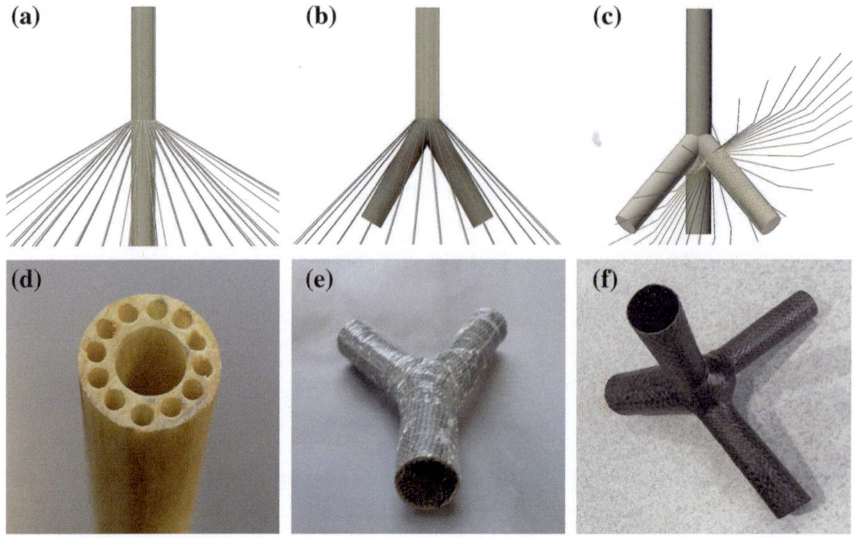

Fig. 3 Braiding techniques and bioinspired products. **a** Braiding technique allowing for a fiber-layup in varying layers of 0°, 30°, 45°, and 60° angle to the stem's axis. **b** Braiding technique allowing for a fiber-layup of approximately 45° on a branched braiding core. **c** Braiding technique with a variable braiding eye, allowing for a close and dense fiber-layup on a branched braiding core forming the legs of a threefold branching. **d–f** Demonstrators manufactured with the techniques shown in **a**, **b,** and **c,** respectively (**f**, courtesy of the Institute of Lightweight Engineering and Polymer Technology of the TU Dresden)

For the use in building construction and architecture, similar fiber-reinforced composites currently are developed in collaboration between biologists, constructional engineers, and architects. These composites are upscaled to a diameter of approximately 200 mm in a first phase and filled with concrete so that they are able to act as pillars for supporting roof constructions. The use of lightweight concrete with bioinspired braided outer hulls could markedly reduce the amount of concrete necessary for high load-bearing branched pillars and by this help to reduce the CO_2 footprint of buildings.

Energy Dissipation and Fracture Toughness in Plants: Inspiration for Biomimetic Impact Protection Systems

Before giving a concise outlook in biomimetic products that are currently under development based on form–structure–function relationships analyzed in highly damping and energy dissipating plant structures, a short introduction is presented into the physical background of impact and impact protection, and on how these principles apply to plants. In the following, different dynamic experiments and selected role models are described, and it is shown how their specific hierarchical structuring provides them with high performance in energy dissipation and/or puncture resistance.

Physical Background—Some Basics for Understanding Energy Dissipation

A simple definition of impact is given by the Oxford Dictionary (2017) as "The action of one object coming forcibly into contact with another". This implies that (at least) the first-mentioned object—or impactor—is moving with a certain velocity $\vec{v}$ and as real objects have a nonzero mass m, it also has a momentum ($\vec{p}$) which is defined as the product of its mass and its velocity:

$$\vec{p} = m \cdot \vec{v} \tag{3}$$

For reasons of clarity, we are only describing the case where an impactor is hitting a nonmoving object (with no velocity and therefore no momentum). According to the law of conservation of momentum, the momentum of the impactor is not lost during impact but transferred to the impacted object. The force ($\vec{F}$) involved thereby is directly proportional to the rate of change of momentum (Newton's second law of motion, with $\vec{a}$ being the acceleration):

$$\vec{F} = m \cdot \vec{a} \tag{4}$$

$$\vec{F} = m \cdot \frac{d\vec{v}}{dt} \tag{5}$$

$$\vec{F} = \frac{d\,(m \cdot \vec{v})}{dt} \tag{6}$$

$$\vec{F} = \frac{d\vec{p}}{dt} \tag{7}$$

This change of momentum is referred to as impulse J which is defined for a constant force F_{const} as

$$J = F_{const} \cdot \Delta t \tag{8}$$

In practice, however, the force F is not constant but rather expressed as function of time t. In this case, J has to be reformulated as the integral of the force with respect to time from the start of the impact (t_1) to its end (t_2):

$$J = \int_{t_1}^{t_2} F\,dt \tag{9}$$

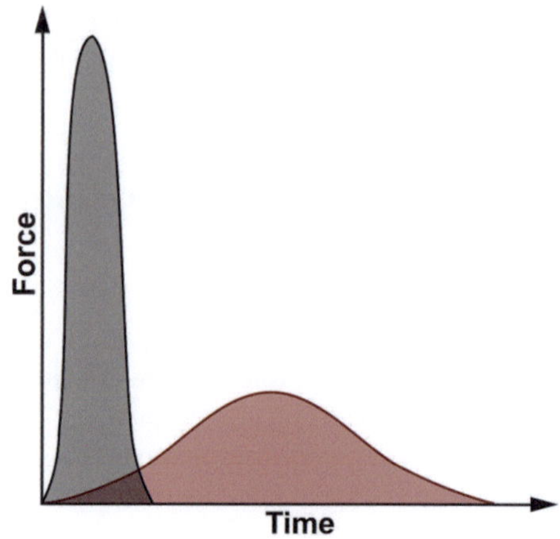

Fig. 4 Different force–time curves showing the same impulse (=shaded area under the curve)

Combining Eq. 5 and Eq. 9 proves that the impulse indeed is the same as the change in momentum:

$$J = \int_{t_1}^{t_2} m\frac{dv}{dt}\,dt = m \cdot v_{t_2} - m \cdot v_{t_1} = p_2 - p_1 = \Delta p \tag{10}$$

From Fig. 4, it becomes clear that the same impulse can result from a high force acting over a short time span (gray area) or from a low force acting over a longer period (red area). It is thus, for a given impulse, of advantage to extend the impact duration in order to keep the forces low and under-critical. This can be achieved by mounting a padding on the impacted body which is compressed during the collision, and thus increases the distance *s* the impactor has to travel and consequently also the time interval during which the momentum of the impactor changes.

Combining the general equations of kinematics Eqs. 11 and 12

$$v_{final}^2 = v_{init.}^2 + 2as \tag{11}$$

$$v_{final} = v_{init.} + a \cdot \Delta t \tag{12}$$

with $v_{init.}$ being the initial velocity and v_{final} being the final velocity shows (at least for uniform deceleration *a*) how Δt increases with increasing distance:

$$\Delta t = \frac{2s}{v_{final} + v_{init.}} \tag{13}$$

In order to come to a complete rest, the impactor must lose all of its kinetic energy ($E_{kin} = \frac{1}{2}mv^2$) by converting it to another form of energy or transferring it to another object (e.g., heat, deformation). This happens at time $t_{m.c.}$ ($t_1 < t_{m.c.} < t_2$) when the compression of the padding is at maximum (marked by suffix: *m.c.*) and when the impactor has a velocity $v_{t_{m.c.}} = 0$. Therefore, $v_{init.} = v_{t_1}$ and $v_{final} = v_{t_{m.c.}} = 0$. The amount of energy that is transformed to another form or transferred to another object by the action of a force is called work (W). When the force of the padding is acting on the impactor over the spatial interval s (i.e., deformation of the padding), the work performed is defined as

$$W = F \cdot s \tag{14}$$

From Eq. 11, we know the relation between the velocity at the start of the impact (v_{t_1}) and the velocity at maximum compression ($v_{t_{m.c.}}$), and thus that the distance ($s_{m.c.}$) when the compression of the padding is at maximum can be expressed as

$$s_{m.c.} = \frac{v_{t_{m.c.}}^2 - v_{t_1}^2}{2a} \tag{15}$$

When we combine Eqs. 4, 14 and 15, it becomes obvious that the work done to stop the impactor is equal to the change in its kinetic energy:

$$W = F \cdot s_{m.c.} \tag{16}$$

$$W = m \cdot a \cdot s_{m.c.} \tag{17}$$

$$W = \frac{m \cdot a \left(v_{t_{m.c.}}^2 - v_{t_1}^2\right)}{2a} \tag{18}$$

$$W = \frac{1}{2} m v_{t_{m.c.}}^2 - \frac{1}{2} m v_{t_1}^2 \tag{19}$$

$$W = \Delta E_{kin} \tag{20}$$

The kinetic energy absorbed by the padding can be stored as potential elastic energy ($E_{pot.,elast.}$) and/or be dissipated as other forms of energy like sound or heat (e.g., due to plastic deformation of the padding). In the case of a perfectly elastic padding, having a spring constant k the relation between deformation and force is given by Hooke's law:

$$F = -k \cdot s \tag{21}$$

which gives us the amount of stored elastic strain energy:

$$E_{pot.,elast.} = \int_0^{s_{m.c.}} ks\, ds = \frac{1}{2} k \cdot s_{m.c.}^2 \tag{22}$$

After the maximal compression of the padding is reached, the stored elastic strain energy is released and accelerates the impactor into the opposite direction. In the more realistic case in which the padding is not perfectly elastic, the proportion of the initial kinetic energy of the impactor that has been stored as elastic strain energy can be determined by comparing kinetic energy of the impactor at the start and the end of the impact, respectively. This ratio expresses the impact's degree of elasticity and is called coefficient of restitution *e*. As the mass of the impactor does not change (and the impactor is considered undeformable) *e* can be reduced to the ratio of rebound velocity v_{t_2} to impact velocity v_{t_1} of the impactor.

$$e = \frac{E_{kin_{t_2}}}{E_{kin_{t_1}}} = \frac{\frac{1}{2} \cdot m \cdot v_{t_2}^2}{\frac{1}{2} \cdot m \cdot v_{t_1}^2} = \frac{v_{t_2}}{v_{t_1}} \quad (23)$$

If the impact is perfectly elastic, the rebound and impact velocities are identical, and thus *e* equals 1. In case of a perfectly plastic impact, the impactor does not rebound at all and consequently *e* equals 0.

Biological Concept Generators

When reflecting on impact situations in nature the first examples that come to one's mind probably are animals that protect themselves from excessive forces involved in impacts (e.g., turtles (Achrai et al. 2015)) or that kill their prey by generating high forces resulting from very short impulses (e.g., mantis shrimps (Patek and Caldwell 2005)). But also plant organs often experience impacts. This occurs, for example, when ripe fruits and other diaspores are shed, accelerated toward the ground due to gravity and eventually hit the ground. It was a falling apple that purportedly led to the establishment of Newton's third law of motion (Stukeley 1752) stating that for every action there is an equal and opposite reaction. Also, the bark of some trees can act—besides other functions, such as fire retardant—as impact protector. During rockfalls in mountainous areas, it prevents the underlying cambium from being physically destroyed by impacting boulders.

In order to understand how different plant structures are evolutionary "optimized" in terms of energy dissipation during impact situations and to derive concepts for the development of biomimetic impact protection systems, we analyzed three biological concept generators with different structural buildups. These were, on the one hand, the thick foam-like peel of the pomelo (*Citrus maxima*) (Fig. 5a) and the multilayered fibrous bark of the giant sequoia (*Sequoiadendron giganteum*) (Fig. 5b) which both can (from the mechanical point of view) be regarded as padding structures that serve to dissipate energy by being deformed upon impact. On the other hand, the pericarp of the coconut (*Cocos nucifera*) (Fig. 5c) was chosen, because it combines two protection systems: the mesocarp and, to a lesser extent also the exocarp, act as padding, whereas the relatively thin endocarp constitutes a strong and tough hull enclosing and protecting the embryo and constituting a stiff backing wall for the padding layer(s) to compress against.

Pomelo peel and the bark of giant sequoia were analyzed with regard to energy dissipation based on deformation mechanisms and the capacity to avoid high force

Fig. 5 **a** *Citrus maxima* fruit. The fruit pulp is surrounded by the thick pericarp whose main part—the whitish albedo—can be regarded as an open pored graded foam in which a three-dimensional network of vascular bundles is embedded. It consists of living parenchymatic cells. **b** Bark sample of *Sequoiadendron giganteum* revealing its layered and fibrous composition. **c** Drupe of *Cocos nucifera*. The pericarp can be distinguished into an outer layer, called exocarp, a bulky fibrous part, called mesocarp and a tough inner shell called endocarp

peaks. In case of the coconut, the focus was set on the endocarp which exhibits a remarkably good compromise between sufficient strength and toughness, properties that typically are contradictory (Ritchie 2011; Wagermaier et al. 2015). One aspect of the latter research is to test if Weibull's theory of failure probability can be applied to quasi-brittle and anisotropic botanic materials. The probability P that a sample of volume V fails at a strength σ is defined as

$$P = 1 - e^{-\frac{V}{V_0}\left(\frac{\sigma}{\sigma_0}\right)^m} \tag{24}$$

where V_0 is a reference volume, σ_0 a reference strength, and m a dispersion index (Bažant 2004; Schmier et al. 2016; Zok 2017). Although Weibull's theory is based on the assumption of homogenous materials, he himself suggested to test the applicability of his probability function on various, non-homogenous materials (Weibull 1951). As the failure probability P is based on the weakest link fracture theory, the investigation of crack nucleation is essential. In technical materials (e.g., ceramics), these weakest links are flaws or imperfections, whereas in quasi-brittle biological materials, crack nucleation points often are functional substructures like intercellular cavities or channels (e.g., vascular vessels providing the surrounding tissues with nutrients). Theoretically, a material's strength is inversely proportional to the occurrence of imperfections or weak links, and thus, a large sample is more likely to fail than a small one made of the same material because the probability of occurrence of imperfections increases with volume (e.g., Weibull 1939; Danzer 2014; Wagermeier et al. 2015). In biological materials, however, the aforementioned "imperfections" can concurrently influence crack propagation and crack deflection and thus—by dissipating deformation energy through the creation of new, often highly structured surfaces—toughen the material. Especially in the context of a transfer of the structure–function properties into biomimetic applications, these mechanisms are of eminent interest and will facilitate upscaling. A complete review of biological

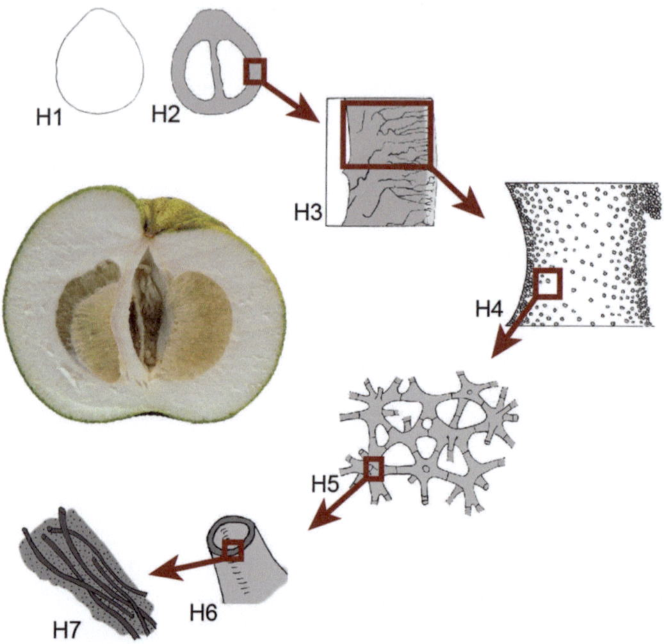

Fig. 6 Schematic illustration of the hierarchical buildup of the pomelo peel (not to scale). Seven hierarchical structural levels, ranging from the macro- to the micro- and molecular level (H1–H7), can be discerned. Adapted from Bührig-Polaczek et al. (2016) with permission

toughening mechanisms is far beyond the scope of this book chapter and can be found in excellent reviews elsewhere (e.g., Launey et al. 2010; Wegst et al. 2015).

All three role models have in common that, like the vast majority of natural materials, they consist of only a very limited number of basic components with comparably poor individual mechanical properties and that, despite this fact, they achieve astonishingly good overall mechanical properties. This remarkable performance is achieved by a hierarchical buildup, ranging from the nano- to the macroscale and concomitant the scale-overarching mechanical properties. In case of the pomelo at least seven hierarchical levels could be identified ranging from the differentiation between fruit pulp and peel over vascular bundles permeating the whole peel which exhibits a density gradient to the composite buildup of the primary cells walls (Bührig-Polaczek et al. 2016; Thielen et al. 2013) (Fig. 6).

The hierarchical structuring of the bark of giant sequoia and the coconut endocarp is similarly complex as was recently shown in Klang et al. (2016) and Schmier et al. (2016). Pomelo peel and bark of giant sequoia can be regarded as low-density materials. While both comprise comparably large air-filled voids, the peel of *C. maxima* mainly consists of living parenchymatic cells constituting the struts of a graded foam-like parenchymatic tissue that is reinforced by only few branched vascular bundles with lignified vessels (Bührig-Polaczek et al. 2016; Thielen et al. 2013), the main parts of the bark of *S. giganteum* consist of dead, highly lignified cells that are arranged in layers connected to each other via thin-walled cells (Klang et al. 2016). In contrast, the coconut endocarp consists to a significant extent of highly lignified sclereid cells whose secondary cell walls nearly completely fill the cell's volume and thus lead to a comparably high density of the whole endocarp. This dense matrix is traversed by a less dense network of branched vascular bundles running almost in parallel with the endocarp's surface (Schmier et al. 2016).

Biomechanical Analyses of Plant Structures

Even though it is often unavoidable to dissect plants or plant organs in order to analyze them, one should always bear in mind that the functionality to its full extent often is only guaranteed if the whole structure remains unaltered. Nevertheless, also tests with excised and cut samples can provide important contributions to the understanding of the form–structure–function relationship and also to comprehend the functional properties of the whole system.

Pomelo Peel and Giant Sequoia Bark

While we analyzed the overall ability of pomelos to withstand drops from several meters height onto a concrete floor and on a platform equipped with force transducers by performing free fall tests with whole fruits (Fischer et al. 2010; Thielen et al. 2012), a more detailed analysis of the deformation behavior and energy dissipation capacity of fresh and freeze-dried peel tissue was only feasible with smaller peel samples. Therefore, we cut out cylindrical peel samples and tested them by performing drop weight tests (Thielen et al. 2015) using a custom-built test rig (Fig. 7a). This test rig allowed to record the force exerted by the impactor and transmitted through the sample to the force sensor over time. By tracking the velocity of the impactor, the coefficient of restitution was determined. The same experimental setup was utilized to analyze cuboid samples of the bark of an approximately 90 years old and 42 m high giant sequoia.

Coconut Endocarp

The impact resistance of coconut endocarp samples was measured using an impact pendulum with an instrumented impact hammer of 7.5 J work capacity (Fig. 7b). The hammer had an impact velocity v_0 of 3.85 m/s when hitting the sample. With this test setup, it was possible to record the force F at the contact point of hammer and sample over the time t. By integrating twice the force with regard to time using Eq. 25, the displacement of the impact hammer while fracturing the sample was calculated (e.g., Thielen et al. 2015):

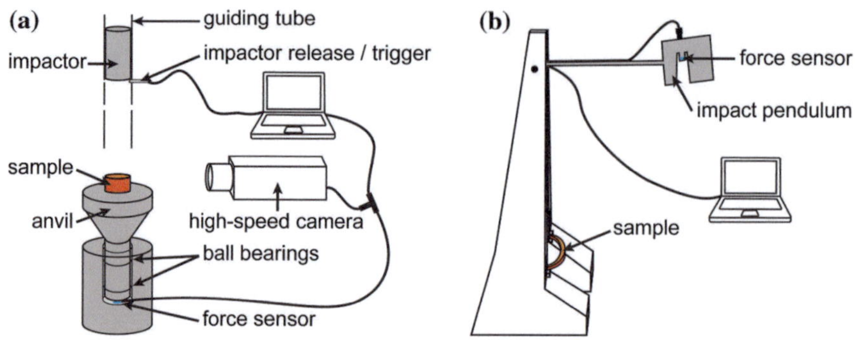

Fig. 7 **a** Drop weight test rig used for testing samples of pomelo peel and giant sequoia bark. **b** Impact pendulum used for testing coconut endocarp samples. **a** Adapted from Thielen et al. (2015) and **b** adapted from Schmier et al. (2016) with permission

$$s(t) = \int_{t_1}^{t} \left(v_0 - \int_{t_1}^{y} \frac{F(x)}{m}\, dx\right) dy \tag{25}$$

with $t_1 < x < y < t < t_2$. For the tests, endocarp samples were used that were arch-shaped due to the curvature of the endocarp (Schmier et al. 2016).

Energy Dissipation Mechanisms

The drop weight tests showed that cell water content and stress distribution play important roles in minimizing peak forces transmitted through the pomelo peel to the fruit pulp (Fig. 6 H2) during impact. While for fresh peel samples the peak force strongly depends on the sample height (i.e., the peel thickness), this correlation is much less pronounced for freeze-dried samples. On the one hand, the liquid filled cells (Fig. 6 H5) of fresh samples are tautened more and more as they are compressed by the impactor. This allows them to transfer stresses to neighboring cells and thus leads to a stiffening effect and eventually to a uniform collapse of cells over the whole height of the sample, whereas freeze-dried cells collapse in a more brittle manner. The latter entails a progressive collapse of the different cell layers and especially for less severe impacts not the whole tissue volume is involved in energy dissipation in freeze-dried samples. On the other hand, cell fluids are squeezed through cell membranes and cell walls (Fig. 6 H6) when these are becoming too taut, a process similar to methods also used in technical dampers for absorbing energy via viscous friction. Another mechanical property that could be shown during these tests is the auxetic behavior of fresh pomelo peel samples. The negative Poisson's ratio leads to a lateral contraction of the samples when compressed, an effect—albeit not verified quantitatively in the described tests—that may lead to a local densification of the impacted area because adjacent tissue volumes are pulled toward the impact location. In order to measure quantitatively this effect, the samples would have to be larger than the impactor's cross section in order to have spare tissue volumes that are not directly submitted to the impact and thus can lead to the abovementioned densification. A similar effect of involving tissue adjacent to the impacted area is caused by the three-dimensional network of relatively stiff vascular bundles permeating the less stiff foam-like parenchymatic tissue (Fig. 6 H3) and thus leading to a distribution of impact stresses to peel volumes that are not directly affected by the impact (Thielen et al. 2013) as well as to a global stiffening of the foamy parenchymatic tissue (Bührig-Polaczek et al. 2016). The density gradient (i.e., number of struts) in radial direction of the peel (Fig. 6 H4) leads to a progressively rising compression curve (i.e., stiffening) and thus prevents bottoming out during more severe impacts.

A minimization of peak forces can also be found in the bark of the giant sequoia. Force–time curves of drop weight tests conducted with the bark samples of the giant sequoia differ from force–time curves of the same test conducted with bark samples of the tree of heaven (*Ailanthus altissima*) (Fig. 8). In contrast to the giant sequoia, the latter tree species is not native to mountainous areas susceptible to rockfall events and thus normally does not face high energy impacts. Its thin and dense bark leads to a high force acting over a short time period, whereas the thick and less dense bark of the giant sequoia is capable of markedly decreasing the magnitude of the force peak while at the same time extending the impact duration (Fig. 8). Thus, the critical factor for the survival of the tree during an impact event (i.e., protection of the cambium) is significantly improved in case of the bark of the giant sequoia by reducing the maximum force transferred through the bark to the cambium. Besides this evaluation of an impact event using a dynamic testing method (i.e., the drop weight test), the deformation behavior of the bark was further analyzed by quasi-static compression tests. During compression of the giant sequoia bark, its intersecting layers of long fibers dissipate energy and lead to a nonuniform compaction behavior. Thereby, a mixture of elastic, viscoelastic, and plastic deformation can be observed even in this dry biological material system. Even though their exact proportion still needs to be further examined in ongoing studies, the existing findings point toward friction between fibers or fiber layers and/or viscoelastic deformation as main components for a storage and dissipation of kinetic energy, whereas plastic deformation is of minor importance (Klang et al. 2016).

While the main biological significance of the pomelo peel and the giant sequoia bark as well as of the coconut mesocarp is the reduction of the maximum force, the coconut endocarp has to act as tough backing wall for the outer layers and as puncture resistant hull to protect the seed from injuries due to the occurring impact. To fulfill these functions, the endocarp has to be sufficiently strong. As it has to further absorb the still occurring impact energies transferred through the exocarp and mesocarp and potential attacks of predators, it has to be in addition sufficiently tough. Analyses of the fracture surfaces after impact pendulum tests allowed for hypothesizing an energy dissipation mechanism for the coconut endocarp. The network of the vascular bundles represents the weakest regions of the endocarp, as they are significantly less dense compared to the sclereid cell matrix. Therefore, crack paths are deviated in a

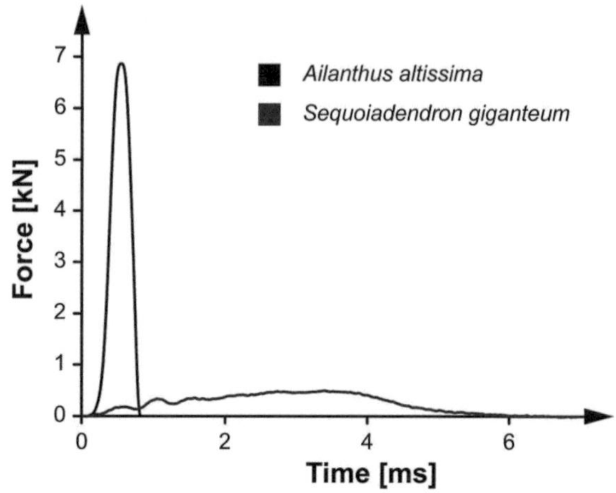

Fig. 8 Force–time curves of drop weight tests conducted on bark samples of *S. giganteum* and *A. altissima* (samples were taken from about 18 m height above ground, trees were of comparable age)

direction more or less parallel to the surface, namely, along the vascular bundles and their interfaces with the surrounding denser sclereid matrix. Through this mechanism, the crack path considerably lengthens before the crack can reach the inner surface of the endocarp. Thus, the probability increases that the impact energy driving the crack is absorbed before the crack reaches the seed. Within the sclereid cell matrix, it has been observed that the connection between neighboring cells, the pits, are highly lignified in ripe coconuts and may prevent the cracks from running between two cells by strongly interconnecting individual cells. If finally broken, the pits will cause extraordinary high friction. We hypothesize that by this structural composition the probability is increased that the cracks run into the cell wall. Here on a smaller hierarchical level, the cracks are further deflected and often come to a stop in the thick cell wall due to its multilayered composition. By involving these three hierarchical levels—fiber bundles, lignified pits, and multilayered cell walls of sclereids—in crack stopping the resulting fracture surface becomes very rough. The rougher a fracture surface the higher is the newly created surface area, indicating that more impact energy has been absorbed (Beismann et al. 2000). Therefore, it can be concluded that the main energy dissipation mechanism of the coconut endocarp is a sophisticated crack path deviation on several hierarchical levels rather than an avoidance of crack nucleation. The result is a (partly) damaged, but most probably still closed shell which allows the coconut for bearing further loads and for protecting the seed or embryo (Schmier et al. 2016, 2017).

Biomimetic Implementation

Summarizing the results of various possible damping mechanics, it is not surprising that the probably most elementary and straightforward method for enhancing energy dissipation and reducing excessive force peaks is also realized in plants. It relates to Eq. 14 and consists in increasing the work performed on the impactor by simply increasing the sheer thickness of the energy dissipating layer(s). However, apart from this, all biological energy dissipation mechanisms described above offer prominent and novel aspects for a transfer into technical applications like impact dampers or other energy dissipating constructional elements.

Several of the structure–function properties found in the pomelo peel were abstracted and implemented into metal foams (Bührig-Polaczek et al. 2016; Fischer et al. 2010, 2014). The density gradient was realized by thickening of struts in selected areas of 10 ppi open-cell Al-7Si-0.3 Mg-foam samples (Fig. 9a). Quasi-static compression tests showed that even a simplified continuous density gradient gives rise to a second—tailorable—plateau stress region which can be interpreted as stiffening effect and thus prevents bottoming out of the impact protector during impacts that are more severe than expected. A stiffening effect could also be realized by introducing stiff ceramic fiber bundles parallel to the loading axis. The increase of the energy dissipation capacity thereby depends on the exact failure mechanism of the fiber bundles (Bührig-Polaczek et al. 2016; Fischer et al. 2014). Analyzing fresh and freeze-dried peel samples indicates that a continuous collapse of foam-like structures seems to be advantageous over a progressive collapse, the latter being a failure mode found in most technical foams and manifested by sharp compaction fronts. This effect is based on the strain-dependent behavior of the fluid-filled struts (i.e., biological cells). Technically, the strain dependency of individual foam struts and by these means the simultaneous involvement of a multitude of struts in energy dissipation was realized by recasting of a 10 ppi aluminum A356 foam with a second aluminum A356 melt using a modified investment casting process (Bührig-Polaczek et al. 2016). As the auxeticity of the pomelo peel can hardly be implemented into metal foams, this effect is currently being transferred into polymer foams. These polymer foams, which in addition to auxeticity will also implement structural gradation of foam cell size, will be used for the inner layer of personal safety equipment such as helmets or knee and shoulder guards (Fig. 9b).

Also, the bark of the giant sequoia is very well capable of dissipating energy and increasing impact duration and thus acts as a highly efficient biological damping system. However, unlike the reinforced open cell foam of the pomelo peel, the bark of the giant sequoia achieves this partly by high thickness, but additionally also by a sophisticated multilayered composition and a high amount of fibers. This unique bark structure results in a mixture of elastic, viscoelastic, and plastic deformation during impact, enabling an outstanding energy dissipation performance during rockfall events compared to the bark of other tree species being not exposed to rockfalls in their natural environments. Thus, the composition of the bark of the giant sequoia serves as a starting point for a transfer into biomimetic concrete-based construction elements that benefit from an improved energy dissipation capability, e.g., for protection against damage by earthquakes or impacts from cars, airplanes, or derailed trains. For that purpose, our project partners from the Institute for Lightweight Structures and Conceptual Design of the University of Stuttgart adapt the composition of functionally graded concrete in order to achieve a hierarchical inner structuring. This

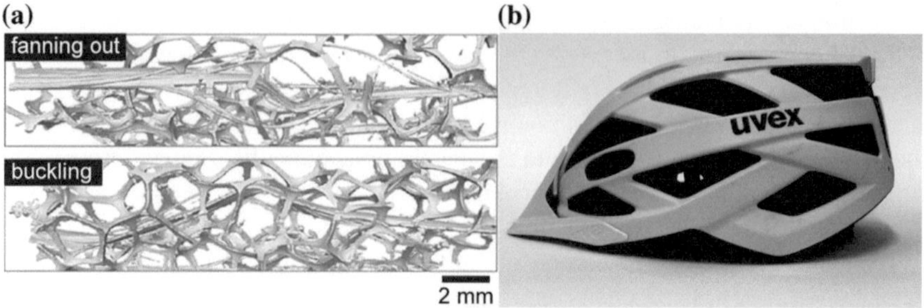

Fig. 9 **a** Microtomographic volume reconstructions of fiber-reinforced Al-foams showing the two typical failure mechanisms under compression, namely, fanning out and buckling of the fibre bundles within the foam matrix (adapted from Bührig-Polaczek et al. 2016) **b** Technical impact protector (i.e., a bicycle helmet) which can be further improved by biomimetic optimization

has already been realized successfully inspired by density gradients of porous sea urchin spines, and a next step will additionally include the bark of the giant sequoia as biological concept generator. The envisaged aim is to develop fiber-reinforced functionally graded lightweight concretes with the fiber arrangement and composition inspired by the bark of the giant sequoia (Klang et al. 2016).

In contrast to the pomelo peel and the giant sequoia bark, the quasi-brittle coconut endocarp exhibits a completely different energy dissipation mechanism, namely, by crack path deviation and elongation and the concomitant consumption of energy. This opens up the possibility to further improve the impact damping capabilities of the abovementioned technical impact dampers. In contrast to using fibers as reinforcement in functionally graded concrete, hollow fibers or channels can also be used as weak spots to redirect cracks into uncritical zones within the concrete structures. For these technical materials, an increase in stability and longevity by dissipating high amounts of energy via directed crack deflection without putting at risk the structural integrity of the main load-bearing part of a construction element is aimed for. Thus, even if partially damaged, such technical load-bearing structures will be able to survive catastrophic events as mentioned above.

Plant Movements: Bioinspired Technical Applications in Architecture

Virtually any plant displays motion without having muscles and hinges. The moving organs (leaves, petals, etc.) represent functionally resilient natural compliant mechanisms which in the last decade became of increasing interest for biomimetic applications (Schleicher et al. 2015; Poppinga et al. 2016, 2017a). During the last years, a multitude of studies showed how plant movements can be used as concept generators for bioinspired kinetic devices, e.g., in the fields of architecture, soft robotics, microfluidics, and micro-electromechanics (reviewed by Burgert and Fratzl 2009; Guo et al. 2015; Li and Wang 2017). The abstraction and technical implementation of the movement actuation, as well as of the kinematics (reviewed by Poppinga et al. 2013), can drive technical innovation in various fields. This requires a thorough understanding of the biological role model, as well as engineering methods for tailoring and fine-tuning the bioinspired product.

Methods

For investigations of the role models, a multitude of biophysical techniques are applicable. Analyses of the functional morphology and anatomy include "classical" methods like light-, (environmental) scanning electron-, confocal laser scanning-, atomic force-, and transmission electron microscopy. The shape of the organs can nowadays be assessed via 3D scanners, and the inner architecture resolved by, e.g., micro-computer tomography (μ-CT) and focused ion beam scanning electron microscopy. Also, noninvasive methods like magnetic resonance imaging (MRI) have emerged as useful tools (Hesse et al. 2016).

Regarding the kinematics, many plant movements are either too slow or too fast to be perceptible by the human eye. Time-lapse cameras allow for accelerated playback of slow motions, whereas high-speed cameras enable fully time-resolved recordings of very rapid processes, e.g., the slow movement of false indusia and ultrafast motions of sporangia in leptosporangiate ferns (Poppinga et al. 2015). By applying stereo recordings with two (or even more) cameras, spatially complex motions can be investigated and the 3D deformation can be quantified, e.g., *Dionaea* trap closure (Forterre et al. 2005). Cinematography in combination with laser-sheet-based fluorescence microscopy allows for visualization of obscured and/or small structures, e.g., the rapidly moving trapdoors of carnivorous bladderworts (Vincent et al. 2011). The flow of water within or around plant organs or cells, as well as the deformation of whole plants, organs, tissues, or even single cells can be analyzed with appropriate imaging methods, as, e.g., X-ray microscopy (Song et al. 2014), in combination with particle image velocimetry (PIV) algorithms. Physical modeling helps to verify the observed movement phenomena, as, e.g., for snap-through transitions observed during the fast motions of carnivorous plant traps (e.g., Poppinga and Joyeux 2011; Vincent et al. 2011; Schleicher et al. 2015).

Measurements of cellular hydrostatic pressure alterations are indispensable for analyses of active hydraulic movements. This becomes possible, for example, via the pressure probe technique (Caliaro et al. 2013b), with the help of special "pico gauges" (Knoblauch et al. 2014), or by AFM indentation techniques (Beauzamy et al. 2015). Also, the tissue pressure (e.g., of collenchyma) can be recorded and tracked over long periods of time for explaining shape changes of plants or plant organs due to water availability alterations (Caliaro et al. 2013a). Visualization of water flows in vivo within plants, i.e., between different tissues or even between cells, is possible, e.g., via MRI (Holbrook et al. 2001).

For the abstraction, scalability evaluation, implementation, and optimization of the bioinspired technical structure, Finite Element Methods (FEM) are commonly used (e.g., Lienhard et al. 2011; Schleicher et al. 2015). FEM allow, e.g., for identifying areas of high stress concentrations when the simulated technical structure is under load. Such weak spots can then be "corrected", e.g., by applying optimized contour lines according to Mattheck and Burkhard (1990).

Biomimetic Applications

Hydraulic plant motion has inspired a multitude of biomimetic morphing structures. Actuators based on differential expansion and contraction of its functional layers can generate a multitude of shape changes like bending and twisting. The responsive layers may consist of expandable cellular structures with tailored geometry (e.g., honeycombs) and specific mechanical properties. Active pressurization (e.g., by pumping) or osmotic actuation can be applied, the latter indeed mimicking the water displacement and cellular pressure generation mechanism of plants (reviewed by Li and Wang 2017).

Plant motions actuated by passive hydraulics, especially the hygroscopic bending behavior of pine cone seed scales, represent "classical" role models in a multitude of biomimetic approaches. The differential passive swelling and shrinking properties of the scale tissues can be similarly achieved with composites containing natural (e.g., wood, flax fibers) and/or technical materials (e.g., hydrogels, polymers, and ceramics). By this, fully autonomous and responsive structures can be produced which require no energy input to move and which are actuator, sensor, and regulator in one (e.g., Ionov 2014; Reichert and Menges 2015; Rüggeberg and Burgert 2015; Erb et al. 2013; Bargardi et al. 2016). 3D printing of such structures has recently become an emerging field (Gladman et al. 2016). Besides hygro-responsiveness, deployable structures which react to light, heat, enzymes, electric voltage, and other stimuli have been developed (e.g., Jiang et al. 2015; Athas et al. 2016). Mechanically robust and large (meter-sized) smart kinetic structures performing complex shape changes with passive-hydraulic actuation at a reasonable timescale are the main future challenges (Poppinga et al. 2017b).

Elastic instabilities like rapid snap-through transitions play a crucial role as speed boosts in many (very) fast plant movements and arouse increasing interest for shape-changing biomimetic devices (reviewed by Guo et al. 2015; Hu and Burgueño 2015). Harnessing such instabilities, which are typically avoided because they are often considered as mechanical failure, can lead to completely new functionalities like microlenses with tunable focal points, induced by snap-buckling (Holmes and Crosby 2007), and jumping hydrogels (Lee et al. 2010). The Venus flytrap has become a widely applied model system for the development of bioinspired actuators featuring snap-through transitions. Forterre et al. (2005) explained the fast closure and bistable behavior of its trap by physical modeling of elastic shells. The amount of elastic energy released during the trap lobe, i.e., shell curvature inversion, and thereby the trapping speed, increases with the dimensionless parameter α:

$$\alpha = \frac{L^4 \cdot K^2}{h^2} \tag{26}$$

with L the lobe size (length), K the mean curvature of the open lobe, and h the lobe thickness. Alpha (α) purely depends on trap morphology and geometry and constitutes the ratio between the energy barrier which separates the two states of lobe curvature (convex or concave), and the bending energy supplied by the change of the rest-state curvature.

Most biomimetic deployable structures as the ones described above possess rather small dimensions, and only few are intended for architectural use (Knippers and Speck 2012). "Failure modes" like torsional buckling phenomena as observed in two partly fused petals of the bird of paradise flower (Lienhard et al. 2011) (Fig. 10a–d), or complex folding processes incorporating kinematic amplification mechanisms, like in the trap closure of the carnivorous waterwheel plant (Poppinga and Joyeux 2011; Schleicher et al. 2015; Poppinga et al. 2016) (Fig. 10e–g), are scalable and proved their applicability as large-scale bioinspired kinetic façade-shading elements. These examples not only show the physical functionality but also a "functional elegance" similar to the aesthetic biological role models (Speck 2015).

Self-Repair in Plants: Analysis and Inspiration for Biomimetic Applications

Definitions: Self-Sealing, Self-Healing, and Self-Repair

Self-repair is a well-known phenomenon in living nature. In the course of life history, a variety of self-repair mechanisms evolved in the plant kingdom. Although various functional principles on different hierarchical levels can be found, self-repair processes in all investigated plants to date can be subdivided into an initial self-sealing phase and a subsequent self-healing phase. During the rapid sealing phase, damages are repaired functionally. Fast sealing of injuries guarantees protection against infection by pathogens and against desiccation by overcritical loss of water. During the long-lasting self-healing phase, the injury is structurally repaired and finally sometimes entirely disappears. These healing processes result in a (partly) restoration of mechanical properties of the injured plant organ. The term self-repair can be used as an umbrella term covering self-sealing and self-healing. Appropriately enough, these terms with their respective definitions describe self-repair processes in plants and animals but also in technical materials, whereas in the latter case, the sealing and the healing phase can also be found separately (Harrington et al. 2015; Speck et al. 2013a, b). Inspired by the self-repair mechanisms found in plants and animals,

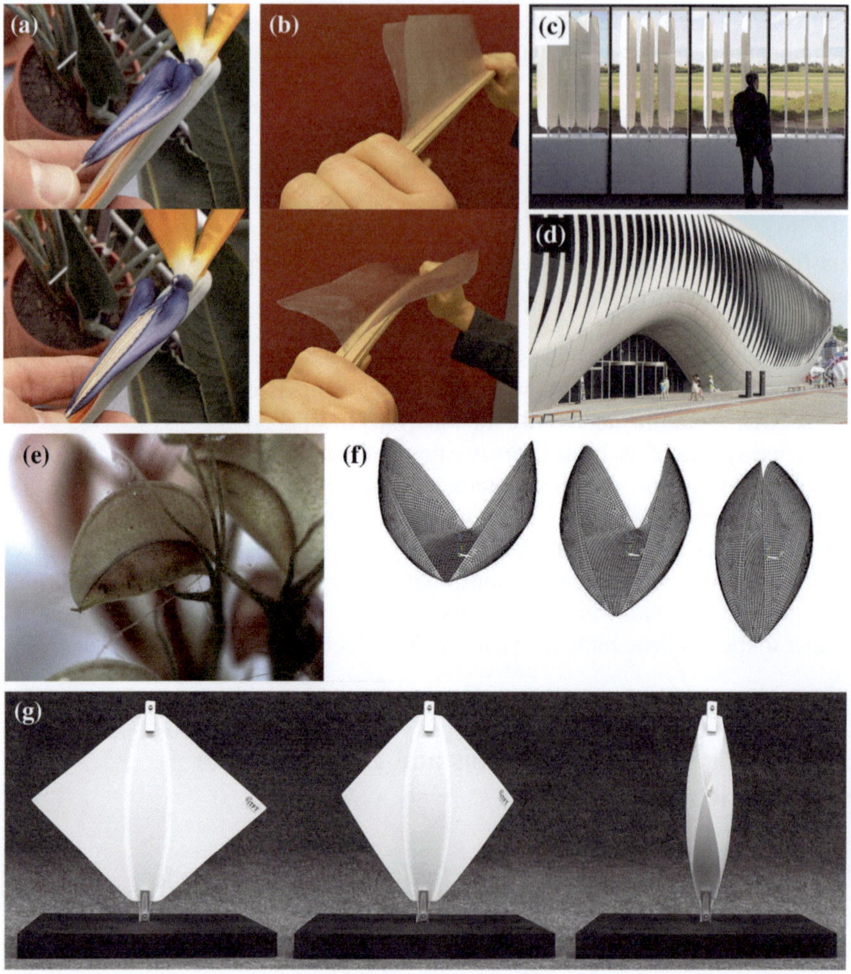

Fig. 10 Two examples of plant movement-inspired, large-scale kinetic structures for architectural purposes. Besides the physical principles of motion, the structures also show the aesthetics and "functional elegance" of the biological role models. **a** The body weight of a pollinator (here "simulated" by hand) naturally causes the closed perch (upper image) in the bird of paradise flower (*Strelitzia reginae*) to open, by which the stamens become exposed and the pollen is released (lower image). **b** During this completely passive movement, the involved petals perform torsional buckling. This principle can be easily transferred into a simple physical hand-model made of a wooden rod-shaped backbone with attached plastic sheets. **c** The resulting biomimetic scalable shading element Flectofin® allows for hinge-free and adjustable covering of façades (Image by Boris Miklautsch, Werkstatt für Photographie, University of Stuttgart). **d** A further abstraction of the Flectofin® principle led to the development of the kinetic façade of the 2012 Expo's thematic pavilion (Image with kind permission from soma architecture). **e** The snap trap of the carnivorous waterwheel plant (*Aldrovanda vesiculosa*) in the closed condition. **f** Simulation of the *Aldrovanda* trap with different steps from open (left) to closed (right). Bending of the middle portion (midrib) of the trap entails the kinematically amplified movement of the two trap lobes. **g** The biomimetic façade-shading element Flectofold shows the same deformation behavior as its biological role model. **f**, **g** Modified from Poppinga et al. (2016)

scientist and engineers developed various synthetic materials and materials systems with self-sealing and self-healing function (Speck et al. 2013a, b; Diesendruck et al. 2015).

Self-Repair Function

The capability of living organisms and technical applications to repair damages represents an essential function of damage management. In contrast to the "damage prevention" concept, where sufficiently strong materials are created that can withstand deformation or extension of micro-cracks, the alternative design philosophy of "damage management" is based on materials that counteract damage with sealing and healing processes to finally regain (at least partly) the original functional performance (Van der Zwaag 2007). Well-known functions are, e.g., the restoration of optical appearance, mechanical properties and tightness (Van Tittelboom and De Belie 2013). The concept of "damage management" is also found in plants and animals in which the occurrence of injuries is (nearly) unavoidable in the life course of each individual. In evolution, various sophisticated ways of dealing with injuries and of keeping them under-critical to ensure functioning and finally survival of the respective living being have evolved and can be tapped for bioinspired developments.

Repairing Efficiency

For the description of self-repair, one has to bear in mind that self-repair is not a property that can be reliably determined on basis of a "repair standard test", but a function that has to be evaluated by its sealing or healing efficiency. Independent of the class of material, the repairing efficiency is assessed by comparing the change in function of virgin and repaired materials (Diesendruck et al. 2015). For that purpose virgin and repaired material undergo identical test protocols. Comparing the test results of the repaired and the virgin samples results in the calculation of a dimensionless quantity between 0 (= no repair) and 1 (= full recovery). Key advantage of a dimensionless value is the comparability of all data derived from various material classes and different test protocols and that in principle can be applied to living and technical structures and materials.

Material Properties

Functional recovery of mechanical properties is an ultimate goal for self-healing materials. Mechanical characterizations of virgin and repaired materials are often carried out by test protocols under fracture conditions. In general, healing efficiency

η_e is calculated by the ratio of mechanical properties of the healed and the virgin materials. An example is the recovery of tear strength T for elastomeric materials as shown in Eq. 27 (Andersson et al. 2007).

$$\eta_e = \frac{T_{healed}}{T_{virgin}} \tag{27}$$

Inspired by latex-based self-healing mechanisms in plants, where the latex is typically stored in cellular micro-tubes (laticifers), synthetic healing agents in technical materials are either encapsulated in microcapsules or micro-tubes or small domains of self-healing agents are engendered directly in the matrix material by phase separation mimicking microcapsules. Mechanical analyses of tensile strength in bark samples of *Ficus benjamina* showed, in addition to wound sealing, a significant increase in tensile strength within 30 min after external injury (Bauer and Speck 2012; Diesendruck et al. 2015). Based on this self-healing mechanism two approaches were used for producing novel bioinspired self-healing elastomers showing significant mechanical restoration after a macroscopic cut. First, inspired by the function of Ca^{2+}-ions, ionomeric self-healing elastomers were designed (Nellesen et al. 2011). In a second approach nitrile butadiene rubber (NBR) blends with microphase separated hyperbranched polyethyleneimine (PEI) were developed (Schüssele et al. 2012). These blends chemically mimic microcapsules with self-healing agents without an outer hull by phase separation. With respect to different PEI molecular architectures, a maximum self-healing efficiency calculated for tensile strength of 44% could be achieved (Schüssele et al. 2012; Speck et al. 2013a, b).

Tightness

Fast regaining of gas tightness is of utmost importance for pneumatic structures. Inspired by the rapid self-sealing process found in two *Aristolochia* species, a bioinspired foam coating for pneumatic structures was developed (Speck et al. 2006). Young stems of these twining vines are characterized by a layer of parenchyma, xylem, and phloem located inside a closed ring of strengthening tissue which is enclosed by a parenchymatic cortex and the epidermis with cuticle (Fig. 11a). During secondary growth, tiny fissures occur in the sclerenchymatous ring due to internal growth stresses and strains caused by the increase of the vascular cylinder. As soon as a lesion arises, parenchyma cells from the surrounding cortex tissue swell—due to their internal pressure (turgor) of several bars—into the lesion and seal it. In this initial fast sealing phase, the cell wall of the sealing cell expands locally through viscoelastic/plastic deformation without observable cell wall biosynthesis (Busch et al. 2010). In the subsequent healing phase typical cell growth and cell division are involved (Fig. 11b). Inspired by the mode of functioning of the sealing cells a bioinspired repair mechanism for pneumatic systems in terms of a self-sealing closed cell polyurethane foam coating for membranes was developed. To determine the gas

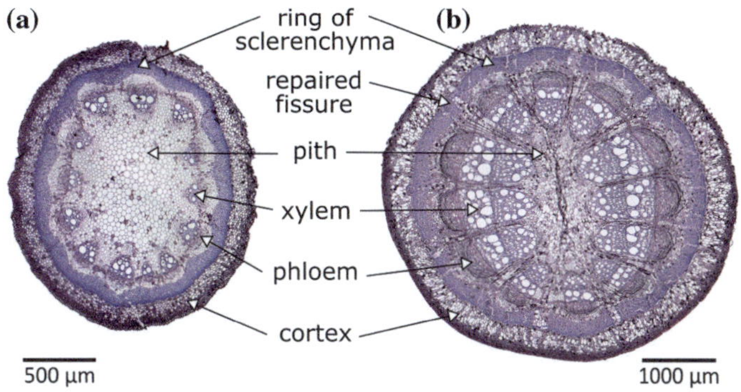

Fig. 11 Cross sections of *Aristolochia macrophylla* **a** young stem, **b** older stem showing self-sealing of fissures

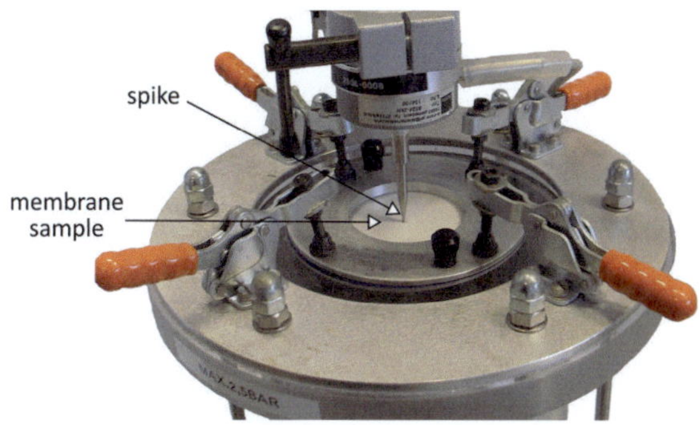

Fig. 12 Experimental setup to measure gas tightness of membranes

tightness R_{flow} achieved by the bioinspired product, a custom-made experimental setup (Fig. 12) had to be developed to measure the air mass flow through the fissure of a foam coated self-repairing membrane $\dot{m}_{coat}$ and the air mass flow through a punctured reference, i.e., an uncoated membrane $\dot{m}_{ref}$. The dimensionless value provides information about the sealing efficiency (Eq. 28), with 0 being no sealing and 1 being perfect sealing of the fissure.

$$R_{flow} = 1 - \frac{\dot{m}_{coat}}{\dot{m}_{ref}} \quad (28)$$

Finally, the sealing mechanism was found to be the result of compressive strains in the coating layer, mostly induced by the curvature of inflated membranes. In consideration of low coating weight and thickness of the foam layer, maximum sealing efficiencies of $R_{flow} > 0.99$ have been obtained for flexible closed cell foam coatings with up to 90% of fissures being entirely sealed (Rampf et al. 2011, 2012, 2013).

For plants growing in arid and semiarid environments, water tightness after damage is essential to survive. Interesting self-sealing mechanisms along different time

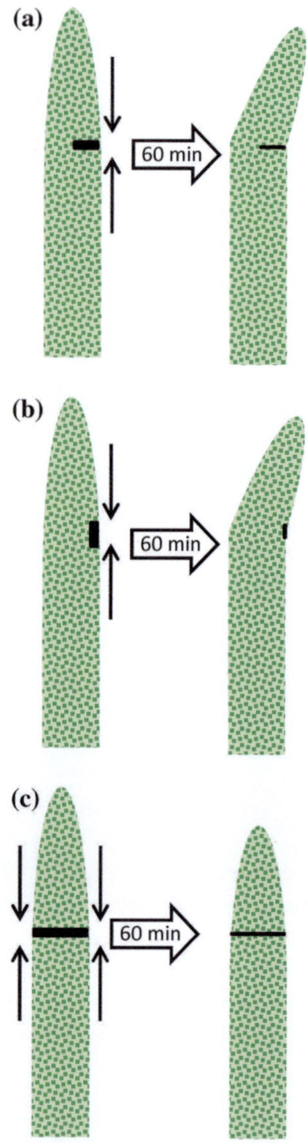

Fig. 13 Schematic drawings of *Delosperma cooperi* leaves after damage. Succulent leaf after a non-axisymmetric incision in **a** transversal and **b** longitudinal direction, and **c** after an axisymmetric ring incision

scales could be found in succulent leaves of *Delosperma* species (Speck et al. 2013b, 2018). In the case of any injury, the wound edges roll inward within a few minutes and reduce the size of the fissure. Additionally, dependent on the injury type, the entire leaf moves until the wound edges touch each other finally sealing the wound. Hydraulic movements and mechanical prestresses are the driving forces. Unilateral damages in transversal or longitudinal direction result in a bending, whereas axisymmetric damages like ring incisions lead to a contraction of the entire leaf (Fig. 13).

Based on this sealing mechanism by internal movement an analytical model has been presented (Konrad et al. 2013) and a numerical model is currently being developed. Technical multi-layer materials with self-sealing effect, e.g., due to internal prestressing are possible applications in future.

Biomimetics and Sustainable Development

Spontaneously thought, biomimetic solutions promise to contribute to sustainable development. Having a closer look, however, reveals that the presumed exceptional qualities like lower levels of risk, fault tolerance, environmental compatibility, energy and material efficiency, and many others are normative aspects. This biomimetic promise cannot be fulfilled because nature—in general—and biological role models—in particular—are not sustainable. They are the result of evolution, which itself—in contrast to the paradigm of sustainability—is neither teleological nor anthropocentric. On the other hand, we can learn a lot from nature for a sustainable world (Horn et al. 2016; Speck et al. 2017). Biomimetic self-repairing materials and materials systems as well as other bioinspired materials systems like the presented resilient and robust façade-shading systems or lightweight and resource-efficient truss systems or pillars and damage-tolerant materials systems may contribute to the "sustainability R's: reduce, reuse, recycle, repair" and more. However, one has to keep in mind that this has to be tested for each biomimetic product separately (Antony et al. 2014, 2016).

Acknowledgements The authors thank the German Research Foundation (DFG) for funding within the Transregional Collaborative Research Centre (SFB/Transregio) 141 "Biological Design and Integrative Structures". S. P. and T. S. additionally thank the JONAS research network (Joint Research Network on Advanced Materials and Systems) for funding. T. S. and M. T. are grateful for financial support from the German Federal Ministry of Education and Research (BMBF) "NanoMatTextil" within the framework of "BISS: Bio Inspired Safety Systems". The research on self-repairing materials was funded by the Ministry of Science, Research and the Arts of Baden-Württemberg, Germany, within the framework of the "Sustainability Center Freiburg", which is gratefully acknowledged by O. S. and T. S.

References

Achrai B, Bar-On B, Wagner HD (2015) Biological armors under impact—effect of keratin coating, and synthetic bio-inspired analogues. Bioinspir Biomim 10:016009. https://doi.org/10.1088/1748-3190/10/1/016009

Andersson HM, Keller MW, Moore JS, Sottos NR, White SR (2007) Self healing polymers and composites. In: van der Zwaag S (ed) Self-healing materials—an alternative approach to 20 centuries of materials sciences. Springer, Dodrecht, pp 19–44

Antony F, Grießhammer R, Speck T, Speck O (2014) Sustainability assessment of a lightweight biomimetic ceiling structure. Bioinspir Biomim 9:016013. https://doi.org/10.1088/1748-3182/9/1/016013

Antony F, Grießhammer R, Speck T, Speck O (2016) The cleaner—the greener? Product sustainability assessment of the biomimetic façade paint Lotusan® in comparison to the conventional façade paint Jumbosil®. Beilstein J Nanotechnol 7:2100–2115. https://doi.org/10.3762/bjnano.7.200

Athas JC, Nguyen CP, Zarket BC, Gargava A, Nie Z, Raghavan SR (2016) Enzyme-triggered folding of hydrogels: toward a mimic of the Venus flytrap. ACS Appl Mater Interfaces 8:19066–19074. https://doi.org/10.1038/ncomms13912

Bargardi FL, Le Ferrand H, Libanori R, Studard AR (2016) Bio-inspired self-shaping ceramics. Nat Commun 7:13912. https://doi.org/10.1038/ncomms13912

Barthlott W, Mail M, Neinhuis C (2016) Superhydrophobic hierarchically structured surfaces in biology: evolution, structural principles and biomimetic applications. Phil Trans R Soc A 374:20160191. https://doi.org/10.1098/rsta.2016.0191

Barthlott W, Mail M, Bhushan B, Koch K (2017) Plant surfaces: structures and functions for biomimetic innovations. Nano Micro Lett 9:23. https://doi.org/10.1007/s40820-016-0125-1

Bauer G, Speck T (2012) Restoration of tensile strength in bark samples of *Ficus benjamina* due to coagulation of latex during fast self-healing of fissures. Ann Bot 109:807–811. https://doi.org/10.1093/aob/mcr307

Bažant ZP (2004) Scaling theory for quasibrittle structural failure. Proc Natl Acad Sci USA 101(37):13400–13407. https://doi.org/10.1073/pnas.0404096101

Beauzamy L, Derr J, Boudaoud A (2015) Quantifying hydrostatic pressure in plant cells by using indentation with an atomic force microscope. Biophys J 108:2448–2456. https://doi.org/10.1016/j.bpj.2015.03.035

Beismann H, Wilhelmi H, Baillères H, Spatz H-C, Bogenrieder A, Speck T (2000) Brittleness of twig bases in the genus *Salix*: fracture mechanics and ecological relevance. J Exp Bot 51:617–633. https://doi.org/10.1093/jexbot/51.344.617

Born L, Jonas FA, Bunk K, Masselter T, Speck T, Knippers J, Gresser G (2016) Branched structures in plants and architecture. In: Knippers J, Nickel K, Speck T (eds) Biomimetic research for architecture and building construction: biological design and integrative structures. Biologically-inspired systems, vol 9. Springer International Publishing Switzerland, pp 195–216. https://doi.org/10.1007/978-3-319-46374-2_10

Bührig-Polaczek A, Fleck C, Speck T, Schüler P, Fischer SF, Caliaro M, Thielen M (2016) Biomimetic cellular metals—using hierarchical structuring for energy absorption. Bioinspir Biomim 11:045002. https://doi.org/10.1088/1748-3190/11/4/045002

Burgert I, Fratzl P (2009) Actuation systems in plants as prototypes for bioinspired devices. Phil Trans R Soc A 367:1541–1557. https://doi.org/10.1098/rsta.2009.0003

Busch S, Seidel R, Speck O, Speck T (2010) Morphological aspects of self-repair of lesions caused by internal growth stresses in stems of *Aristolochia macrophylla* and *Aristolochia ringens*. Proc R Soc B 277:2113–2120. https://doi.org/10.1098/rspb.2010.0075

Caliaro M, Flues F, Speck T, Speck O (2013a) Novel method for measuring tissue pressure in herbaceous plants. Int J Plant Sci 174:161–170. https://doi.org/10.1086/668791

Caliaro M, Schmich F, Speck T, Speck O (2013b) Effect of drought stress on bending stiffness in petioles of *Caladium bicolor* (Araceae). Am J Bot 100:2141–2148. https://doi.org/10.3732/ajb.1300158

Danzer R (2014) On the relationship between ceramic strength and the requirements for mechanical design. J Eur Ceram Soc 34(15):3435–3460. https://doi.org/10.1016/j.jeurceramsoc.2014.04.026

Diesendruck CE, Sottos NR, Moore JS, White SR (2015) Biomimetic self-healing. Angew Chem Internat Ed 54(36):10428–10447. https://doi.org/10.1002/anie.201500484

Erb RM, Sander JS, Grisch R, Studard AR (2013) Self-shaping composites with programmable bioinspired microstructures. Nat Commun 4:1712. https://doi.org/10.1038/ncomms2666

Falk A, von Buelow P (2008) Exploration and optimization of combined timber plate and branching column systems using evolutionary computation. In: Gerardo OLJ (ed) Proceedings of the international symposium of the international association for shell and spatial structures (IASS). New materials and structures. Grupo Editorial Formato, Mexico pp 27–31. https://doi.org/10.13140/2.1.3947.7124

Fischer SF, Thielen M, Loprang RR, Seidel R, Fleck C, Speck T, Bührig-Polaczek A (2010) Pummelos as concept generators for biomimetically inspired low weight structures with excellent damping properties. Adv Eng Mater 12(12):B658–B663. https://doi.org/10.1002/adem.201080065

Fischer SF, Thielen M, Weiß P, Seidel R, Speck T, Bührig-Polaczek A, Bünck M (2014) Production and properties of a precision-cast bio-inspired composite. J Mat Sci 49:43–51. https://doi.org/10.1007/s10853-013-7878-4

Forterre Y, Skotheim JM, Dumais J, Mahadevan L (2005) How the Venus flytrap snaps. Nature 433:421–542. https://doi.org/10.1038/nature03185

Fratzl P, Speck T, Gorb S (2016) Function by internal structure—preface to the special issue on bioinspired hierarchical materials. Bioinspir Biomim 11:60301. https://doi.org/10.1088/1748-3190/11/6/060301

Gladman AS, Matsumoto EA, Nuzzo RG, Mahadevan L, Lewis JA (2016) Biomimetic 4D printing. Nat Mater 15:413–419. https://doi.org/10.1038/nmat4544

Gorb SN (ed) (2009) Functional surfaces in biology: little structures with big effects, vol 1. Springer, Dordrecht. https://doi.org/10.1007/978-1-4020-6697-9

Guo Q, Dai E, Han X, Xie S, Chao E, Chen Z (2015) Fast nastic motion of plants and bioinspired structures. J R Soc Interface 12:20150598. https://doi.org/10.1098/rsif.2015.0598

Harrington MJ, Speck O, Speck T, Wagner S, Weinkamer R (2015) Biological archetypes for self-healing materials. Adv Polym Sci 273:307–344. https://doi.org/10.1007/12_2015_334

Haushahn T, Speck T, Masselter T (2014) Branching morphology of decapitated arborescent monocotyledons with secondary growth. Am J Bot 101:754–763. https://doi.org/10.3732/ajb.1300448

Hesse L, Masselter T, Leupold J, Spengler N, Speck T, Korvink JG (2016) Magnetic resonance imaging reveals functional anatomy and biomechanics of a living dragon tree. Sci Rep 6:32685. https://doi.org/10.1038/srep32685

Holbrook NM, Ahrens ET, Burns MJ, Zwieniecki MA (2001) In vivo observation of cavitation and embolism repair using magnetic resonance imaging. Plant Physiol 126:27–31. https://doi.org/10.1002/adma.200700584

Horn R, Gantner J, Widmer L, Sedlbauer KP, Speck O (2016) Bio-inspired sustainability assessment—a conceptual framework. In: Knippers J, Nickel K, Speck T (eds) Biomimetic research for architecture and building construction: biological design and integrative structures. Biologically-inspired systems, vol 9. Springer International Publishing Switzerland, pp 361–377. https://doi.org/10.1007/978-3-319-46374-2_18

Holmes DP, Crosby AJ (2007) Snapping surfaces. Adv Mater 19:3589–3593. https://doi.org/10.1002/adma.200700584

Hu N, Burgueño R (2015) Buckling-induced smart applications: recent advances and trends. Smart Mater Struct 24:063001. https://doi.org/10.1088/0964-1726/24/6/063001

Hufenbach W, Gruhl A, Lepper M, Renner O (2013) Verfahren für die Fertigung komplexer Faserverbund-Hohlstrukturen. Lightw Des 6:44–48. https://doi.org/10.1007/978-3-658-04025-3_11

Ionov L (2014) Hydrogel-based actuators: possibilities and limitations. Mater Today 17:494–503. https://doi.org/10.1016/j.mattod.2014.07.002

Jiang S, Liu F, Lerch A, Ioniv L, Agarwal S (2015) Unusual and superfast temperature-triggered actuators. Adv Mater 27:4865–4870. https://doi.org/10.1002/adma.201502133

Klang K, Bauer G, Toader N, Lauer C, Termin K, Schmier S, Kovaleva D, Haase W, Berthold C, Nickel KG, Speck T, Sobek W (2016) Plants and animals as source of inspiration for energy dissipation in load bearing systems and facades. In: Knippers J, Nickel K, Speck T (eds) Biomimetic research for architecture and building construction: biological design and integrative structures, biologically-inspired systems, vol 9. Springer International Publishing Switzerland, pp 10–133. https://doi.org/10.1007/978-3-319-46374-2_7

Knippers J, Nickel K, Speck T (eds) (2016): Biomimetic research for architecture and building construction: biological design and integrative structures. Biologically-inspired systems, vol 9, Springer, Heidelberg, Berlin. https://doi.org/10.1007/978-3-319-46374-2

Knippers J, Speck T (2012) Design and construction principles in nature and architecture. Bioinspir Biomim 7:015002. https://doi.org/10.1088/1748-3182/7/1/015002

Knoblauch J, Mullendore DL, Jensen KH, Knoblauch M (2014) Pico gauges for minimally invasive intracellular hydrostatic pressure measurements. Plant Physiol 166:1271–1279. https://doi.org/10.1104/pp.114.245746

Konrad W, Flues F, Schmich F, Speck T, Speck O (2013) An analytic model of the self-sealing mechanism of the succulent plant *Delosperma cooperi*. J Theor Biol 336:96–109. https://doi.org/10.1016/j.jtbi.2013.07.013

Launey ME, Buehler MJ, Ritchie RO (2010) On the mechanistic origins of toughness in bone. Annu Rev Mater Res 40:25–53. https://doi.org/10.1146/annurev-matsci-070909-104427

Lee H, Xia C, Fang NX (2010) First jump of microgel; actuation speed enhancement by elastic instability. Soft Matter 6:4342–4345. https://doi.org/10.1039/C0SM00092B

Li S, Wang KW (2017) Plant-inspired adaptive structures and materials for morphing and actuation: a review. Bioinspir Biomim 12:011001. https://doi.org/10.1088/1748-3190/12/1/011001

Lienhard J, Schleicher S, Poppinga S, Masselter T, Milwich M, Speck T, Knippers J (2011) Flectofin: a hinge-less flapping mechanism inspired by nature. Bioinspir Biomim 6:045001. https://doi.org/10.1088/1748-3182/6/4/045001

Masselter T, Eckert S, Speck T (2011) Functional morphology, biomechanics and biomimetic potential of stem-branch-connections in *Dracaena reflexa* and *Freycinetia insignis*. Beilstein J Nanotechnol 2:173–174. https://doi.org/10.3762/bjnano.2.21

Masselter T, Haushahn T, Schwager H, Milwich M, Müller L, Böhm H, Gude M, Gruhl A, Hufenbach W, Neinhuis C, Speck T (2013) From natural branchings to technical joints: branched plant stems as inspiration for biomimetic fibre-reinforced composites. J Des Nat Ecodyn 8:144–153. https://doi.org/10.2495/DNE-V8-N2-144-153

Masselter T, Haushahn H, Fink S, Speck T (2016a) Biomechanics of selected arborescent and shrubby monocotyledons. Beilstein J Nanotechnol 7:1602–1619. https://doi.org/10.3762/bjnano.7.154

Masselter T, Hesse L, Böhm H, Spengler N, Speck T, Korvink JG (2016b) Biomimetic optimisation of branched fibre-reinforced composites in engineering by detailed analyses of biological concept generators. Bioinspir Biomim 11:055005. https://doi.org/10.1088/1748-3190/11/5/055005

Mattheck C (2011) Thinking tools after nature. Karlsruher Institute of Technology, Germany

Mattheck C, Burkhardt S (1990) A new method of structural shape optimization based on biological growth. Int J Fatigue 12:185–190. https://doi.org/10.1016/0142-1123(90)90094-U

Milwich M, Speck T, Speck O, Stegmaier T, Planck H (2006) Biomimetics and technical textiles: solving engineering problems with the help of nature's wisdom. Am J Bot 93:1455–1465. https://doi.org/10.3732/ajb.93.10.1455

Müller L, Gruhl A, Böhm H, Gude M, Haushahn T, Masselter T, Schwager H, Neinhuis C, Speck T (2013) Biomimetisch optimierte verzweigte Faserverbundstrukturen mit hoher Tragfähigkeit. Melliand Textilberichte 2:88–93

Nellesen A, Tapavicza Mv, Bertling J, Schmidt A, Bauer G, Speck T (2011) Pflanzliche Selbstheilung als Vorbild für selbstreparierende Elastomerwerkstoffe. GAK—Gummi, Fasern, Kunststoffe 64/8:472–475

Oxford dictionary (2017) https://en.oxforddictionaries.com/definition/impact. Accessed 22 Feb 2017

Patek SN, Caldwell RL (2005) Extreme impact and cavitation forces of a biological hammer: strike forces of the peacock mantis shrimp *Odontodactylus scyllarus*. J Exp Biol 208:3655–3664. https://doi.org/10.1242/jeb.01831

Poppinga S, Joyeux M (2011) Different mechanics of snap-trapping in the two closely related carnivorous plants *Dionaea muscipula* and *Aldrovanda vesiculosa*. Phys Rev E 84:041928. https://doi.org/10.1103/PhysRevE.84.041928

Poppinga S, Masselter T, Speck T (2013) Faster than their prey: new insights into the rapid movements of active carnivorous plants traps. BioEssays 35:649–657. https://doi.org/10.1002/bies.201200175

Poppinga S, Haushahn T, Warnke M, Masselter T, Speck T (2015) Sporangium exposure and spore release in the Peruvian maidenhair fern (*Adiantum peruvianum*, Pteridaceae). PLoS ONE 10:e0138495. https://doi.org/10.1371/journal.pone.0138495

Poppinga S, Nestle N, Šandor A, Reible B, Masselter T, Bruchman B, Speck T (2017a) Hygroscopic motions of fossil conifer cones. Sci Rep 7:40302. https://doi.org/10.1038/srep40302

Poppinga S, Zollfrank C, Prucker O, Rühe J, Menges A, Cheng T, Speck T (2017b) Towards a new generation of smart biomimetic actuators for architecture. Adv Mater (online before print):1703653. https://doi.org/10.1002/adma.201703653

Poppinga S, Körner A, Sachse R, Born L, Westermeier A, Hesse L, Knippers J, Bischoff M, Gresser G, Speck T (2016) Compliant mechanisms in plants and architecture. In: Knippers J, Nickel K, Speck T (eds) Biomimetic research for architecture and building construction: biological design and integrative structures. Biologically-inspired systems, vol 9. Springer, Heidelberg, Berlin, pp 169–193. https://doi.org/10.1007/978-3-319-46374-2_9

Prüm B, Seidel R, Bohn HF, Rubach S, Speck T (2013) Microscopical surface roughness: a relevant factor for slipperiness of plant surfaces with cuticular folds and their replica. Acta Biomater 9:6360–6368. https://doi.org/10.1016/j.actbio.2013.01.030

Prüm B, Seidel R, Bohn HF, Speck T (2012) Plant surfaces with cuticular folds are slippery for beetles. J R Soc Interface 9:127–135. https://doi.org/10.1098/rsif.2011.0202

Rampf M, Speck O, Speck T, Luchsinger RH (2011) Self-repairing membranes for inflatable structures inspired by a rapid wound sea-ling process of climbing plants. J Bionic Eng 8:242–250. https://doi.org/10.1016/S1672-6529(11)60028-0

Rampf M, Speck O, Speck T, Luchsinger RH (2012) Structural and mechanical properties of flexible polyurethane foams cured under pressure. J Cell Plast 48:49–65. https://doi.org/10.1177/0021955X11429171

Rampf M, Speck O, Speck T, Luchsinger RH (2013) Investigation of a fast mechanical self-repair mechanism for inflatable structures. Int J Eng Sci 63:61–70. https://doi.org/10.1016/j.ijengsci.2012.11.002

Reichert S, Menges A (2015) Meteorosensitive architecture: Biomimetic building skins based on materially embedded and hygroscopically enabled responsiveness. Comput Aided Des 60:50–69. https://doi.org/10.1016/j.cad.2014.02.010

Ren S, Galjaard S (2015) Topology optimisation for steel structural design with additive manufacturing. In: Thomsen MR et al (eds) Modelling behaviour. Springer International Publishing, Switzerland, pp 35-44. https://doi.org/10.1007/978-3-319-24208-8_3

Richardson J, Adriaenssens S, Filomeno Coelho R, Bouillard P (2014) Discrete topology optimization—connectivity for gridshells. In: Adriaenssens S, Block P, Veenendaal D, Williams C (eds) Shell structures for architecture: form finding and optimization. Routledge, USA, pp 171–180

Ritchie RO (2011) The conflicts between strength and toughness. Nat Mater 10:817–822. https://doi.org/10.1038/nmat3115

Rüggeberg M, Burgert I (2015) Bio-inspired wooden actuators for large scale applications. PLoS ONE 10:e0120718. https://doi.org/10.1371/journal.pone.0120718

Schleicher S, Lienhard J, Poppinga S, Speck T, Knippers J (2015) A methodology for transferring principles of plant movements to elastic systems in architecture. Comput Aided Des 60:105–117. https://doi.org/10.1016/j.cad.2014.01.005

Schmier S, Lauer C, Schäfer I, Klang K, Bauer G, Thielen M, Termin K, Berthold C, Schmauder S, Speck T, Nickel KG (2016) Developing the experimental basis for an evaluation of scaling properties of brittle and 'quasibrittle' biological materials. In: Knippers J, Nickel K, Speck T (eds) Biomimetic research for architecture and building construction: biological design and integrative structures. Biologically-inspired systems, vol 9. Springer International Publishing Switzerland, pp 277–294. https://doi.org/10.1007/978-3-319-46374-2_14

Schmier S, Masic A, Schwaiger R, Speck T (2017) Functional morphology of the endocarp of *Cocos nucifera*, in prep

Schüssele AC, Nübling F, Thomann Y, Carstensen O, Bauer G, Speck T, Mülhaupt R (2012) Self-healing rubbers based on NBR blends with hyperbranched polyethylenimines. Macromol Mater Eng 297:411–419. https://doi.org/10.1002/mame.201100162

Schwager H, Masselter T, Speck T, Neinhuis C (2013) Functional morphology and biomechanics of branch-stem junctions in columnar cacti. Proc R Soc B 280:20132244. https://doi.org/10.1098/rspb.2013.2244

Song K, Yeom E, Lee SJ (2014) Real-time imaging of pulvinus bending in *Mimosa pudica*. Sci Rep 4:6466. https://doi.org/10.1038/srep06466

Spatz H-C, Beismann H, Brüchert F, Emanns A, Speck T (1997) Biomechanics of the giant reed *Arundo donax*. Phil Trans R Soc B 352:1–10. https://doi.org/10.1098/rstb.1997.0001

Speck O, Schlechtendahl M, Borm F, Kampowski T, Speck T (2018) Humidity-dependent wound sealing in succulent leaves of *Delosperma cooperi* – An adaptation to seasonal drought stress. Beilstein J Nanotechnol 9:175–186. https://doi.org/10.3762/bjnano.9.20

Speck O, Speck D, Horn R, Gantner J, Sedlbauer KP (2017) Biomimetic bio-inspired biomorph sustainable? an attempt to classify and clarify biology-derived technical developments. Bioinspir Biomim 12:011004. https://doi.org/10.1088/1748-3190/12/1/011004

Speck T (2015) Approaches to bio-inspiration in novel architecture. In: Imhof B, Gruber B (eds) Built to grow—blending architecture and biology. Birkhäuser Verlag, Basel, pp 145–149

Speck T, Bauer G, Flues F, Oelker K, Rampf M, Schüssele AC, von Tapavicza M, Bertling J, Luchsinger R, Nellesen A, Schmidt AM, Mülhaupt R, Speck O (2013a) Bio-inspired self-healing materials. In: Fratzl P, Dunlop JWC, Weinkamer R (eds) Materials design inspired by nature: function through inner architecture, RSC smart materials no. 4. The Royal Chemical Society, London, pp 359–389

Speck T, Luchsinger R, Busch S, Rüggeberg M, Speck O (2006) Self-healing processes in nature and engineering: self-repairing biomimetic membranes for pneumatic structures. In: Brebbia CA (ed) Design and nature III. WIT Press, Southampton, pp 105–114. https://doi.org/10.2495/DN060101

Speck T, Mülhaupt R, Speck O (2013b) Self-healing in plants as bio-inspiration for self-repairing polymers. In: Binder W (ed) Self-healing materials. Wiley-VCH, Weinheim, pp 69–97. https://doi.org/10.1002/9783527670185.ch2

Speck T, Speck O, Emanns A, Spatz H-C (1998) Biomechanics and functional anatomy of hollow stemmed sphenopsids: III. *Equisetum hyemale*. Bot Acta 111:366–376. https://doi.org/10.1111/j.1438-8677.1998.tb00721.x/

Speck T, Speck O (2008) Process sequences in biomimetic research. In: Brebbia CA (ed) Design and nature IV. WIT Press, Southampton, pp 3–11. https://doi.org/10.2495/DN080011

Stukeley W (1752) Memoirs of Sir Isaac Newton's life. http://ttp.royalsociety.org/ttp/ttp.html?id=1807da00-909a-4abf-b9c1-0279a08e4bf2&type=book. Accessed 22 Feb 2017

Thielen M, Speck T, Seidel R (2012) The ecological relevance of the pomelo (*Citrus maxima*) peel acting as an effective impact protection. In: Moulia B, Fournier M (eds) Proceedings of the 7th plant biomechanics international conference, Clermont-Ferrand, France, pp 20–24

Thielen M, Schmitt CNZ, Eckert S, Speck T, Seidel R (2013) Structure–function relationship of the foam-like pomelo peel (*Citrus maxima*)—an inspiration for the development of biomimetic damping materials with high energy dissipation. Bioinspir Biomim 8:025001. https://doi.org/10.1088/1748-3182/8/2/025001

Thielen M, Speck T, Seidel R (2015) Impact behaviour of freeze-dried and fresh pomelo (*Citrus maxima*) peel: influence of the hydration state. R Soc Open Sci 2:140322. https://doi.org/10.1098/rsos.140322

Van der Zwaag S (2007) An introduction to material design principles: damage prevention versus damage management. In: van der Zwaag S (ed) Self-healing materials—an alternative approach to 20 centuries of materials sciences. Springer, Dodrecht, pp 1–18. https://doi.org/10.1007/978-1-4020-6250-6_1

Van Tittelboom K, De Belie N (2013) Self-healing in cementitious materials—a review. Materials 6:2182–2217. https://doi.org/10.3390/ma6062182

Vincent O, Weißkopf C, Poppinga S, Masselter T, Speck T, Joyeux M, Quilliet C, Marmottant P (2011) Ultra-fast underwater suction traps. Proc R Soc B 278:2909–2914. https://doi.org/10.1098/rspb.2010.2292

Wagermaier W, Klaushofer K, Fratzl P (2015) Fragility of bone material controlled by internal interfaces. Calcif Tissue Int 97:201–212. https://doi.org/10.1007/s00223-015-9978-4

Wegst UG, Bai H, Saiz E, Tomsia AP, Ritchie RO (2015) Bioinspired structural materials. Nat Mater 14:23–36. https://doi.org/10.1038/nmat4089

Weibull W (1939) A statistical theory of the strength of materials. Generalstabens litografiska anstalts förlag, Stockholm

Weibull W (1951) A statistical distribution function of wide applicability. J Appl Mech 18:293–297

Zok FW (2017) On weakest link theory and Weibull statistics. J Am Ceram Soc 100:1265–1268. https://doi.org/10.1111/jace.14665

Index

A
Acetabularia, 212, 213
Acetobacter, 315, 331
Adiantum, 217, 220, 222, 228
Aerenchyma, 214, 220, 222
Agriculture, 391–392
Alga, *see* Chara, Nitella
Allium (onion), 217, 311, 316, 324–328, 336, 338–339, 342
Amorphous, Non-crystalline, 4–15, 241, 248–250, 254–265
Angiosperm (Broad-leaved), 4, 95, 211, 224, 353, 355
Anisotropy
  growth, 110, 117, 120, 121, 129, 132, 182, 194–203, 211, 226–227, 288, 299, 315–316, 339, 386
  mechanical, 4, 11, 22, 46, 49, 52, 129, 134, 158, 178, 184, 195–199, 200, 202, 214, 248–250, 253, 259, 261, 266, 286, 312, 322–324, 328–330, 339, 342, 411
Apple (*Malus*), 362
Arabidopsis, 115, 127, 129–132, 142, 148, 167, 172, 173, 177, 179–180, 197, 200, 218–219, 226–229, 309–317, 324, 334, 335, 342, 365, 378–380, 382–384, 386–391
Arabinogalactan Protein (AGP), 148–152
Aristolochia (pipevine), 424–425
Apoplasmic (apoplast), 171, 175, 226, 237, 277, 308, 310, 364, 366, 367, 368
Aquaporin, 150, 199, 355, 365, 368, 372
Asplenium (fern), 217, 222
Atomic Force Microscopy (AFM), 5, 199, 306–317, 419
Auxin, 115, 116, 128, 150, 168–178, 196–197, 201, 299, 311, 376, 386, 390, 391

B
Bark, 22, 401, 410–418, 424
Barley (*Hordeum*), 213, 215, 391
Beam/beam theory, 25–26, 38, 45–46, 49–52, 69–70, 90–97
Bean, *see* Glycine, Phaseolus, Vigna
Bending, 21–23, 26, 34, 37–47, 49–75, 83–92
Biomimetic, 315, 399–427
Bitter melon (*Momordica*), 128
Brassinosteroid, 298
Broad-leaved, *see* Angiosperm
Buckling/Elastic stability, 14, 26, 38, 49, 74, 83, 113–114, 122, 182–183, 209–229, 324, 418, 420–422
Bundle sheath, 351, 356, 361–370, 372

C
Calcium, 273–300, 310
Callose, 150, 219
Cambium/cambial, 22, 23, 83, 142–147, 410, 415
Cannabis (hemp), 142, 143, 147, 151, 153
Canopy, 88–89
Cardamine (cardamine), 237
Carrot (*Daucus*), 241
Cell division, 111, 129, 132, 166, 168, 171, 176, 187, 198, 199, 376, 382, 390, 424
Cellular Force Microscopy (CFM), 306, 308, 314, 317
Cellularity, 338, 339
Cellulose, 3–15, 116, 129, 137, 141, 151–160, 175, 178, 195–200, 211, 224, 228–244,

A. Geitmann and J. Gril (eds.), *Plant Biomechanics*,
https://doi.org/10.1007/978-3-319-79099-2

247–267, 273–276, 279, 286–288, 299, 300, 322–325, 328–331, 338, 341, 364, 365, 382, 386
Cellulose microfibril (CMF), *see* Microfibril
Cellulose synthase/CESA, 129, 195, 228, 386
Cell wall
creep, 305–317
loosening, 115, 129, 176, 180, 197, 227, 274–277, 292, 306–308, 313–316, 386
primary, 5–6, 152–153, 213, 228, 248–249, 273–275, 298, 321–324, 386, 412
secondary, 3–15, 151–153, 159, 238–239, 247–249, 273–274, 413
tertiary, *see* G-layer
Channel, 170, 201, 282, 378, 379, 381–385, 389
Chara, 277 *See also* Alga
Citrus (pomelo), 401, 410–418
Clover (*Trifolium*), 211
Cocos (coconut), 401, 410–418
Coil, 237, 239, 242–244, 331, 382
Collenchyma, 213, 215, 322, 419
Composite, 4, 53, 135–137, 178, 238, 243, 248, 255–262, 266–267, 315, 328–329, 338, 399–406, 412, 420
Compression, 14, 21, 30, 38, 43, 46, 51, 72, 74, 97, 134, 156, 157, 181–182, 193, 195, 198, 210–221, 228, 239, 243, 321, 322, 403–404, 408–409, 415, 417–418
Conifer, *see* Gymnosperm
Contraction/contractile, 27, 43, 111, 151, 155, 159, 199, 202, 209–211, 235–244, 314, 326–328, 414, 420, 426
Cortical microtubules (cMT), 116, 195–202, 211, 221–229, 287
Cotyledon, 167, 199, 210, 213, 226, 228
Crack, 21, 169, 170, 237, 326, 411, 415–418, 423
Creep, 12, 121, 289, 306–317, 324, 325, 331, 341
Cross-link, 229, 238, 243, 274–278, 284, 293–297
Crystallinity/Crystallite, 6–15, 155, 158, 195, 228, 238–239, 248, 254, 255–266
Cucumis (cucumber), 372
Cycad, 211
Cytoskeleton, 226, 375, 378, 385

D

Damage, 45, 79, 80–83, 90, 93, 96–98, 145, 389, 399, 401, 405, 406, 410, 420, 422, 437, 309, 316, 321, 417, 423, 425, 426, 427
Damping, 94, 400, 406, 416–418
Dandelion (*Taraxacum*), 213–215
Darcy's law, 358
Death, 80, 211, 235, 237–239, 361, 381, 385, 413
Delosperma (iceplant), 426
Density, 3, 14, 24, 28, 72, 83, 85, 88, 220, 250, 258, 277, 278, 284, 296, 297, 310, 324, 339, 368, 370, 390, 402, 405, 412–418
Development
cell/tissue, 142–152, 171, 209, 218–226, 236–237, 309, 324, 342, 383
organ, 175–177, 186–187, 200–203, 213, 384
plant, 22, 137, 167–168, 193–194, 209, 210, 273, 277, 280, 299, 322, 375–378
Differentiation, 117, 129, 171, 322, 377, 382, 386
Diffusion, 167, 170–173, 251–253, 257, 260–261, 282, 293, 335, 351, 352, 358, 364, 366, 369
Dissipation, 6, 381, 399, 401, 406–418
Drag, 79, 85–86, 89, 94–95
Drying, 43, 159, 169, 170, 235, 238, 241, 243

E

Earlywood, 250–251
Elasticity, 6–15, 23–24, 34, 38, 45, 53, 59, 74, 92, 94, 96, 111–113, 118–120, 127, 129–136, 153, 157, 177, 184–185, 198, 239, 264, 284, 291–293, 296, 314–315, 321, 325–330, 336–337, 339, 360, 386, 399–400
hyper, 155–159, 286 *See also* Hyperelasticity
limit/range, 80, 83, 326–327, 403 *See also* Yield stress
nonlinear, 336
poro, 118, 264–265 *See also* Poroelasticity
strain/deformation, 24, 135–136, 180–181, 185–186, 288, 292–297, 308, 311, 314, 325, 330, 336 *See also* Strain
visco, 6, 43, 45, 115–121, 176, 278, 293, 314, 325, 328, 331, 337, 416, 417, 425 *See also* Viscoelasticity
Elastic modulus/Young's modulus, 13–14, 22, 24, 27, 29, 32–34, 38, 43, 47, 68, 83, 90, 96, 109, 120, 134–136, 261, 279, 294, 329–330, 332, 336–337
Elastic stability, *see* Buckling
Endocarp, 242, 410–416, 418
Energy
absorption, 403–404
chemical/bonding, 237, 250–251, 260–261, 265, 366

deformation/strain, 59, 111–113, 118, 132, 135–136, 182, 288, 326, 421
dissipation/loss, 6, 94, 337, 399, 401–418
kinetic, 95, 408–410, 415
thermal/activation, 171, 289
Enzyme, 148, 153, 176, 180, 229, 274–279, 293–295, 314, 324, 368, 375, 387, 420
Epidermis, 127, 129, 130, 135–137, 170, 172–174, 194, 196–203, 209–237, 309–311, 315–316, 324–328, 336, 338, 342, 386, 424
Erodium (geranium), 243, 244
Eucalyptus (eucalypt), 159
Evaporation, 352, 361, 364–366, 372
Expansive growth, 289–290, 305–307, 314, 317 *See also* Growth
Extensibility, 134, 273–280, 297, 299, 303–317, 324, 337

F

Façade shading, 399, 401, 421, 422, 427
Feedback, 137, 167, 170, 172–174, 186–187, 199–202, 333
Fern, *see* Adiantum, Asplenium
Fiber
plant, 4–6, 11–15, 22, 83, 141–160, 211, 235, 236, 239, 241, 242, 244, 274, 415–418, 420
macromolecular, 3–15, 116, 129, 131–133, 141, 151–159, 195–197, 199, 211, 214–229, 235, 238–244, 248–250, 255–258, 262, 273, 275–276, 279, 286–299, 324, 328–329, 338, 386 *See also* Microfibril
(reinforced) composite, 286–290, 329, 399, 399, 401, 405–406, 418
Finite element (FEM), 49–75, 92–98, 141, 156–159, 181, 194, 226, 264, 322, 340, 420
Flax (*Linum)*, 142–160, 239, 329, 420
Flow
air, 85, 333, 425
auxin, 174
sugar, 351, 372
water, 170, 351–372, 419
wind, 89, 96, 98
Fluid, 133, 176, 248, 264, 267, 331, 357–359, 370, 417
Foam, 399, 401, 410–418, 424–425
Fold, 197, 200, 209–229, 336, 338, 421
Forest, 79–80, 83, 88–90, 96, 98
Fracture, 6, 49, 111, 243, 325–327, 337–340, 401, 403–404, 406, 411, 415–416, 423
freeze, 221, 310

G

Galacturonan, *see* Rhamnogalacturonan
Gas exchange, 235–236, 361, 391
Gas space, 209, 220–225, 228–229
Gas tightness, 424–425
Gelatinous fiber, 211, 236
Gelatinous layer/G-layer/Tertiary cell wall, 141, 151–160, 236
Gene/genetic, 109, 113, 115, 127, 128, 147–150, 166–167, 173, 175, 177–178, 182–183, 186, 193–194, 199–200, 202, 298, 365, 375, 378–382, 386, 389, 392
Geranium, 239, 241, 242
Germination, 235, 237, 381, 384
Glutamate receptor, 389
Glycine (soybean), 220
Gravity, 129, 236, 310, 321, 375–377, 386–392, 410
Growth
allometric, 28–29, 41–42
differential, 110–112, 127, 130, 137, 186, 193, 197–203, 209–210, 214–229, 236, 316
expansive, 289–290, 305–307, 314, 317
leaf, 109–123, 167–169, 172, 177–183, 186, 194, 353, 365
primary, 22, 129
secondary, 22, 424
Growth stress, *see* Stress
Gymnosperm/Conifer, 32–34, 86, 94, 353–360, 366

H

Hagen–Poiseuille, 358
Handedness, 127, 128, 130, 133
Helianthus (sunflower), 182, 198, 211–213, 386
Helical/helicoidal, 110, 127–137, 145, 248
Hemicellulose, 3–14, 186, 238, 243, 248–249, 254–258, 262, 273, 275, 286–287, 300, 364–365
Hemp (*Cannabis)*, 142, 143, 147, 151, 153
Hevea, 95
Hoffmeister's rule, 168–169
Homogalacturonan (HG), 152, 274–277, 297–299
Hooke's law, 24, 329, 409
Hordeum (barley), 213, 215, 391
Hull, 406, 410, 415, 424
Humidity, 11, 155, 238, 239, 241, 249–251, 255, 257, 263, 334, 335, 352
Hyacinthus (hyacinth), 211
Hydration, 238, 275, 310, 328, 339, 381, 384
Hydrogel, 155, 159, 284, 420

Hydrogen bond, 158, 247, 250–251, 253–255, 260–263, 266
Hygromechanics, 248, 265
Hygroscopicity, 11, 235, 238–244, 248, 420
Hyperelasticity, 155–159, 286
Hypocotyl, 127, 129, 167, 194, 197–199, 211, 213, 220, 221, 309–313, 316–317, 324, 335, 339
Hysteresis, 119, 247–263, 337–338

I
Iceplant (*Delosperma)*, 426
Instability, 89, 170, 420
Intercalary/intercellular, 147, 220–229, 274, 338–339, 364–365, 411
Intracellular, 279, 365–366, 375, 378, 381, 384–385, 387, 390

K
Kedem-Katchalsky, 369–370
Kinase, 149, 298, 375, 378, 384, 387, 388–389, 391
Kübler's model, 21–22, 29–34

L
Larix (larch), 213
Latewood, 250–252
Leaf development, 177
Leaf/Leaflet, 80, 88, 109–123, 128, 167–174, 177–186, 214, 217–222, 226–229, 236, 342, 351–357, 360–366, 372, 377, 388, 426
Leaf growth, 109–123, 167–169, 172, 177–183, 186, 194, 353, 365 *See also* Growth
Lignin, 3–5, 7, 10–14, 147, 151, 155, 237–239, 248–249, 253–257, 262, 273, 322, 382
Lilium (lily), 384
Linum (flax), 142–160, 239, 329, 420
Lockhart, 129, 185, 279–284, 306–307
Loosening, 115, 129, 147, 151, 153, 175–176, 197, 201, 227, 274–277, 292, 298, 305–317

M
Magnetic resonance imaging (MRI), 402–404, 419
Malus (apple) 362
Maize (*Zea)*, 71–72, 75, 217, 222, 226–228, 375, 379, 382, 386, 391
Matrix, 3, 7, 10–11, 52–53, 151, 155, 158, 159, 175, 176, 210, 214, 217, 219–225, 228, 238–241, 243, 248–249, 255–259, 264, 266, 274, 276–277, 286–297, 329, 338, 386, 413, 415–418, 424
Maturation
  cell-wall, 21–47, 158–159, 217–219
  pollen, 384
Mechano transduction, sensing, response, 116, 375–392
Meristem, 115, 116, 168–169, 171–173, 186, 194, 196–202, 209, 212–213, 236, 299, 323, 386
Mesophyll, 170, 220–225, 228, 229, 351–372
MicroCT, *see* X-ray tomography
Microfibril/cellulose microfibril (CMF), 3–15, 116, 129, 131–133, 141, 151–159, 195–197, 199, 211, 214–229, 235, 238–244, 248–250, 255–258, 262, 273, 275–276, 279, 286–299, 324, 328–329, 338, 386
Micromechanics, 13, 194, 199, 201, 264, 331, 336, 340
Microtubule, 116, 127, 129–133, 136–138, 145–146, 195–202, 211, 221, 276, 287, 375, 378, 386
Microscopy, 5–7, 153, 173, 194, 195, 215, 225, 239, 240, 248, 306, 316, 323, 329, 332–336, 342, 351, 362, 365, 402, 419
Middle lamella (ML), 11, 13–14, 141–142, 147–148, 152–153, 176, 214–225, 249–250, 273–274, 287, 295–297, 339–340
Mimosa, 236
Molecular Dynamic (MD), 12–13, 247–266
Momordica (bitter melon), 128
Morphogenesis, 115, 127–130, 137, 165–187, 193–203, 209–226, 273–274, 297, 299–300, 321–322, 377, 386
Movement/motion
  cell, 167, 200, 235–244, 323
  external fluid, 85, 89, 169
  internal fluid, 328, 351–372
  plant/organ, 38–47, 79, 94–95, 123, 141, 142, 159, 176, 211, 236, 238–243, 323, 376, 418–422, 426–427
  signal/molecule, 171–175, 250, 252, 260, 279, 293, 296, 386
Multiscale, 122, 197, 264, 266, 296, 340
Münch, 353–359

N
Nanoindentation, 14–15, 306, 314, 323, 334, 419
Needle, 85, 222–224, 347, 351, 356–360
Nicotiana (tobacco), 120, 382

Nitella, 315 *See also* Alga
Nonlinearity, 43, 46, 95, 174, 258, 260, 286, 294, 332, 336

O
Onion (*Allium*), 217, 311, 316, 324–328, 336, 338–339, 342
Organelle, 361, 378, 395, 390
Organogenesis, 194, 202, 322, 323
Orthotropy, 45, 52–53, 68, 70, 74, 96–97, 248, 258
Oryza (rice), 148, 375, 379, 382
Oscillation, 94, 96, 122, 260, 281, 283, 286–287, 387
Osmotic, 109, 115, 144, 279, 284, 305, 340, 351–357, 364, 368–369, 378, 381–385, 420

P
Parenchyma, 3–4, 72, 146, 151, 211, 220, 241–242, 322, 366, 411–412, 424
Patterning, 165–187, 200, 299, 382
Pavement cell, 195, 197, 225–229, 240
Pea (*Pisum*), 211, 217, 220
Pectate lyase, 276
Pectin, 121, 129, 148, 149, 175, 176, 178, 199, 219–220, 238, 243, 273–300, 323, 325, 331, 353, 364–365
Pectin Methyl Esterase (PME), 274–280, 293–300
Peel, 315, 401, 410, 412–418
Petal, 129, 130, 182–183, 224–225, 229, 418, 421–422
pH, 275–278, 308, 310–311, 313, 324
Phaseolus (bean), 220, 222, 388
Phloem, 142–144, 146–148, 153, 159, 211, 236, 351, 353, 355, 356, 361, 362, 364–369, 372, 424
Photosynthesis, 351, 352, 361, 364
Phyllotaxis, 168, 169, 201, 386
Picea (spruce), 87, 93, 248, 252
Pine (*Pinus*) 95, 155, 222–223, 238, 241, 359, 420
PIN protein, 172–175, 196–197, 201, 299
Pinus (pine), 95, 155, 222–223, 238, 241, 359, 420
Pipevine (*Aristolochia*), 424–425
Pisum (pea), 211, 217, 220
Pith, 25, 34–35, 38, 43, 46, 65–66, 71–73
Plasma membrane, 7, 129, 149, 150, 152, 157, 171–173, 279, 283, 296, 365, 366, 368, 380–385, 387–389
Plasmodesmata, 145, 146, 171, 172, 362, 364, 366–370, 372
Plasmolysis/plasmolyzed, 213, 339, 384
Plasticity/plastic deformation, 11, 44, 45, 83, 121, 135, 153, 176, 181, 184–186, 202, 286, 289, 308, 314–315, 325–328, 330, 336–338, 409–410, 415, 417, 424
Poisson's ratio, 52, 155–157, 282, 285, 326, 329, 331, 414
Pollen tube, 142, 147, 148, 150, 175, 177, 180, 194, 280–284, 286, 287, 384
Pomelo (*Citrus*), 401, 410–418
Populus (poplar) 42–44, 142, 158, 362, 365, 379
Pore/porosity, 157, 158, 209, 217, 219, 222, 239, 250, 256, 259, 260, 263, 264, 265, 266, 323, 328, 353, 354, 361, 368, 369–372, 379, 380, 389, 411, 418
Poroelasticity 118, 264–265
Pressure
  drag, 85
  osmotic, 109, 353, 354
  hydraulic 115, 330, 331, 356, 357, 358, 360, 372, 389
  traction, 75
  turgor, 118, 123, 129, 134, 136, 144, 155–158, 176, 178, 180, 181, 184–186, 194, 195, 197, 199, 210, 212, 279–296, 306, 322, 328, 329, 330, 331, 337, 352, 369, 372, 419, 420, 424
Pressure vessel, 195, 197, 199, 328, 329, 386
Primary wall, *see* Cell wall
Proprioception, 193

Q
Quantitative imaging, 193

R
Reactive oxygen species (ROS), 201, 308, 309, 376
Relaxation, 46, 132, 176, 229, 253, 278, 297, 306, 321, 324, 325, 341
Remodeling, 117, 129, 195, 199
Repair, 400, 401, 421, 423, 424, 425, 427
Rhamnogalacturonan (RG), 141, 151–159, 274–275
Rice (*Oryza*), 148, 375, 379, 382
Root, 22, 79, 80, 82, 83, 92, 93, 97, 127-137, 142, 145, 167, 171, 173, 177, 180, 197, 199, 209, 210, 211, 212, 220, 222, 229, 236, 277, 309, 316, 317, 324, 328, 334, 351-354, 356, 359, 343, 364, 378, 381, 382, 383, 384, 389, 390, 391
  contractile, 209, 210, 229

S

S2 layer/S2 wall/S2 cell wall layer, 3–15, 247–267
Sclereid, 413, 415, 416
Sclerenchyma, 242, 248, 322, 424
Seed, 199, 235–244, 415, 416, 420
Seed dispersal, 235–244
Self-
  cleaning, 400
  healing, 421, 423, 424
  repair, 399–401, 421, 423
  sealing, 421–427
  weight, 21–47, 92
Sequoia, 401, 410–418
Sesamum (sesame), 424
Shoot apical meristem, 116, 169, 194, 197, 198–201, 212–213, 299, 323, 386
Shrinkage, 7, 111, 122, 159, 249, 326, 364
Sieve element, 351–372
Sieve tube, 351–372
Signal/long-distance signal/signaling, 115, 116, 137, 150, 168, 171, 193, 200, 201, 203, 226, 274, 297–300, 306, 361, 375–392
Softwood, 3, 9–13, 250, 254, 255, 264
Sorption, 247–267
Spruce (*Picea*), 87, 93, 248, 252
Stem, 21–47, 49–75, 79–98, 137, 141–160, 186, 197, 222, 258, 328, 353–355, 389, 391, 401, 404–406, 425
Stiffness, 4–15, 24, 30, 45, 53, 54, 57, 58, 60, 63, 65–69, 72–74, 83, 116, 157, 158, 180, 194, 240, 248, 265, 279, 299, 321–342
Stochastic, 98, 202, 203
Stomate, 176, 209, 217–220, 222, 236, 323, 352, 361, 364, 365, 372, 383, 387
Strain
  bending, 54, 61
  compressive, 114, 425
  elastic, 24, 185, 409
  failure, 13
  plastic, 83
  maturation, growth 21–47, 180, 227, 284–297, 424
  release, 21, 34, 43, 95, 109, 119, 159, 185
  swelling, volumetric 251, 258–260, 263, 265, 283
  tensile, 8, 118, 156–159, 321–343
Strain gauge, 43, 94, 332–336
Strelitzia (bird of paradise), 422
Strength
  anchoring, 93
  bending, flexural 21, 50, 74
  fracture, 325, 326
  tensile, 11, 239, 273, 321–343, 403
  tissue, 72, 118, 159, 411, 424
  wood, stem, stalk, 3, 4, 21–23, 26, 46, 47, 50, 67, 71, 72, 83, 92
  yield, 325
Stress-strain, 95, 109, 119, 120, 155, 325, 326, 336–338
Stress
  bending, 21, 22, 38, 39, 42, 46, 47, 49–75, 92, 321, 322
  breaking, 13
  cartesian, 58
  compression, 51, 72, 92, 193
  growth, 21–47, 97, 132, 200, 236, 424
  maturation, 21–47
  salt, 388
  shear, 51–53, 55–58, 60, 67, 68, 74, 88, 89
  tensile, 8, 44, 51, 120, 195, 197, 198, 200, 209–229
  tissue, 137, 176
  twisting, 132, 135
  yield, 45
Stretch, 51, 111–113, 117, 119–121, 134, 137, 176, 180, 197, 222, 282, 295, 296, 309, 315–316, 324–338, 378, 383, 385
Sucrose, 226, 236, 366, 368, 369, 372
Sugar
  cell wall, 158, 274
  transport, 150, 351–372
Sunflower (*Helianthus*), 182, 198, 211–213, 386
Swelling, 11, 111, 122, 155, 158, 210, 214, 216, 219–229, 235–251, 253–267, 276, 420
Symplasmic/symplastic, 143–146, 351, 361–369

T

Taraxacum (dandelion), 213–215
Temperature, 10, 11, 85, 109, 110, 155, 185, 212, 252–253, 257, 289, 291, 292, 324, 334, 335, 385
Tendril, 128, 236
Tensile, tension, 8, 11, 14, 21, 30, 38, 43, 44, 46, 47, 49, 51, 54, 92, 117, 120, 129, 132, 137, 141-160, 176, 181, 182, 184, 195, 197, 198, 200, 209-229, 236, 240, 243, 261, 273, 277, 306, 310, 316, 321–328, 330–342, 403, 404, 424
Tension wood, 22, 35, 83, 151, 153, 155, 156, 158–160, 211
Thigmomorphogenesis, 50, 79–85, 92, 97
Tissue tension, 127, 132, 135–137, 159
Tobacco (*Nicotiana*), 120, 377–378, 382

Toughness, 4, 326, 400–411
Tracheid, 3–5, 142, 352–353
Transcription, 147, 151, 298, 375, 389
Transpiration, 236, 279–280, 351, 352, 364, 383
Transport
  auxin/hormone, 115, 166–168, 172–177, 299, 390
  gas, 220
  pectin, 295
  sugar, 150, 351, 366–372
  water, 2, 170, 351–372
Tree, 3, 21–47, 79–98, 127, 155, 211, 321, 351–356, 376–377, 402, 415–417
Trifolium (clover), 211
Triticum (wheat), 155, 220, 238–240, 391
Turgor, 118, 123, 129, 134, 144, 155–159, 176–185, 194, 210, 212, 213, 219, 220, 222, 229, 236–237, 278–300, 305–309, 312, 315, 322, 324, 328–331, 337, 339, 365, 372, 424
Twist, 8, 127–137, 239–244

V
Vallisneria, 128
Vascular system/strand, 116–117, 180, 351–372
Vein, 116–117, 167–174, 180, 351–372
Vibration, 94, 95
Vigna (azuki bean) 221, 386
Viscoelasticity, 6, 43, 45, 115–121, 176, 278, 321, 293, 314, 325, 328, 331, 337, 416, 417, 425
Viscosity, 85, 109, 118–120, 155, 176, 194, 293, 280–294 328, 330, 335, 337, 358, 414

W
Wall, *see* cell wall
  anticlinal, 197, 211, 215, 217, 219, 224–229
  periclinal, 215, 217, 219, 224–229
Water
  transport/flux, 3, 4, 150, 170, 351–372, 402, 419
  cell wall/apoplast, 12, 154–155, 159, 211, 216, 237, 244, 247–267, 273, 322, 328
  environment/external/humidity, 109, 169, 170, 220, 236, 425
  structural role/turgor, 73–74, 123, 235–237, 279–280, 305, 280, 414, 420
Weight, 21–47, 92, 236, 291, 399–427
Wheat (*Triticum*), 155, 220, 238–241, 391
Wind, 21, 79–98, 238, 321, 355, 376–378, 385, 391
Wood, woody, 3–15, 21–47, 57, 68–69, 79–98, 146, 151, 153, 155, 158, 211, 236, 247–267, 274, 322, 422
Wood
  reaction, 22, 83, 236
  properties, 3–15, 21–47, 52–75, 79, 92, 94–97, 247–267
  tension, 35, 151, 153, 155, 156, 158–160, 211
Wrinkling, 111–113, 209–211, 214, 229

X
X-ray diffraction, 15, 242
X-ray tomography/computed (micro, nano) tomography/microCT/μCT, 15, 71, 250, 402–403, 419
Xylem, 3–15, 142–147, 211, 274, 322, 352, 353, 361–367, 372, 424 *See also* Wood
Xyloglucan, 152, 229, 275–276, 308, 331

Y
Yeast, 378–382
Yield stress, 80, 83, 326–327, 403
Young's modulus, 46, 68, 155, 157, 213, 261, 288, 291, 325, 329, 336–339

Z
Zea (maize), 71–72, 75, 217, 222, 226–228, 375, 379, 382, 386, 391

MIX
Papier aus verantwortungsvollen Quellen
Paper from responsible sources
FSC® C105338

If you have any concerns about our products,
you can contact us on
ProductSafety@springernature.com

In case Publisher is established outside the EU,
the EU authorized representative is:
Springer Nature Customer Service Center GmbH
Europaplatz 3, 69115 Heidelberg, Germany

Printed by Libri Plureos GmbH
in Hamburg, Germany